NONLINEAR AND NONSTATIONARY
SIGNAL PROCESSING

The Isaac Newton Institute of Mathematical Sciences of the University of Cambridge exists to stimulate research in all branches of the mathematical sciences, including pure mathematics, statistics, applied mathematics, theoretical physics, theoretical computer science, mathematical biology and economics. The research programmes it runs each year bring together leading mathematical scientists from all over the world to exchange ideas through seminars, teaching and informal interaction.

NONLINEAR AND NONSTATIONARY
SIGNAL PROCESSING

edited by

W.J. Fitzgerald

University of Cambridge

Richard L. Smith

University of North Carolina, Chapel Hill

A.T. Walden

Imperial College of Science, Technology and Medicine

Peter Young

Lancaster University

CAMBRIDGE
UNIVERSITY PRESS

PUBLISHED BY THE PRESS SYNDICATE OF THE UNIVERSITY OF CAMBRIDGE
The Pitt Building, Trumpington Street, Cambridge, United Kingdom

CAMBRIDGE UNIVERSITY PRESS
The Edinburgh Building, Cambridge CB2 2RU, UK
40 West 20th Street, New York, NY 10011–4211, USA
10 Stamford Road, Oakleigh, VIC 3166, Australia
Ruiz de Alarcón 13, 28014 Madrid, Spain
Dock House, The Waterfront, Cape Town 8001, South Africa

http://www.cambridge.org

First published 2000

Printed in the United Kingdom at the University Press, Cambridge

Typeset in 12pt Computer Modern

A catalogue record for this book is available from the British Library

ISBN 0 521 80044 7 hardback

CONTENTS

Contributors

Christophe Andrieu, Signal Processing Laboratory, Department of Engineering, University of Cambridge, Cambridge CB2 1PZ, UK
ca226 @eng.cam.ac.uk

Richard Baraniuk, Department of Electrical and Computer Engineering, Rice University, 6100 Main Street, Houston, TX 77005, USA
richb@ece.rice.edu

Metin Bayram, Department of Electrical and Computer Engineering, Rice University, 6100 Main Street, Houston, TX 77005, USA
mebay@ece.rice.edu

Keith Beven, Institute of Environmental and Natural Sciences, Lancaster University, Lancaster LA1 4YQ, UK
k.beven@lancaster.ac.uk

Richard A. Davis, Department of Statistics, Colorado State University, Fort Collins, CO 80523-1877, USA
rdavis@stat.colostate.edu

Anthony Davison, Department of Mathematics, Swiss Federal Institute of Technology Lausanne, CH-1015 Lausanne EPFL, Switzerland
anthony.davison@epfl.ch

Arnaud Doucet, Signal Processing Laboratory, Department of Engineering, University of Cambridge, Cambridge CB2 1PZ, UK
ad2@eng.cam.ac.uk

Bill Fitzgerald, Signal Processing Laboratory, Department of Engineering, University of Cambridge, Cambridge CB2 1PZ, UK
wjf@eng.cam.ac.uk

Patrick Flandrin, Laboratoire de Physique (UMR 5672 CNRS), Ecole Normale Supérieure de Lyon 46, allée d'Italie, F-69364 Lyon Cedex 07, France
flandrin@ens-lyon.fr

Jim Freer, Institute of Environmental and Natural Sciences, Lancaster University, Lancaster LA1 4YQ, UK
j.freer@lancaster.ac.uk

Neil Gordon, Signal Processing and Imagery Department, DERA, Malvern, Worcs, WR14 3PS, UK
N.Gordon@signal.dera.gov.uk

Barry Hankin, Institute of Environmental and Natural Sciences, Lancaster University, Lancaster, LA1 4YQ UK
b.hankin@lancaster.ac.uk

Alan Marrs, Signal Processing and Imagery Department, DERA, Malvern, Worcs, WR14 3PS, UK
A.Marrs@signal.dera.gov.uk

Alistair Mees, Department of Mathematics and Statistics, University of Western Australia, Nedlands, Western Australia 6907, Australia
`alistair@maths.uwa.edu.au`

Thomas Mikosch, Department of Mathematics and Computing Science, University of Groningen, PO Box 800, 9700 AV Groningen, The Netherlands
`T.Mikosch@math.rug.nl`

José Miguel Pérez, Centro de Estadística y Software Matemático, Universidad Simón Bolívar, Caracas 1080-A Venezuela
`jperez@cesma.usb.ve`

Don Percival, Applied Physics Laboratory, Box 355640, University of Washington, Seattle, WA 98195–5640, USA; *and* MathSoft Inc., 1700 Westlake Avenue North, Suite 500, Seattle, WA 98109–3044, USA
`dbp@apl.washington.edu`

Tina Rydberg, Paribas, 10 Harewood Avenue, London NW1 6AA, UK; *and* Nuffield College, Oxford OX1 1NF, UK
`tina.rydberg@paribas.com`

David Salmond, DERA, Farnborough, Hants, GU14 OLX, UK
`djsalmond@dera.gov.uk`

Sylvain Sardy, Department of Mathematics, Swiss Federal Institute of Technology Lausanne, CH-1015 Lausanne EPFL, Switzerland
`sardy@epfl.ch`

Karsten Schulz, Institute of Environmental and Natural Sciences, Lancaster University, Lancaster LA1 4YQ, UK

Neil Shephard, Nuffield College, Oxford OX1 1NF, UK
`neil.shephard@nuf.ox.ac.uk`

Richard L. Smith, Department of Statistics, University of North Carolina, Chapel Hill, NC 27599-3260, USA
`rls@email.unc.edu`

Vasily Strela, Department of Mathematics & Computer Science, Drexel University, Philadelphia, PA 19104, USA
`vstrela@mcs.drexel.edu`

David Thomson, Room 2C-360, Bell Laboratories, Murray Hill, NJ 07974, USA
`djt@research.bell-labs.com`

Andrew Walden, Department of Mathematics, Imperial College of Science, Technology and Medicine, Huxley Building, 180 Queen's Gate, London SW7 2BZ, UK
`a.walden@ic.ac.uk`

Peter Young, Institute of Environmental and Natural Sciences, Lancaster University, Lancaster LA1 4YQ, UK
`P.Young@lancaster.ac.uk`

Introduction

The programme "Nonlinear and Nonstationary Signal Processing" was held at the Isaac Newton Institute from July to December 1998. This programme was motivated by the observation that the whole field of signal processing and time series analysis has by now moved far beyond its roots in the theory of linear stationary processes, but many of the new techniques to handle nonlinear and nonstationary processes have developed in individual areas of statistics, engineering or more specialised fields such as environmental science or mathematical finance, with limited interaction between different groups. This programme brought together researchers from many different areas and with a wide range of expertise, resulting in a very successful synthesis of ideas. Particularly noteworthy achievements were new methodological developments and applications of wavelets, a wider appreciation of Bayesian methods, the interaction between nonlinear time series analysts and dynamical systems experts, and the development of new areas of application such as risk management in insurance and finance.

As part of the programme activities, five open workshops and a host of more informal meetings were held. The open workshops were on Bayesian statistics in signal processing, environmental modelling, the interaction between time series analysis and dynamical systems, statistical methods in finance, and data analysis with a particular emphasis on wavelet methods.

The chapters of this volume were all originally presented as talks at one of the workshops (and are divided into subsections according to the themes of those workshops), or are research contributions by long-stay participants in the programme.

Following the development of fast computers and sophisticated Monte Carlo simulation methods, the Bayesian community has been able to address more complex problems of Bayesian inference than was possible before. The chapter by Christophe Andrieu, Arnaud Doucet, Bill Fitzgerald and José Miguel Pérez, on Bayesian Computational Approaches to Model Selection, which is co-authored by the Programme's main organiser, Bill Fitzgerald, discusses some of the computational issues of this new approach and provides two examples which demonstrate how the resulting methodology can be used in stochastic model selection. The chapter first considers the state-of-the-art approaches to the problem of choosing prior distributions in the context of model selection. And it then goes on to describe numerical methods for computing Bayes factors and posterior model probabilities, concentrating on Markov Chain Monte Carlo (MCMC) methods and, in particular, the reversible jump MCMC method. The chapter concludes with two interesting examples that illustrate well the application of the methods: the detection of sinusoids in noise and the identification of the components in a Gaussian mixtures model.

There is much current interest in the development of sequential simulation methods with particular applications to Bayesian inference. Many of these ideas can be traced back to the control literature of the 1960s but due to computational restrictions at this time the methods were not really developed. The chapter by Neil Gordon, Alan Marrs and David Salmond, on sequential analysis of dynamic sytems using particle filters and mixtures, describes some of the theory, within a Bayesian framework, and introduces the ideas associated with particle filters and sequential MCMC methods. The chapter then goes on to consider three applications, concerning target tracking, in-situ ellipsometry for monitoring the growth of SiGe alloys and finally, the monitoring of chemical and nuclear agents as a function of space and time. These applications show the power of particle filters and sequential methods.

The current worldwide interest in environmental issues was reflected in the first workshop of the programme on Environmental Modelling and Data Analysis. The chapter by another of the programme organisers, Peter Young, gives excellent coverage of dynamic modelling for time variable and state dependent parameter estimation. This chapter starts by reviewing some of the author's previous and fundamental work concerning time variable and recursive parameter estimation and it is shown that if changes in parameters are functions of the state or input variables then the system can exhibit complex nonlinear behaviour, and this then requires state dependent modelling techniques, which are then covered in great detail in the chapter. The ideas associated with transfer function models and fixed interval smoothing for instrumental variable equations are introduced and the methodology is supported by various simulation examples starting with a simple first order time varying parameter model, then a forced logistic growth equation model, a chaotic logistic growth model and finally the so-called cosine map model, before moving onto the application of the methodology to the famous Nicholson blowfly data set where a totally new interpretation of the data is given.

One feature of the environmental workshop was the wide-ranging background of the participants. The chapter by Keith Beven, Jim Freer, Barry Hankin and Karsten Schulz is typical of the presentations made by scientists from different disciplines at the Workshop. Less mathematical than most of the chapters in the book, the authors concentrate on the important issues of uncertainty and over-parameterisation in large, mechanistic simulation models, with particular emphasis on hydrological systems. In relation to such large models, the chapter questions the concept of an optimum parameter set and proposes that this should be replaced by a concept they call 'equifinality'. This is intended to focus attention on the multiple possibilities for producing simulations that are, in some sense, 'acceptable', rather than optimum. The authors suggest that Beven's Bayesian Generalised Likelihood Uncertainty Estimation (GLUE) methodology allows the model builder and user to live with the problems of equifinality, and they present two practical examples which illustrate its application: the modelling of land surface to atmosphere fluxes, and dispersion in open channel flows.

The chapter by Richard Smith provides a very useful, partly tutorial, review of spatial statistics, but it has particular emphasis on the use of such methods in environmental applications. The first part of the chapter provides a general overview of the subject that covers specific methods of current interest. These include geostatistics and kriging, nonstationary models, models defined by conditional probabilities, the design of spatial experiments and spatio-temporal data issues. The second part of the chapter introduces a simplified, hierarchical model for spatially varying temporal trends and presents two applications of this approach. The first is concerned with spatial analysis of trends in global temperature series, and the second is based on sulphur dioxide measurements at 35 locations in the eastern US over the period 1989–1995.

The workshop on dynamical systems is represented by the chapter written by Alistair Mees. Mees's chapter addresses the problem of determining which components of an apparently unpredictable time series can be identified with a deterministic nonlinear system as opposed to stochastic noise. A particular feature of his approach is the use of criteria based on information theory. The concept of code length, originally due to Kolmogorov but developed independently by Wallace and Rissanen, is introduced as a unifying theme. This is first applied in the case of symbol strings, where the whole system can be represented as series of symbols from a finite alphabet. A key step here is the use of context trees to allow the system to be represented as a Markov chain, followed by estimation of the transition probabilities of that chain. The discussion then moves on to real-valued time series, including the use of embedding to allow this kind of system to be represented as a Markov chain, then estimating the transition relations through either local linear approximations or a radial basis function representation. As an example, a time series of firing voltage measurements from a squid giant axon is analysed using both the finite-alphabet and real-value methods. For finite-alphabet methods, Mees concludes that information theory is the key tool allowing efficient identification of the system. For real-valued time series, information-theoretic methods are also valuable but appreciation of the geometry of the system is another essential part of the reconstruction.

The workshop on statistical methods in finance is represented by two chapters. Tina Rydberg and Neil Shephard focus on the use of compound Poisson processes to model trade-by-trade data. Such data consist of jumps between discrete levels and so render traditional models for financial time series, such as geometric Brownian motion, inappropriate. Their proposal is to model the point process of jump times as a Cox process, together with a random process for the price changes which occur at the jumps. To illustrate their methods, they use electronically recorded IBM share price data from the New York Stock Exchange. Their basic model assumes an Ornstein–Uhlenbeck process to define the random intensity of the point process of trading times, combined with a first-order moving average process for the price changes. A particular feature is the use of signal-extraction methods to infer the unobserved random intensity process from the observed sequence of trading times. They also consider the consequences of their model for the distribution of returns over various time periods. They conclude that there are many features of the observed data which are well described by their model but they also highlight some aspects where more sophisticated modelling is required.

The second chapter on finance, by Richard Davis and Thomas Mikosch, focusses on the behaviour of sample autocorrelations in time series which may have both non-linear and heavy-tailed features. Examples such as the GARCH model, well-known in econometrics, often have both these features. For linear heavy-tailed processes, the sample autocorrelation function generally has attractive properties, for example, converging rapidly to a well-defined quantity even when the population autocorrelation does not exist. For nonlinear processes, however, there is no guarantee that the sample autocorrelation converges to anything as the sample size tends to infinity. The authors illustrate this possibility especially for ARCH and GARCH models, which they treat as a special case of the more general class of models satisfying stochastic recurrence equations. On the other hand, for stochastic volatility models, the behaviour of the sample autocorrelations is far more satisfactory. The results may provide some grounds for preferring the stochastic volatility models in practice, though the authors are careful to stress that conclusions of this nature need further research.

Time-frequency and wavelet methods formed an important theme in the programme.

Long-stay participant Patrick Flandrin illustrates in a clear and interesting way how tools for non-parametric time-frequency analysis can be motivated in a large number of different contexts, by considering representations of the signal, its energy or power, distribution or correlation, or by making use of probabilistic or geometrical properties. There are few existing time-frequency methodologies that are not covered in this comprehensive survey.

Metin Bayram and Richard Baraniuk (elected as Rosenbaum Fellow for the programme) give a new non-parametric method for estimating the time-varying spectrum of a nonstationary random process. The method extends Thomson's multiple window spectrum estimation method and works by averaging over sets of orthogonal, optimally concentrated windows – the Hermite functions for time-frequency analysis and the Morse wavelets for time-scale analysis. Nonstationary line components are detected and extracted by approximating them as piecewise linear chirps.

David Thomson, originator of the multiple window (or multitaper) spectrum estimation method, made two visits to the programme.

His chapter gives a number of multitaper approaches to the analysis of nonlinear and nonstationary time series with a viewpoint between those of statistics and signal processing; included are detailed analyses of the Central England temperature series and space physics data (magnetic fields and electron fluxes).

The data analysis workshop is further represented by two chapters on wavelet methods. The first, by Vasily Strela and Andrew Walden, looks at signal and image denoising via wavelet thresholding in the context of scalar and multiple wavelet transforms, using both scalar and vector thresholding. Multiwavelets outperform scalar wavelets for three out of four noisy 1D test signals, and are similarly generally preferable for the four 2D image denoising problems. Chui–Lian scaling functions and wavelets combined with repeated row preprocessing appears to be a good general method.

Don Percival, Sylvain Sardy and Anthony Davison consider the problem of decorrelating random processes using wavelet methods. They achieve this by adaptively selecting an orthonormal transform from a wavelet packet table using a series of white noise tests. Having thus obtained transform coefficients with very low correlations, they create new sets of coefficients by bootstrapping; these are then inverse transformed to create bootstrapped time series, from which distributions of statistical quantities of interest can be calculated. This innovative procedure is given the name 'wavestrapping'.

Acknowledgements

The editors wish to thank Professor H.K. Moffatt and all of the staff at the Isaac Newton Institute for making our six month programme such a success. The editors are also very grateful for the sponsorship of the programme by EPSRC, TSUNAMI, the NSF, BP, Barclaycard, Schlumberger Cambridge Research, and Autonomy Ltd.

Bayesian Computational Approaches to Model Selection

C. Andrieu, A. Doucet, W.J. Fitzgerald and J.-M. Pérez

1   Introduction

A fundamental task in signal and data processing and science is, in general, to develop models for signals which are observed and to determine whether the model function that one is using to describe the data is actually appropriate for the particular problem under investigation. Except in artificial problems, a model is just an approximation that, up to some degree, describes the process which generates a particular signal or set of observations. In this way, one usually entertains several plausible models, realizing that in terms of the real data generation process, the correct model may not be within the set chosen. Multiple model selection, therefore, appears naturally in the analysis of trying to determine which of the entertained models best describes the data at hand. Further, in parametric models, one is also interested in extracting values for the free parameters of the model.

The problems of parameter estimation and model selection can be coherently approached within a Bayesian framework. In Bayesian analysis, the statistical inference is obtained in the form of *posterior distributions*, which incorporate both the scientist's beliefs and the observations, in a well founded probabilistic framework. In particular, the model selection problem can be summarized by the posterior probability of each model. This distribution is meaningful, and certainly easier to interpret than, say, classical P-values.

Bayesian analysis is not without problems, however. In practice, one is forced to establish prior beliefs, in the form of *prior probability distributions*, on the models under consideration and, often more difficult, on the parameters in each model. The latter could prove a daunting task, as the characteristics of many model parameters may well not be known precisely. One other problem one faces in the Bayesian framework is the computation of the quantities that lead to Bayesian model selection – typically integrals of large dimension that do not admit any closed-form analytical solution.

The problem of choosing prior distributions for the parameters is a very delicate one in the framework of model selection as illustrated in Section 2, where we briefly present state-of-the-art approaches to address these difficulties. In Section 3, we describe numerical methods that address the practical computation of the quantities of interest, namely the Bayes Factors and posterior model probabilities. We focus on Markov chain Monte Carlo (MCMC)

1

methods and especially the reversible jump MCMC method introduced by
Green [33]. In Section 4 we present two applications of the methodology to
the detection of sinusoids in noise and the determination of the components
of Gaussian mixture models.

2   Bayesian model selection

2.1   Bayesian methodology for model selection

Assume that we are analyzing data[1] $\mathbf{y}$ and we believe that the data arise
from one of a set of possible models $\mathcal{M}_0, \ldots, \mathcal{M}_{k_{\max}}$ ($k_{\max}$ can be infinite),
where under model $\mathcal{M}_i$, $\mathbf{y}$ has density $p_i(\mathbf{y} | \theta_i)$, conditional on $\theta_i \in \Theta_i$. The
parameter vectors θ_i are unknown and are typically of different dimension. Let
$p_i(\theta_i)$ denote the prior density for θ_i (with respect to a dominating measure,
usually Lebesgue), and let p_i denote the prior probability of the model $\mathcal{M}_i$.
For the sake of convenience, we introduce a random variable $k \in \{0, \ldots, k_{\max}\}$
such that $\Pr(k = i) = \Pr(\mathcal{M}_i) = p_i$. The prior probability distribution for the
random parameters (k, θ) is defined on a space of the form $\Theta \triangleq \bigcup_{i=0}^{k_{\max}} \{i\} \times \Theta_i$
and can be written

$$p(k, d\theta) = \sum_{i=0}^{k_{\max}} p_i(i, d\theta_i) \, \mathbb{I}_{\{i\} \times \Theta_i}(k, \theta), \tag{2.1}$$

where the integrable and distinct functions p_i admit the form

$$p_i(i, d\theta_i) = p_i(\theta_i) \, d\theta_i p_i \tag{2.2}$$

and

$$\mathbb{I}_{\{i\} \times \Theta_i}(k, \theta) = \begin{cases} 1, & \text{if } (k, \theta) \in \{i\} \times \Theta_i, \\ 0, & \text{otherwise,} \end{cases} \tag{2.3}$$

i.e. (k, θ) is in one of the spaces $\{i\} \times \Theta_i$, and the prior probability of k
being equal to i and for θ being in an infinitesimal set centered around θ_i is
$p_i(i, \theta_i) \, d\theta_i$.

After observing $\mathbf{y}$, one obtains the posterior distribution using Bayes' theorem

$$p(k, d\theta | \mathbf{y}) = \sum_{i=0}^{k_{\max}} p(i | \mathbf{y}) \, p_i(d\theta_i | \mathbf{y}) \, \mathbb{I}_{\{i\} \times \Theta_i}(k, \theta) \tag{2.4}$$

where $p(i | \mathbf{y})$ is the posterior probability of model $\mathcal{M}_i$ and is given by

$$p(i | \mathbf{y}) \triangleq p(\mathcal{M}_i | \mathbf{y}) = \frac{m_i(\mathbf{y}) \, p_i}{\sum_{j=0}^{k_{\max}} m_j(\mathbf{y}) \, p_j}, \tag{2.5}$$

---

[1]We do not distinguish between random variables and their realizations.

where

$$m_i\left(\mathbf{y}\right) \triangleq p\left(\mathbf{y}\,|\,i\right) = \int_{\Theta_i} p\left(\mathbf{y}\,|\,\theta_i\right) p_i\left(\theta_i\right) d\theta_i \tag{2.6}$$

is called the *marginal* distribution of $\mathbf{y}$ under model $\mathcal{M}_i$. Assuming $\mathcal{M}_i$ is the *true* model, $p\left(\mathbf{y}\,|\,i\right)$ is the density according to which $\mathbf{y}$ will actually occur. For this reason, $m_i\left(\mathbf{y}\right)$ is also called the *predictive* density of $\mathbf{y}$. Under a 0-1 loss function, the optimal model is that $\mathcal{M}_k$ which maximizes the posterior model probability $p\left(k\,|\,\mathbf{y}\right)$, $k = 0, \ldots, k_{\max}$.

Note that $p\left(k\,|\,\mathbf{y}\right)$ can be written as

$$p\left(k\,|\,\mathbf{y}\right) = \left(1 + \sum_{i \neq k} \frac{p_i}{p_k} B_{ik}\right)^{-1}, \tag{2.7}$$

where the factor

$$B_{ik} = \frac{m_i\left(\mathbf{y}\right)}{m_k\left(\mathbf{y}\right)} \tag{2.8}$$

is called the *Bayes Factor* of model $\mathcal{M}_i$ against $\mathcal{M}_k$. Intuitively, the Bayes Factor can be interpreted as the odds of $\mathcal{M}_i$ against $\mathcal{M}_k$ given by the observations. Note that Bayes Factors can be used to summarize the analysis independently of the model prior beliefs, p_i.

The Bayesian approach to model selection can be applied to a wide variety of problems, including multiple comparisons and the testing of non-nested hypotheses. The results are easily interpreted (as opposed to frequentist P-values) and automatically penalize overparametrizations [12], [61]. For a detailed discussion of the advantages and applications of Bayes Factors see [7], [36].

We present two classical model selection problems arising in signal processing and statistics.

Example 2.1 *Sinusoids in noise.* We would like to model the data $\mathbf{y} \triangleq \{y_0, \ldots, y_{T-1}\}$ with one of the following models:

$$\begin{aligned} \mathcal{M}_0 &: y_t = w_{0,t}, & \text{if } k = 0, \\ \mathcal{M}_k &: y_t = \sum_{j=1}^{k} \left(a_{c_{j,k}} \cos\left[\omega_{j,k} t\right] + a_{s_{j,k}} \sin\left[\omega_{j,k} t\right]\right) + \varepsilon_{k,t} & \text{if } k \geq 1, \end{aligned} \tag{2.9}$$

where $\varepsilon_{k,t} \overset{iid}{\sim} \mathcal{N}\left(0, \sigma_k^2\right)$. The model $\mathcal{M}_k$ describes the data in terms of k sinusoids in white Gaussian noise. The unknown parameters for $\mathcal{M}_k$ are $\theta_k = \left(a_{c_{1,k}}, a_{s_{1,k}}, \omega_{1,k}, \ldots, a_{c_{k,k}}, a_{s_{k,k}}, \omega_{k,k}, \sigma_k^2\right)$. We added a subscript k to emphasize that these parameters depend on the model $\mathcal{M}_k$. Bayesian inference is performed on the parameter space $\Theta = \bigcup_{i=0}^{k_{\max}} \{i\} \times \Theta_i$ where $\Theta_i = \left(\mathbb{R}^2 \times (0, \pi)\right)^i \times \mathbb{R}^+$, *i.e.* if $k = i$, the unknown parameters θ_k belongs to $\Theta_{i=k}$. The space **Theta** is a union of disjoint spaces. In this case, one says that the models are nested since $\textbf{Theta}_{i+1} = \mathbb{R}^2 \times (0, \pi) \times \textbf{Theta}_i$ for $i = 0, \ldots, k_{\max-1}$.

Example 2.2 *Mixture of normals.* The data $\mathbf{y} \triangleq \{y_1, \ldots, y_T\}$ are assumed to be distributed according to one of the following models:

$$\mathcal{M}_k : \mathbf{y}_t \overset{iid}{\sim} \sum_{j=1}^{i} w_{j,k} \mathcal{N}\left(\mu_{j,k}, \Sigma_{j,k}\right).$$

The model $\mathcal{M}_k$ describes the data in terms of a mixture of k Gaussian distributions, where $k \in \{1, \ldots, k_{\max}\}$ represents the unknown number of components. Conditional on k, the weights of the mixture are given by $\mathbf{w}_k = (w_{1,k}, \ldots, w_{k,k})$, where $w_{j,k}$ is the probability of an observation coming from component j and $\sum_{j=1}^{i} w_{j,k} = 1$. The parameters of the model components are given for $\mathcal{M}_k$ by $\phi_k = (\mu_k, \Sigma_k)$ where $\mu_k = (\mu_{1,k}, \ldots, \mu_{k,k})$ and $\Sigma_k = (\Sigma_{1,k}, \ldots, \Sigma_{k,k})$, with $\mu_{j,k}$ and $\Sigma_{j,k}$ being respectively the mean and covariance matrix of the kth component. The unknown parameters for $\mathcal{M}_k$ are $\theta_k = (\mathbf{w}_k, \phi_k)$. Bayesian inference is performed on the parameter space $\Theta = \bigcup_{i=1}^{k_{\max}} \{i\} \times \Theta_i$. In this case, the models are also nested.

In practice, such as in the two previous examples, Bayesian analysis requires specification of prior distributions on the parameter spaces. Unfortunately, it is usually impossible, or sometimes not desirable, to specify distributions on some or all model parameters. In such cases, one may alternatively[2] consider 'non-informative' or 'default' priors, which are discussed in the next section.

2.2 Specification of default prior distributions

The need for automatic or *default* approaches has been recognized for a long time [35]. In estimation problems, the use of vague or 'non-informative' prior distributions, including sometimes improper prior distributions[3], is typically a satisfactory solution [5], [55]. When performing model selection, however, one has to be much more careful because default priors are typically improper, and, thus, depend on arbitrary multiplicative constants, *i.e.*, $p_i^N(\theta_i) = c_i f_i^N(\theta_i)$. (We use the superscript N to indicate the use of a *non-informative* or *default prior* for the model parameters.) Hence, the resultant Bayes Factor

$$B_{ji}^N = \frac{c_j}{c_i} \frac{\int_{\Theta_j} p\left(\mathbf{y} \mid \theta_j\right) p_j^N(\theta_j) d\theta_j}{\int_{\Theta_i} p\left(\mathbf{y} \mid \theta_i\right) p_i^N(\theta_i) d\theta_i} \tag{2.10}$$

is indeterminate, and cannot be used for model selection.

A number of proposals to overcome this problem have been made. Approaches using conventional priors were studied in [35], [67]. In the case of nested models, proper hierarchical robust prior models have been successfully developed for various applications [2], [53], [58]. Other approaches include the Intrinsic Bayes Factor [7], the Fractional Bayes Factor [45] and the method

---

[2] Another alternative is given by Robust Bayesian Analysis (see, for example, [4]).
[3] A density $p(\theta)$ is improper iff $\int p(\theta) d\theta = +\infty$.

suggested in [61], among others. Most of these later methods deal with the problem by rescaling the Bayes Factor by a *correction factor* in such a way that any normalizing constants would be canceled. We briefly discuss these methods in the following sections as well as a new alternative method called Expected Posterior Prior distributions [47, 48].

2.2.1 Conventional priors

One possible way to obtain a default model selection criterion is to choose a *conventional* prior according to the problem at hand. The term 'conventional prior' is used to refer to a prior which is agreed upon as reasonable for the problem [37].

Conventional prior approaches for tests concerning the mean of a normal model have been proposed in [35] and [67]. For the comparison of models $\mathcal{M}_1$: $\mathcal{N}(0, \sigma_1^2)$ vs. $\mathcal{M}_2$: $\mathcal{N}(\mu, \sigma_2^2)$, Jeffreys utilizes the priors $p_1^N(\sigma_1) \propto \sigma_1^{-1}$ and $p_2^N(\mu, \sigma_2) = p_1^N(\sigma_2)p_{22}(\mu|\sigma_2)$ for models $\mathcal{M}_1$ and $\mathcal{M}_2$, respectively, where p_{22} is a proper density function. Jeffreys argued that p_{22} should be an even function such that $\int p_{22}(\mu|\sigma_2)\mu^{n-1}d\mu$ diverges for any $n > 1$. He proposed the Cauchy density with scale σ_2 as the simplest density to meet these requirements.

For tests of hypotheses concerning multivariate normal means, Zellner and Siow [67], largely based on Jeffreys' work, suggested using a multivariate Cauchy density in place of p_{22}. The authors applied this idea to typical linear regression testing problems with orthogonal covariates, and provided Laplace approximations for the Bayes Factor in such cases.

2.2.2 Hierarchical models

Another possible solution consists of using a hierarchical model. Indeed, in many cases the prior distribution of the parameter θ_k further depends on a set of hyperprior parameters η. It is possible to add an extra layer in the prior distribution structure, which therefore models the uncertainty on the distribution of θ_k through a prior on η. The model for the prior density now takes the form

$$p(k, \theta_k, \eta) = p(k, \theta_k|\eta) p(\eta) \tag{2.11}$$

and the parameters η either become part of the inference problem, or may be integrated out as nuisance parameters, defining the prior for θ_k as

$$p(k, \theta_k) = \int p(k, \theta_k|\eta) p(d\eta). \tag{2.12}$$

Typically the prior for η will be a *vague* or *non-informative* (proper) prior so that its influence on the model selection is minimum, leading to a robust estimation. This approach has been applied for example in [53] for Gaussian mixture models, where vague, data dependent hyperpriors are introduced to model the uncertainty on the unknown component variances and locations of the mixture.

The hierarchical prior approach will be further illustrated in the present chapter in Section 4 for the problem of detection of sinusoids in noise.

2.2.3   Spiegelhalter–Smith Bayes Factor

Spiegelhalter and Smith [61] used the device of *imaginary training samples* in the context of linear model comparisons to choose the constant c_j/c_i in (2.10).

Assume that an imaginary training sample, $\mathbf{y}_0^*$, is available such that

1. it is of *minimal size*, *i.e.*, it involves the smallest possible sample size that allows comparing $\mathcal{M}_i$ against $\mathcal{M}_j$ under non-informative priors;

2. it provides a maximum possible support for the simpler model, say, $\mathcal{M}_i$.

The authors argued that, for such a training sample, the Bayes Factor of $\mathcal{M}_j$ against $\mathcal{M}_i$ should be equal to one. Under this assumption, it follows that

$$\frac{c_i}{c_j} = \left\{ \frac{\int_{\Theta_i} p\left(\mathbf{y}_0^*|\,\theta_i\right) p_i^N\left(\theta_i\right) d\theta_i}{\int_{\Theta_j} p\left(\mathbf{y}_0^*|\,\theta_j\right) p_j^N\left(\theta_j\right) d\theta_j} \right\}^{-1}. \tag{2.13}$$

The Spiegelhalter–Smith Bayes Factor for observations $\mathbf{y}$ is now given by

$$B_{ji}^{SS} = B_{ji}^N \frac{\int_{\Theta_i} p\left(\mathbf{y}_0^*|\,\theta_i\right) p_i^N\left(\theta_i\right) d\theta_i}{\int_{\Theta_j} p\left(\mathbf{y}_0^*|\,\theta_j\right) p_j^N\left(\theta_j\right) d\theta_j}. \tag{2.14}$$

Note, however, that the choice of $\mathbf{y}_0^*$ depends on the models under comparison, and so there is no guarantee that B^{SS} is coherent for multiple model comparisons, *i.e.*, that $B_{ij}^{SS} \neq B_{ik}^{SS} B_{kj}^{SS}$.

2.2.4   Fractional Bayes Factors

The Fractional Bayes Factor, developed by O'Hagan [45], is based on using a fraction b of the likelihood, denoted here $L^b\left(\theta_i\right)$ for θ_i, to update non-informative priors. The Fractional Bayes Factor for model $\mathcal{M}_j$ against $\mathcal{M}_i$ is given by

$$B_{ji}^F = B_{ji}^N\left(\mathbf{y}\right) \frac{\int_{\Theta_i} L^b\left(\theta_i\right) p_i^N\left(\theta_i\right) d\theta_i}{\int_{\Theta_j} L^b\left(\theta_j\right) p_j^N\left(\theta_j\right) d\theta_j}. \tag{2.15}$$

One common choice of b is $b = m/T$, where m is the minimal training sample size and T the number of available data. The Fractional Bayes Factors are typically easier to compute than the Intrinsic Bayes Factors.

2.2.5   Intrinsic Bayes Factors

Consider two models $\mathcal{M}_1$ and $\mathcal{M}_2$ with data $\mathbf{y}$ having density $p\left(\mathbf{y}|\,\theta_i\right)$ and prior $p_i^N\left(\theta_i\right)$ under model $\mathcal{M}_i$. The *Arithmetic Intrinsic Bayes Factor* developed by Berger and Pericchi [7] is given by

$$B_{21}^{\mathrm{AIBF}} = B_{21}^N \frac{1}{L} \sum_{l=1}^{L} B_{12}^N\left(\mathbf{y}\left(l\right)\right), \tag{2.16}$$

where $B_{21}^N = B_{21}^N(\mathbf{y})$ is the usual Bayes Factor between models $\mathcal{M}_2$ and $\mathcal{M}_1$ and

$$m_i^N(\mathbf{y}) \triangleq \int_{\Theta_i} p(\mathbf{y}|\theta_i) p_i^N(\theta_i) d\theta_i, \tag{2.17}$$

the marginal under model $\mathcal{M}_i$. The sum is over all possible *minimal training samples*, $\mathbf{y}(l)$, of the observed data $\mathbf{y}$. A minimal training sample is defined as the minimum number of observations such that all marginals $m_i^N(\mathbf{y}(l))$ are positive and finite. Other versions of the IBF can be constructed by using averages rather than the mean in (2.16). For instance, using the geometric mean yields the *Geometric IBF* (B^{GIBF}). Choosing the median of Bayes Factors $B_{12}^N(\mathbf{y}(l))$ gives the *Median IBF* (B^{MIBF}), which appears to be more stable in cases where the sample size is relatively small. For a discussion of the Median IBF, see [10].

For the Arithmetic IBF, it is typically the case that $B_{ij}^{\mathrm{IABF}} \neq 1/B_{ji}^{\mathrm{IABF}}$. Thus, the AIBF is not suitable for multiple comparisons in its pure form. In [7] the authors suggest placing the more complex model, say $\mathcal{M}_j$, in the numerator and defining $B_{ij}^{\mathrm{IABF}} = 1/B_{ji}^{\mathrm{IABF}}$. Alternatively, one might define an *encompassing* model $\mathcal{M}_E$ such that every other model under consideration is nested in it. One can then obtain B_{Ei}^{IABF} for all i, and define the *Encompassing* IBF (based on $\mathcal{M}_E$) by $B_{ji}^{EIBF} = B_{Ei}^{\mathrm{AIBF}}/B_{Ej}^{\mathrm{AIBF}}$.

One appealing property of the arithmetic IBF is its asymptotic equivalence with a 'proper' Bayes Factor arising from *Intrinsic Priors*. By using a Schwartz approximation to the Bayes Factor one obtains

$$B_{21} = B_{21}^N \frac{p_2\left(\hat{\theta}_2\right) p_1^N\left(\hat{\theta}_1\right)}{p_2^N\left(\hat{\theta}_2\right) p_1\left(\hat{\theta}_1\right)} (1 + o(1)). \tag{2.18}$$

Equating this with (2.16) yields

$$\frac{p_2\left(\hat{\theta}_2\right) p_1^N\left(\hat{\theta}_1\right)}{p_2^N\left(\hat{\theta}_2\right) p_1\left(\hat{\theta}_1\right)} (1 + o(1)) = \frac{1}{L} \sum_{l=1}^L B_{12}^N(\mathbf{y}(l)) = \tilde{B}_{12}^N. \tag{2.19}$$

Assume
- Under $\mathcal{M}_1$: $\hat{\theta}_1 \to \theta_1$, $\hat{\theta}_2 \to \psi_2(\theta_1)$ and $\tilde{B}_{12}^N \to B_1^*(\theta_1)$;
- Under $\mathcal{M}_2$: $\hat{\theta}_2 \to \theta_2$, $\hat{\theta}_1 \to \psi_1(\theta_2)$ and $\tilde{B}_{12}^N \to B_2^*(\theta_2)$;
- For $i = 1$ or 2, the following exists:

$$B_i^*(\theta_i) = \lim_{L \to \infty} \mathbb{E}_{\mathcal{M}_i} \left[\frac{1}{L} \sum_{l=1}^L B_{12}^N(\mathbf{y}(l)) \middle| \theta_i \right], \tag{2.20}$$

where $\mathbb{E}_{\mathcal{M}_i}(\cdot)$ is the expectation with respect to the observations $\mathbf{y}(l)$ under model $\mathcal{M}_i$ and conditional upon θ_i.

If $\mathbf{y}(l)$ is exchangeable then the limit and average over l can be removed. Passing to the limit in (2.19), first under $\mathcal{M}_2$, and then under $\mathcal{M}_1$, results in the *Intrinsic Equations* for the Intrinsic Priors

$$\frac{p_2^I(\theta_2)\,p_1^N(\psi_1(\theta_2))}{p_2^N(\theta_2)\,p_1^I(\psi_1(\theta_2))} = B_2^*(\theta_2), \qquad \frac{p_2^I(\psi_2(\theta_1))\,p_1^N(\theta_1)}{p_2^N(\psi_2(\theta_1))\,p_1^I(\theta_1)} = B_1^*(\theta_1). \quad (2.21)$$

In the case where $\mathcal{M}_1$ is nested in $\mathcal{M}_2$, it can be shown that the Intrinsic Equations (2.21) have a solution of the form

$$\begin{aligned} p_2^I(\theta_2) &= p_2^N(\theta_2)\ \mathbb{E}_{\mathcal{M}_2}\left[\left.\frac{m_1^N(\mathbf{y})}{m_2^N(\mathbf{y})}\right|\theta_2\right], \\ p_1^I(\theta_1) &= p_1^N(\theta_1) \end{aligned} \qquad (2.22)$$

(see, for example, [24]).

2.2.6   Expected Posterior Priors

Expected Posterior Priors have recently been proposed by Pérez and Berger [47], [48]. The method consists of adjusting the initial priors for each model by

$$p_i^*(\theta_i) = \int p_i^N(\theta_i|\mathbf{y}^*)\,m^*(\mathbf{y}^*)d\mathbf{y}^*, \qquad (2.23)$$

where m^* is a suitable *predictive measure* on the *imaginary training sample* space and, as in the Intrinsic Bayes Factor, $\mathbf{y}^*$ is of minimal size. Several choices for m^* are possible. One choice for m^* that is attractive arises from selecting a *base model* $\mathcal{M}_*$ for the training sample and defining $m^*(\mathbf{y}^*) = m_*^N(\mathbf{y}^*)$. In the case of nested models, the Expected Posterior Priors resulting from this choice of m^* correspond to the Intrinsic Priors for the Arithmetic IBF. An empirical version of m^* can also be considered, where training samples are obtained by resampling from the observations.

The updated prior, p^*, is called the *Expected Posterior Prior* (or EP Prior) under m^*. The Bayes Factor of $\mathcal{M}_i$ against $\mathcal{M}_j$ resulting from the *Expected Posterior Priors* is given by

$$B_{ij}^*(\mathbf{y}) = \frac{m_{p_i^*}(\mathbf{y})}{m_{p_j^*}(\mathbf{y})}, \qquad (2.24)$$

where

$$m_{p_i^*}(\mathbf{y}) \triangleq \int_{\Theta_i} p(\mathbf{y}|\theta_i)\,p_i^*(\theta_i)\,d\theta_i, \qquad (2.25)$$

which can be written as

$$B_{ij}^*(\mathbf{y}) = \frac{m_{p_i^*}(\mathbf{y})}{m_{p_j^*}(\mathbf{y})} = \frac{\int m_i^N(\mathbf{y}|\mathbf{y}^*)\,m^*(\mathbf{y}^*)\,d\mathbf{y}^*}{\int m_j^N(\mathbf{y}|\mathbf{y}^*)\,m^*(\mathbf{y}^*)\,d\mathbf{y}^*}. \qquad (2.26)$$

Therefore, the resulting Bayes Factor does not depend on arbitrary multiplicative constants and, therefore Bayesian model selection based on B_{ij}^* is coherent and allows for multiple comparisons.

The EP Prior approach admits several extensions. The key insight for the EP Prior is the use of $\mathbf{y}^*$ to update model parameters with improper priors. Note, however, that the improper priors p_i^N could be updated with any statistic $\psi_i(\cdot)$ of $\mathbf{y}^*$ and its associated 'likelihood', *i.e.*

$$p_i^N(\theta_i|\psi_i(\mathbf{y}^*)) = \frac{f_i(\psi_i(\mathbf{y}^*)|\theta_i)p_i^N(\theta_i)}{m^N(\psi_i(\mathbf{y}^*))}. \tag{2.27}$$

This gives rise to a more general definition of EP Priors. Suppose there exist statistics, $\psi_i(\mathbf{y}^*)$, for each model $\mathcal{M}_i$, such that $p_i^N(\theta_i|\psi_i(\mathbf{y}^*))$ is a proper density function and such that

$$0 < \mathbb{E}_{\mathcal{M}_i}\left[\left.\frac{m^*(\mathbf{y}^*)}{m^N(\psi_i(\mathbf{y}^*))}\right|\theta_i\right] < \infty, \tag{2.28}$$

for each model. Then EP Priors can be more generally defined as

$$p_i^*(\theta_i) = \int p_i^N(\theta_i|\psi_i(\mathbf{y}^*))m^*(\mathbf{y}^*)d\mathbf{y}^*. \tag{2.29}$$

Some other interesting properties of this approach are

- The resulting Bayesian inference is coherent and allows for multiple comparisons.
- In many cases, it is possible to find m^* such that, for a sample of minimal size, there is *predictive matching* for the comparisons of model $\mathcal{M}_i$ against $\mathcal{M}_j$, *i.e.*, the Bayes Factor $B_{ij} = 1$.
- In the case of nested models, where $\mathcal{M}_1$ is nested in every other model, choosing $m^*(\mathbf{y}^*)$ to be the marginal of $\mathbf{y}^*$ under $\mathcal{M}_1$ is asymptotically equivalent to the arithmetic Intrinsic Bayes Factor [7].

2.3 Practical issues

The previous subsections have outlined some of the 'theoretical' problems related to the definition of a Bayesian model to perform model selection, specially in the cases where improper, default priors are used. One further problem in the Bayesian framework relates to the computation of integrals, such as the Bayes Factors, which do not admit any closed-form expression, except for simple examples. Although numerical approximations were available, their application was often constrained by the high dimensionality of the integral involved. This problem had severely limited the application of Bayesian inference until the beginning of the 90's. Since the introduction of simulation methods, the Bayesian community has been able to computationally address more complex and large problems that were out of the scope before. The next section focusses on Monte Carlo and MCMC simulation methods and their uses in solving the numerical problems often encountered in Bayesian analysis. We will discuss the applications of these methods to perform Bayesian detection of sinusoids in noise and Bayesian analysis of finite mixtures of normals in Section 4.

3 MCMC algorithms for Bayesian model selection

This section first presents the principles of Monte Carlo approximations for integrals of the type $\int f(\phi) p(d\phi)$. Here, $p(d\phi)$ is some probability distribution from which it is possible to draw i.i.d. samples. This can be difficult in practice. An often easier approach is the MCMC approach. The MCMC methodology for approximation of integrals is shown in detail in subsection 3.2, after recalling some results on the convergence of Markov chains admitting $p(d\phi)$ as invariant distribution.

3.1 Principle of Monte Carlo methods for integration

Consider a probability distribution $p(d\phi)$ defined on a general state space Φ. Assume that $M \gg 1$ samples $\{\phi^{(j)}; j = 1, ..., M\}$ distributed according to the distribution $p(d\phi)$ are available. Then, a Monte Carlo approximation $\widehat{P}_M(d\phi)$ of this distribution is given by the empirical estimate

$$\widehat{P}_M(d\phi) = \frac{1}{M} \sum_{j=1}^{M} \delta_{\phi^{(j)}}(d\phi), \tag{3.1}$$

where $\delta_{\phi^{(j)}}(d\phi)$ is the Dirac delta measure located at $\phi^{(j)}$. That is, the concentration of the samples in a given zone of the space Φ is assumed sufficiently representative of the probability of this zone under the distribution $p(d\phi)$.

Using this approximation for $p(d\phi)$, one can propose the following estimate f_M of $\mathbb{E}_{p(d\phi)}[f(\phi)]$, where $f : \Phi \to \mathbb{R}^{n_f}$ is a $p(d\phi)$-integrable function:

$$f_M = \int_{\Phi} f(\phi) \widehat{P}_M(d\phi) = \frac{1}{M} \sum_{j=1}^{M} f\left(\phi^{(j)}\right). \tag{3.2}$$

This estimate is unbiased and if the samples $\{\phi^{(j)}; j = 1, ..., M\}$ are statistically independent, then

$$\lim_{M \to +\infty} f_M \overset{a.s.}{\to} \mathbb{E}_{p(d\phi)}(f(\phi)) \tag{3.3}$$

from the strong law of large numbers. Moreover if

$$var[f_M] = \frac{1}{M}\left[\mathbb{E}_{p(d\phi)}\left(f^2(\phi)\right) - \mathbb{E}^2_{p(d\phi)}(f(\phi))\right] \triangleq \frac{\sigma_f^2}{M} < +\infty, \tag{3.4}$$

then the central limit theorem yields

$$\lim_{M \to +\infty} \sqrt{M} \left(f_M - \mathbb{E}_{p(d\phi)} \left(f\left(\phi\right)\right)\right) \stackrel{dist.}{\Rightarrow} \mathcal{N}\left(0, \sigma_f^2\right). \tag{3.5}$$

In the case of model selection, *i.e.* $\Phi = \Theta$ and $p\left(d\phi\right) = p\left(k, d\theta\right)$, this type of integration method will allow us to obtain approximations of quantities of the type

$$p\left(i | \mathbf{y}\right) = \int_{\Theta_i} p_i\left(i, d\theta_i | \mathbf{y}\right) = \int_{\Theta} \mathbb{I}_{\{i\} \times \Theta_i}\left(k, \theta\right) p\left(k, d\theta | \mathbf{y}\right) \tag{3.6}$$

using estimators of the form

$$\widehat{p}\left(i | \mathbf{y}\right) = \frac{\sum_{j=1}^{M} \mathbb{I}_{\{i\}}\left(k^{(j)}\right)}{M} \tag{3.7}$$

when i.i.d. samples $\left\{\left(k^{(j)}, \theta^{(j)}\right) ; j = 1, ..., M\right\}$ distributed according to $p\left(k, d\theta_k | \mathbf{y}\right)$ are available. This in turn allows for the Bayes Factors to be estimated as

$$\widehat{B}_{ji} = \frac{\widehat{p}\left(j | \mathbf{y}\right) / p_j}{\widehat{p}\left(i | \mathbf{y}\right) / p_i}. \tag{3.8}$$

This approach has two major advantages. Contrary to classical numerical integration methods, the empirical distribution (3.1) enables estimates of $\mathbb{E}_{p(d\phi)}\left[f\left(\phi\right)\right]$ to be obtained easily for any function f, regardless of how complex this function may be. Moreover, when the samples are statistically independent, the dimension of the integration space n_ϕ does not appear in the convergence rate of the estimate towards its theoretical value. This stems from the fact that stochastic methods cleverly explore the space, according to the importance of the different areas, while deterministic methods mainly rely on a regular discretization and exploration of the space Φ. However, it might be extremely difficult to obtain i.i.d. samples $\left\{\phi^{(j)}; j = 1, ..., M\right\}$ distributed according to a given distribution, especially for non-standard and high dimensional distributions.

In the next subsections, we present iterative techniques, Markov chain Monte Carlo (MCMC), that allow for samples to be asymptotically drawn from almost any given distribution. As we shall see in the next subsection, although the samples obtained at the output of the MCMC methods are not independent, inherent to the Markov chain, the ergodic properties (3.3) and (3.5) still hold under mild conditions.

3.2 Markov chains with a given invariant distribution

Throughout this subsection, we will assume that the state space of interest is Φ and that the target distribution of the Markov chain is $p\left(d\phi\right)$. The basic idea of MCMC methods is to simulate an ergodic Markov chain whose samples are asymptotically distributed according to the distribution $p(d\phi)$. Thus, non-independent samples distributed according to $p(d\phi)$ can be obtained and still allow for Monte Carlo integration to be performed. Before presenting practical mechanisms for building such Markov chains, we need to present some basic results on the convergence of Markov chains which are the rationale of the method.

Let $K(\phi, A) \triangleq \Pr\left(\phi^{(j+1)} \in A \mid \phi^{(j)} = \phi\right)$ be the transition kernel of the Markov chain and define

$$K^1(\phi, A) = K(\phi, A), \tag{3.9}$$

$$K^n(\phi, A) = \int_\Theta K^{n-1}(\phi, d\gamma) K(\gamma, A) \text{ for } n > 1. \tag{3.10}$$

Definition 3.1 Invariance. *Let $p(d\phi)$ be a probability distribution. We say that $p(d\phi)$ is an invariant or stationary distribution for the transition kernel K if for any $A \in \mathcal{B}(\Phi)$*

$$p(A) = \int_\Phi p(d\phi) K(\phi, A) = \int_\Phi p(d\phi) \int_A K(\phi, d\gamma). \tag{3.11}$$

The notion of invariance implies that, if a state of the Markov chain ϕ_i is distributed according to $p(d\phi)$, then ϕ_{i+1} and all the following states are distributed marginally according to $p(d\phi)$.

Definition 3.2 Irreducibility. *Let φ be a probability measure on $(\Phi, \mathcal{B}(\Phi))$. A Markov chain is φ-irreducible if for any $\phi \in \Theta$ and for any $A \in \mathcal{B}(\Phi)$,*

$$\varphi(A) > 0 \Rightarrow \exists n \in \mathbb{N}^* \quad \text{such that } K^n(\phi, A) > 0. \tag{3.12}$$

On a φ-irreducible chain, all sets of non-null φ measure can be reached with a non-null probability in a finite number of iterations from any starting point. A sufficient condition to verify the irreducibility of K with respect to φ is that there exists $n \in \mathbb{N}$ such that K^n has a strictly positive density $K^n_\varphi(\phi, \gamma) > 0$ with respect to φ, i.e. $K^n(\phi, A) = \int_A K^n_\varphi(\phi, d\gamma) \varphi(d\gamma)$ [29, p. 63].

Definition 3.3 Aperiodicity. *A Markov chain is called aperiodic if there does not exist a partition of $\Phi = (A_1, ..., A_d)$ for $d \geq 2$ such that $K(\phi, A_{i+1 \bmod d}) = 1$ for any $\phi \in A_i$.*

The aperiodicity condition eliminates kernels which induce a periodic behavior in the trajectories of the Markov chain. If the transition kernel $K(\phi, A)$ is such that the probability of staying at ϕ is non-null, or if this kernel has a positive density in the neighborhood of ϕ, then the chain is aperiodic.

Before stating the main result of convergence, it is necessary to introduce a distance between two probability distributions μ_1 and μ_2 on $(\Phi, \mathcal{B}(\Phi))$.

Definition 3.4 *The total variation norm is defined by*

$$\|\mu_1 - \mu_2\|_{TV} \triangleq \sup_{A \in \mathcal{B}(\Phi)} |\mu_1(A) - \mu_2(A)|. \tag{3.13}$$

If $\mu_1(d\phi) = \mu_1(\phi)\,d\phi$ and $\mu_2(d\phi) = \mu_2(\phi)\,d\phi$, then

$$\|\mu_1 - \mu_2\|_{TV} = \frac{1}{2}\int_{\Phi} |\mu_1(\phi) - \mu_2(\phi)|\,d\phi. \tag{3.14}$$

Let $\mathbb{E}_p[f] \triangleq \int_{\Phi} f(\phi)\,p(d\phi)$. The estimate of this quantity obtained by averaging the M first simulated values of the Markov chain will be denoted by

$$f_M \triangleq \frac{1}{M}\sum_{j=0}^{M-1} f\left(\phi^{(j)}\right). \tag{3.15}$$

We are now able to state the main convergence result.

Theorem 3.1 [41], [65, Theorem 1*, p. 1758] *Let $\{\phi^{(j)}; j \in \mathbb{N}\}$ be a Markov chain with transition kernel K and invariant distribution p. Assume that the Markov chain is φ-irreducible, where φ is a probability distribution on $(\Phi, \mathcal{B}(\Phi))$. Let $f : \Phi \to \mathbb{R}$ be such that $\mathbb{E}_p[|f|] < +\infty$; then*

$$P_{\phi_0}\left[f_M \stackrel{M \to +\infty}{\rightarrow} \mathbb{E}_p[f]\right] = 1$$

for p-almost all points $\phi^{(0)}$. Moreover, if K is aperiodic

$$\lim_{M \to +\infty} \left\|K^M\left(\phi^{(0)}, \cdot\right) - p(\cdot)\right\|_{TV} = 0$$

for p-almost all $\phi^{(0)}$.

Remark 3.1 *The aperiodicity condition is not necessary if one is only interested in estimates of $\mathbb{E}_p[f]$.*

The conditions on the transition kernel which lead to this theorem are weak and intuitive: the Markov chain must be able to explore the support of the target distribution and this exploration must not be periodic. The estimates obtained by averaging the simulated samples are asymptotically consistent and these samples are asymptotically distributed according to $p(d\phi)$. However, this theorem allows some starting points (of null p measure) for which convergence is not ensured. The theorem is valid for all starting points ϕ_0 if the Markov chain is in addition Harris recurrent [41], [65], [57].

The convergence theorem theoretically validates under weak assumptions the basic principle of MCMC algorithms, that is, the idea that constructing a Markov chain of invariant distribution $p(d\phi)$, when simulating directly from

$p\,(d\phi)$ is not feasible. Some care must be taken, however, on the construction of the Markov chain, as this theorem does not give any information on the rate of convergence of the chain towards its invariant distribution. For some Markov chains, the convergence can be so slow that the approach would be of no practical interest. It is possible to obtain, under additional assumptions on the transition kernels, better convergence results that ensure a geometric or even a uniform geometric convergence rate of the Markov chain [65], [41] and [57].

The following subsection presents a general method that allows for transition kernels for Markov chain that admit a given distribution $p\,(d\phi)$ as invariant distribution to be built.

3.3   The general Metropolis–Hastings algorithm: reversible jump MCMC

This section describes how to build MCMC algorithms for model selection, *i.e.* how to construct ergodic Markov chains admitting $p_k\,(k,d\theta_k|\,\mathbf{y})$ as their invariant distribution. To ease notation in this section, we will write $p_k\,(k,d\theta_k)$ for $p_k\,(k,d\theta_k|\,\mathbf{y})$. In the case of model selection, the main difficulty for the Markov chain is to be able to jump from one subspace Θ_n to another subspace Θ_m while preserving the correct invariant distribution. Green [33] has developed a general methodology that addresses this problem.

The Metropolis–Hastings (MH) algorithm is a method of constructing a reversible transition kernel for a Markov chain with a specified invariant distribution $p_k\,(k,d\theta_k)$. The algorithm requires a proposal distribution $q_{k,k^*}\,(k^*,d\theta_{k^*}^*|\;k,\theta_k)$ and an acceptance probability $\alpha\,(k,\theta;k^*,\theta_{k^*}^*)$. If the current state is $(k,d\theta_k)$ then a candidate (k^*,θ_k^*) for the next state is proposed from $q_{k,k^*}\,(k^*,d\theta_{k^*}^*|k,\theta_k)$ and accepted with probability $\alpha\,(k,\theta;k^*,\theta_{k^*}^*)$. Otherwise, the candidate is rejected and the process remains in the state (k,θ). The transition kernel of the Markov chain can be written

$$K\,(k,\theta_k;k^*,d\theta_{k^*}^*) \;\;=\;\; q_{k,k^*}\,(k^*,d\theta_{k^*}^*|\,k,\theta_k)\,\alpha\,(k,\theta_k;k^*,\theta_{k^*}^*) \qquad (3.16)$$

$$+\;\mathbb{I}_{\{k\}}\,(k^*)\,\delta_{\theta_k}\,(d\theta_{k^*}^*)\int_\Theta (1-\alpha\,(k,\theta_k;l,\gamma))\,q_{k,l}\,(l,d\gamma|\,k,\theta_k)\,.$$

A very simple condition for ensuring that a given distribution $p_k\,(k,d\theta_k)$ is an invariant distribution for a given transition kernel K is to impose *reversibility*, *i.e.*

$$p_k\,(k,d\theta_k)\,K\,(k,\theta_k;k^*,d\theta_{k^*}^*) = p_{k^*}\,(k^*,d\theta_{k^*}^*)\,K\,(k^*,\theta_{k^*}^*;k,d\theta_k)\,. \qquad (3.17)$$

The two sides of this equality are measures on the product space $\Theta \times \Theta$, and the equality ensures that they are the same. Integrating $(k^*,d\theta_{k^*}^*)$ over the space Θ yields

$$\int_\Theta p_{k^*}\,(k^*,d\theta_{k^*}^*)\,K\,(k^*,d\theta_{k^*}^*;k,d\theta_k) = p_k\,(k,d\theta_k)\,, \qquad (3.18)$$

which is precisely the invariance property. Ensuring reversibility when the dimension of the problem is fixed, $p_{k^*}(k^*, d\theta_{k^*}^*)$ and $q_{k,k^*}(k^*, d\theta_{k^*}^* \mid k, \theta_k)$ admit densities $p_{k^*}(k^*, \theta_{k^*}^*)$ and $q_{k,k^*}(k^*, \theta_{k^*}^* \mid k, \theta_k)$ with respect to the same dominating measure is well known, and the following choice for $\alpha(k, \theta_k; k^*, \theta_{k^*}^*)$,

$$\alpha(k, \theta_k; k^*, \theta_{k^*}^*) = \min\left\{1, \frac{p_{k^*}(k^*, \theta_{k^*}^*) \, q_{k^*,k}(k, \theta_k \mid k^*, \theta_{k^*}^*)}{p_k(k, \theta_k) \, q_{k,k^*}(k^*, \theta_{k^*}^* \mid k, \theta_k)}\right\}, \qquad (3.19)$$

is satisfactory.

Returning to the model selection problem, a Markov chain can be constructed using the above methodology, where each state of the chain is proposed from the model spaces Θ_m. However, any information common to the different subspaces might be discarded, leading to an inefficient algorithm. A scenario for which this seems to be crucial is that of nested models in which components need to be added or removed more or less independently. This type of approach raises measure theoretic problems, and Green [33] has described a very general methodology for solving this problem. It requires, for each pair of communicating spaces Θ_m and Θ_n, firstly the definition of extended versions of these subspaces into $\overline{\Theta}_{m,n} \triangleq \Theta_m \times \Psi_{m,n}$ and $\overline{\Theta}_{n,m} \triangleq \Theta_n \times \Psi_{n,m}$, secondly the definition of a deterministic invertible mapping $f_{n,m}$ everywhere differentiable between $\overline{\Theta}_{m,n}$ and $\overline{\Theta}_{n,m}$,

$$\begin{aligned}(\theta_m, \varphi_{m,n}) &= f_{n,m}(\theta_n, \varphi_{n,m}) \\ &= \left(f_{n,m}^\theta(\theta_n, \varphi_{n,m}), f_{n,m}^\varphi(\theta_n, \varphi_{n,m})\right); \qquad (3.20)\end{aligned}$$

(we define $f_{m,n}$ such that $f_{m,n}(f_{n,m}(\theta_n, \varphi_{n,m})) = (\theta_n, \varphi_{n,m})$), and thirdly the choice of proposal densities for $\varphi_{n,m}$ and $\varphi_{m,n}^*$, respectively $q_{n,m}(\cdot \mid n, \theta_n)$ and $q_{m,n}(\cdot \mid m, \theta_m)$. The choice of the extended spaces, deterministic transformation $f_{m,n}$ and proposal distributions for $q_{n,m}(\cdot \mid n, \theta_n)$ and $q_{m,n}(\cdot \mid m, \theta_m)$ is problem dependent and needs to be addressed on a case by case basis.

The proposal distribution can now be written

$$\begin{aligned}q(k^*, d\theta_{k^*}^* \mid n, \theta_n) &= \sum_{m=1}^{k_{\max}} \delta_{f_{n,m}^\theta(\theta_n, \varphi_{n,m})}(d\theta_m) \qquad (3.21) \\ &\quad \times q_{n,m}(m, d\varphi_{n,m} \mid k, \theta_k) \, \mathbb{I}_{\{m\} \times \Theta_m}(k^*, \theta_{k^*}^*).\end{aligned}$$

With the assumption that all probability distributions admit a density with respect to the Lebesgue measure, the acceptance probability of a move from $\overline{\Theta}_n$ to $\overline{\Theta}_m$ satisfies

$$\begin{aligned}r((n, \theta_n), (m, \theta_m^*)) &= \qquad\qquad\qquad\qquad\qquad\qquad\qquad (3.22) \\ &\frac{p_m\left(m, f_{n,m}^\theta(\theta_n, \varphi_{n,m})\right) q_{m,n}\left(f_{n,m}^\varphi(\theta_n, \varphi_{n,m}) \mid m, \theta_m^*\right)}{p_n(n, d\theta_n) \, q_{n,m}(\varphi_{n,m} \mid n, \theta_n)} \times \mathcal{J}_{f_{n,m}},\end{aligned}$$

where $\mathcal{J}_{f_{n,m}}$ is the Jacobian, when only continuous variables are involved in the transformation, of the invertible mapping $f_{n,m}$ between the spaces Θ_n and Θ_m,

$$\mathcal{J}_{f_{m,n}} = \left| \det \frac{\partial f_{n,m}(\theta_m, \varphi_{m,n})}{\partial(\theta_m, \varphi_{m,n})} \right|. \tag{3.23}$$

The procedure can be summarized with the following pseudo-code.

Reversible jump MCMC algorithm

1. Initialization: set $\left(k^{(0)}, \theta_k^{(0)}\right) \in \Theta$, and $i = 1$.

2. Iteration i.

   - Sample m from the discrete distribution $q_{k^{(i-1)},m}\left(m \mid k^{(i-1)}, \theta_{k^{(i-1)}}^{(i-1)}\right)$.
   - Sample $\varphi_{k^{(i-1)},m}^{(i-1)} \sim q_{k^{(i-1)},m}\left(d\varphi_{k^{(i-1)},m}^{(i-1)} \mid m, k^{(i-1)}, \theta_{k^{(i-1)}}^{(i-1)}\right)$ and perform the invertible transformation

     $$\left(\theta_m^*, \varphi_{m,k^{(i-1)}}^*\right) = f_{k^{(i-1)},m}\left(\theta_{k^{(i-1)}}^{(i-1)}, \varphi_{k^{(i-1)},m}^{(i-1)}\right) \tag{3.24}$$

   - Accept the move with probability

     $$\alpha\left(\left(k^{(i-1)}, \theta_{k^{(i-1)}}^{(i-1)}\right), (m, \theta_m^*)\right)$$
     $$= \min\left\{1, r\left(\left(k^{(i-1)}, \theta_{k^{(i-1)}}^{(i-1)}\right), (m, \theta_m^*)\right)\right\} \tag{3.25}$$

     i.e. $\left(k^{(i)}, \theta_{k^{(i)}}^{(i)}\right) = (m, \theta_m^*)$, otherwise stay at $\left(k^{(i-1)}, \theta_{k^{(i-1)}}^{(i-1)}\right)$.

Note that this algorithm is not guaranteed to produce, even asymptotically, samples from the correct distributions, as the two extra properties, namely irreducibility and aperiodicity (See Subsection 3.2), of the Markov chain need to be checked.

Remark 3.2 One could object that it is possible to use a classical MCMC within each subspace Θ_n and then propose the following naive estimate $\widehat{m}_i(\mathbf{y})$:

$$\widehat{m}_i(\mathbf{y}) = p_i\left\{\frac{1}{M}\sum_{j=1}^{M}\frac{\mathbb{I}_{\{i\}}\left(k^{(j)}\right)}{p_i\left(\mathbf{y} \mid \theta_i^{(j)}\right)}\right\}^{-1}. \tag{3.26}$$

Despite being consistent, $\widehat{m}_i(\mathbf{y})$ often has infinite variance [44]. Several attempts have been made to improve this estimator (3.26) (see for example [44], [66]) but the reversible jump Markov chain approach still allows for the simultaneous estimation of all model parameters and model posterior probabilities in the same simulation run.

3.4 Other approaches

3.4.1 The method of subspace extension and the algorithm of Carlin and Chib

One can, of course, generalize what has been presented in the previous subsection by extending the probabilistic models corresponding to the two subspaces Θ_i and Θ_j. More precisely, we can introduce extended parameters $\bar{\theta}_i \triangleq (\theta_i, \varphi_i)$ defined on extended sets $\overline{\Theta}_i \triangleq \Theta_i \times \Psi_i$ and associated with probabilistic models $\bar{p}_i(i, d\bar{\theta}_i)$ such that

$$\int_{\Psi_i} \bar{p}_i(i, d\bar{\theta}_i) = p_i(i, d\theta_i), \qquad (3.27)$$

which means that there exists a distribution $\bar{p}_i(d\varphi_i | \theta_i)$ such that

$$\bar{p}_i(i, d\bar{\theta}_i) = \bar{p}_i(d\varphi_i | \theta_i) p_i(i, d\theta_i). \qquad (3.28)$$

Assuming that there exists a deterministic invertible mapping $f_{i,j}$ between $\overline{\Theta}_i$ and $\overline{\Theta}_j$, one can apply the strategy described in the previous subsection.

A particular case of interest is the method proposed by Carlin and Chib [16], in which the following choice for the family of spaces $(\Psi_i)_{i=1,\dots,k_{\max}}$ was made:

$$\Psi_i = \Theta_1 \times \dots \times \Theta_{i-1} \times \Theta_{i+1} \times \dots \times \Theta_{k_{\max}}. \qquad (3.29)$$

This requires the definition of the distributions

$$\bar{p}_i(d\varphi_i | \theta_i), \qquad (3.30)$$

called 'pseudo-priors', as they do not have any meaning in the true statistical sense, despite the fact that φ_i is composed of parameters similar to θ_l for $l \neq i$. Note that the complete state space can then be written as

$$\begin{aligned}
\overline{\Theta} &= \bigcup_{i=1}^{k_{\max}} \{i\} \times \Theta_1 \times \dots \times \Theta_{k_{\max}} \\
&= \{1, \dots, k_{\max}\} \times \Theta_1 \times \dots \times \Theta_{k_{\max}} \qquad (3.31)
\end{aligned}$$

and that a fixed number, $k_{\max}$, of probabilistic models must be defined at the beginning of the procedure; $k_{\max}$, therefore, cannot change while the procedure is carried out.

The algorithm proposed by Carlin and Chib is a Gibbs sampler on $\overline{\Theta}$ which, contrary to Green's reversible jump MCMC, allows $\bar{\theta}_i$ to be drawn first and then the new model j conditional upon $\bar{\theta}_i$. A summary of the algorithm is as follows.

Carlin and Chib's algorithm

1. Iteration i.

   - Sample $\varphi^{(i)}_{k^{(i-1)}}$ and $\theta^{(i)}_{k^{(i-1)}}$,

   $$\varphi^{(i)}_{k^{(i-1)}} \sim \overline{p}_{k^{(i-1)}}\left(\cdot \,|\, \theta^{(i-1)}_{k^{(i-1)}}\right), \qquad (3.32)$$

   $$\theta^{(i)}_{k^{(i-1)}} \sim \overline{p}_{k^{(i-1)}}\left(\cdot \,|\, \varphi^{(i)}_{k^{(i-1)}}\right). \qquad (3.33)$$

   - Sample the new index $k^{(i)}$,

   $$k^{(i)} \sim \overline{p}_k\left(\cdot \,|\, \theta^{(i)}_k, \varphi^{(i)}_k\right) \propto \overline{p}_k\left(d\varphi^{(i)}_k \,\big|\, \theta^{(i)}_k\right) p_k\left(k, d\theta^{(i)}_k\right). \qquad (3.34)$$

2. Go to 1.

∎

Note that Carlin and Chib make further assumptions that, with $\varphi_k = (\varphi_{1,k}, \ldots, \varphi_{k-1,k}, \varphi_{k+1,k}, \ldots, \varphi_{k_{\max},k})$, the $\varphi_{j,k}$ are independent among themselves and of θ_k, conditional upon k. More precisely, they assume that

$$\overline{p}_k\left(k, d\left(\theta_k, \varphi_k\right)\right) = p_k\left(k, d\theta_k\right) \prod_{j=1, j\neq k}^{k_{\max}} \overline{p}_k\left(d\varphi_{j,k}\right) \qquad (3.35)$$

and that the pseudo-priors do not depend on the current index, that is, for any k and distinct j and i such that $j \neq k$ and $i \neq k$ then

$$\overline{p}_j\left(d\varphi_k\right) = \overline{p}_i\left(d\varphi_k\right). \qquad (3.36)$$

This algorithm has several drawbacks:

- $k_{\max}$ has to be finite,

- simulation of φ_j is required at each iteration, although they are used neither for estimation purposes nor for proposing the θ_k in a 'clever' way,

- pseudo-priors must be carefully chosen in order for the exploration of the different indices to be efficient,

- it requires a Gibbs sampler, and hence the availability of full conditional distributions, which seriously limits the range of applications, even if Metropolized versions can be proposed.

Interesting remarks and comments on this algorithm can be found in [19] and [30], where connections between the 'Metropolized' Carlin and Chib type algorithm and reversible jump MCMC were independently made.

3.4.2 Jump-diffusion sampling

Much of the research on sampling from posterior distributions defined on $\Theta \triangleq \bigcup_{i=0}^{k_{max}} \{i\} \times \Theta_i$ was initiated by the work of Grenander and Miller published in 1994 [32]. The proposed method of inference is based on a random process which follows jump-diffusion dynamics, and whose samples are drawn from a posterior of the form (2.1).

The main idea in [32] is to construct a single posterior distribution over the union of all considered parameter spaces and then sample from it using a Markov process that has jump-diffusion dynamics. In particular, at random times the Markov process jumps from one of the parameter spaces to another, and in between the jumps, the process follows Langevin stochastic differential equations. Early work on simulations from a given probability density by using Langevin equations appears in [31].

The jump dynamics can be defined in various ways, and always must satisfy certain regularity and balance conditions as well as reversibility. The jump times are obtained from marginal jump intensities, which are computed from jump intensities chosen to satisfy balance equations that ensure sampling from the desired posterior [32]. Once the jump time is determined, a decision where to jump is made by using transition kernels, which are conditional probability densities of jumping from the current to a new parameter space. Two useful jump dynamics approaches are Gibbs and Metropolis-Hastings jumps [50].

The continuous-time diffusion process $\theta^{(t)}$ within a fixed parameter space and between jump times satisfies the Langevin stochastic differential equation

$$d\theta_k^{(t)} = \frac{dt}{2}\left(\frac{d}{d\theta_k}\log p_k(k,\theta_k)\right)_{\theta_k^{(t)}} + d\mathbf{W}^{(t)}, \qquad (3.37)$$

where $\mathbf{W}^{(t)}$ is a standard Brownian motion process whose dimension is the same as that of θ_k. In practice, (3.37) is approximated by a discrete-time version.

4 Applications

We address in this section two applications of Bayesian model selection and Bayesian computation using MCMC. These models have been briefly described in Section 2. In Subsection 4.1, the classical signal processing problem of detection of sinusoids in noise is addressed. A proper hierarchical and robust prior distribution is defined in the spirit of [53], [58], see [2] for further details. Bayesian computation is addressed using reversible jump MCMC and the implementation issues are carefully detailed.

In Subsection 4.2, the classical statistical problem of analysis of finite mixture of Gaussians is addressed. The model selection is carried out using an extension of the EP Priors approach. In particular, it is shown how the EP Priors can be embedded into the MCMC algorithm in a natural way. A sketch of the reversible jump MCMC procedure is given, see [47], [49] for further details. The method is applied to the analysis of the third BATSE gamma ray burst data [39]. This data set is an example of a mixture of bivariate normals, with the additional complication that the observations are associated with a known error of measurement.

Both applications show how both the model selection and the model parameter inference can be performed in a Bayesian framework. The power of the simulation methods described above allows us to explore fairly complex problems in a natural and coherent way.

4.1 Analysis of sinusoids in noise

4.1.1 Data models

We recall that the problem addressed is the selection of one of the following models

$$
\begin{aligned}
\mathcal{M}_0 &: y_t = w_{0,t}, & \text{if } k = 0, \\
\mathcal{M}_k &: y_t = \sum_{j=1}^k \left(a_{c_{j,k}} \cos\left[\omega_{j,k} t \right] + a_{s_{j,k}} \sin\left[\omega_{j,k} t \right] \right) + \varepsilon_{k,t}, & \text{if } k \geq 1,
\end{aligned} \tag{4.1}
$$

for representing the data for $t = 0, \ldots, T-1$, where $\varepsilon_{k,t} \overset{iid}{\sim} \mathcal{N}\left(0, \sigma_k^2\right)$. In vector-matrix form, we have

$$
\mathbf{y} = \mathbf{D}\left(\omega_k\right) \mathbf{a}_k + \varepsilon_k, \tag{4.2}
$$

where $\varepsilon_k \triangleq \left(\varepsilon_{k,0}, \ldots, \varepsilon_{k,T-1} \right)^{\mathsf{T}}$, $[\mathbf{a}_k]_{2i-1,1} \triangleq a_{c_{i,k}}$, $[\mathbf{a}_k]_{2i,1} \triangleq a_{s_{i,k}}$ and $[\omega_k]_{i,1} \triangleq \omega_{i,k}$ for $i = 1, \ldots, k$. The $T \times 2k$ matrix $\mathbf{D}\left(\omega_k\right)$ is defined as

$$
\begin{aligned}
\left[\mathbf{D}\left(\omega_k\right)\right]_{i+1,2j-1} &= \cos\left[\omega_{j,k} i\right], \text{ for } i = 0, \ldots, T-1, \ j = 1, \ldots, k, \\
\left[\mathbf{D}\left(\omega_k\right)\right]_{i+1,2j} &= \sin\left[\omega_{j,k} i\right], \text{ for } i = 0, \ldots, T-1, \ j = 1, \ldots, k, \tag{4.3}
\end{aligned}
$$

with $\omega_k \in \Omega_k \triangleq \left\{ \omega_k; \omega_k \in (0, \pi)^k / \omega_{j_1,k} \neq \omega_{j_2,k} \text{ for } j_1 \neq j_2 \right\}$. We assume here that the number k of sinusoids and their parameters $\theta_k \triangleq \left(\mathbf{a}_k, \omega_k, \sigma_k^2 \right)^{\mathsf{T}}$ are unknown. Given the data set $\mathbf{y}$, our objective is to estimate k and θ_k.

4.1.2 Prior distributions

We will introduce here hyperparameters, δ^2 and Λ, which can be respectively interpreted as an expected signal to noise ratio and the expected number of sinusoids. Then we assume for the prior of $\left(\Lambda, \delta^2, k, \theta_k \right)$ the following structure:

$$
p\left(\Lambda, \delta^2, k, \theta_k \mid \delta^2 \right) = p\left(k, \mathbf{a}_k, \omega_k \mid \Lambda, \delta^2, \sigma_k^2 \right) p\left(\sigma_k^2 \right) p\left(\delta^2 \right) p\left(\Lambda \right), \tag{4.4}
$$

where σ_k^2 is a scale parameter, assumed to be distributed according to a conjugate inverse-Gamma prior distribution, *i.e.* $\sigma_k^2 \sim \mathcal{IG}\left(\frac{v_0}{2}, \frac{\gamma_0}{2} \right)$. When $v_0 = 0$

and $\gamma_0 = 0$, we obtain Jeffreys' non-informative prior $p\left(\sigma_k^2\right) \propto 1/\sigma_k^2$ [5]. For $(k, \mathbf{a}_k, \omega_k)$ we introduce the following prior distribution:

$$p\left(k, \mathbf{a}_k, \omega_k \middle| \Lambda, \delta^2, \sigma_k^2\right) \propto \frac{\Lambda^k}{k!} \frac{1}{\left|2\pi\sigma_k^2 \Sigma_k\right|^{1/2}} \exp\left[-\frac{\mathbf{a}_k^{\mathsf{T}} \Sigma_k^{-1} \mathbf{a}_k}{2\sigma_k^2}\right] \frac{\mathbb{I}_\Omega\left(k, \omega_k\right)}{\pi^k}, \quad (4.5)$$

where $\Sigma_k^{-1} = \delta^{-2} \mathbf{D}^{\mathsf{T}}\left(\omega_k\right) \mathbf{D}\left(\omega_k\right)$. The prior probability model distribution p_k is a truncated Poisson distribution. Conditional upon k, the frequencies are assumed uniformly distributed in Ω_k. Finally, conditional upon (k, ω_k), the amplitudes are assumed zero-mean Gaussian with covariance $\sigma_k^2 \Sigma_k$. Proportionality in (4.5) comes from the fact that $k \leq k_{\max}$.

The prior distributions for the hyperparameters will now be specified. The values of these hyperpriors could be fixed by the user if prior information is available. However, we choose here to include these hyperparameters in the estimation and thus assign them prior distributions. This is the approach adopted for example in [53], which proves to be robust in practice. As δ^2 is a scale parameter, we ascribe a vague conjugate prior density (which means that the variance is infinite [5]) to it

$$\delta^2 \sim \mathcal{IG}\left(\alpha_{\delta^2}, \beta_{\delta^2}\right) \ \left(\alpha_{\delta^2} = 2 \ \text{and} \beta_{\delta^2} > 0\right). \quad (4.6)$$

We apply the same method to Λ by assigning a non-informative conjugate prior [5]

$$\Lambda \sim \mathcal{G}\left(1/2 + \varepsilon_1, \varepsilon_2\right) \ \left(\varepsilon_i \ll 1 \ \text{for} \ i = 1, 2\right). \quad (4.7)$$

From the description of the model given above, the parameter space is $\mathbb{R}^{+2} \times \Theta$ where Θ can be written as a finite union of subspaces $\Theta = \bigcup_{k=0}^{k_{\max}} \{k\} \times \Theta_k$, $\Theta_0 \triangleq \mathbb{R}^+$, and $\Theta_k \triangleq \Omega_k \times \mathbb{R}^k \times \mathbb{R}^+$ with $\Omega_k \triangleq (0, \pi)^k$ for $k \in \{1, \dots, k_{\max}\}$ and[4] $k_{\max} \triangleq T - 1$. By convention $\Omega_0 \triangleq \{\omega_0\} = \emptyset$ and we denote $\Omega \triangleq \bigcup_{k=0}^{k_{\max}} \{k\} \times \Omega_k$.

4.1.3 Integration of the nuisance parameters

With the prior distributions as defined above, one can integrate out the amplitudes and the variance of the noise, conditional upon the hyperparameters, to obtain the following conditional posterior distribution for the number of sinusoids and their frequencies:

$$p\left(k, d\omega \middle| \Lambda, \delta^2, \mathbf{y}\right) \propto$$
$$\sum_{i=0}^{k_{\max}} \left(\mathbf{y}^{\mathsf{T}} \mathbf{P}_i \mathbf{y}\right)^{-\frac{T}{2}} \times \frac{\left(\Lambda/\left(\left(\delta^2 + 1\right)\pi\right)\right)^i}{i!} d\omega_i \mathbb{I}_{\{i\} \times \Omega_i}\left(k, \omega\right), \quad (4.8)$$

---

[4] The following constraint $k < T$ is added because otherwise the columns of $\mathbf{H}_k$ are necessarily linearly dependent and the parameters θ_k may not be uniquely defined in terms of the data (see (4.2)).

where the hyperparameters Λ and δ^2 can be interpreted as the mean number of expected sinusoids and the expected signal to noise ratio, respectively. The matrix $\mathbf{P}_i$ is defined as

$$\begin{aligned}
\mathbf{P}_i &= \mathbf{I}_T - \mathbf{D}\left(\omega_i\right)\mathbf{M}_i\mathbf{D}^{\mathsf{T}}\left(\omega_i\right) \\
\text{with } \mathbf{M}_i^{-1} &= \tfrac{1+\delta^2}{\delta^2}\mathbf{D}^{\mathsf{T}}\left(\omega_i\right)\dot{\mathbf{D}}\left(\omega_i\right).
\end{aligned}$$

We will also define $\mathbf{m}_i = \left[\mathbf{D}^{\mathsf{T}}\left(\omega_i\right)\mathbf{D}\left(\omega_i\right)\right]^{-1}\mathbf{D}^{\mathsf{T}}\left(\omega_i\right)\mathbf{y}$. To perform model selection, one is interested in evaluating $p\left(k\mid\mathbf{y}\right)$ in order to evaluate the Bayes Factors, for which one cannot obtain a closed-form expression.

4.1.4   Overview of the algorithm

For our problem, the following moves have been selected:

1. Birth of a new sinusoid, *i.e.* proposing a new sinusoid with frequency ω^* at random on $(0,\pi)$.

2. Death of an existing sinusoid, *i.e.* removing a sinusoid chosen randomly.

3. Update of the parameters of all the sinusoids, when $k \neq 0$, and the variance of the observation noise.

The birth and death moves represent changes from k to $k+1$ and k to $k-1$, respectively. These moves are defined by heuristic considerations, the only condition to be fulfilled being to maintain the correct invariant distribution. A particular choice will only have an influence on the convergence rate of the algorithm. Other moves may be proposed, but we have found that the ones suggested here lead to satisfactory results [1].

The resulting transition kernel of the simulated Markov chain is then a mixture of the different transition kernels associated with the moves described above. This means that at each iteration one of the candidate moves – birth, death or update – is randomly chosen. The probabilities for choosing these moves are $q_{k,k+1}\left(k+1\mid k,\vartheta_k\right)$, $q_{k,k-1}\left(k-1\mid k,\vartheta_k\right)$ and $q_{k,k}\left(k\mid k,\vartheta_k\right)$ respectively, where $q_{k,k+1}\left(k+1\mid k,\vartheta_k\right) + q_{k,k-1}\left(k-1\mid k,\vartheta_k\right) + q_{k,k}\left(k\mid k,\vartheta_k\right) = 1$ for all $0 \leq k \leq k_{\max}$ and $\vartheta_k \triangleq \left(\Lambda,\delta,\theta_k\right)$. The move is performed if the algorithm accepts it. For $k = 0$ the death move is impossible, so that $q_{0,-1}\left(-1\mid 0,\vartheta_0\right) \triangleq 0$. For $k = k_{\max}$ the birth move is impossible and thus $q_{k_{\max},k_{\max}+1}\left(k_{\max}+1\mid k_{\max},\vartheta_{k_{\max}}\right) \triangleq 0$. Except in the cases described above, we take the following probabilities

$$\begin{aligned}
q_{k+1,k}\left(k+1\mid k,\vartheta_k\right) &= q_{k+1,k} \triangleq c\,\min\left\{1,\tfrac{p_{k+1}}{p_k}\right\}, \\
q_{k,k+1}\left(k\mid k+1,\vartheta_{k+1}\right) &= q_{k,k+1} \triangleq c\,\min\left\{1,\tfrac{p_k}{p_{k+1}}\right\},
\end{aligned} \tag{4.9}$$

where $p_k \propto \left(\Lambda^k/k!\right)\mathbb{I}_{\{0,\dots,k_{\max}\}}\left(k\right)$ is the prior probability of model $\mathcal{M}_k$ and c is a parameter which tunes the 'relative frequencies' of dimension change and update moves. As pointed out in [33, p. 719], this choice ensures that

$$q_{k,k+1}p_k\left[q_{k+1,k}p_{k+1}\right]^{-1} = 1, \tag{4.10}$$

which means that an MH algorithm on the sole dimension in the case of no observation would have 1 as acceptance probability. We take $c = 0.5$ and then $q_{k+1,k} + q_{k-1,k} \in [0.5, 1]$ for all k [33]. For this algorithm, $\varphi_{k,k+1} = \omega^*$ and $\omega_{k+1} = f_{k,k+1}(\omega_k, \omega^*)$ is just any concatenating function such as $\omega_{k+1} = (\omega_k, \omega^*)$.

One can then describe the main steps of the algorithm as follows.

Reversible jump MCMC algorithm

1. Initialization: set $\left(\Lambda^{(0)}, \delta^{2(0)}, k^{(0)}, \theta_{k^{(0)}}^{(0)} \right) \in \Theta$.

2. Iteration i, $i \geq 1$.

    - If $\left(u \sim \mathcal{U}_{[0,1]} \right) \leq q_{k^{(i-1)}, k^{(i-1)}+1}$ then 'birth' move
    else if $u \leq q_{k^{(i-1)}, k^{(i-1)}+1} + q_{k^{(i-1)}, k^{(i-1)}-1}$ then 'death' move
    else update the frequencies one at a time, using MH steps (see details in [2]).

    - Sample the nuisance parameters,

    $$
    \begin{aligned}
    \sigma_{k^{(i)}}^{2(i)} \Big| \left(\mathbf{y}, k^{(i)}, \omega_{k^{(i)}}^{(i)} \right) &\sim \mathcal{IG}\left(\frac{v_0 + T}{2}, \frac{\gamma_0 + \mathbf{y}^{\mathsf{T}} \mathbf{P}_{k^{(i)}} \mathbf{y}}{2} \right), \\
    \mathbf{a}_{k^{(i)}}^{(i)} \Big| \left(\mathbf{y}, k^{(i)}, \omega_{k^{(i)}}^{(i)}, \sigma_{k^{(i)}}^2 \right) &\sim \mathcal{N}\left(\mathbf{m}_{k^{(i)}}, \sigma_{k^{(i)}}^2 \mathbf{M}_{k^{(i)}} \right).
    \end{aligned}
    \tag{4.11}
    $$

    - Sample the hyperparameters $\delta^{2(i)}$,

    $$
    \delta^{2(i)} \Big| \left(k^{(i)}, \omega_{k^{(i)}}^{(i)}, \sigma_{k^{(i)}}^{2(i)} \right) \sim \mathcal{IG}\left(k^{(i)} + \alpha_{\delta^2}, \frac{\mathbf{a}_{k^{(i)}}^{\mathsf{T}} \mathbf{D}^{\mathsf{T}}\left(\omega_{k^{(i)}}^{(i)} \right) \mathbf{D}\left(\omega_{k^{(i)}}^{(i)} \right) \mathbf{a}_{k^{(i)}}}{2\sigma_{k^{(i)}}^2} + \beta_{\delta^2} \right),
    \tag{4.12}
    $$

    and then sample $\Lambda^{(i)}$ according to an MH step with proposal $\mathcal{G}\left(1/2 + k^{(i)} + \varepsilon_1, 1 + \varepsilon_2 \right)$ and target distribution

    $$
    p\left(\Lambda | k^{(i)} \right) \propto \frac{\Lambda^k / k^{(i)}!}{\sum_{l=0}^{k_{\max}} (\Lambda^l / l!)} \mathcal{G}\left(\Lambda; 1/2 + \varepsilon_1, \varepsilon_2 \right)
    \tag{4.13}
    $$

    End If.

3. Set $i \leftarrow i + 1$ and go to 2.

We describe more precisely these different reversible jump moves below. In what follows, in order to simplify notation, we drop the superscript $\cdot^{(i)}$ from all variables at iteration i.

4.1.5   The birth and death moves

Suppose that the current state of the Markov chain is in $\{k\} \times \Theta_k$, then

---

Birth move

- Propose a new frequency at random on $(0, \pi)$: $\omega^* \sim \mathcal{U}_{(0,\pi)}$ and set $\omega_{k+1} = (\omega_k, \omega^*)$.

- Evaluate $\alpha_{k,k+1}$, see (4.15), and sample $u \sim \mathcal{U}_{[0,1]}$.

- If $u \leq \alpha_{k,k+1}$ then the state of the Markov chain becomes $(k+1, \omega_{k+1})$, else it remains at (k, ω_k).

---   ∎

Assume that the current state of the Markov chain is in $\{k+1\} \times \Theta_{k+1}$, then

---

Death move

- Choose a sinusoid with label l at random among the $k+1$ existing sinusoids: $l \sim \mathcal{U}_{\{1,\dots,k+1\}}$.

- Evaluate $\alpha_{k+1,k}$, see (4.15), and sample $u \sim \mathcal{U}_{[0,1]}$.

- If $u \leq \alpha_{k+1,k}$ then the state of the Markov chain becomes (k, ω_k), else it remains at $(k+1, \omega_{k+1})$.

---   ∎

The acceptance ratio for the proposed moves is deduced from the expression (3.22):

$$r\left((k, \vartheta_k), (k+1, \vartheta_{k+1})\right)$$

$$= \frac{p_{k+1}\left(k+1, \omega_{k+1} | \Lambda, \delta^2, \mathbf{y}\right) q_{k+1,k} q_{k+1,k}\left(\omega_k | k, k+1, \vartheta_{k+1}\right)}{p_k\left(k, d\omega_k | \Lambda, \delta^2, \mathbf{y}\right) q_{k,k+1} q_{k,k+1}\left(\omega^* | k+1, k, \vartheta_k\right)}$$

$$= \frac{p_{k+1}\left(k+1, \omega_{k+1} | \Lambda, \delta^2, \mathbf{y}\right) q_{k+1,k} \frac{1}{k+1}}{p_k\left(k, \omega_k | \Lambda, \delta^2, \mathbf{y}\right) q_{k,k+1} \pi^{-1}}$$

which yields, after simplifications,

$$r\left((k, \vartheta_k), (k+1, \vartheta_{k+1})\right) = \left(\frac{\mathbf{y}^{\mathsf{T}} \mathbf{P}_k \mathbf{y}}{\mathbf{y}^{\mathsf{T}} \mathbf{P}_{k+1} \mathbf{y}}\right)^{\frac{T}{2}} \frac{1}{(k+1)(1+\delta^2)}. \qquad (4.14)$$

Note that here the Jacobian is equal to one as the invertible deterministic transformation is the identity transformation and the acceptance probabilities corresponding to the described moves are

$$\alpha_{k,k+1} \triangleq \alpha\left((k, \vartheta_k), (k+1, \vartheta_{k+1})\right) = \min\left\{1, r\left((k, \vartheta_k), (k+1, \vartheta_{k+1})\right)\right\}$$

$$\alpha_{k+1,k} \triangleq \alpha\left((k+1, \vartheta_{k+1}), (k, \vartheta_k)\right) = \min\left\{1, r^{-1}\left((k, \vartheta_k), (k+1, \vartheta_{k+1})\right)\right\}$$

$$(4.15)$$

The update moves consist of standard MH steps and are not detailed here, see [2] for details.

4.1.6 Example of simulation

We present here results for the case of two sinusoids ($k = 2$) with parameters given in Table 1, where $E_i = a_{s_i}^2 + a_{c_i}^2$. The number of observed samples was $T = 64$ and the standard deviation of the noise $\sigma = 1.0$.

i	E_i	$-\arctan\left(a_{s_i}/a_{c_i}\right)$	$\omega_i/2\pi$
1	20	0	0.2
2	20	$\pi/3$	$0.2 + 2/T$

Table 1: Parameters for the experiment.

We ran the algorithm for $M = 20,000$ iterations. Then the posterior model probabilities were computed using

$$\widehat{p}(k = j|\mathbf{y}) = \frac{1}{M}\sum_{i=1}^{M}\mathbb{I}_{\{j\}}\left(k^{(i)}\right) \tag{4.16}$$

The results are shown in Figures 1 and 2 where we present the observed data and the posterior distribution of the models and the posterior distributions of the frequencies for the model with two sinusoids. Note that in practice it has been found that this approach outperforms classical model selection rules such as AIC, MDL and BIC, and that the prior on the hyperparameter δ^2 has a limited effect on the results, see [2] for detailed experiments.

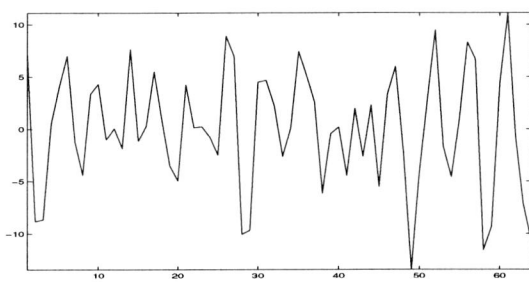

Figure 1: Noisy observations

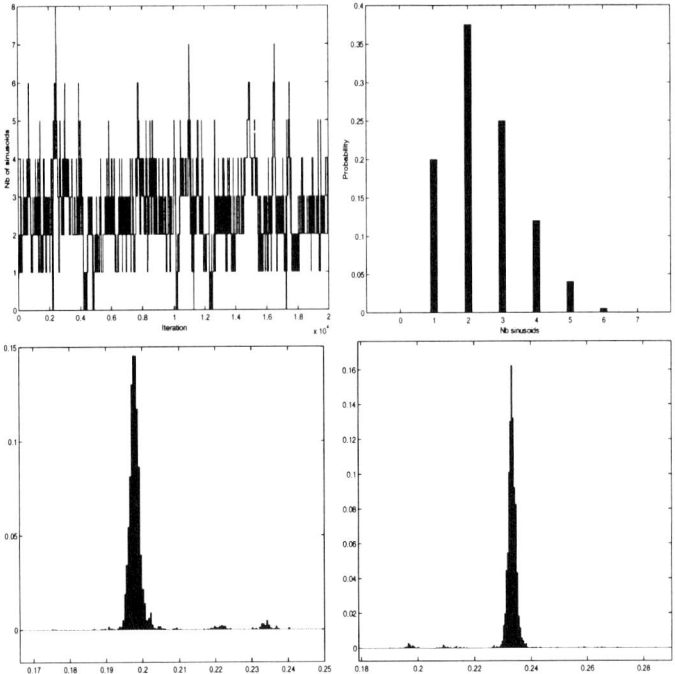

Figure 2: Top: the component of the Markov chain corresponding to the dimension, and an estimate of $p_k(k|\mathbf{y})$. Bottom: estimation of $p_k(\omega/2\pi|\mathbf{y})$

4.2   Model selection for finite mixtures of Gaussians
4.2.1   Data model

Consider the finite mixture model, where observations $\mathbf{y} = (\mathbf{y}_1, \dots, \mathbf{y}_T)$ are distributed independently as

$$\mathcal{M}_k : \mathbf{y}_t \overset{iid}{\sim} \sum_{j=1}^{k} w_{j,i} \mathcal{N}(\mu_{j,k}, \Sigma_{j,k}). \qquad (4.17)$$

Conditional on k, the weights of the mixture are given by $\mathbf{w}_k = (w_{1,k}, \dots, w_{k,k})$, where $w_{j,k}$ is the probability of an observation coming from component j and $\sum_{j=1}^{k} w_{j,k} = 1$. The parameters of the model components are given for $\mathcal{M}_k$ by $\phi_k = (\mu_k, \Sigma_k)$ where $\mu_k = (\mu_{1,k}, \dots, \mu_{k,k})$ and $\Sigma_k = (\Sigma_{1,k}, \dots, \Sigma_{k,k})$, with $\mu_{j,k}$ and $\Sigma_{j,k}$ being respectively the mean and standard deviation of the jth component. The unknown parameters for $\mathcal{M}_k$ are $\theta_k = (\mathbf{w}_k, \phi_k)$. Mixture models can provide a general framework for statistical inference and, indeed, have been used in situations ranging from pseudo-parametric density estimation to clustering analysis, changepoint problems and image analysis.

4.2.2 Bayesian model

Define the latent variables $\mathbf{z} = (z_1, \dots, z_T)$ where $z_i = j$ indicates that observation i comes from component j. The joint (Bayesian) density for all variables, based on the mixture model in (4.17), can be written in a hierarchical form as

$$p(k, \theta_k, \mathbf{z}, \mathbf{y}) = p_k p_k(\mathbf{w}_k) p_k(\mathbf{z} \mid \mathbf{w}_k) p_k(\phi_k) p_k(\mathbf{y} \mid \phi_k, \mathbf{z}). \qquad (4.18)$$

The following structure for the various probability distributions is assumed

- p_k is the probability for the number of components; this is chosen here to be uniform on $\{1, \dots, k_{\max}\}$;

- $p_k(\mathbf{w}_k)$ is a Dirichlet distribution with known parameter $\alpha_k = (\alpha_0, \dots, \alpha_0)$;

- the allocation variables z_i, $i = 1, \dots, T$, are i.i.d. with

$$p_k(z_i = j \mid \mathbf{w}_k) = w_{j,k}; \qquad (4.19)$$

- the likelihood $p_k(\mathbf{y} \mid \phi_k, \mathbf{z}) = \prod_{i=1}^{T} \mathcal{N}(\mathbf{y}_i; \mu_{z_i,k}, \Sigma_{z_i,k})$;

- the initial prior for the component parameters is given by $p_k(\phi_k) = \prod_{j=1}^{k} \pi(\mu_{j,k}, \Sigma_{j,k})$ for a specified prior $\pi(\cdot)$.

Note that further hyperparameters might be introduced in this hierarchical structure, allowing for greater flexibility of the model [53].

Default analysis of mixture models Due to the large number of parameters involved in the mixture model, it is very difficult to provide detailed prior information for each component. Therefore, it will be desirable to provide methods that will allow the estimation of mixture model parameters in an automatic way, or with minimal intervention. Note, however, that default priors are typically not proper, and, therefore, the marginal for the mixture model is not finite. Indeed, the marginal for any number of observations under an improper prior, p^N, is not finite.

Several approaches have been proposed in order to solve this problem ([21], [55],[40], [59], [60].) Some of these approaches utilize a prior structure in which the components are linked to a common parameter with a flat prior, so that all observations contribute to its estimation. Shui [60] proposes use of resampling from the observations in order to produce versions of the Intrinsic Bayes Factor [7] and the Fractional Bayes Factor [45] which then can be used for model selection. Note that the use of 'vague priors' does not solve this problem, as pointed out by many [21], [37], [60]. While vague priors are proper densities, the results can be very sensitive to the way in which these priors are made 'vague'.

For the normal univariate case, [53] suggests the use of weak, data dependent, hierarchical priors for the component parameters. The hyperparameters of these priors are estimated from the observations, according to a criterion designed to minimize their influence on the final estimates. The results produced by this approach are sensible. However, the method seems to be sensitive to the choice of hyperparameters in their simulations.

Provided there exist training samples to update each component, it is possible to use standard improper priors in the context of mixture models, while avoiding compromises concerning the prior distribution structure of the model. When lacking a suitable training sample, one could use imaginary training samples arising from a predictive measure, m^*, to update the non-informative priors of the components. This is the basis of the definition of the *Expected Posterior Prior* developed in the next subsection.

Expected Posterior Priors for mixture models Consider the mixture model in (4.17). Assume that the initial prior for the component parameters, conditional on the number of components, k, is given by a non-informative, improper prior, $p_k(\phi_k) = \prod_{j=1}^{k} p^N(\mu_{j,k}, \Sigma_{j,k})$. Let $k_{\max}$ be the maximum number of components and consider imaginary training samples $\mathbf{y}^* = (\mathbf{y}_1^*, \dots, \mathbf{y}_{k_{\max}}^*)$, where the dimension of each $\mathbf{y}_j^*$ is chosen as the minimal size such that $p^N\left(\mu_{j,k}, \Sigma_{j,k} | \mathbf{y}_j^*\right)$ exists. The Expected Posterior Prior for (μ_k, Σ_k), conditional on $k \leq k_{\max}$, is given by

$$p_k^*(\mu_k, \Sigma_k) = \int \prod_{j=1}^{k} p^N\left(\mu_{j,k}, \Sigma_{j,k} | \mathbf{y}_j^*\right) m^*\left(\mathbf{y}_1^*, \dots, \mathbf{y}_{k_{\max}}^*\right) d\mathbf{y}^*. \qquad (4.20)$$

As discussed in [47], [48], [49], one might choose the predictive measure m^* to be the marginal from a *base model* $\mathcal{M}_*$. A natural choice for $\mathcal{M}_*$ in this situation is given by the 'mixture' from a single component. The resulting predictive density for the training samples is

$$m^*\left(\mathbf{y}_1^*, \dots, \mathbf{y}_{k_{\max}}^*\right) = \int p\left(\mathbf{y}_1^*, \dots, \mathbf{y}_{k_{\max}}^* | \mu_{1,1}, \Sigma_{1,1}\right) p^N\left(\mu_{1,1}, \Sigma_{1,1}\right) d\mu_{1,1} d\Sigma_{1,1}.$$
$$(4.21)$$

A second possible choice for the predictive measure of the training samples is given by an empirical version, where $m^*\left(\mathbf{y}_1^*, \dots, \mathbf{y}_{k_{\max}}^*\right)$ is the probability mass function arising from resampling, with or without replacement, $k \leq k_{\max}$ training samples from the observations. In this case, $p_k^*(\mu_k, \Sigma_k | k)$ is a proper density.

For m^* in (4.21), one can easily see that the resulting prior $p_k^*(\mu_k, \Sigma_k)$ is marginally coherent, in the sense that by integrating this prior over $\mu_{k,k}$,

$\Sigma_{k,k}$, we obtain the same functional form as $p_{k-1}^* (\mu_{k-1}, \Sigma_{k-1})$. In particular $p_1^* (\mu_1, \Sigma_1) = p^N (\mu_{1,1}, \Sigma_{1,1})$. Therefore, components with no allocated data are no longer an issue in estimation of the model and hence, there is no need to impose conditions on the number of allocated observations in each component, as done in [21]. This implies that the latent allocation variables, z_i, are now independent, making the statistical inference easier. Furthermore, for one observation, the EP Prior p_k^* produces *predictively matched* Bayes Factors for the comparison of models with different components, *i.e.* for any k_1, k_2

$$B_{k_1,k_2}^* = \frac{m_{k_2}(\mathbf{y}_1)}{m_{k_1}(\mathbf{y}_1)} = 1. \tag{4.22}$$

In this sense, the information content given by m^* is balanced among all models and the problem of selecting the number of components is 'unbiased'.

4.2.3 MCMC computation for mixture models under EP Priors

Reversible jump MCMC algorithm We propose here a reversible MCMC method to produce inferences from the posteriors arising from the Expected Posterior Priors. The key idea is to use $m^* (\mathbf{y}_1^*, \ldots, \mathbf{y}_{k_{\max}}^*)$ as a prior density for the imaginary training samples, $\mathbf{y}_1^*, \ldots, \mathbf{y}_{k_{\max}}^*$, and so incorporate the generation of the training samples alongside the other parameters in the simulation.

With some abuse of notation, the joint density for the observations $\mathbf{y}$, the parameters $(\mathbf{w}_k, \phi_k, \mathbf{z})$, and the training samples $\mathbf{y}^* = (\mathbf{y}_1^*, \ldots, \mathbf{y}_{k_{\max}}^*)$ can be written as

$$p_k (k, \mathbf{w}_k, \phi_k, \mathbf{z}, \mathbf{y}, \mathbf{y}^*) = p_k p_k (\mathbf{w}_k) p_k (\mathbf{z} | \mathbf{w}_k) \times \tag{4.23}$$

$$m^* (\mathbf{y}^*) \left(\prod_{j=1}^{k} p^N (\mu_{j,k}, \Sigma_{j,k} | \mathbf{y}_j^*) \right) p_k (\mathbf{y} | \phi_k, \mathbf{z}).$$

To improve the performance of the algorithm, an allocation variable for the training samples is introduced. Let $\mathbf{l} = (l_1, \ldots, l_k)$, such that $l_j = i$ indicates that training sample $\mathbf{y}_i^*$ is (uniquely) assigned to component j. We will denote by $\mathbf{y}^* (\mathbf{l}) = \left(\mathbf{y}_{l_1}^*, \ldots, \mathbf{y}_{l_k}^* \right)$ and by $p(\mathbf{l} | k)$ the uniform distribution over all possible assignments of $\mathbf{y}_1^*, \ldots, \mathbf{y}_{k_{\max}}^*$ into k components.

For m^* chosen as in (4.21), the reversible jump MCMC algorithm proceeds as follows at iteration i.

Reversible jump MCMC algorithm

1. Sample weights from

$$\mathbf{w}_{k^{(i-1)}}^{(i)} \Big| \left(k^{(i-1)}, \mathbf{z}^{(i-1)}, \mathbf{y} \right) \sim \mathcal{D} \left(\alpha_0 + n_1^{(i-1)}, \ldots, \alpha_0 + n_k^{(i-1)} \right),$$

where $n_j^{(i-1)} = \sum_{l=1}^{T} \mathbb{I}_{\{j\}}\left(z_l^{(i-1)}\right)$ is the number of observations in component j.

2. Sample component parameters from $\phi_{k^{(i-1)}}^{(i)}\Big|\left(k^{(i-1)}, \mathbf{z}^{(i-1)}, \mathbf{y}^*\left(\mathbf{l}^{(i-1)}\right), \mathbf{y}\right)$ with density

$$\prod_{j=1}^{k} p_{k^{(i-1)}}^N\left(\mu_{j,k^{(i-1)}}^{(i)}, \Sigma_{j,k^{(i-1)}}^{(i)}\Big| \mathbf{z}^{(i-1)}, \mathbf{y}_{l_j^*}^*{}_{(i-1)}, \mathbf{y}\right).$$

3. Sample allocations from $\mathbf{z}^{(i)}\Big|\left(k^{(i-1)}, \mathbf{w}_{k^{(i-1)}}^{(i)}, \phi_{k^{(i-1)}}^{(i)}, \mathbf{y}\right)$, with probability

$$p_{k^{(i-1)}}\left(z_i = j\Big| \phi_{k^{(i-1)}}^{(i)}, \mathbf{y}\right) \propto w_{j,k^{(i-1)}}^{(i)} \mathcal{N}\left(\mathbf{y}_i; \mu_{j,k^{(i-1)}}^{(i)}, \Sigma_{j,k^{(i-1)}}^{(i)}\right).$$

4. Sample $\mathbf{y}^{*(i)}\left(-\mathbf{l}^{(i-1)}\right) \triangleq \left\{\mathbf{y}_m^* : m \neq l_j^{(i-1)}; j = 1, \ldots, k\right\}$, conditional on $\mathbf{y}^{*(i-1)}\left(\mathbf{l}^{(i-1)}\right)$, with density

$$m^*\left(\mathbf{y}^*\left(-\mathbf{l}\right)\Big| \mathbf{y}^*\left(\mathbf{l}\right)\right) = m^*\left(\mathbf{y}^*\right)/m^*\left(\mathbf{y}^*\left(\mathbf{l}\right)\right)$$

5. Sample $\mathbf{y}^{*(i)}\left(\mathbf{l}^{(i-1)}\right)\Big|\left(\phi_{k^{(i-1)}}^{(i)}, \mathbf{l}^{(i-1)}\right)$.

6. Sample one-at-a-time assignments for $j = 1, \ldots, k^{(i-1)}$

$$l_j^{(i)}\Big|\left(\phi_{k^{(i-1)}}^{(i)}, l_1^{(i)}, \ldots, l_{j-1}^{(i)}, l_{j+1}^{(i-1)}, \ldots, l_k^{(i-1)}\right)$$

with probabilities

$$p\left(l_j = i\Big| \phi_{k^{(i-1)}}^{(i)}, l_1^{(i)}, \ldots, l_{j-1}^{(i)}, l_{j+1}^{(i-1)}, \ldots, l_k^{(i-1)}\right)$$
$$\propto p^N\left(\mu_{j,k^{(i-1)}}^{(i)}, \Sigma_{j,k^{(i-1)}}^{(i)}\Big| \mathbf{y}_i^*\right) p\left(l_1^{(i)}, \ldots, l_{j-1}^{(i)}, i, l_{j+1}^{(i-1)}, \ldots, l_k^{(i-1)}\Big| k^{(i-1)}\right).$$
$$(4.24)$$

7. Sample $k^{(i)}\Big|\left(\mathbf{w}_{k^{(i-1)}}^{(i)}, \phi_{k^{(i-1)}}^{(i)}, \mathbf{z}^{(i)}, \mathbf{y}\right)$.

$\blacksquare$

Steps 1 and 3 can be carried out using the Gibbs kernels described above. In Step 2, new component parameters are generated from the posterior of p^N, updated with the training samples assigned to each component.

Steps 4–6 are the main modifications due to the introduction of the imaginary training samples. In Step 5, a new set of training samples, $\mathbf{y}_{l_1}^*, \ldots, \mathbf{y}_{l_{k^{(i-1)}}}^*$, is generated at iteration i, using a Metropolis–Hastings move, with proposal distributions $p\left(\mathbf{y}_{l_j}^*\Big| \phi_{j,k^{(i-1)}}^{(i)}\right)$, $j = 1, \ldots, k^{(i-1)}$. The remaining training samples $\mathbf{y}^{*(i)}\left(-\mathbf{l}^{(i-1)}\right)$ are generated in Step 4 from their 'posterior'

$m^*\left(\cdot \quad \mathbf{y}^*\left(\mathbf{l}^{(i-1)}\right)\right)$. Note that, since $m^* = m^N$, this can be done by generating a dummy parameter $\left(\tilde{\mu}, \tilde{\Sigma}\right) \sim p^N\left(\mu, \Sigma | \mathbf{y}^*\left(\mathbf{l}^{(i-1)}\right)\right)$, and then taking $\mathbf{y}^{*(i)}\left(-\mathbf{l}^{(i-1)}\right) \sim p\left(\cdot | \tilde{\mu}, \tilde{\Sigma}\right)$. Step 6 uniquely assigns the training samples at iteration i to the components. Step 7 involves changing k by adding or deleting one component, and updating $(\theta_k, \mathbf{z})$ accordingly. At this point the reversible jump mechanism is needed. As in [53], we consider splitting or combining with probabilities c_k and $s_k = 1 - c_k$ respectively. Of course, $s_{k_{\max}} = c_1 = 0$. A split move will produce components j_1 and j_2 by splitting a single component j_0 into two. The weights and component parameters for these two components are obtained by independently generating a random variable $\mathbf{u}$, of the same dimension as $\theta_{j_0,k}$, and deterministically assigning $(\theta_{j_1,k+1}, \theta_{j_2,k+1}) = T\left(\theta_{j_0,k}, \mathbf{u}\right)$ where T is an invertible, everywhere differentiable function. When the mixture is univariate, the mapping selected is similar to the one in [53]. In the bivariate case, details are given in [47], [48]. To preserve the reversible property, the jump from k to $k + 1$ components in Step 7 is done conditional on all the training samples $\mathbf{y}_1^*, \ldots, \mathbf{y}_{k_{\max}}^*$. In this way, splitting a component j_0 into j_1 and j_2 only involves choosing training samples $\mathbf{y}_{l_{j_1}}^*$ and $\mathbf{y}_{l_{j_2}}^*$ from the available pool of training samples $\left(\mathbf{y}^*\left(-\mathbf{l}\right), \mathbf{y}_{l_{j_0}}^*\right)$. A combine move entails merging two randomly chosen components j_1 and j_2 into a new component j_0. The allocation variables are simply updated by setting $z_i = j_0$ for all y_i such that z_i is either j_1 or j_2. For the training samples it only entails choosing $\mathbf{y}_{l_{j_0}}^*$ from the pool of unassigned training samples $\left(\mathbf{y}^*\left(-\mathbf{l}\right), \mathbf{y}_{l_{j_1}}^*, \mathbf{y}_{l_{j_2}}^*\right)$. This effectively avoids introducing unknown constants in the acceptance/rejection probabilities for the move. Indeed, the acceptance probability for the split move from k to $k + 1$ components is given by

$$\min\left\{1, C_w L \frac{p^N\left(\theta_{j_1,k+1} | \mathbf{y}_{l_{j_1}}^*\right) p^N\left(\theta_{j_2,k+1} | \mathbf{y}_{l_{j_2}}^*\right)}{p^N\left(\theta_{j_0,k} | \mathbf{y}_{l_{j_0}}^*\right)} \frac{p(l|k+1)}{p(l|k)} \frac{c_{k+1} d_{k+1}^1 \mathcal{J}_T\left(\theta_{j_0,k}, \mathbf{u}\right)}{s_k d_k^2 q(\mathbf{u}) C_z}\right\},$$

$$(4.25)$$

where $\mathcal{J}_T$ is the Jacobian of the mapping $T(\cdot)$ and

$$L = \prod_{\{l; z_l = j_1\}} p\left(\mathbf{y}_l | \theta_{j_1,k+1}\right) \prod_{\{l; z_l = j_2\}} p\left(\mathbf{y}_l | \theta_{j_2,k+1}\right) \Big/ \prod_{\{l; z_l = j_0\}} p\left(\mathbf{y}_l | \theta_{j_0,k}\right), \quad (4.26)$$

$$C_w = \frac{w_{j_1,k+1}^{\alpha_0 - 1 + n_{j_1}} w_{j_2,k+1}^{\alpha_0 - 1 + n_{j_2}}}{w_{j_0}^{\alpha_0 - 1 + n_{j_0}} Be\left(\alpha_0, \alpha_0 k\right)}, \quad (4.27)$$

C_z is the probability of the allocations in the split move, and d_k^i is the probability of choosing i training samples from a pool of $k_{\max} - (k + i)$ samples.

The same algorithm can be used for the empirical version of the predictive density m^*. Note that, even though in theory it is possible to generate the training samples in Steps 4 and 5 with a resampling mechanism, it is typically computationally infeasible to do so when the number of observations is large. The Metropolis-Hastings move is thus preferred.

Identifiability The mixture model components are permutation invariant, and hence their parameters are not identifiable. For inference concerning individual components, one usually imposes constraints on the parameter space, *e.g.*, by taking the location of the components to be ordered: $\mu_1 < \cdots < \mu_k$. This, however, is not a completely satisfactory solution, as different parameter constraints could yield radically different inferences (see, for example, [56], [40], [53].) If, for example, a constraint of the form $\mu_1 < \mu_2$ is imposed, the posterior estimates of the parameters resulting from the MCMC simulation will typically produce multimodal marginal posteriors when $\Pr(\mu_1 > \mu_2 | \mathbf{y}) > 0$. The multimodality in the estimated posteriors is especially noticeable when the posterior variances are not kept small. This is, unfortunately, the case when Expected Posterior Priors are used, as the posterior variance for empty components is not controlled. However, if the number of observations, T, is large, the probability of observing empty components is reduced and use of a set of constraints is reasonable.

We view the identifiability problem as being a separate issue, arising only when individual components are of interest. For many inferences, constraints are not necessary, and the MCMC simulation can then be run without constraints. Indeed, Diebolt and Robert [21] report that their results are improved when the identifiability constraints are removed. Furthermore, the MCMC algorithm can be greatly improved by introducing a random ordering move which shuffles all components. In this way, every realization of the chain will approximately come from

$$p_k(\theta_{1,k}, \ldots, \theta_{k,k} | \mathbf{y}) = \frac{1}{S} \sum_s p_k(\theta_{s(1),k}, \ldots, \theta_{s(k),k} | \mathbf{y}),$$

where s corresponds to a permutation of the components, S the number of possible permutations s, and $p_k(\cdot | \mathbf{y})$ is the posterior density of interest.

Note that permutation invariant statistics are not affected by the lack of identifiability. For example, the predictive density of future observations given k and the data,

$$p_k(\cdot | \mathbf{y}) = \mathbb{E}\left[p_k(\cdot | \theta_k, \mathbf{y}) | \mathbf{y}\right],$$

where $p_k(\cdot | \theta_k, \mathbf{y}) = \sum_{j=1}^k w_{j,k} \mathcal{N}(\cdot; \mu_{j,k}, \Sigma_{j,k})$, can be estimated by

$$\widehat{p}_k(\cdot | \mathbf{y}) = \frac{1}{M} \sum_{j=1}^M p_k\left(\cdot | \theta_k^{(i)}, \mathbf{y}\right)$$

from an MCMC run of length M over the states with k components. Another important example of inference that is not affected by the lack of identifiability is the use of Bayes Factors for comparisons between models with different numbers of components.

Estimating the parameter components from a run of (4.17) is indeed a difficult problem that requires further study. Inspection of the modes of $p_k(\cdot\,|\,\mathbf{y})$ can suggest a suitable set of constraints (which may depend on k) that could be used in estimation of the component parameters. Alternatively, under the assumption that the posterior marginals are unimodal densities, one could approximate $p_k(\cdot\,|\,\mathbf{y})$ by fitting a family of densities $g(\cdot\,|\,\eta_1,\dots,\eta_k)$ with unimodal marginals to the MCMC run $\theta^{(i)}$.

BATSE gamma ray burst data   The third BATSE catalog contains 1122 measurements performed by the Compton Gamma Ray Observatory between 1991 and 1994 [39]. We will study the relationship of the duration (T90) and hardness ratio (HR) of the burst for a subset of 745 observations, consisting of observations with no missing data and which have not been flagged with possible acquisition errors of any kind. Hurley [34] observed a bimodal distribution for the duration (T90), on the log-scale. A correlation between the hardness ratio and the duration had also been observed. Figure 3 shows the duration and hardness ratio on the log-scale. The figure also shows the standard deviation of the measurement errors provided in the BATSE catalog, adjusted to the log-scale.

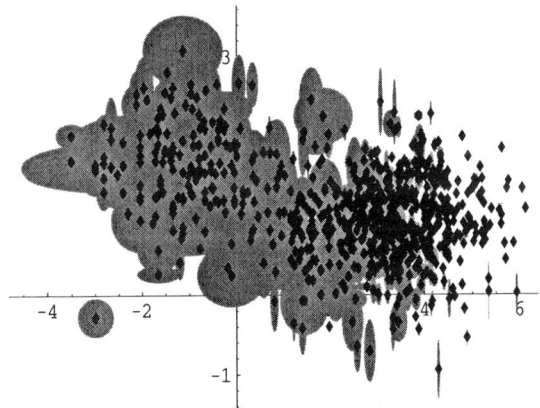

Figure 3: BATSE: log(T90) vs. log(HR). Ellipse radii indicate associated measurement errors.

For the analysis, a bivariate normal mixture model was fitted to $\log(T90)$

and $\log(HR)$ as follows. Let $\mathbf{x}_i = (x_{i1}, x_{i2}) = (\log{(T90)}_i, \log{(HR)}_i)$ denote the ith observation and $\sigma_i = (\sigma_{i1}, \sigma_{i2}) = (\sigma_{T90_i}, \sigma_{HR_i})$ the standard deviation of the associated observational errors. Then x_{i1} and x_{i2} are assumed to have independent normal distributions, with means y_{i1} and y_{i2} and variances σ_{i1}^2 and σ_{i2}^2, respectively. The mean parameter $\mathbf{y}_i = (y_{i1}, y_{i2})$ is assumed to have a density given by a mixture of k bivariate normal distributions, with means $\mu_{j,k}$ and covariance matrices $\Sigma_{j,k}$, $j = 1, \dots, k$. The initial prior distributions for the component parameters were chosen as in Subsection 4.2.2, with $\alpha_0 = 1$ and $p^N(\mu, \Sigma) = |\Sigma|^{-3/2}$.

Analysis under the base model EP Priors and under the empirical version of the EP Priors was considered. The reversible jump MCMC algorithm was used in both cases. An additional step was added to the MCMC algorithm to deal with the generation of $\mathbf{y}_i$ conditional on all other parameters and on $\mathbf{x}_i$. The length of the run for the MCMC simulation was 100,000 sweeps, and only the last 60,000 were used for inference. Identifiability was obtained by ordering the first coordinate of the means.

The posterior probabilities for k strongly indicate that only two components are present $(\Pr(k = 2|\mathbf{y}) = .99)$. Tables 2 and 3 give the corresponding estimates of the location and covariance matrices for the components under the 'base model' and the empirical versions of the EP Prior. Both appear to agree in the results. Notice that no evidence of correlation between y_{i1} and y_{i2} is found within groups.

Figures 4 and 5 show the allocation distribution $p(\mathbf{z}|\mathbf{y}, k = 2)$ for the gamma ray bursts, together with predictive confidence sets of levels 90%, 95% and 99% for each of the two components.

Component 1		
$\widehat{w}_1 = 0.2440$		
$\widehat{\mu}_{T90} = -0.8451$	$\widehat{\mu}_{HR} = 1.6071$	
$\widehat{\sigma}_{T90} = 1.0431$	$\widehat{\sigma}_{HR} = 0.4966$	$\widehat{\rho} = -0.03098$
Component 2		
$\widehat{w}_{2,2} = 0.7560$		
$\widehat{\mu}_{T90} = 3.3103$	$\widehat{\mu}_{HR} = 0.9478$	
$\widehat{\sigma}_{T90} = 1.1036$	$\widehat{\sigma}_{HR} = 0.4924$	$\widehat{\rho} = 0.01358$

Table 2: BATSE: MMSE Estimates for $\log(T90)$ and $\log(HR)$ under base model EP Priors.

5   Conclusions

Bayesian analysis is a powerful methodology which has seldom been exploited in the signal processing community. With the recent advances of the theory

Component 1		
$\widehat{w}_1 = 0.2396$		
$\widehat{\mu}_{T90} = -0.9154$	$\widehat{\mu}_{HR} = 1.6203$	
$\widehat{\sigma}_{T90} = 0.9847$	$\widehat{\sigma}_{HR} = 0.4950$	$\widehat{\rho} = -0.02255$
Component 2		
$\widehat{w}_{2,2} = 0.7604$		
$\widehat{\mu}_{T90} = 3.3097$	$\widehat{\mu}_{HR} = 0.9474$	
$\widehat{\sigma}_{T90} = 1.1044$	$\widehat{\sigma}_{HR} = 0.4918$	$\widehat{\rho} = 0.01363$

Table 3: BATSE: MMSE Estimates for $\log(T90)$ and $\log(HR)$ under empirical EP Priors.

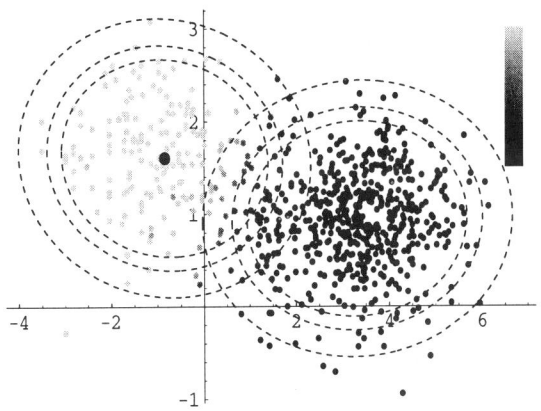

Figure 4: BATSE: classification probabilities under base model EP priors. Bar indicates value of $p(z_i = 2|\mathbf{y}, k = 2)$.

of MCMC computations, this methodology has been generalized to allow for simultaneous selection of models and the estimation of their parameters. This has become possible once algorithms for sampling from target distributions defined over joint sample spaces of models and their parameters had been developed.

The main objective of this article was to provide a summary of the state-of-the-art theory on Bayesian model selection and the application of MCMC algorithms. It has been shown how applications of considerable complexity can be handled successfully within this framework. Several methods for dealing with the use of default, improper priors in the Bayesian model selection framework have been shown. Special care has been taken to pinpoint the subtleties of jumping from one parameter space to another, and in general, to show the construction of MCMC samplers in such scenarios. The focus in the

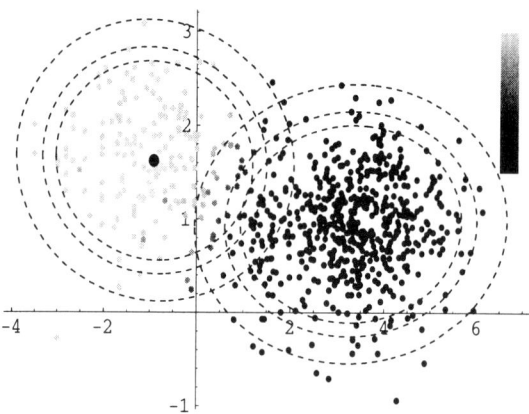

Figure 5: BATSE: classification probabilities under empirical EP priors. Bar indicates value of $p(z_i = 2|\mathbf{y}, k = 2)$

article was on the reversible jump MCMC algorithm as this is the most widely used of all existing methods; it is easy to use, flexible and has nice properties. Many references have been cited, with the emphasis being given to articles with signal processing applications.

Appendix. Notation

- $\mathbb{R}$ is the set of real numbers.

- $\mathbb{R}^+$ is the set of positive real numbers.

- $[\mathbf{A}]_{i,j}$: i^{th} row, j^{th} column of matrix $\mathbf{A}$.

- $|\mathbf{A}|$: determinant of matrix $\mathbf{A}$.

- $\mathbf{A}^{\mathrm{T}}$ is the matrix $\mathbf{A}$ transposed

- If $\mathbf{z} \triangleq (z_1, ..., z_{j-1}, z_j, z_{j+1}, ..., z_k)^{\mathrm{T}}$, then $\mathbf{z}_{-j} \triangleq (z_1, ..., z_{j-1}, z_{j+1}, ..., z_k)^{\mathrm{T}}$.

- $\mathbf{0}_{n \times p}$: null matrix of dimension $n \times p$.

- $\mathbf{I}_n$: identity matrix of dimension $n \times n$.

- $\mathbb{I}_E(\mathbf{z})$: indicator function of the set E (1 if $\mathbf{z} \in E$, 0 otherwise).

- $\delta_{\mathbf{x}}(d\mathbf{z})$ is the Dirac delta measure such that $\int_A \delta_{\mathbf{x}}(d\mathbf{z}) = 1$ if $\mathbf{x} \in A$ and 0 otherwise.

- $\lfloor z \rfloor$: highest integer strictly less than z.

- $\mathbf{z} \sim p(\mathbf{z})$: $\mathbf{z}$ is distributed according to $p(\mathbf{z})$.

- $\mathbf{z}|\mathbf{y} \sim p(\mathbf{z})$: the conditional distribution of $\mathbf{z}$ given $\mathbf{y}$ is $p(\mathbf{z})$.

Distribution	Symbol	Density
Beta	$\mathcal{B}(\alpha,\beta)$	$\frac{\Gamma(\alpha)\Gamma(\beta)}{\Gamma(\alpha+\beta)} z^{\alpha-1}(1-z)^{\beta-1} \mathbb{I}_{[0,1]}(z)$
Dirichlet	$\mathcal{D}(\alpha_1,\dots,\alpha_s)$	$\frac{\Gamma(\sum_{j=1}^s \alpha_j)}{\prod_{j=1}^s \Gamma(\alpha_j)} \prod_{j=1}^s z_j^{\alpha_j-1} \mathbb{I}_{\{\sum_{j=1}^s z_j=1\}}(\mathbf{z})$
Gamma	$\mathcal{G}(\alpha,\beta)$	$\frac{\beta^\alpha}{\Gamma(\alpha)} z^{\alpha-1} \exp(-\beta z) \mathbb{I}_{[0,+\infty)}(z)$
Gaussian	$\mathcal{N}(\mathbf{m},\Sigma)$	$\frac{\exp(-\frac{1}{2}(\mathbf{z}-\mathbf{m})^{\mathsf{T}}\Sigma^{-1}(\mathbf{z}-\mathbf{m}))}{\|2\pi\Sigma\|}$
Inverse Gamma	$\mathcal{IG}(\alpha,\beta)$	$\frac{\beta^\alpha}{\Gamma(\alpha)} z^{-\alpha-1} \exp(-\beta/z) \mathbb{I}_{[0,+\infty)}(z)$
Uniform	$\mathcal{U}_A$	$\left[\int_A d\mathbf{z}\right]^{-1} \mathbb{I}_A(\mathbf{z})$

References

[1] C. Andrieu, *Méthodes MCMC pour l'analyse bayésienne de modèles de régression paramétrique non-linéaires. Applications à l'analyse de raies et à la déconvolution impulsionnelle* (in French), PhD Thesis, Université Paris XV – Cergy-Pontoise, 1998.

[2] C. Andrieu and A. Doucet, 'Joint Bayesian model selection and estimation of noisy sinusoids via reversible jump MCMC', *IEEE Trans. Sig. Proc.* **47**, (10) 2667–2676, 1999.

[3] S.A. Barker and P.J.W. Rayner, 'Unsupervised image segmentation', *Proceedings of ICASSP*, Seattle, 1998.

[4] M.J. Bayarri and J. Berger, 'Robust Bayesian analysis of selection models', *Ann. Stat.* **26**, 645–659, 1998.

[5] J.M. Bernardo and A.F.M. Smith, *Bayesian Theory*, Wiley, 1994.

[6] Y.C. Bechtel, C. Bonaiti-Pellié, N. Poisson, J. Magnette, P.R. Bechtel, 'A population and family study on N-acetyltransferase using caffeine urinary meteabolites', *Clin. Pharm. Therap.* **54**, 134–141, 1993.

[7] J.O. Berger and L.R. Pericchi, 'The intrinsic Bayes factor for model selection and prediction', *J. Am. Stat. Assoc.* **91**, 109–122, 1996.

[8] J.O. Berger, L.R. Pericchi and J. Varshavsky, 'Bayes factors and marginal distributions in invariant situations', *Sankhya: Ind. J. Stat. A* **60**, 307–321, 1998.

[9] J.O. Berger, L.R. Pericchi and J. Varshavsky, 'Bayes factors and marginal distributions in invariant situations', Purdue University, Department of Statistics, TR95-7C, 1996.

[10] J.O. Berger and L.R. Pericchi, 'The median intrinsic Bayes factor for model selection', Duke University, ISDS, TR97–45, 1997.

[11] J.O. Berger and L.R. Pericchi, 'On the justification of default and intrinsic Bayes factors', in *Modeling and Prediction*, J.C. Lee *et al.* (eds.), Springer Verlag, 276–293, 1997.

[12] J.O. Berger and T. Sellke, 'Testing a point null hypothesis: the irreconciability of P-values and evidence', *J. Am. Stat. Assoc.* **82**, 112–122, 1987.

[13] J.M. Bernardo, 'Reference prior distributions for Bayesian inference', *J. Roy. Stat. Soc. B* **41**, 113–147, 1979.

[14] P. Billingsley, *Probability and Measure*, Wiley, 1985.

[15] S.P. Brooks, 'Markov chain Monte Carlo method and its application', *Statistician* **47**, 69–100, 1998.

[16] B.P. Carlin and S. Chib, 'Bayesian model choice via Markov Chain Monte Carlo', *J. Roy. Stat. Soc. B* **57**, 473–484, 1995.

[17] S. Chib and E. Greenberg, 'Understanding the Metropolis–Hastings algorithm', *Am. Statistician* **49**, 327–335, 1995.

[18] R.A. Cook, C. Andrieu, A. Doucet and W.J. Fitzgerald, 'Bayesian detection and estimation of gaussian peaks via reversible jump MCMC', Tech. Rep. CUED/F–INFENG/TR. 350, 1999.

[19] P. Dellaportas, J.J. Forster and I. Ntzoufras, 'On Bayesian model and variable selection using MCMC', paper based upon a talk presented at the HSSS Workshop on Variable Dimension MCMC, New Forest, September 1997.

[20] L. Devroye, *Non-Uniform Random Variate Generation*, Springer Verlag, New York, 1986.

[21] J. Diebolt and C. P. Robert, 'Estimation of finite mixture distributions through Bayesian sampling',*J. Roy. Stat. Soc. B* **56**, 363–375, 1994.

[22] P.M. Djurić, 'Bayesian methods for signal processing', *IEEE Signal Processing Magazine* **15**, (5) 26–28, 1998.

[23] P.M. Djurić, 'Variable selection by a reversible jump MCMC approach', *Proceedings of EUSIPCO*, **4**, 2013–2016, Rhodes, Greece, 1998.

[24] J. Domochowski, 'Intrinsic priors via Kulback–Leibler geometry', Purdue University, Department of Statistics, TR94–15, 1994.

[25] J.F.G. de Freitas, *Bayesian Methods for Neural Networks*, PhD Cambridge University, January 2000.

[26] D. Gamerman, *Markov Chain Monte Carlo*, Chapman & Hall, 1997.

[27] A. Gelman, J.B. Carlin, H.S. Stern, and D.B. Rubin, *Bayesian Data Analysis*, Chapman & Hall, 1995.

[28] C. Geyer, 'Likelihood inference for spatial point processes', in *Stochastic Geometry: Likelihood and Computation*, O.E. Barndorff-Nielsen, W.S. Kendall and M.N.M. Van Lieshout (eds.), Chapman & Hall, 1998.

[29] W.R. Gilks, S. Richardson and D.J. Spiegelhalter (eds.), *Markov Chain Monte Carlo in Practice*, Chapman & Hall, 1996.

[30] S. Godsill, 'Some new relationships between MCMC model uncertainty methods', University of Cambridge, CUED/F–INFENG/TR. **305**, December 1997.

[31] U. Grenander, 'Tutorial in Pattern Theory', Division of Applied Mathematics, Brown University, Providence, RI, 1983.

[32] U. Grenander and M.I. Miller, 'Representation of knowledge in complex systems' (with discussion), *J. Roy. Stat. Soc. B* **56**, 549–603, 1994.

[33] P.J. Green, 'Reversible jump MCMC computation and Bayesian model determination', *Biometrika* **82**, 711–732, 1995.

[34] K. Hurley, 'Gamma-Ray Bursts', in AIP Conf. Proc. **265**, Paciesas, W.S. and Fishman, G.J. (eds.), American Institute of Physics, New York, 1991.

[35] H. Jeffreys, *Theory of Probability*, Oxford University Press, 1961.

[36] R.E. Kass and A.E. Raftery, 'Bayes factors', *J. Am. Stat. Assoc.* **90**, 773–796, 1995.

[37] R.E. Kass and L. Wasserman, 'The Selection of prior distributions by formal rules', *J. Am. Stat. Assoc.*, 91, 1343–1369, 1996.

[38] A.D. Lanterman, M.I. Miller, and D.L. Snyder, 'General Metropolis–Hastings jump-diffusions for automatic target recognition in infrared scenes', *Optical Engineering* **36**, 1123–1137, 1997.

[39] C.A. Meegan, G.N. Pendleton, M. S. Briggs, C. Kouveliotou, T.M. Koshut, J.P. Lestrade, W.S. Paciesas, M.L. McCollough, J.J. Brainerd, J.M. Horack, J. Hakkila, W. Henze, R.D Preece, R.S. Mallozzi, and G.J. Fishman, 'The third BATSE Gamma-Ray Burst Catalog', *Astrophys. J. Supplement* **106**, 65, 1995.

[40] K. Mengersen and C.P. Robert, 'Testing for mixtures: a Bayesian entropy approach', in *Bayesian Statistics* **5**, Oxford University Press, 1996.

[41] S.P. Meyn and R.L. Tweedie, *Markov Chains and Stochastic Stability*, Springer Verlag, 1993.

[42] M.I. Miller, A. Srivastava, and U. Grenander, 'Conditional-mean estimation via jump-diffusion processes in multiple target tracking/recognition', *IEEE Trans. Sig. Proc.* **43**, 2678–2690, 1995.

[43] M.I. Miller, U. Grenander, J.A. O'Sullivan, and D.L. Snyder, 'Automatic target recognition organized via jump-diffusion algorithms', *IEEE Trans. Image Proc.* **6**, 157–174, 1997.

[44] M.A. Newton and A.E. Raftery , 'Approximate Bayesian inference by the weighted likelihood bootstrap', *J. Roy. Stat. Soc. B* **56**, 1–48, 1994.

[45] A. O'Hagan, 'Fractional Bayes factors for model comparison', *J. Roy. Stat. Soc. B* **57**, 99–138, 1995.

[46] J.J.K. O'Ruanaidh and W.J. Fitzgerald, *Numerical Bayesian Methods Applied to Signal Processing*, Springer Verlag, 1996.

[47] J.M. Pérez, *Development of Expected Posterior Prior Distributions for Model Comparisons*, PhD Thesis, Department of Statistics, Purdue University, 1998.

[48] J.M. Pérez and J.O. Berger, 'Expected posterior prior distributions for model selection', CESMA, University Simon Bolivar, TR99–09, 1999.

[49] J.M. Pérez and J.O. Berger, 'Default analysis of mixture models using expected posterior priors', CESMA, University Simon Bolivar, TR99–12, 1999.

[50] D.B. Philips and A.F.M. Smith, 'Bayesian model comparison via jump diffusions', in *Markov Chain Monte Carlo in Practice*, Chapman & Hall, 214–239, 1996.

[51] C.J. Preston, 'Spatial birth-and-death processes', *Bull. Inst. Int. Stat.* **39**, 177–212, 1976.

[52] E. Punskaya, C. Andrieu, A. Doucet and W.J. Fitzgerald, 'Bayesian segmentation of piecewise constant autoregressive processes using MCMC', CUED/F–INFENG/TR 344, University of Cambridge, 1999.

[53] S. Richardson and P.J. Green, 'On Bayesian analysis of mixtures with unknown number of components', *J. Roy. Stat. Soc. B* **59**, (4) 731–792, 1997.

[54] B.D. Ripley, 'Modelling spatial patterns', (with discussion) *J. Roy. Stat. Soc. B* **39**, 172–212, 1977.

[55] C.P. Robert, *The Bayesian Choice*, Springer Verlag, 1996.

[56] C.P. Robert, 'Mixtures of distributions: inference and estimation', in *Markov Chain Monte Carlo in Practice*, Chapman and Hall, 1996.

[57] C.P. Robert and G. Casella, *Monte Carlo Statistical Methods*, Springer Verlag, 1999.

[58] C.P. Robert, T. Rydén and D.M. Titterington, 'Bayesian inference in hidden Markov models through reversible jump Markov chain Monte Carlo', *J. Roy. Stat. Soc. B* **62**, (1) 57–75.

[59] K. Roeder and L. Wasserman, L., 'Practical Bayesian density estimation using mixtures of normals', *J. Am. Stat. Assoc.* **92**, 894–902, 1997.

[60] C. Shui, *Default Bayesian Analysis of Mixture Models*, PhD Thesis, Dept. of Statistics, Purdue University, 1996.

[61] D.J. Spiegelhalter and A.F.M. Smith, 'Bayes factor for linear and for log-linear models with vague prior information', *J. Roy. Stat. Soc. B* **44**, 377–387, 1982.

[62] A. Srivastava, M. I. Miller, and U. Grenander, 'Multiple target direction of arrival tracking', *IEEE Trans. Sig. Proc.* **43**, 1282–1285, 1995.

[63] D. Sun and J.O. Berger, 'Reference priors with partial information', *Biometrika* **85**, 55–71, 1998.

[64] D. D. Sworder and J. E. Boyd, 'Jump-diffusion in tracking/recognition', *IEEE Trans. Sig. Proc.* **46**, 235–239, 1998.

[65] L. Tierney, 'Markov chains for exploring posterior distributions', (with discussion) *Ann. Stat.* **22**, 1701–1762, 1994.

[66] I. Verdinelli and L. Wasserman, 'Computing Bayes factors using a generalization of the Savage–Dickey density ratio', *J. Am. Stat. Assoc.* **90**, 614–617, 1995.

[67] A. Zellner and A. Siow, 'Posterior odds ratios for selected regression hypotheses', in *Bayesian Statistics*, J.M. Bernardo, M.H. DeGroot, D.V. Lindley and A.F.M. Smith (eds.)., University of Valencia Press, 585–603, 1980.

Sequential Analysis of Nonlinear Dynamic Systems using Particles and Mixtures

Neil Gordon, Alan Marrs and David Salmond

1   Introduction

Many problems in science require estimation of the state of a system that changes over time using a sequence of noisy measurements made on the system. In this chapter we shall concentrate on the state space approach to modelling dynamic systems and the focus will be on the discrete time formulation of the problem. Thus, difference equations are used to model the evolution of the system with time, and measurements are assumed to be available at discrete times. For dynamic state estimation, the discrete time approach is widespread and convenient.

The state space approach to time series modelling focuses attention on the state vector of a system. The state vector contains all relevant terms required to describe the system under investigation. For example, in tracking problems the terms could be kinematic characteristics of the target. Alternatively, in an econometrics problem the terms could be related to monetary flow, interest rates, inflation etc. The measurement vector represents (noisy) observations that are related to the state vector. The measurement vector is generally (but not necessarily) of lower dimension than the state vector. The state space approach is convenient for handling multivariate data and non-stationary/non-Gaussian processes and it provides a significant advantage over traditional time series techniques for these problems. A full description is provided in [50]. Also, many varied examples illustrating the application of nonlinear/non-Gaussian state space models are given in [29].

In order to analyse and make inference about a dynamic system at least two models are required: firstly a model describing the evolution of the state with time (the system model), and secondly a model relating the noisy measurements to the state (the measurement model). We shall assume that these models are available in a probabilistic form. The probabilistic state space formulation and the requirement for the updating of information on receipt of new measurements are ideally suited for the Bayesian approach. This provides a rigorous general framework for dynamic state estimation problems.

In the Bayesian approach to dynamic state estimation one attempts to construct the probability density function (pdf) of the state based on all available information including the set of received measurements. Since this pdf embodies all available statistical information, it may be said to be the complete solution to the estimation problem. In principle, an optimal (with respect to any criterion) estimate of the state may be obtained from the pdf. A measure of the accuracy of the estimate may also be obtained. For many problems an estimate is required every time that a measurement is received. In this case a recursive filter is a convenient solution. A recursive filtering approach means that received data can be processed sequentially rather than as a batch, so that it is not necessary to store the complete data set or to reprocess existing data if a new measurement becomes available. Such a filter consists of essentially two stages: prediction and update. The prediction stage uses the system model to predict the state pdf forwards from one measurement time to the next. Since the state is usually subject to unknown disturbances (modelled as random noise) prediction generally translates, deforms and spreads the state pdf. The update operation uses the latest measurement to modify the prediction pdf. This is achieved using Bayes' theorem which is the mechanism for updating knowledge about the target state in the light of extra information from new data.

For the linear-Gaussian estimation problem, the required pdf remains Gaussian at every iteration of the filter, and the Kalman filter relations [26, 21] propagate and update the mean and covariance of the distribution. However, such assumptions are rarely appropriate and for nonlinear or non-Gaussian problems there is no general analytic (closed form) expression available for the required pdf.

Recently there has been a surge of interest in methods utilising a random sample based representation of the state pdf [19, 23, 28]. These approaches have come to be called 'particle filters' and they have the advantage of being able to handle any functional nonlinearity, and system or measurement noise of any distribution. As the number of random samples used in the filters becomes very large, they effectively provide an exact, equivalent, representation of the required pdf. Estimates of moments (such as mean and covariance) or percentiles of the state vector pdf can be obtained directly from the samples. If necessary, a functional estimate of the pdf could also be constructed from the samples and from this estimates of highest posterior density (HPD) intervals or the mode can be obtained.

In this chapter we shall outline the simplest form of particle filter for dynamic state estimation and review some of the available procedures for improving the efficiency of the method. However, our key aim is to demonstrate the utility of the methods so we then provide three examples illustrating how particle filters have been applied in complex on-line estimation problems: tracking of targets in the presence of intermittent decoys, estimation of semi-conductor composition via *in situ* ellipsometry and estimation and monitoring of chemical release sites. See [13] for a more complete overview of the area including many interesting applications of sequential Monte Carlo methods.

2   Recursive Bayesian estimation

Here we are concerned with the discrete time estimation problem. The state vector, $x_k \in \Re^n$, is assumed to evolve according to the following system model:

$$x_{k+1} = f_k(x_k, w_k) \tag{2.1}$$

where $f_k : \Re^n \times \Re^m \to \Re^n$ is the system transition function and $w_k \in \Re^m$ is a white noise sequence independent of past and current states. The pdf of w_k is assumed known. At discrete times, measurements $y_k \in \Re^p$ become available. These measurements are related to the state vector via the observation equation

$$y_k = h_k(x_k, v_k) \tag{2.2}$$

where $h_k : \Re^n \times \Re^r \to \Re^p$ is the measurement function and $v_k \in \Re^r$ is another white noise sequence of known pdf, independent of past and present states and the system noise. It is assumed that the initial pdf $p(x_1 \mid D_0) \equiv p(x_1)$ of the state vector is available together with the functional forms f_i and h_i for $i = 1, \ldots, k$. The available information at time step k is the set of measurements $D_k = \{y_i : i = 1, \ldots, k\}$.

The requirement is to construct the pdf of the current state x_k, given all the available information: $p(x_k \mid D_k)$. In principle this pdf may be obtained recursively in two stages: prediction and update. Suppose that the required pdf $p(x_{k-1} \mid D_{k-1})$ at time step $k - 1$ is available. Then using the system model it is possible to obtain the prior pdf of the state at time step k:

$$p(x_k \mid D_{k-1}) = \int p(x_k \mid x_{k-1}) p(x_{k-1} \mid D_{k-1}) dx_{k-1}. \tag{2.3}$$

Here the probabilistic model of the state evolution, $p(x_k \mid x_{k-1})$, which is a Markov model, is defined by the system equation (2.1) and the known statistics of w_{k-1}.

Then at time step k a measurement y_k becomes available and may be used to update the prior via Bayes' rule:

$$p(x_k \mid D_k) = \frac{p(y_k \mid x_k) p(x_k \mid D_{k-1})}{p(y_k \mid D_{k-1})} \tag{2.4}$$

where the normalising denominator is given by

$$p(y_k \mid D_{k-1}) = \int p(y_k \mid x_k) p(x_k \mid D_{k-1}) dx_k. \tag{2.5}$$

The conditional pdf of y_k given x_k, $p(y_k \mid x_k)$, is defined by the measurement model (2.2) and the known statistics of v_k. In the update equation (2.4), the measurement y_k is used to modify the predicted prior from the previous time step to obtain the required posterior of the state.

The recurrence relations (2.3) and (2.4) constitute the formal solution to the Bayesian recursive estimation problem. Analytic solutions to this problem are only available for a relatively small and restrictive choice of system and measurement models, the most important being the Kalman filter, which assumes f_k and h_k are linear and w_k and v_k are additive Gaussian noise of known variance. Considerations of realism imply that these assumptions are unreasonable for many applications.

3   Some approximate methods of implementation

The search for improved approximate implementation procedures for general recursive Bayesian filters has been an active area of research for many years. Many alternative approaches have been suggested and these can broadly be split into three groups: analytic approximations where the aim is to use some form of distributional approximation to estimate the pdf; numerical approximations where a grid of nodes is used as the basis for numerical integration strategies and functional approximations; and most recently Monte Carlo methods where random samples are utilised to effectively give filtering by simulation.

3.1   Analytic approximations

The most well known approach to implementing recursive nonlinear filters is the Extended Kalman Filter (EKF) [24]. This is applicable to nonlinear models with additive Gaussian noise and uses a linearisation technique based on a first order Taylor series expansion of the nonlinear system and measurement functions. However, the posterior pdf is still modelled by a Gaussian which may be a significant distortion of the true underlying pdf and for many important practical applications, such as bearings-only tracking for instance, the EKF is well known to be unstable and unreliable, leading to gross errors which are notoriously difficult to detect in an on-line environment.

An obvious development from the EKF is to extend the Taylor series expansion to include higher order terms. A filter based on second order expansions is described in [24]; however, the Gaussian assumption is still implicit in the approximation and divergence problems can still occur. An attempt to remove the Gaussian assumption via the use of series expansions was made by [45] who used an Edgeworth expansion. However, when the infinite series is truncated for practical implementation the result can become negative over portions of the state space. This is particularly problematic when significant probability mass is contained in the region where the density estimate is non-positive.

The idea of replacing the normal distribution framework with one composed of a convex linear combination of normal distributions was proposed in [44, 2]. Here the nonlinear/ non-Gaussian filtering problem is essentially approximated by a convex combination of EKF's. The two problems with this approach are the exponential increase in the number of mixture terms as measurements are processed and the inherent inaccuracy of the EKF.

3.2   Numerical approximations

Here the posterior pdf is estimated on a (prescribed) grid in state space and some form of interpolating function used to create the full pdf approximation. Research has focused on the choices of grid location, methods of approximating the density at those grid points and different interpolation functions. For instance, [27] used a large fixed set of equally spaced nodes with a linear interpolation between grid points, whereas [31] chose instead to have a constant interpolation function, resulting in a 'histogram' style pdf approximation. These methods all require that the grid of nodes be defined to encompass the region of high posterior probability. A quadrature technique with dynamic grid allocation [40] and a Monte Carlo grid placement strategy [47, 48] both address this problem. However, the exponential increase in required number of grid points as the dimension increases and the non-trivial calculation required at each grid point still remain as problems for these approaches.

A different style of numerical approximation has recently been proposed by [25]. He uses points derived on the Gaussian constant probability ellipse to effectively perform numerical differentiation for the EKF. This results in significantly improved performance over the EKF but is still only really valid for straightforward unimodal problems.

3.3   Monte Carlo approximations

The idea of using Monte Carlo methods in dynamic estimation problems originated in [20] but has only reappeared in recent years. The resurgence in interest has led to the development of 'particle filters' which are effectively a sequential Monte Carlo strategy for on-line filtering with random sample based representations of the posterior pdf. Whereas the goal of current approaches is to approximate the full posterior pdf, [20] were only concerned with estimating the mean and covariance. It can only be assumed that this was due to the limited computational power available at the time.

Other previous works involving the use of Monte Carlo methods include [7] who use MCMC methods to analyse the data as a batch (rather than recursively). Also, [36] which is based on rejection sampling for the key update stage and so results in a random, decreasing sample set size and [49] who used iterative kernel fitting to improve the efficiency of an importance sampling technique.

A related algorithm involving branching of particle streams is proposed and analysed in some detail in [10, 11]. This technique is referred to as interacting particle filters. The birth/death process is arranged so that particles having higher likelihood support will tend to spawn more birth particles than those having lower likelihood support.

4 Particle Filters

In a particle filter the underlying posterior pdf is represented by a cloud of random samples in state space. This differs from grid based methods in that there is no requirement to specify the sample locations and the samples automatically congregate in regions of high posterior probability. Also, the rate of convergence of particle methods is theoretically not sensitive to the size of the state space [9]. A recent review of the field is given in [12].

4.1 Why random samples?

The first question which must be answered is why use such a computationally spend-thrift approach where pdf's are represented by thousands of random samples rather than a low dimensional parametric summary? The key advantages which make the computational sacrifice worth while include routine use of (practically) any nonlinear/non-Gaussian model, approximation to the full pdf rather than a limited summary, coordinate transformations are easily accomplished, physical constraints and other prior information can be easily incorporated, HPD regions and pdf moments easily approximated, independence of association hypotheses over time and the scalability (i.e. the sample representation will approach the exact solution as computational capabilities improve).

4.2 A basic particle filter

In this section we outline the simplest approach to implementing a particle filter [19, 28, 23]. Suppose we have a set of independent random samples $\{x_{k-1}(i) : i = 1, \ldots, N\}$ drawn from the pdf $p(x_{k-1} \mid D_{k-1})$. The particle filter is an algorithm for propagating and updating these samples to obtain a set of values $\{x_k(i) : i = 1, \ldots, N\}$, which are approximately distributed as independent random samples from $p(x_k \mid D_k)$. Thus the filter is an approximate mechanisation (simulation) of the prediction and update stages of a recursive Bayesian filter.

Prediction   Noting the approximate result

$$\begin{aligned} p(x_k \mid D_{k-1}) &= \int p(x_k \mid x_{k-1})p(x_{k-1} \mid D_{k-1})dx_{k-1} \\ &\approx N^{-1}\sum_{i=1}^{N} p(x_k \mid x_{k-1} = x_{k-1}(i)) \end{aligned}$$

where $\{x_{k-1}(i) : i = 1, \ldots, N\}$ are N iid samples from $p(x_{k-1} \mid D_{k-1})$, the following algorithm is suggested [18]:

- uniformly resample (with replacement) from the values $\{x_{k-1}(i)\}$;

- pass each resampled value through the system model.

Repeated N^* times, this generates the required sample set $\{x_k^*(i) : i = 1, \ldots, N^*\}$. Note that N^* is not necessarily equal to N. These are (approximately) independent random samples from the pdf $p(x_k \mid D_{k-1})$.

SIR update On receipt of the measurement y_k, evaluate the likelihood of each prior sample and so obtain a normalised weight for each sample:

$$q_i = \frac{p(y_k \mid x_k^*(i))}{\sum_{j=1}^{N^*} p(y_k \mid x_k^*(j))}. \qquad (4.1)$$

Thus define a discrete distribution over $\{x_k^*(i) : i = 1, \ldots, N^*\}$, with probability mass q_i associated with element $x_k^*(i)$. Now resample (with replacement) N times from the discrete distribution to generate samples $\{x_k(i) : i = 1, \ldots, N\}$, so that for any j, $\Pr(x_k(j) = x_k^*(i)) = q_i$.

The above steps of prediction and update form a single iteration of the recursive algorithm. To initiate the algorithm, N^* samples $\{x_1^*(i) : i = 1, \ldots, N^*\}$ are drawn from the known prior $p(x_1)$. These samples feed directly into the update stage of the filter. The samples $x_k(i)$ are approximately distributed as the required pdf $p(x_k \mid D_k)$ – see discussion in [19, 16, 28] for further details. For multi-modal problems the following alteration to the basic update stage was suggested [17].

Hybrid filter update As before, define a discrete distribution over $\{x_k^*(i) : i = 1, \ldots, N^*\}$, with probability mass q_i associated with element $x_k^*(i)$. However, instead of resampling directly from this discrete sample set we now use the set of weighted points to generate a mixture approximation to the posterior pdf at time step k:

$$\hat{p}(x_k \mid D_k) = \sum_{i=1}^{C(k)} \pi_i d(x_k \mid m_i, V_i), \qquad (4.2)$$

where $d(x \mid m, V)$ denotes a density function of random variable x, centred at mode m and scaled by some positive definite matrix V. A key example has $d(x \mid m, V) = N(x; m, V)$, the density of a multivariate Gaussian with mean m and covariance matrix V, [44]. Other possibilities include choosing d to be the multivariate Student-t distribution [49]. Here we shall only consider Gaussian mixture approximations. This gives

$$\hat{p}(x_k \mid D_k) = \sum_{i=1}^{C(k)} \pi_i N(x_k; m_i, V_i), \qquad (4.3)$$

where $C(k)$ is the number of components required to approximate the N^* point discrete sample set, m_i is the mean of the ith component, V_i is the covariance of the ith component and π_i is the weight attached to the ith component with $\pi_1 + \ldots + \pi_{C(k)} = 1$. Details of the clustering algorithm used are given in [41]. We then draw N samples $\{x_k(i) : i = 1, \ldots, N\}$ from this mixture to complete the update stage.

4.3 Improvements

The basic approach to particle filtering described above is not particularly efficient and many techniques have been proposed leading to improved computational performance. In this section we provide a brief review.

[39] recently proposed an auxiliary variable method (ASIR) which rolls the prediction and update stages of the filter into a single process so that the measurement can cause 'good' (where good means going to produce high likelihood) values in the prior sample set to be sampled more often. The additional (or auxiliary) variable is an indicator variable pointing to the prior sample to be used as parent in the system model. Their method also allows for aspects of the calculation to be carried out analytically (if possible) so reducing the Monte Carlo variation in the approach and increasing the efficiency. In [6] a similar approach is used but the auxiliary variable mixture component selected is then used to define a proposal distribution for a Metropolis-Hastings step.

A related idea on kernel based importance sampling was presented by [22]. They also outline a method of approximating the likelihood which is very useful for model selection.

A totally different perspective on the particle filtering problem is presented in [8, 14, 15]. Here they consider the aim to be that of providing a weighted discrete approximation to the pdf rather than trying to generate a cloud of random samples. Some impressive results have been obtained on a range of problems.

[33, 32] develop a sequential importance sampling strategy. Here, resampling is only carried out when the variance of the weighted particle sets is deemed to be too large (or equivalently the effective sample set size too small). Otherwise the non-uniformly weighted samples are propagated through the filter. [8] highlights some of the problems with the choice of measure of effective sample set size.

Contrasting the essentially discrete nature of the sample set with the continuous parameters they represent in state space motivates the use of a smoothing (or regularising) technique to enhance filter performance. Techniques for implementing smoothing were suggested in [19, 16, 17, 18, 46, 1].

5 Tracking in the presence of intermittent decoys

This section is concerned with the problem of maintaining track on a target in the presence of intermittent spurious objects. These spurious objects or sources are spawned in close proximity to the required target at random times and persist for random periods. Only one such source is present at any instant and its birth/death process is modelled by a Markov process. The problem is complicated by the presence of dense random clutter. Position and discrimination (signature) information is available on the objects and clutter from a scanning sensor (e.g. radar). It is further assumed that the resolution of the sensor is finite, so when in close proximity, the target and spurious object may generate only a single common return. There is a high degree of measurement association uncertainty.

The usual (Bayesian) approach to problems of this type involves some form of multiple hypothesis filter. For optimality, it is necessary to take account of the history of every possible measurement association hypothesis. As measurements accumulate, this rapidly becomes infeasible, and some hypothesis pruning and/or merging scheme, such as Probabilistic Data Association [5], or some mixture reduction method [37], must be imposed. In particular, a successful application of the IMMJPDAF (Interacting Multiple Model Joint Probabilistic Data Association Filter) to the problem of tracking splitting targets with finite sensor resolution is reported by [4]. Also a multiple hypothesis approach to tracking two closely spaced targets with finite sensor resolution is presented by [30]. If system or measurement nonlinearities are present, further approximations (typically a version of the extended Kalman filter) must be employed.

Here we consider an application of particle filters to this problem. It was noted by [3] that particle filters can be applied directly to problems involving measurement association uncertainty, provided that the measurement likelihood can be specified (also see [17]). The main advantage of this approach is that measurement association hypotheses from previous time steps do not need to be explicitly stored - they are contained implicitly in the sample set. In this case, suboptimality is due to the number of samples being finite, and as the sample size tends to infinity, the exact (optimal) result is obtained.

5.1 Scenario

The dynamics of the primary target (T) are described by the following (known) discrete system model:

$$x_{Tk+1} = f_{Tk}(x_{Tk}, w_{Tk}) \qquad (5.1)$$

where x_{Tk} is the target state vector, w_{Tk} is system driving noise, and f_{Tk} describes the dynamics of the target.

At some random time step the target may spawn a secondary object D in the vicinity of T. Thereafter the secondary object moves independently of T according to the following dynamic model:

$$x_{Dk+1} = f_{Dk}(x_{Dk}, w_{Dk}), \tag{5.2}$$

where x_{Dk} is the state vector of D. The initial distribution of x_{Dk} at birth is a (known) function of x_{Tk}. The secondary object disappears after a random period and later (following another random period) another object D may be produced. The birth/death sequence of the object D is described by a Markov process. If $\gamma_k = 0$ indicates that D does not exist at time t_k and $\gamma_k = 1$ indicates that D is in existence, the transitions $0 \rightarrow 1$ and $1 \rightarrow 0$ depend only on the probabilities:

$$p_{01} = \Pr\{\gamma_k = 1 \mid \gamma_{k-1} = 0\} \quad \text{and} \quad p_{10} = \Pr\{\gamma_k = 0 \mid \gamma_{k-1} = 1\}. \tag{5.3}$$

Clearly, $\Pr\{\gamma_k = 1 \mid \gamma_{k-1} = 1\} = 1 - p_{10}$ and $\Pr\{\gamma_k = 0 \mid \gamma_{k-1} = 0\} = 1 - p_{01}$. If the time step Δt is constant, the average period between the death of one secondary object and the birth of another is $\Delta t / p_{01}$, while the average lifetime of D is $\Delta t / p_{10}$. Note that with this model at most two objects may be present at any instant.

At $k = 0$, it is assumed that only the primary target T is present ($\gamma_0 \equiv 0$). The prior distribution of x_{T0} is also assumed to be known.

5.2 Sensor model

At each time step k, N_k position measurements z_{ik} are received from a sensor whose position is precisely known. If only the primary target T is present (i.e. $\gamma_k = 0$), the probability of detecting the target is P_{TD}. If the secondary object D is also present (i.e. $\gamma_k = 1$), depending on its proximity to the primary target and the relative geometry, the sensor may be capable of resolving two objects or it may only be able to resolve a single composite object. If the two objects can be resolved, then the probability of receiving a measurement from T is P_{TD} and the probability of receiving a measurement from D is P_{DD} (and these are independent events). If the objects cannot be resolved, the probability of receiving a single composite measurement is P_{JD}. These probabilities may be functions of the appropriate T or D states. Additionally, other spurious or clutter measurements (independent of the two objects) may also be produced by the sensor. We assume that these clutter measurements are uniformly distributed over the measurement space and that they are not subject to resolution limitations (although this would only be significant in exceedingly dense clutter). The number of clutter measurements received at a given time step follows a Poisson distribution with mean m.

Associated with each position measurement z_{ik} is a classification flag or signature parameter c_{ik} which may provide an indication of the type of object from which the measurement originated (target, secondary object, composite or clutter), but gives no direct information on object position; c_{ik} could be a discrete output (for example, target, secondary object or clutter) or a continuous parameter such as a measure of target amplitude. The N_k measurements and classifications received at time t_k are denoted

$$Z'_k = \{(z_{1k}, c_{1k}), (z_{2k}, c_{2k}), \ldots, (z_{N_k k}, c_{N_k k})\}, \tag{5.4}$$

and the set of all data received up to and including time t_k is denoted

$$Z_k = \{Z'_1, Z'_2, \ldots, Z'_k\}. \tag{5.5}$$

It is assumed that the association between measurements and the objects is *a priori* unknown. An association hypothesis $\mathcal{H}$ defines a mapping λ from the subscripts of the measurements to their source (target (T), secondary (D), composite (J) or clutter(C)): $\lambda : \{1, 2, \ldots, N_k\} \rightarrow \{T, D, J, C\}$. Given λ, the conditional pdfs of the measurements z_{ik} are denoted by $p_T(z_{ik} \mid x_{Tk})$ if $\lambda(i) = T$, $p_D(z_{ik} \mid x_{Dk})$ if $\lambda(i) = D$, $p_J(z_{ik} \mid x_{Tk}, x_{Dk})$ if $\lambda(i) = J$, and $p_C(z_{ik})$ if $\lambda(i) = C$. The performance of the classifier is denoted similarly by $p_T(c_{ik} \mid x_{Tk})$ for $\lambda(i) = T$, etc. Note that the classifier performance may be state dependent. It is assumed that these conditional distributions are known.

Although the specification of the classifier output is essentially identical in form to that of the measurements, it is convenient to make the distinction between z and c to emphasise that two quite different types of information are available, one of which is strongly indicative of object position while the other is primarily dependent on object type. However, if the classifier output is state dependent (albeit only weakly), then this also provides some information on the object state vector. The particle filter is able to exploit this.

5.3   Measurement/classification likelihood

For this problem it is convenient to define a system state vector $X_k = (x_{Tk}, x_{Dk}, \gamma_k)$ which evolves with time according to equations (5.1) to (5.3). Note that if $\gamma_k = 0, x_{Dk}$ is redundant. In this case it is convenient to set $x_{Dk} = x_{Tk}$.

5.3.1   General form of likelihood

By careful consideration of the possible measurement-state associations, the likelihood of the state vector X given the data set Z' (dropping the time subscript k) may be shown to be (for $m > 0$)

$$
\begin{aligned}
p(Z' \mid X) \quad \propto \quad & (1 - \gamma) f_T(x_T \mid Z') + \gamma (1 - P_{\text{res}}) f_J(x_T, x_D \mid Z') \\
& + \gamma P_{\text{res}} [f_T(x_T \mid Z') f_D(x_D \mid Z') \\
& \qquad - P_{TD} P_{DD} \sum_{j=1}^{N} \ell_T(x_T \mid z_j, c_j) \ell_D(x_D \mid z_j, c_j)]. \tag{5.6}
\end{aligned}
$$

Here, the likelihood ratio is

$$\ell_T(x_T \mid z, c) = \frac{1}{m} \frac{p_T(z \mid x_T)}{p_C(z)} \frac{p_T(c \mid x_T)}{p_C(c)}; \qquad (5.7)$$

ℓ_D is similar, and

$$\ell_J(x_T, x_D \mid z, c) = \frac{1}{m} \frac{p_J(z \mid x_T, x_D)}{p_C(z)} \frac{p_J(c \mid x_T, x_D)}{p_C(c)}. \qquad (5.8)$$

The function $f_T(x_T \mid Z')$, which may be interpreted as the likelihood of x_T for the case of a single target in clutter, is given by

$$f_T(x_T \mid Z') = (1 - P_{TD}) + P_{TD} \sum_{j=1}^{N} \ell_T(x_T \mid z_j, c_j); \qquad (5.9)$$

$f_D(x_D \mid Z')$ is similar, and

$$f_J(x_T, x_D \mid Z') = (1 - P_{JD}) + P_{JD} \sum_{j=1}^{N} \ell_J(x_T, x_D \mid z_j, c_j). \qquad (5.10)$$

Also P_{res} is the probability that T and D can be resolved (if $\gamma = 1$) and this is a function of x_T and x_D.

The likelihood (5.6) has three principal terms. The first of these (for $\gamma = 0$) corresponds to cases when the secondary object is not present so that only measurements from the prime target or clutters are available. The second term (for $\gamma = 1$ and $P_{\mathrm{res}} = 0$) represents the case where the secondary object is present but it is not resolved so that only measurements from the composite object or clutter are available. The third term (for $\gamma = 1$ and $P_{\mathrm{res}} = 1$) corresponds to the case where again the secondary object is present but it can be resolved from the target. Various special cases follow directly from this expression, for example if $\gamma \equiv 0$ the likelihood for a single target in clutter is obtained. Also for $\gamma \equiv 1$ and $P_{TD} = P_{TD} = P_{\mathrm{res}} \equiv 1$, we recover the special case presented by [42] where the decoy is always present, detected and resolved.

5.4 Solution for Gaussian measurements and uniform clutter

The above solution is valid for any form of (time independent) measurement error characteristic (including, for example, quantisation effects and skewed or truncated distributions). Likewise, any form of clutter distribution may be employed provided it is independent of the state X. However, consider the common assumptions of Gaussian measurement errors and uniformly distributed clutter: $p_T(z \mid x) = p_D(z \mid x) = N(z; h(x), R)$, $p_J(z \mid x_T, x_D) = N(z; h_J(x_T, x_D), R_J)$ and $p_C(z) = 1/V$. The measurements z are independent, zero mean Gaussian perturbations about a function of the state. For the resolved objects the (possibly nonlinear) measurement function is h and the covariance of the Gaussian perturbation or error is R. For the composite return, the measurement function h_J depends on both x_T and x_D. This function could depend on some centroid of x_T and x_D such as $(x_T + x_D)/2$. Also, the measurement error has covariance R_J which may be different to that of the resolved objects. For example, in the case of a radar sensor, the composite measurement might have a larger error due to glint type effects. The parameter V in the clutter distribution is the volume of the sensor surveillance region (which is assumed to 'comfortably' encompass T and D). Note that the spatial density of clutter measurements in this case is $\mu = m/V$.

5.5   Simulation example

5.5.1   Scenario

Both objects T and D (if it exists) are assumed to obey the common second order, linear, Cartesian model. The X-coordinate for this model is

$$\left.\begin{aligned}
x_{ik+1} &= x_{ik} + \Delta t \dot{x}_{ik} + (\Delta t^2/2)w_{iXk}, \\
\dot{x}_{ik+1} &= \dot{x}_{ik} + \Delta t w_{iXk},
\end{aligned}\right\} \qquad (5.11)$$

for $i = T$ and D. The Y-coordinate is similar. The driving noise sequences w_{TXk}, w_{TYk}, w_{DXk} and w_{DYk} are independent Gaussian random processes of variance q. Note that since the models for T and D are the same in this example, trajectory characteristics cannot be used to distinguish between T and D. If the dynamics of T and D were known to be different and this could be modelled (as allowed in the general formulation), this would aid the discrimination process. The state vector for the system is

$$X_k = (x_T, \dot{x}_T, y_T, \dot{y}_T, x_D, \dot{x}_D, y_D, \dot{y}_D, \gamma)_k \,.$$

In the example below, the system driving noise variance q is set to 0.015^2 and the time step $\Delta t = 0.25$. Also, the probability of D being born at time $k+1$ given that it does not exist at k is $p_{01} = \Pr\{\gamma_{k+1} = 1 \mid \gamma_k = 0\} = 0.1$, and the probability of it dying is $p_{10} = \Pr\{\gamma_{k+1} = 0 \mid \gamma_k = 1\} = 0.2$. Thus the average lifetime of a D object is $5\Delta t$, and the average interval between the death of one D object and the birth of another is $10\Delta t$.

Measurements of range and bearing $(z_{ik} = (z_{irk}, z_{i\theta k}))$ are taken from a sensor located at the origin. Thus for measurements originating from T or D, the nonlinear measurement function is defined by

$$z_{iRk} = r_{\lambda(i)k} + v_{iRk} \quad \text{and} \quad z_{i\theta k} = \theta_{\lambda(i)k} + v_{i\theta k}, \qquad (5.12)$$

where for resolved objects $(\lambda(i) = T$ or $D)$

$$r_{\lambda(i)k} = \sqrt{x_{\lambda(i)k}^2 + y_{\lambda(i)k}^2} \quad \text{and} \quad \theta_{\lambda(i)k} = \tan^{-1}\left(\frac{y_{\lambda(i)k}}{x_{\lambda(i)k}}\right), \qquad (5.13)$$

and similarly for the case of an unresolved composite measurement ($\lambda(i) = J$) using the centroid of the individual states. The measurement errors v_{irk} and $v_{i\theta k}$ are independent, zero mean, Gaussian processes of variance σ_r^2 and σ_θ^2, respectively (for $\lambda(i) = T$, D or J). We can resolve T and D provided they do not fall into the same range/bearing resolution cell (Δr, $\Delta\theta$), i.e.

$$P_{\text{res}}(x_T, x_D) = \begin{cases} 0 & \text{if } |r_T - r_D| < \Delta r \quad \text{and} \quad |\theta_T - \theta_D| < \Delta\theta, \\ 1 & \text{otherwise}; \end{cases} \quad (5.14)$$

see also the resolution model suggested by [30]. For the example results presented below: $\sigma_r = 0.01$, $\Delta r = 0.05$, $\sigma_\theta = 0.01$ radians and $\Delta\theta = 0.05$ radians. Note that the measurement error is substantially less than the sensor resolution, a common feature of radar systems. Also the detection probabilities for T, D and the composite return are all 0.99 (i.e. $P_{TD} = P_{DD} = P_{JD} = 0.99$).

Clutter measurements are uniformly distributed in range and bearing over the field of view of the sensor. In the simulation experiments, an acceptance gate in the measurement space is defined to reject any measurements that clearly originate from clutter and do not assist in the estimation of the state vector. This gate is defined by the maximum extent of the predicted T and D position samples plus four standard deviations of the measurement error. In the simulation experiment below, the clutter density is high: the average number of clutter returns in a $\Delta r \times \Delta\theta$ resolution cell is 0.25 and the average number of returns in a $\sigma_r \times \sigma_\theta$ cell is 0.01.

Associated with each measurement is a discrete classification flag which takes the values T, D or C. The classifier performance of the sensor against the target is a function of the target aspect presented. Assuming that the target's axis is directed along its velocity vector, the classification performance is a function of $\psi = \theta_{\text{heading}} - \theta_T$ (where $-180° < \psi < 180°$) where θ_T is the sightline angle between the sensor and the target with respect to the X-axis and

$$\theta_{\text{heading}} = \tan^{-1}(\dot{y}_T/\dot{x}_T) \quad (5.15)$$

is the target heading relative to the X-axis. The classification probabilities for the simulation example are given in Table 5.1. Thus it is assumed that for D and clutter, performance is independent of aspect. Note that when the subtended target aspect is within 10° of the target's axis, the classifier output is equally likely to be T or D. At other aspects, performance is useful. Also the classifier has a 10% chance of mistaking the T, D or J for clutter, and correctly recognises clutter with a probability of 70%. The composite return J is equally likely to be classified as T or D – there is no classification J in this example. The state dependency of the target classification performance considerably complicates the estimation problem. The accuracy of the state information affects the degree to which the filter can rely on the classifications, while, conversely, it is possible to learn about the direction of the target

Origin of actual measurement	Classifier output		
	T	D	C
T for $|\psi| \in (10°, 170°)$	0.60	0.30	0.10
T for $|\psi| \notin (10°, 170°)$	0.45	0.45	0.10
D	0.30	0.60	0.10
J	0.45	0.45	0.10
C	0.15	0.15	0.70

Table 5.1: Classification probabilities

velocity vector from the sequence of classifications. The particle filter is able to accommodate (and exploit) this state dependency.

At $k = 0$, $\gamma \equiv 0$ so that only T is present. The prior distribution of the position of T is Gaussian with mean (x_{T0}, y_{T0}) and covariance

$$\begin{pmatrix} \sigma^2_{TX0} & 0 \\ 0 & \sigma^2_{TY0} \end{pmatrix}.$$

The prior distribution of the initial target velocity is defined in terms of direction and magnitude: the direction being uniformly distributed over $[0, 2\pi)$ and the magnitude being uniformly distributed over $[0, V_{T\,\mathrm{max}}]$. In the example below, $\bar{x}_{T0} = 0.1, \bar{y}_{T0} = 1.9, \sigma_{TX0} = \sigma_{TY0} = 0.25$ and $V_{T\,\mathrm{max}} = 0.05$. The prior distribution of the position and velocity of each D at birth is assumed to Gaussian with mean x_T and a diagonal covariance matrix. In the example below, the standard deviation of the x and y positions is $0.03(= 3\sigma_r = 0.6\Delta r)$ and the standard deviations of the x and y velocity components is $0.01(= 0.2V_{T\,\mathrm{max}})$. Thus, D type objects are generated in close proximity and with similar velocities to the target (relative to the problem parameters).

5.6   Illustrative result

Figures 5.1 and 5.2 show the object paths, measurements and track for the hybrid particle filter. In Figure 5.2, the actual path of the target is shown as a continuous line joining diamonds which indicate the target positions at each time step. The actual paths of the secondary D type objects are shown as continuous lines connecting crosses. There are two of these D type paths (alive at different times) in this example. The remaining individual symbols indicate the positions of those measurements that actually originate from the target. Those measurements correctly classified as type T are shown as diamonds, those incorrectly classified as type D are shown as crosses and those incorrectly classified as clutter are shown as asterisks. Those measurements which are the product of an unresolved T/D pair are surrounded by a square box. The

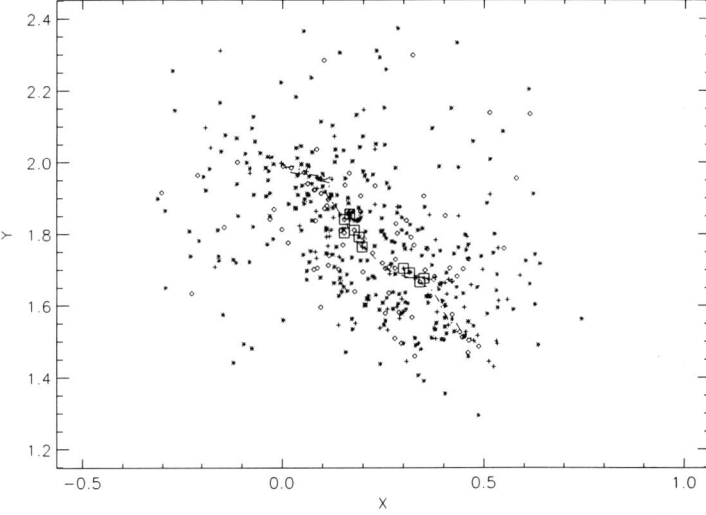

Figure 5.1. Measurements, paths and tracks

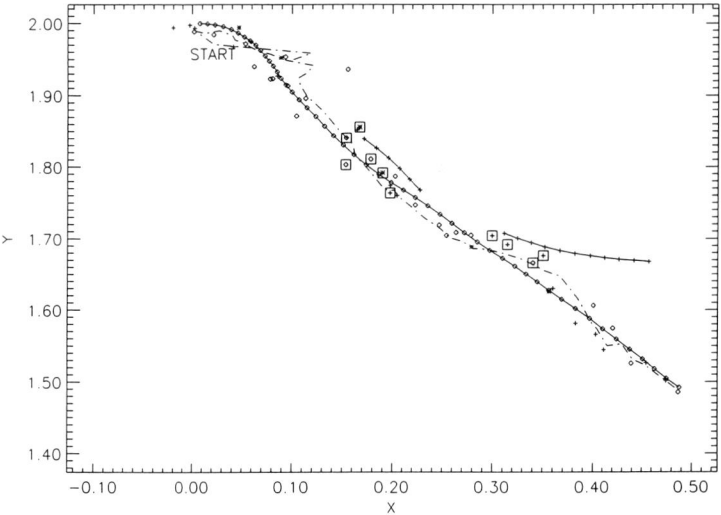

Figure 5.2. Estimated and true target position

position estimate (i.e. the mean of the posterior, $E[x_{Tk}, y_{Tk} \mid Z_k]$) from the tracking filter is shown as a dashed line. Figure 5.1 shows all measurements accepted by the filter over the 50 time steps. The notation is as above and the track from the filter is repeated. The actual value of γ_k is also shown in Figure 5.3 together with the estimated value (continuous line) of γ_k from the filter, i.e. $E[\gamma_k \mid Z_k]$. Figure 5.4 shows the number of measurements that pass the filter's acceptance test.

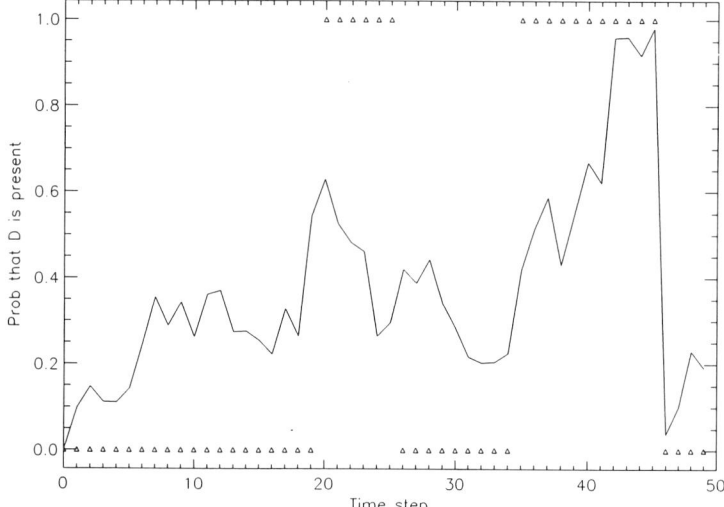

Figure 5.3. Estimated and true values of γ_k

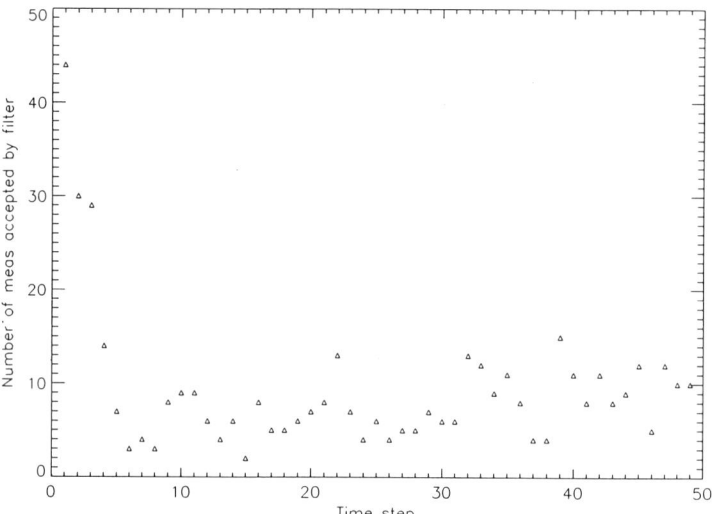

Figure 5.4. Number of gated measurements

The particle filter has successfully accommodated the release of two D type objects (at different times) produced in close proximity to the target and moving with similar velocities. In the case of the first release at time step 20, the object only exists for six time steps and is never resolved. This has little effect on the filters although there is some indication of an increase in the estimate of γ_k (possibly due to a perceived jump in the target measurements when the decoy is launched and a change in the dynamics - which is effectively smoothed by the averaging of the two object paths). In the case of the second release at time step 35, the object gradually separates from the target and is resolved as a separate entity after four time steps. This presents a more difficult problem. The dynamics of D are quite similar to T. Furthermore, during the seven time steps when T and D are resolved, the target is incorrectly classified as a D type object on four occasions while D is correctly classified on only three occasions. Note that soon after the two objects are resolved, the estimate of γ_k rises to above 0.9 showing that the filter is confident that two objects are present. When D dies, the estimates of γ_k rapidly fall to about 0.2.

The number of clutter measurements accepted by the filter is generally between about 5 and 10 for each time step after the transient has decayed. Although this is quite high, the acceptance gate is generous so that these false returns are often fairly remote from the objects. Also, the classifier correctly recognises clutter on 80% of occasions. Nevertheless, the filter does successfully accommodate this level of disturbance.

6 In situ ellipsometry

This section describes the use of a particle filter to perform *in situ* monitoring of growing silicon–germanium (SiGe) alloys [35]. Silicon (Si) technology is revolutionising the design of both civil and military systems through the production of complex integrated circuits at low unit cost. The electronic and optical properties of SiGe devices are sensitive to the composition, thickness and strain-state of the alloys, and if the additional gains of lower power and higher speed are to be realised, these layer parameters must be accurately controlled in manufacture.

Real-time spectroscopic ellipsometry (RTSE) is currently used for quantitative determination of alloy composition and thickness. Ellipsometry is a surface analytical technique used to determine the optical properties and morphology of a surface from measurements of the change in polarisation state of light reflected by it (parameterised by the ellipsometric polarisation angles Ψ and Δ). Until recently, the main weakness of the technique has been the large time lag between real-time data acquisition and computation of film optical properties and thickness [38].

The main problem underlying the direct calculation of film properties from raw measurements lies in the transcendental nature of the equations relating polarisation state change to the physical attributes of a given surface. Their solution requires time consuming numerical techniques. The most commonly used technique is the Levenberg-Marquardt algorithm for nonlinear model optimisation which, given an initial guess at film properties, updates the parameters of the ellipsometry equations such that the predicted ellipsometric spectra will match the measured profiles as closely as possible. This method typically requires several tens of iterations to converge to a local minimum. Rate of convergence to a solution is, of course, dependent upon the initial guess. In addition, each measurement is considered in isolation from the previous measurements, which can lead to large fluctuations in predicted composition from one measurement to the next. The use of iterative model fitting techniques also makes it difficult to incorporate prior knowledge regarding the growth conditions in a coherent manner.

The analysis of RTSE data relies upon a 'virtual interface' approximation which uses a 3-layer model for the growing material: pseudo-substrate/surface layer/gas. The surface layer is defined as the material deposited in a chosen time interval; the pseudo-substrate represents the true substrate and all material deposited prior to the surface layer.

6.1   Ellipsometry sensor model

Making use of the layer model, the ellipsometric polarisation angles for a given wavelength of incident light can be calculated from the reflection coefficients for s and p polarisation:

$$\tan(\Psi)\exp(j\Delta) = \frac{R_p}{R_s} = \frac{S_{21_p}}{S_{11_p}} \times \frac{S_{11_s}}{S_{21_s}}, \tag{6.1}$$

where the reflection coefficients are defined in terms of the elements of the s and p polarisation structure matrices **S** which are in turn defined by combining interface matrices (**I**) and layer matrices (**L**);

$$\mathbf{S} = \mathbf{I}_{\text{gas/surface}}\mathbf{L}_{\text{surface}}\mathbf{I}_{\text{surface/pseudo}}. \tag{6.2}$$

The interface matrices are defined using the interface Fresnel reflection and transmission coefficients for s and p polarisation:

$$\mathbf{I_{a/b}} = \frac{1}{\mathbf{t_{a/b}}}\begin{pmatrix} 1 & r_{a/b} \\ r_{a/b} & 1 \end{pmatrix}, \tag{6.3}$$

where the Fresnel coefficients for the interface between layer a and layer b are calculated from the layer refractive indices N at that wavelength and the incident angle ϕ_a and refraction angle ϕ_b,

$$r_{a/b_p} = \frac{N_b \cos\phi_a - N_a \cos\phi_b}{N_b \cos\phi_a + N_a \cos\phi_b}, \tag{6.4}$$

$$r_{a/b_s} = \frac{N_a \cos\phi_a - N_b \cos\phi_b}{N_b \cos\phi_a + N_a \cos\phi_b}, \tag{6.5}$$

$$t_{a/b_p} = \frac{2N_a \cos\phi_a}{N_b \cos\phi_a + N_a \cos\phi_b}, \tag{6.6}$$

$$t_{a/b_s} = \frac{2N_a \cos\phi_a}{N_a \cos\phi_a + N_b \cos\phi_b}. \tag{6.7}$$

The layer matrices **L** represent the effect of propagation of light of wavelength λ through a homogeneous layer of refractive index N and thickness d:

$$L_a = \begin{pmatrix} e^{j\frac{2\pi dN}{\lambda}cos\phi_a} & 0 \\ 0 & e^{-j\frac{2\pi dN}{\lambda}cos\phi_a} \end{pmatrix}. \qquad (6.8)$$

Equation (6.1) only partially fulfils the requirements of a measurement model. In fact only one element of the state, namely layer depth d, appears explicitly in this model. The use of ellipsometric measurements to monitor composition is only possible through the ability to relate the alloy fraction of a SiGe layer, x, to its refractive index N. This is achieved through the recording of a set of material reference files where a set of reference structures are grown in the laboratory and analysed off-line. The material reference files then consist of a set of optical properties for a range of alloy fractions. The optical properties for intermediate values of alloy fraction can be derived using established interpolation methods.

The optical properties of the pseudo-substrate can be calculated from previous ellipsometric measurements using the following equation for the complex pseudo-dielectric constant at wavelength λ:

$$\epsilon_\lambda = \sin^2\phi\left\{1 + \tan^2\phi\left(\frac{1 - \tan\Psi_\lambda e^{j\Delta_\lambda}}{1 + \tan\Psi_\lambda e^{j\Delta_\lambda}}\right)^2\right\}. \qquad (6.9)$$

The pseudo-dielectric constant can then be easily converted to a value for the refractive index.

The combination of reference files and an optical model permit easy computation of the forward model, i.e. calculate the ellipsometric angles for a series of layers of prescribed composition and depth. Unfortunately the inverse problem is analytically intractable. The widely accepted method for generating estimates of layer composition and depth utilises the Levenberg–Marquardt algorithm to iteratively fit the parameters of the optical model to the ellipsometric measurements. The accuracy of these estimates can be improved by performing the fit for spectroscopic measurements.

Unfortunately, the iterative nature of such model fitting techniques makes them unsuitable for *in situ* monitoring. In addition, each model fit is performed in isolation with no sense of the sequential nature of the growth being incorporated into the solutions. A further key feature of the growth of such alloys is that it is performed upon a substrate of known composition with the concentration of gases from which growth occurs being externally controlled. This amounts to a high degree of prior knowledge which could be further extended as complex models of the growth process are developed.

6.2 Monitoring via a particle filter

The sequential nature of the growth process combined with the availability of prior knowledge led naturally to the proposal of a sequential Bayesian approach to monitoring of composition and depth. The complex measurement model ruled out standard filtering techniques such as the EKF. However, the ability to evaluate the measurement model for a given alloy fraction and layer depth lent itself immediately to the use of a particle filter.

The system evolution model was represented as a simple random walk. However, in recognition of the three common growth regimes (increasing, constant and decreasing alloy fraction), a multiple model approach was taken with a separate system model for each of the regimes and a switching matrix representing the probability of the system moving between regimes. The system models were

$$
\begin{aligned}
x_{t+1} &= x_t + Ga(\alpha_{x_+}, \beta_{x_+}), \\
d_{t+1} &= d_t + N(\mu_{d_+}, \sigma_{d_+}^2),
\end{aligned} \quad \Big\} \text{Increasing}
$$

$$
\begin{aligned}
x_{t+1} &= x_t + N(0.0, \sigma_{x_c}^2), \\
d_{t+1} &= d_t + N(\mu_{d_c}, \sigma_{d_c}^2),
\end{aligned} \quad \Big\} \text{Constant} \qquad (6.10)
$$

$$
\begin{aligned}
x_{t+1} &= x_t - Ga(\alpha_{x_-}, \beta_{x_-}), \\
d_{t+1} &= d_t + N(\mu_{d_-}, \sigma_{d_-}^2),
\end{aligned} \quad \Big\} \text{Decreasing.}
$$

The switching matrix reflects the probability of the growing alloy moving from a regime where alloy fraction is constant, increasing or decreasing into another regime or remaining in the current regime. This matrix can also be used to incorporate prior knowledge into the particle filter. For example, as the relative concentrations of gases entering the growth chamber are changed, the switching matrix can be modified to reflect the higher probability of alloy fraction increasing or decreasing.

6.3   Results

In this section we present some results for the laboratory growth of SiGe on a Si substrate via chemical vapour deposition. The structure comprised three layers with alloy fraction stepping down for each layer. The rate of growth was approximately one mono-layer ($\sim$ 1 Angstrom) per second. The optical model is essentially a bulk model and is only applicable to layers which are a few mono-layers thick ($\sim$ 10 Angstroms). This determines the time-scale within which the particle filter must perform a single update. In this example 100 samples were used in the update with a boosting factor of 2 during the predictive step of the filter. The evaluation of the ellipsometry sensor model accounted for the bulk of the computational burden masking any gains to be obtained using efficient sampling schemes.

The resulting state estimate from the particle filter is shown in Figure 6.1. The solid line is the mean posterior estimate formed by taking the posterior mean of x for each incremental layer and estimate of total depth formed by summing the posterior mean estimates of d for all layers grown up to that time. The dotted lines use the same posterior mean estimate for x but sum

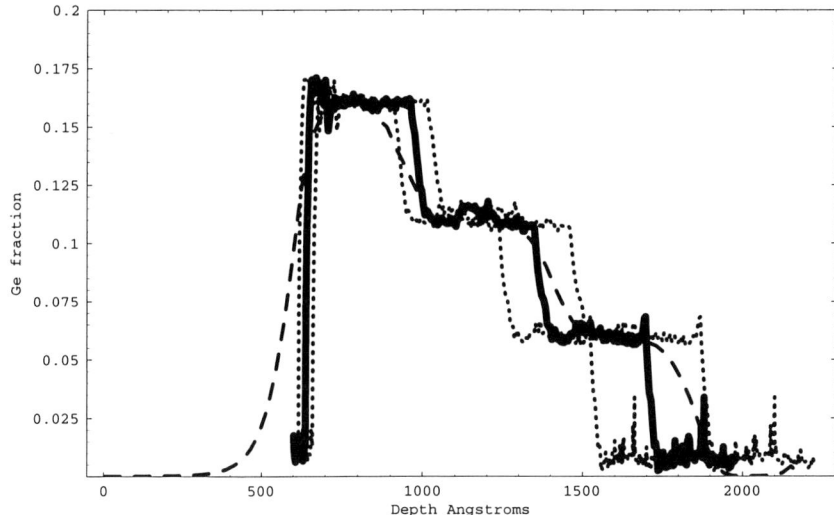

Figure 6.1. Estimates of alloy fraction and layer depth for SiGe structure (Solid line is posterior mean from filter, dashed line is result from SIMS analysis of structure).

the posterior estimates of the $\pm 90\%$ confidence limits for d to form a measure of the uncertainty in the estimate of total depth. For comparison the results of independent off-line analysis using secondary ion mass spectroscopy (SIMS) are included as the dashed line. SIMS analysis is a destructive procedure and the smoothing of the SIMS profile is characteristic of this procedure. The stepped layer structure is evident in both the filter estimate and the SIMS estimate with the filter estimate of alloy fraction x in good agreement with the SIMS analysis. The depth estimate shows poorer agreement. However, the discrepancy is well within the error bounds for both the filter and SIMS (typically 5% error in the SIMS depth estimates).

7 NBC Hazard assessment

The threat of Nuclear, Biological and Chemical (NBC) attack is a constant feature of the modern battlefield. Faced with such a threat the commander on the battlefield needs to be able to make a rapid assessment of its current status and determine an optimum plan of action to counteract it.

Over the next 25 years the quality and quantity of NBC-related information available on the battlefield are likely to improve radically. New detection equipment will be capable of identifying and quantifying any NBC agent released during an attack. Combined with information from local vehicle mounted meteorological sensors and improved communications, the commander will be faced with an ever increasing deluge of information.

An important capability of a future NBC hazard assessment system will be the ability to exploit multiple detector responses and utilise all available information to develop a picture of the current NBC threat. In particular, the number, type and location of agent releases would indicate areas of the battlefield under immediate threat.

Such a system will need to combine information from various detection methods with intelligence assessments regarding the likely threat agents. All of this information will contain varying degrees of uncertainty. For example, intelligence estimates may be able to indicate which agents are likely to be used by the enemy yet it may not be possible to completely exclude other agents entirely. In addition, NBC detectors will have some inherent probability of giving a false alarm. Perhaps the most important area of uncertainty is the determination of source location.

Given a single observation of an agent's presence and the associated meteorological conditions, it would be very difficult to accurately determine the release site. Bayesian methods provide a structured methodology for combining such information to update our uncertainty regarding an event.

This problem of source location extends beyond the work of [43] who use Bayesian methods to improve the prediction of atmospheric dispersion models, in cases where the source location is known, by incorporating information from measurements.

In this section we present the development of such a hazard assessment system which uses a particle filtering approach to perform the necessary inferences [34].

7.1   Tracking agent release

The inference of agent release position from a set of sensor observations represents a scenario similar to that seen in multiple target tracking; the agent type, concentration, release position and time being the state ($\theta = [A, x, y, t, \text{Conc.}]$) which we observe indirectly through the effect of atmospheric dispersion upon the original agent release. In reality some types of observation could be caused by a number of different agent types, so inference of agent type and release position could be treated separately. Inference of agent type and concentration can be treated using a multiple model approach where each agent type and concentration is represented by a separate model. The prior samples for our particle filter are then drawn from this set of models in proportions which reflect the prior probability of each agent type.

Inferring the position of a single release would be insufficient for an operational system since many releases could occur at several positions over a period of time. Following the tracking analogy, recognition that a new release has occurred can be regarded as equivalent to track initiation. Thus, each distinct release would have its own track which could be deleted when the expected concentration of the associated agent has fallen below lethal levels.

7.1.1 Observation model

The observation likelihood is based upon an atmospheric dispersion model. In this case a simple 2-dimensional Gaussian puff model [43] was used, with a slight modification to model the volatile break-up of an agent with time (this could easily be extended to include temperature dependence),

$$
c_{x_t, y_t} = c_{x_0, y_0} e^{-t/\tau} \exp\left(-\frac{\left(\left(\delta_t \sum_{j=0}^{t-1} U_x(j) - x_0 \right) - x_t \right)^2}{2\sigma_x^2} \right)
$$
$$
\times \exp\left(-\frac{\left(\left(\delta_t \sum_{j=0}^{t-1} U_y(j) - y_0 \right) - y_t \right)^2}{2\sigma_y^2} \right), \tag{7.1}
$$

where c_{x_0, y_0} represents the agent concentration upon release at position (x_0, y_0) at time $t = 0$. The predicted agent concentration at the observation position (x_t, y_t) at time t is then c_{x_t, y_t}. The agent break-up is modelled as an exponential decay. Each agent type has an associated *half-life* τ which can be used to model persistent and non-persistent agents. This equation essentially describes a 2-dimensional Gaussian puff with mean $((\delta_t \sum_{j=0}^{t-1} U_x(j) - x_0), (\delta_t \sum_{j=0}^{t-1} U_y(j) - y_0))$ and a diagonal covariance matrix whose diagonal elements are σ_x, σ_y, given by the equations,

$$
\sigma_x = \kappa_x \sum_{j=0}^{t-1} U_x(j) + \sigma_{x_0}, \tag{7.2}
$$
$$
\sigma_y = \kappa_y \sum_{j=0}^{t-1} U_y(j) + \sigma_{y_0}. \tag{7.3}
$$

These equations implicitly make use of the local air-flow (represented by the wind velocity components $U_x(t)$ and $U_y(t)$ for time t). The assumption inherent in this model is that the wind field over the region of interest is available for each discrete time t with a time resolution of δ_t. In an operational system this information would be supplied by a model of air-flow over complex terrain linked to regular meteorological observations and could be easily extended to a more complex dispersion model.

Given this estimate of the concentration at the observation location due to the current proposed release location, agent type and quantity, the likelihood of the observation given the release can be written for two different types of observation. Firstly, we may have symptom type observations which do not provide any information about the local agent concentration (other than the fact that it exceeds some threshold), but do perhaps provide some information about the agent type, i.e. certain agents give rise to particular symptoms. In these cases the likelihood of an observation $\mathbf{O}$ is defined in terms of an indicator function $I_{C_{x_t, y_t}} > C_{\text{thr}}$:

$$
f(\mathbf{O}|\theta) \propto \mathbf{I}_{\mathbf{C}_{\mathbf{x_t, y_t}}} > \mathbf{C}_{\text{thr}} \mathbf{p}(\mathbf{A}), \tag{7.4}
$$

where C_{thr} is the concentration threshold for the agent and $p(A)$ is the probability of the agent producing the observed symptom.

For observations of agent type and concentration the likelihood is defined by the noise model for the sensor system giving rise to the observation. In this example the sensor was assumed to exhibit a Gaussian error on its measurement of concentration and to have a small probability of false classification of agent type. If the measurement error is zero mean Gaussian with variance σ^2, the likelihood is given by

$$f(\mathbf{O}|\theta) \propto \exp\left(-\frac{(\mathbf{C}_{x_o,y_o,t_o} - \mathbf{C}_{x,y,t})^2}{2\sigma^2}\right)\mathbf{p}(\mathbf{A_o}|\mathbf{A}), \qquad (7.5)$$

where C_{x_o,y_o,t_o} is the observed concentration and $C_{x,y,t}$ is the true concentration. As for the symptoms, $p(A_o|A)$ represents the probability that the observed agent classification is obtained given the true agent type, making it possible to model classification errors.

7.1.2   System model

Although a particle filtering approach was used to perform the hazard assessment, the dynamic nature of the problem derives mainly from the influence of meteorological conditions on the agent after release. The system evolution model was essentially non-existent since, once a release has occurred, its state is fixed.

It is more helpful to view the hazard assessment system as a graphical model, and the use of a particle filter as a means of performing inference upon that model. Since the system is static for a particular release, we require some means of projecting the state samples forward in time, avoiding sample degeneracy, as new observations arrive. This can be achieved through the use of the hybrid filter utilising a Gaussian mixture model.

In the case of the hazard assessment system, since the prior samples are drawn from a number of models corresponding to different agent types, the posterior samples are modelled by using a separate Gaussian mixture for each agent type. These mixtures are then combined to form the posterior pdf by constructing a mixture of mixture models.

7.2   Simulated example

7.2.1   Scenario

A simulated scenario was generated with four releases of different agent types at separate locations within an area 18km×10km. Monitoring was conducted for a period of 1 hour during which the prevailing wind direction veered from south-south-westerly to westerly.

The prior set of agent types was limited to four which we shall label GA, H, L and VX. The probability tables for the various observations are shown below.

Index	Release Time (mins)	Position (km)	Agent	Delivery	Concentration	Half-Life
✓ A	15	(3.00, 1.35)	H (Blister)	Delivery B	1000.00	1700.00
✓ B	18	(2.55, 4.70)	VX (Nerve)	Delivery A	200.00	300.00
✓ C	24	(1.00, 3.10)	H (Blister)	Delivery B	1000.00	1700.00
✓ D	30	(4.50, 7.50)	GA (Nerve)	Delivery A	200.00	500.00

Figure 7.1 Sequence of agent releases

The sequence of agent releases is shown in Figure 7.1 while the sequence of observations is shown in Figure 7.2. As each new release is made it is labelled A, B, C etc, while the observations are numbered in sequence.

Index	Arrival Time (Mins)	Observation Time (Mins)	Position (Km)	Observation
1	22	22	(4.28, 2.31)	Blisters
2	24	24	(4.74, 2.57)	Death
3	24	24	(3.65, 5.02)	Convulsions
4	26	26	(5.08, 2.78)	Mustard
5	26	26	(4.08, 5.10)	VX (Nerve) Conc=4.10847e-...
6	30	30	(2.04, 3.61)	Mustard
7	40	40	(6.68, 7.79)	Convulsions

Figure 7.2 Sequence of observations

7.2.2 Results

At any point in the processing of observations, the joint posterior can be marginalised to answer several questions. For example, after the first observation has been processed the posterior pdf of agent type (Figure 7.3) reflects the lack of discrimination between agents 2 and 3 provided by an observation of symptom 3. Inference regarding possible release position could be performed by marginalising over release time (Figure 7.4). As the observations are processed, a new release track is created if the current release tracks do not support the observation (analogous to gating in a target tracking context). Examination of the final joint posterior pdf of release position shows that four separate release

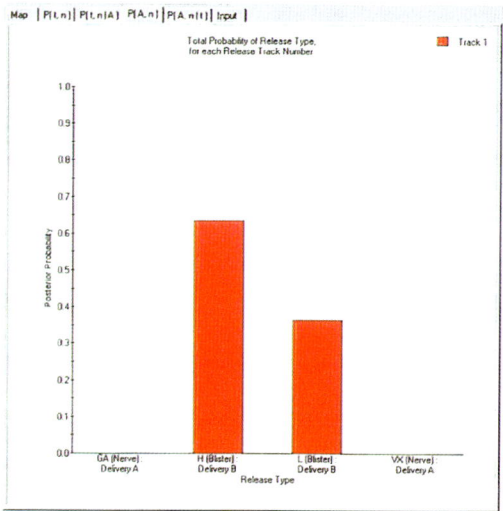

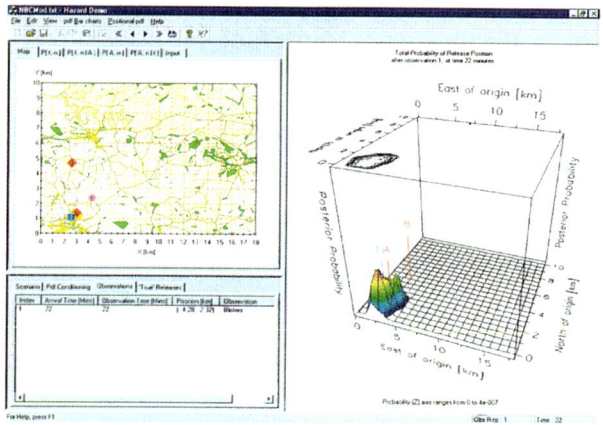

Figure 7.4 Joint posterior pdf of release position after observation 1 with posterior mean release position (blue), true release position (red) and observation location (magenta) shown.

tracks have been identified (Figure 7.5). Inference of release position is not limited to the joint pdf but can be conditioned upon agent type and release time. It is possible to marginalise the posterior pdf to obtain a probability of release time for each track (Figure 7.6).

These results represent an illustrative subset which serve to highlight the success of using a particle filtering approach for such a complex inference task. A further advantage of adopting a particle filtering approach is that an operational hazard assessment system need not be restricted to simple Gaussian puff models of atmospheric dispersion since it is only necessary to be able to evaluate the dispersion model for a given state sample.

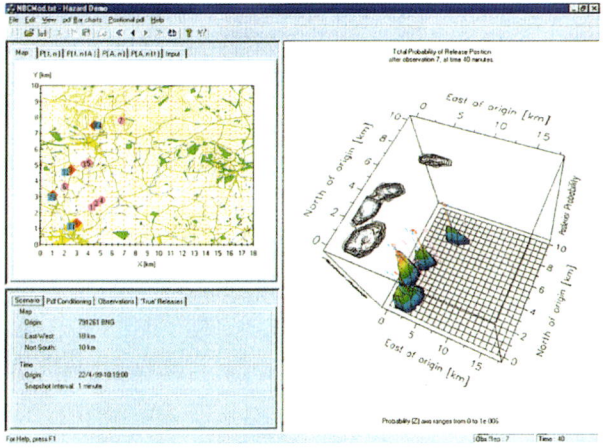

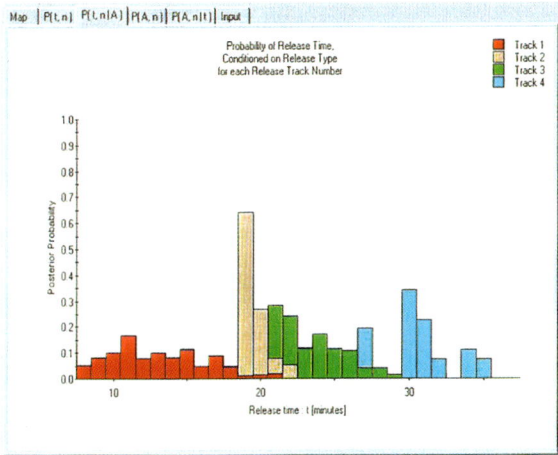

Figure 7.6 Posterior pdf of release time after all observations.

8 Concluding Remarks

In this article we have given an introductory review and tutorial guide to the emerging world of particle filters for solving nonlinear non-Gaussian dynamic estimation problems. The techniques were then demonstrated on three examples. The first example presented was of a complex learning process involving integration of discrete classification data and continuous positional information for tracking in the presence of dense clutter. The second example was of a highly nonlinear measurement model involving only indirect reference to the state parameters and the final example illustrated a data fusion problem with measurements arriving from a diverse set of sensors. In all cases use of a particle filter with mixtures allowed easy incorporation of realistic modelling assumptions and prior physical knowledge into an on-line estimation environment.

Acknowledgements

This research was sponsored by the UK MOD Corporate Research Programmes TG3 and TG9.

References

[1] Acklam, P.J. (1996). *Monte Carlo Methods in State Space Estimation*, Master's thesis, University of Oslo.

[2] Alspach, D.L. and Sorenson, H.W. (1972). Nonlinear Bayesian estimation using Gaussian sum approximation, *IEEE Transactions on Automatic Control* **AC-17**: 439–447.

[3] Avitzour, D. (1995). A stochastic simulation Bayesian approach to multitarget tracking, *IEE Proceedings on Radar and Signal Processing* **142**(2): 41–44.

[4] Bar-Shalom, Y., Chang, K.C. and Blom, H.A.P. (1992). Tracking splitting targets in clutter by using an Interacting Multiple Model Joint Probabilistic Data Association Filter, *in* Y. Bar-Shalom (ed.), *Multitarget multisensor tracking: applications and advances volume II*, Artech House.

[5] Bar-Shalom, Y. and Li, X. R. (1995). *Multitarget-Multisensor Tracking: Principles and Techniques*, YBS Publishing, Storrs, CT.

[6] Berzuini, C., Best, N.G., Gilks, W.R. and Larizza, C. (1997). Dynamic conditional independence models and Markov chain Monte Carlo methods, *Journal of the American Statistical Association* **92**: 1403–1412.

[7] Carlin, B.P., Polson, N.G. and Stoffer, D.S. (1992). A Monte-Carlo approach to non-normal and nonlinear state space modelling, *Journal of the American Statistical Association* **87**(418): 493–500.

[8] Carpenter, J., Clifford, P. and Fearnhead, P. (1999). Building robust simulation-based filters for evolving data sets, *Technical report*, University of Oxford.

[9] Crisan, D., Moral, P. Del and Lyons, T. (2000). Discrete filtering using branching and interacting particle systems, *Markov Processes and Related Fields* **To Appear:**.

[10] Del Moral, P. (1996a). Nonlinear filtering: Interacting particle solution, *Markov Processes and Related Fields* **2**: 555–580.

[11] Del Moral, P. (1996b). Nonlinear filtering using random particles, *Theory of Probability and its Applications* **40**(4): 690–701.

[12] Doucet, A. (1998). On sequential simulation-based methods for Bayesian filtering, *Technical Report CUED/F-INFENG/TR 310*, Cambridge University Engineering Department.

[13] Doucet, A., de Freitas, J.F.G. and Gordon, N.J. (2000). *Sequential Monte Carlo Methods in Practice*, To appear: Springer-Verlag.

[14] Fearnhead, P. (1999a). Particle filters for change-point detection, *Technical report*, University of Oxford.

[15] Fearnhead, P. (1999b). *Sequential Monte Carlo Methods in Filter Theory*, PhD thesis, University of Oxford.

[16] Gordon, N.J. (1993). *Bayesian Methods for Tracking*, PhD thesis, Imperial College, University of London.

[17] Gordon, N.J. (1997). A hybrid bootstrap filter for tracking in clutter, *IEEE Transactions on Aerospace and Electronic Systems* **33**(1): 353–358.

[18] Gordon, N.J., Salmond, D.J. and Ewing, C.M. (1995). Bayesian state estimation for tracking and guidance using the bootstrap filter, *Journal of Guidance, Control and Dynamics* **18**(6): 1434–1443.

[19] Gordon, N.J., Salmond, D.J. and Smith, A.F.M. (1993). A novel approach to nonlinear/non-Gaussian Bayesian state estimation, *IEE Proceedings on Radar and Signal Processing* **140**(2): 107–113.

[20] Handschin, J.E. and Mayne, D.Q. (1969). Monte Carlo techniques to estimate the conditional expectation in multi-stage nonlinear filtering, *International Journal of Control* **9**: 547–559.

[21] Ho, Y.C. and Lee, R.C.K. (1964). A Bayesian approach to problems in stochastic estimation and control, *IEEE Transactions on Automatic Control* **AC-9**: 222–229.

[22] Hürzeler, M. and Künsch, H.R. (1998). Monte Carlo approximations for general state-space models, *Journal of Computational and Graphical Statistics* **7**: 175–193.

[23] Isard, M. and Blake, A. (1998). Condensation - conditional density propagation for visual tracking, *International Journal of Computer Vision* **28**(1): 5–28.

[24] Jazwinski, A.H. (1970). *Stochastic Processes and Filtering Theory*, Academic Press, New York.

[25] Julier, S. (1998). A skewed approach to filtering, *in* O.E. Drummond (ed.), *Signal and Data Processing of Small Targets 1998*, Vol. 3373, SPIE, pp. 271–282.

[26] Kalman, R.E. (1960). A new approach to linear filtering and prediction problems, *Journal of Basic Engineering* **82**: 35–45.

[27] Kitagawa, G. (1987). Non-Gaussian state space modelling of non-stationary time series (with discussion), *Journal of the American Statistical Association* **82**: 1032–1063.

[28] Kitagawa, G. (1996). Monte Carlo filter and smoother for non-Gaussian nonlinear state space model, *Journal of Computational and Graphical Statistics* **5**(1): 1–25.

[29] Kitagawa, G. and Gersch, W. (1996). *Smoothness priors analysis of time series*, Vol. 116 of *Lecture notes in statistics*, Springer-Verlag, New York.

[30] Koch, W. and van Keuk, G. (1997). Multiple hypothesis track maintenance with possibly unresolved measurements, *IEEE Transactions on Aerospace and Electronic Systems* **33**(3): 883–892.

[31] Kramer, S.C. and Sorenson, H.W. (1988). Recursive Bayesian estimation using piece-wise constant approximations, *Automatica* **24**(6): 789–801.

[32] Liu, J.S. and Chen, R. (1998). Sequential Monte Carlo methods for dynamic systems, *Journal of the American Statistical Association* **93**(443): 1032–1044.

[33] Liu, J.S., Chen, R. and Wong, W.H. (1998). Rejection control and sequential importance sampling, *Journal of the American Statistical Association* **93**(443): 1022–1031.

[34] Marrs, A.D. (1999). NBC hazard assessment using a graphical modelling approach, *Technical Report DERA/S&P/SPI/TR980109/1.0*, DERA, Malvern, UK.

[35] Marrs, A.D. and Copsey, K. (1999). In-situ tracking of the surface layer composition of growing SiGe/Si using a particle filter, *Technical Report DERA/S&P/SPI/TR990141/1.0*, DERA, Malvern, UK.

[36] Muller, P. (1991). Monte Carlo integration in general dynamic models, *Contemporary Mathematics* **115**: 145–163.

[37] Pao, L. (1994). Multisensor multitarget mixture reduction algorithms for tracking, *Journal of Guidance, Control and Dynamics* **17**(6): 1205–1211.

[38] Pickering, C., Russell, J., Hope, D., Carline, R., Marrs, A., Robbins, D. and Dann., A. (1998). Instrumental and computational advances for real-time process control using spectrosopic ellipsometry, *Int. Conf. on Metrology and Characterisation for VLSI Technology* .

[39] Pitt, M. and Shephard, N. (1999). Filtering via simulation: auxiliary particle filters, *Journal of the American Statistical Association* **94**(446): 590–599.

[40] Pole, A. and West, M. (1990). Efficient Bayesian learning in dynamic models, *Journal of Forecasting* **9**: 119–136.

[41] Salmond, D.J. (1990). Mixture reduction algorithms for target tracking in clutter, *in* O. E. Drummond (ed.), *Signal and Data Processing of Small Targets 1990*, Vol. 1305, SPIE, pp. 434–445.

[42] Salmond, D.J., Fisher, D. and Gordon, N.J. (1997). Tracking and identification for closely spaced objects in clutter, *European Control Conference.*

[43] Smith, J.Q. and French, S. (1993). Bayesian updating of atmospheric dispersion models for use after an accidental release of radioactivity, *The Statistician* **42**(5): 501–512.

[44] Sorenson, H.W. and Alspach, D.L. (1971). Recursive Bayesian estimation using Gaussian sums, *Automatica* **7**: 465–479.

[45] Srinivasan, K. (1970). State estimation by orthogonal expansion of probability distribution, *IEEE Transactions on Automatic Control* **15**: 3–10.

[46] Stavropolous, P. (1998). *Computational methods for the Bayesian analysis of dynamic models*, PhD thesis, University of Glasgow.

[47] Tanizaki, H. (1996). *Nonlinear filters: estimation and applications*, second edn, Springer-Verlag, New York.

[48] Tanizaki, H. and Mariano, R.S. (1998). Nonlinear and non-Gaussian state space modelling with Monte Carlo simulations, *Journal of Econometrics* **83**(1): 263–290.

[49] West, M. (1993). Mixture models, Monte Carlo, Bayesian updating and dynamic models, *Computing Science and Statistics* **24**: 325–333.

[50] West, M. and Harrison, P.J. (1997). *Bayesian Forecasting and Dynamic Models*, second ed., Springer-Verlag, New York.

Stochastic, Dynamic Modelling and Signal Processing: Time Variable and State Dependent Parameter Estimation

Peter Young

1 Introduction

Previous publications (e.g. Young, 1978, 1983, 1993a,b, 1998a,b, 1999a,b; Young and Runkle, 1989; Young and Minchin, 1991; Young *et al.*, 1991; Young and Lees, 1993; Young and Beven, 1994; Young and Pedregal, 1997, 1998, 1999) have discussed an approach to nonstationary and nonlinear signal processing based on the identification and estimation of stochastic models with time variable (TVP) or state dependent (SDP) parameters. Here the term 'nonstationarity' is assumed to mean that the statistical properties of the signal, as defined by the parameters in an associated stochastic model, are changing over time at a rate which is 'slow' in relation to the rates of change of the stochastic state variables in the system under study. Although such nonstationary systems exhibit nonlinear behaviour, this can often be approximated well by TVP (or piece-wise linear) models, the parameters of which can be estimated using recursive methods of estimation in which the parameters are assumed to evolve in a simple stochastic manner (e.g. Young, 1984, 1999a). On the other hand, if the changes in the parameters are functions of the state or input variables (i.e. they actually constitute stochastic state variables), then the system is truly nonlinear and likely to exhibit severe nonlinear behaviour. Normally, this cannot be approximated in a simple TVP manner; in which case, recourse must be made to the alternative, and more powerful, SDP modelling methods that are the main topic of this chapter.

The extension of the TVP estimation methods to allow for state dependency, as described here, involves two statistical stages.

- First, the non-parametric *identification* of the state dependency using recursive methods of time variable parameter estimation which allow for rapid (state dependent) parametric change. As we shall see, the standard methods of TVP estimation developed previously for nonstationary time series analysis need to be modified considerably in this SDP setting to allow for the much more rapid temporal changes that arise from the state dependency.

- Second, the parameterization of the identified non-parametric relationships, followed by the *statistically efficient estimation* of the (now normally constant) parameters that characterize these nonlinearities.

74

The first identification stage in this process exploits recursive *Fixed Interval Smoothing* (FIS) algorithms, combined with special data re-ordering and 'backfitting' procedures, to obtain estimates of any state dependent parameter variations. These state dependencies are estimated in the form of non-parametric relationships (graphs) between the estimated rapid parameter variation and the associated state or input variable(s). Parameterization of these non-parametric relationships can be accomplished in various ways, from simple curve fitting based on weighted least squares methods (Young, 1993a; Young and Beven, 1994) to the use of neural networks or radial basis functions.

Having identified a structural form for the nonlinear model of the system based on the parameterized nonlinear relationships, this model is converted into a stochastic state space form. The final estimation phase of the nonlinear modelling then exploits *Maximum Likelihood* (ML) methods of estimation, based on Gaussian assumptions for the stochastic disturbances and the application of *Prediction Error Decomposition* (Schweppe, 1965). If successful, this yields statistically efficient estimates of the constant parameters in the identified nonlinear state space model. The resulting model should then provide a parametrically efficient representation of the stochastic, nonlinear system that has considerable potential for use in subsequent signal processing, time series analysis, forecasting and automatic control system design. For example, the methodology described here exploits recursive estimation in an *off-line* manner but this sequential processing of the data facilitates the development of related *on-line adaptive* methods of signal processing, forecasting and control.

Although primarily concerned with nonlinear state dependent parameter models, as outlined above, the chapter also provides a sequel to a previous paper (Young, 1999a) that discusses the simpler class of 'linear' TVP regression relationships. These include the *Dynamic AutoRegressive eXogenous variables* (DARX) model, the constant parameter version of which is often used in the modelling of linear stochastic, dynamic systems. As a prelude to our discussion of SDP estimation, therefore, the next section 2 considers the alternative *Dynamic Transfer Function* (DTF) model, using a new instrumental variable method of FIS, and shows how this is much superior to the DARX model when measurement noise is present (the *errors-in-variables* situation). This leads naturally to the definition of the more complex SDP transfer function models of truly nonlinear, stochastic dynamic systems and the associated methods of statistical identification and estimation.

In order to illustrate the practical application and utility of both the TVP and SDP methods, the chapter also contains a number of simulation examples, as well as a practical study involving a re-analysis of the famous Nicholson blowfly data (Nicholson, 1954). Other, practical examples cited in the references cover a variety of application areas from the environment through engineering to economics.

2   TVP Transfer Function Models: DTF Model Estimation

A previous paper (Young, 1999a) has discussed the estimation of time variable parameters in the various kinds of 'linear' regression model. One of these, the *Dynamic*[1] *Auto-Regressive eXogenous variables* (DARX) model, is capable of modelling the input-output behaviour of stochastic, dynamic systems. The DARX model relating a single input variable u_t to an output variable y_t, can be written in the following form:

$$y_t = -a_{1,t}y_{t-1} - \cdots - a_{n,t}y_{t-n} + b_{0,t}u_{t-\delta} + b_{1,t}u_{t-\delta-1} + \cdots + b_{m,t}u_{t-\delta-m} + e_t$$
(1)

or, in transfer function terms,

$$y_t = \frac{B(L,t)}{A(L,t)}u_{t-\delta} + \frac{1}{A(L,t)}e_t, \qquad e_t = N(0,\sigma^2).$$
(1a)

In these equations, L is the backward shift operator, i.e., $L^r y_t = y_{t-r}$, $A(L,t)$ and $B(L,t)$ are *time variable coefficient polynomials* in L of the following form:

$$
\begin{aligned}
A(L,t) &= 1 + a_{1,t}L + a_{2,t}L^2 + \cdots + a_{n,t}L^n, \\
B(L,t) &= b_{0,t} + b_{1,t}L + b_{2,t}L^2 + \cdots + b_{m,t}L^m.
\end{aligned}
$$
(1b)

The term δ is a pure time delay, measured in sampling intervals, which is introduced to allow for any temporal delay that may occur between the incidence of a change in u_t and its first effect on y_t; and e_t is a zero mean, white noise input with Gaussian normal amplitude distribution and variance σ^2.

Unfortunately, the DARX model is limited in practical terms since it depends on the assumption of the above, rather specific, signal topology, with the noise entering the model through a restricted AR process with a polynomial $A(L,t)$ equal to that of the denominator polynomial in the main DTF between $u_{t-\delta}$ and y_t. A more general Dynamic Transfer Function (DTF) model, without the restrictions of the DARX, is the following:

$$y_t = \frac{B(L,t)}{A(L,t)}u_{t-\delta} + \xi_t$$
(2a)

where ξ_t represents uncertainty in the relationship arising from a combination of measurement noise, the effects of other unmeasured inputs and modelling error. Normally, ξ_t is assumed to be independent of u_t and is modelled as an Auto-Regressive (AR) or Auto-Regressive-Moving Average (ARMA) stochastic process (see e.g. Box and Jenkins, 1970; Young, 1984), although even this restriction can be avoided by the use of instrumental variable methods, as discussed below.

---

[1]The term 'dynamic' is used here for historical reasons (see Young, 1999a) to mean a time variable parameter ARX model.

Equation (2a) can be written in the following vector equation form:

$$y_t = \mathbf{z}_t^T \mathbf{p}_t + \mu_t \tag{2b}$$

where,

$$\begin{aligned}
\mathbf{z}_t^T &= [-y_{t-1} \ -y_{t-2} \ \cdots \ -y_{t-n} \ u_{t-\delta} \ \cdots \ u_{t-\delta-m}] \\
\mathbf{p}_t &= [a_{1,t} \ a_{2,t} \ \cdots \ a_{n,t} \ b_{0,t} \ \cdots \ b_{m,t}]^T
\end{aligned} \tag{2c}$$

and $\mu_t = A(L,t)\xi_t$. For convenience of notation, let $\mathbf{p}_t$ be defined as follows,

$$\mathbf{p}_t = [p_{1,t} \ p_{2,t} \ \cdots \ p_{n+m+1,t}]^T \tag{2d}$$

with $p_{i,t}$, $i = 1, 2, \ldots, n+m+1$, relating to the TF model parameters $a_{i,t}$ and $b_{j,t}$ through (2c).

In order to estimate the assumed time variable model parameters in $\mathbf{p}_t$, it is necessary to make some assumptions about the nature of their temporal variability. Reflecting the statistical setting of the analysis and referring to previous research on this topic, it seems desirable if this is characterized in some stochastic manner. Normally, when little is known about the nature of the time variability, this model needs to be both simple and flexible. One of the simplest and most generally useful classes of stochastic, state space models involves the assumption that the ith parameter, $p_{i,t}$, $i = 1, 2, \ldots, n+m+1$, is defined by a two-dimensional stochastic state vector $\mathbf{x}_{i,t} = [l_{i,t} \ d_{i,t}]^T$, where $l_{i,t}$ and $d_{i,t}$ are, respectively, the changing level and slope of the associated TVP. This selection of a two-dimensional state representation of the TVPs is based on practical experience over a number of years. Initial research by the author and others in the 1960s (Young, 1969, 1970a) tended to use a simple scalar random walk (RW) model for the parameter variations. However, later work in the 1980s (see above references) showed the value of modelling not only the level changes in the TVPs but also their rates of change, as in the definition of $\mathbf{x}_{i,t}$, above.

The stochastic evolution of each $\mathbf{x}_{i,t}$ (and, therefore, each of the $n+m+1$ parameters in $\mathbf{p}_t$) is assumed to be described by a Generalized Random Walk (GRW) process defined in the following State Space (SS) terms:

$$\mathbf{x}_{i,t} = \mathbf{F}_i \mathbf{x}_{i,t-1} + \mathbf{G}_i \boldsymbol{\eta}_{i,t}, \qquad i = 1, 2, \ldots, m+n+1, \tag{2e}$$

where

$$\mathbf{F}_i = \begin{bmatrix} \alpha & \beta \\ 0 & \gamma \end{bmatrix}, \qquad \mathbf{G}_i = \begin{bmatrix} \delta & 0 \\ 0 & \varepsilon \end{bmatrix}$$

and $\boldsymbol{\eta}_{i,t} = [\eta_{1i,t} \ \eta_{2i,t}]^T$ is a 2×1, zero mean, white noise vector that allows for stochastic variability in the parameters and is assumed to be characterized by a (normally diagonal) covariance matrix $\mathbf{Q}_{\eta i}$. This general model comprises as special cases the *Integrated Random Walk* (IRW: $\alpha = \beta = \gamma = \varepsilon = 1$;

$\delta = 0$); the scalar *Random Walk* (RW: scalar but equivalent to (2e) with $\beta = \gamma = \varepsilon = 0$; $\alpha = \delta = 1$: i.e. just the first equation in (2e), see below); the intermediate case of *Smoothed Random Walk* (SRW: $0 < \alpha < 1$; $\beta = \gamma = \varepsilon = 1$; and $\delta = 0$); the first order autoregressive process (AR(1): again scalar with $\beta = \gamma = \varepsilon = 0$; $0 < \alpha < 1$; $\delta = 1$; and, finally, both the *Local Linear Trend* (LLT: $\alpha = \beta = \gamma = \varepsilon = \delta = 1$), [2] and *Damped Trend* (DT: $\alpha = \beta = \delta = \varepsilon = 1$; $0 < \gamma < 1$): see Harvey (1984, 1989). The various, normally constant, parameters in this GRW model (α, β, γ, δ, ε and the elements of $\mathbf{Q}_{\eta i}$) are often referred to as *hyper-parameters*. This is to differentiate them from the TVPs that are the main object of the estimation analysis. However, the hyper-parameters are also assumed to be unknown *a priori* and need to be estimated from the data, as we shall see in the subsequent discussion.

The full GRW model (2e) was introduced in Jakeman and Young (1979, 1984): further discussion and practical examples appear in Young (1988), Young *et al.* (1989), Young and Ng (1989), Ng and Young (1990). Note that, in the case of the RW model, i.e.,

$$l_{i,t} = l_{i,t-1} + \eta_{1i,t}, \qquad l_{i,t} = p_{i,t}, \qquad (3)$$

each parameter can be assumed to be time-invariant if the variance of the white noise input $\eta_{1i,t}$ is set to zero. Then, the stochastic TVP setting reverts to the more normal, constant parameter TF model situation. In other words, if RW models with zero variance white noise inputs are specified for the model parameters, the recursive *Instrumental Variable* (IV) estimation algorithm described below for the general stochastic TVP case will provide recursive estimates that are identical to those obtained with the normal recursive IV estimation algorithm for TF models with constant parameters (Young, 1984). Of course, there is some added value to the recursive solution even in this situation, since the user is provided with the recursive estimates over the whole interval $t = 1, 2, \ldots, N$. These can provide additional useful information on the model: for example, they show how the estimates are converging and can be used (e.g. Brown *et al.*, 1975; Young, 1984) to detect both the presence of potential parametric change and possible over-parameterization (i.e. the model contains too many parameters to provide unambiguous estimation results).

Clearly other, more general and higher order stochastic processes could be used to model the stochastic TVPs, provided such models can be identified satisfactorily from the data. For example the higher order IRWs (Double and Triple Integrated Random Walk (DIRW, TIRW), etc.), the Integrated or Double Integrated AutoRegressive (IAR, DIAR: see Young, 1994) model, and even more general processes (e.g. Pedregal and Young, 1996, 1998). However, the

---

[2]Interestingly, the LLT model can be considered simply as the combination of the simpler RW and IRW models.

more complex models introduce additional hyper-parameters that would have to be well identified from the data and optimized, thus introducing potential practical difficulties.

The idea of assuming that the model parameters evolve over time as nonstationary stochastic variables may seem complex at first sight but it is, in fact, just a statistical device to allow for the estimation of parametric change. After all, the assumption of the RW model is simply a means of introducing into the estimation problem the freedom for the associated parameter to vary at each sample in time by a small random amount defined by the variance of the white noise input $\eta_{1i,t}$. And the more complex GRW models in (2e) are just a way of refining and adding to this freedom. In fact, it can be shown (Young and Pedregal, 1998) that the GRW assumptions on the parameter variations have an *implicit* but physically interpretable effect: they make the recursive parameter estimates, at any sample time t, depend only on the local data in the vicinity of this sample, with the selected GRW model defining the local weighting effect. In the case of the RW model, for instance, this weighting effect or 'kernel' has a Gaussian-like shape that applies maximum weight to the current data with declining weight each side. And the 'bandwidth' of the kernel is defined by the ratio of the variance of the white noise input $\eta_{1i,t}$ to the residual variance σ^2 (the Noise Variance Ratio (NVR): see later). This can be related to the more *explicit* use of localized data weighting in methods such as locally weighted kernel regression (e.g. Holst *et al.*, 1996; Young and Pedregal, 1996) and 'wavelet' methods (e.g., Daubechies, 1988) that are currently receiving so much attention in the statistical and signal processing literature.

Having introduced the GRW models for the parameter variations, an overall SS model can then be constructed straightforwardly by the aggregation of the subsystem matrices defined in (2e), with the 'observation' equation defined by the model equation (2b): i.e.,

$$\text{Observation Equation:} \quad y_t = \mathbf{H}_t\mathbf{x}_t + \mu_t \quad \text{(i)},$$
$$\text{State Equations:} \quad \mathbf{x}_t = \mathbf{F}\mathbf{x}_{t-1} + \mathbf{G}\boldsymbol{\eta}_t \quad \text{(ii)}. \tag{4a}$$

where

$$\mathbf{x}_t = \begin{bmatrix} x_{1,t}^T & x_{2,t}^T & \cdots & x_{n+m+1,t}^T \end{bmatrix}^T. \tag{4b}$$

If $p = 2(n+m+1)$, then $\mathbf{F}$ is a $p \times p$ block diagonal matrix with blocks defined by the $\mathbf{F}_i$ matrices in (2e); $\mathbf{G}$ is a $p \times p$ block diagonal matrix with blocks defined by the corresponding subsystem matrices $\mathbf{G}_i$ in (2e); and $\boldsymbol{\eta}_t$ is a p-dimensional vector containing, in appropriate locations, the white noise input vectors $\boldsymbol{\eta}_{i,t}$ ('system disturbances' in normal SS terminology) to each of the GRW models in (2e). These white noise inputs, which provide the stochastic stimulus for parametric change in the model, are assumed to be independent of the observation noise e_t and have a covariance matrix $\mathbf{Q}$ formed from the combination of the individual covariance matrices $\mathbf{Q}_{\eta,i}$. Finally, $\mathbf{H}_t$ is a $1 \times p$ vector of the following form,

$$\mathbf{H}_t = [-y_{t-1}\; 0\; -y_{t-2}\; 0\; \ldots\; y_{t-n}\; 0\; u_{t-\delta}\; 0\; \ldots\; u_{t-\delta-m}\; 0]\,, \qquad (4c)$$

that relates the scalar observation y_t to the state variables defined by (4a)(ii), so that it represents the DTF model (2b), with each parameter defined as a GRW process. In the case of the scalar RW and AR(1) models, the alternate zeros are simply omitted.

The SS formulation in equations (4) is particularly well suited for optimal recursive estimation in which the time variable parameters (acting as surrogate 'states' in this SS formulation) are estimated sequentially whilst working through the data in temporal order (usually termed 'forward-pass filtering'). In the off-line situation, where all the time series data are available for analysis, this filtering operation is accompanied by optimal recursive smoothing (see e.g. Young, 1984; Young *et al.*, 1998). Here, the estimates obtained from the forward-pass, filtering algorithm are updated sequentially whilst working through the data in *reverse* temporal order (usually termed 'backward-pass smoothing') using a backward-recursive *Fixed Interval Smoothing* (FIS) algorithm, where the 'fixed interval' is the interval covered by the total sample size N.[3]

The reason for this two-pass approach is easy to understand. The forward-pass filtering estimate of $\mathbf{x}_t$, which defines the estimated TVPs, can be denoted by $\hat{\mathbf{x}}_{t|t}$ (or simply $\hat{\mathbf{x}}_t$, for convenience) since it represents the estimate at sample t given only the data up to and including sampling instant t. However, under our assumption that each of the parameters evolves stochastically according to the equation (2e), a superior 'smoothed' estimate $\hat{\mathbf{x}}_{t|N}$ exists and can be generated by the FIS algorithm, in which the estimate at t is based on all the data over the sampling interval $t = 1, 2, \ldots, N$. As a result, the phase lag associated with the forward-pass filtering estimate (since it cannot anticipate any change until the evidence for change in the series has been processed) is eliminated on the backward smoothing pass. Thus, any variation in the parameters is estimated as it occurs, without any lag effect (indeed, it may even be anticipated if the smoothing effect is substantial, as it can be in high noise situations). This proves particularly useful in operations such as interpolation over gaps in the data, estimation and removal of individual components from the data (signal extraction), and seasonal adjustment.

In our previous publications on this topic, a standard algorithmic approach to the problem has been utilized. This involves a two-pass (prediction-correction) version of the forward-pass recursive filtering algorithm; followed by a

---

[3]On-line 'Fixed Lag Smoothing' is also possible, where the recursive estimation works in a forward-pass, filtering mode but with smoothed estimates provided at every sampling instant t over a finite interval of l samples into the past (i.e. over the interval $t - l$ to t), but this is not discussed here.

version of the FĪS algorithm which is stable numerically and which has a structure allowing for the handling of missing observations and outliers. More specifically, it is an adapted version (e.g. Young, 1988; Young *et al.*, 1998 and the references therein) combining Bryson and Ho's recursion for the Lagrange multipliers (Bryson and Ho, 1969) with the state update recursion of Norton (1975, 1986). It should be noted that the recursive filtering algorithm is closely related to the Kalman Filter (KF: 1960) and is often referred to as such. The difference is that the $\mathbf{H}_t$ matrix in the present, recursive TVP estimation context for the TF model (2b), is based on measured variables. In particular, the output variables y_{t-i}, $i = 1, 2, \ldots, n$, in $\mathbf{H}_t$ are affected by the noise ξ_t (the *errors-in-variables* problem); whereas, strictly, $\mathbf{H}_t$ in the KF has to be composed of exactly known (but, if necessary, time variable), deterministic coefficients.

This difference is important in the present TF context since it can be shown that the TVP estimates obtained from the standard recursive filtering/smoothing algorithm (see later, section 3.1) will be asymptotically biased away from their 'true' values. The level of this bias is dependent on the magnitude of the measurement noise and it can be problematic in high noise situations, particularly if the parameters are physically meaningful (see e.g. Young, 1984 for a discussion of this problem).[4] For this reason, it is necessary to modify the standard algorithm to avoid these biasing problems. This can be achieved by attempting to model the noise μ_t in some manner (e.g. Norton, 1975, 1986). However, since μ_t is a complex, nonstationary, noise process, its complete estimation is not straightforward. A new, alternative approach, which does not require modelling μ_t, provided it is independent of the input u_t, is the recursive *TVP Instrumental Variable* (TVPIV) method.

In relation to the time series y_t, $t = 1, 2, \ldots, N$, the recursive TVPIV filtering/smoothing algorithm has the following form:

1. Forward-Pass Symmetric IV Equations (iterative)

Iterate the following recursive equations (5a)–5(c) for $j = 1, 2, \ldots, J$, with $\hat{\mathbf{H}}_t = \mathbf{H}_t$ for $j = 1$:

Prediction:

$$\hat{\mathbf{x}}_{t|t-1} = \mathbf{F}\hat{\mathbf{x}}_{t-1},$$
$$\hat{\mathbf{P}}_{t|t-1} = \mathbf{F}\hat{\mathbf{P}}_{t-1}\mathbf{F}^T + \mathbf{G}\mathbf{Q}_r\mathbf{G}^T.$$

Correction:

$$\hat{\mathbf{x}}_t = \hat{\mathbf{x}}_{t|t-1} + \hat{\mathbf{P}}_{t|t-1}\hat{\mathbf{H}}_t^T \left[1 + \hat{\mathbf{H}}_t\hat{\mathbf{P}}_{t|t-1}\hat{\mathbf{H}}_t^T\right]^{-1} \{\, y_t - \mathbf{H}_t\hat{\mathbf{x}}_{t|t-1}\,\},$$
$$\hat{\mathbf{P}}_t = \hat{\mathbf{P}}_{t|t-1} + \hat{\mathbf{P}}_{t|t-1}\hat{\mathbf{H}}_t^T \left[1 + \hat{\mathbf{H}}_t\hat{\mathbf{P}}_{t|t-1}\hat{\mathbf{H}}_t^T\right]^{-1} \hat{\mathbf{H}}_t\hat{\mathbf{P}}_{t|t-1}$$

$$\text{(5a)}$$

---

[4]It is less important if the model is to be used within a forecasting context since the forecasts produced by the model are not biased (although they may not be statistically efficient).

where

$$\hat{\mathbf{H}}_t = [-\hat{x}_{t-1}\ 0\ -\hat{x}_{t-2}, 0\ \ldots, \hat{x}_{t-n}\ 0\ \ldots\ u_{t-\delta}\ 0\ \ldots\ u_{t-\delta-m}\ 0] \tag{5b}$$

$$\hat{x}_t = \frac{\hat{B}_{j-1}(L,t)}{\hat{A}_{j-1}(L,t)} u_{t-\delta}. \tag{5c}$$

The FIS algorithm is in the form of a backward recursion operating from the end of the sample set to the beginning.[5]

2. Backward-Pass Fixed Interval Smoothing IV (FISIV) Equations (single-pass)

$$\hat{\mathbf{x}}_{t|N} = \mathbf{F}^{-1}\left[\hat{\mathbf{x}}_{t+1|N} + \mathbf{G}\mathbf{Q}_r\mathbf{G}^T\mathbf{L}_t\right],$$

$$\mathbf{L}_t = \left[\mathbf{I} - \hat{\mathbf{P}}_{t+1}\hat{\mathbf{H}}_{t+1}^T\hat{\mathbf{H}}_{t+1}\right]^T\left[\hat{\mathbf{F}}^T\hat{\mathbf{L}}_{t+1} - \hat{\mathbf{H}}_{t+1}^T\left\{y_{t+1} - \hat{\mathbf{H}}_{t+1}\hat{\mathbf{x}}_{t+1}\right\}\right], \tag{5d}$$

$$\hat{\mathbf{P}}_{t|N} = \hat{\mathbf{P}}_t + \hat{\mathbf{P}}_t\hat{\mathbf{F}}^T\hat{\mathbf{P}}_{t+1|t}^{-1}\left[\hat{\mathbf{P}}_{t+1|N} - \hat{\mathbf{P}}_{t+1|t}\right]\hat{\mathbf{P}}_{t+1|t}^{-1}\mathbf{F}\hat{\mathbf{P}}_t,$$

with $\mathbf{L}_N = 0$.

In these algorithms, the $p \times p$ Noise Variance Ratio (NVR) matrix $\mathbf{Q}_r$ and the $p \times p$ matrix $\hat{\mathbf{P}}_t$ are defined as follows,

$$\mathbf{Q}_r = \frac{\mathbf{Q}}{\sigma^2}, \qquad \hat{\mathbf{P}}_t = \frac{\mathbf{P}_t^*}{\sigma^2}, \tag{5e}$$

where $\mathbf{P}_t^*$ is the error covariance matrix associated with the state estimates which, in the present TVP context, define the estimated uncertainty in the parameters. For simplicity, it is normally assumed that the NVR matrix $\mathbf{Q}_r$ is diagonal, although this is not essential. The NVR parameters that characterize $\mathbf{Q}_r$ (as well as any other unknown hyper-parameters in the SS model (4)) are unknown prior to the analysis and clearly need to be estimated on the basis of the time series data y_t and u_t before the filtering and smoothing algorithms can be utilized. The optimization of these hyper-parameters is discussed in the next sub-section.

The main difference between the above algorithm (5a)–(5e) and the standard filtering/smoothing algorithms (see later section 3.1) is the introduction of 'hats' on the $\hat{\mathbf{H}}_t$ vector and the $\hat{\mathbf{P}}_t$ matrix, and the use of an iterative IV solution in the forward-pass algorithm. In (5b) $\hat{\mathbf{H}}_t$ is the *IV vector*, which is used by the algorithm in the generation of all the $\hat{\mathbf{P}}_t$ terms and is the main vehicle in removing the bias from the TVP estimates. The subscript $j-1$ on $\hat{A}_{j-1}(L,t)$ and $\hat{B}_{j-1}(L,t)$ indicates that the estimated DTF polynomials in the *Auxiliary Model*, (5c), which generates the *instrumental variables* $\hat{x}_t$ that appear in the definition of $\hat{\mathbf{H}}_t$, are updated in an iterative manner, starting

---

[5]An alternative FIS algorithm is available in which, at each backward recursion, the estimate $\hat{\mathbf{x}}_{t|N}$ is based on an update of the filtering estimate $\hat{\mathbf{x}}_t$ (see Young, 1984). This can be specified as an alternative to (5d)

with the least squares estimates of these polynomials. Iteration is continued for T iterations, until the forward-pass (filtered) IV estimates of the TVPs are no longer changing significantly: normally only 3 or 4 iterations are required.

This new recursive-iterative IV approach to time variable parameter estimation is based on the IV algorithm for *constant* parameter TF models (e.g. Young, 1984 and the prior references therein), except that the *symmetric gain* version of the IV algorithm (Young, 1970b; 1984, p. 183) is used, rather than the more usual asymmetric version. This is necessary in order that the standard recursive FIS algorithm in (5d) can be used to generate the smoothed estimates of the TVPs.

Maximum Likelihood (ML) Optimization of Hyper-Parameters

The approach to ML optimization based on *Prediction Error Decomposition* (PED) derives originally from the work of Schweppe (1965), who showed how to generate likelihood functions for Gaussian signals using the Kalman filter (see also Bryson and Ho, 1969; p. 389). Its importance in the present UC context was probably first recognized by Harvey (1981) and Kitagawa (1981). Since then, it has become one of the two standard approaches to the problem (the other being the *Expectation and Minimization* (EM) algorithm: Dempster *et al.*, 1977).

In the case of the simpler DARX model (Young, 1999a), $\hat{\mathbf{H}}_t$ and $\hat{\mathbf{P}}_t$ in the recursive TVP least squares estimation algorithm (5a) are replaced by $\mathbf{H}_t$ and $\mathbf{P}_t$, respectively, and iteration is not required. With given initial values for the hyper-parameters, this algorithm will yield the one-step-ahead prediction errors (also termed the 'innovations' or 'recursive residuals') ε_t, where

$$\varepsilon_t = y_t - \mathbf{H}_t \hat{\mathbf{x}}_{t|t-1}, \qquad t = 1, 2, \ldots, N. \tag{6}$$

If the first p observations are regarded as fixed, the log-likelihood function of $y_{p+1}, \ldots, y_N$ can be defined in terms of the standard 'regression' form of prediction error decomposition, i.e.,

$$\log L = \frac{-(N-k)}{2} \log(2\pi) - \frac{1}{2} \log(\sigma^2) - \frac{1}{2} \sum_{t=p+1}^{N} \log(1 + \mathbf{H}_t \mathbf{P}_{t|t-1} \mathbf{H}_t^t)$$

$$- \frac{1}{2\sigma^2} \sum_{t=p+1}^{N} \frac{\varepsilon_t^2}{1 + \mathbf{H}_t \mathbf{P}_{t|t-1} \mathbf{H}_t^T}, \tag{7}$$

where it can be shown that $\sigma^2(1 + \mathbf{H}_t \mathbf{P}_{t|t-1} \mathbf{H}_t^T)$ is the variance of ε_t, so that the last term in (7) is based on the sum of squares of the normalized one-step-ahead prediction errors. Now the ML estimate of σ^2, conditional on the hyper-parameters, is given by

$$\hat{\sigma}^2 = \frac{1}{N-k} \sum_{t=p+1}^{N} \frac{\varepsilon_t^2}{1 + \mathbf{H}_t \mathbf{P}_{t|t-1} \mathbf{H}_t^T}, \tag{8}$$

so that it can be estimated in this manner and 'concentrated out' of the expression (7) by substituting (8) into (7), to yield the following expression for the 'concentrated likelihood'

$$\log(L_c) = -\frac{N-k}{2} \log(2\pi+1) - \frac{1}{2} \sum_{t=p+1}^{N} \log(1 + \mathbf{H}_t \mathbf{P}_{t|t-1} \mathbf{H}_t^T) - \frac{N-k}{2} \log(\hat{\sigma}^2) \tag{9}$$

which needs to be maximized with respect to the unknown hyper-parameters in order to obtain their ML estimates.

Since (9) is nonlinear in the hyper-parameters, the likelihood maximization needs to be carried out numerically. Consequently, it is more convenient to remove the constant term (since it will play no part in the minimization) and multiply (9) by -2, to yield

$$\log(L_c) = \sum_{t=p+1}^{N} \log(1 + \mathbf{H}_t \mathbf{P}_{t|t-1} \mathbf{H}_t^T) + (N-k) \log(\sigma^2), \tag{10}$$

which then needs to be *minimized*. This minimization is accomplished by initiating the optimization with the hyper-parameter estimates either selected by the user or set to some default values (in both cases, ensuring that the resulting optimization does not converge on a local minimum). The recursive TVP estimation algorithm is used repeatedly to generate the one-step-ahead prediction errors ε_t and, thence, the log-likelihood value in (10) associated with the latest selection of hyper-parameter estimates made by the optimization algorithm. The optimization algorithm then adjusts its selection of hyper-parameter estimates in order to converge on those estimates that minimize (10). Further details of this and alternative ML optimization procedures are given, for example, in Harvey (1989) and Harvey and Peters (1990). Typical methods that can be used for numerical optimization are the '*fmins*' and '*fminu*' functions available in the MATLAB software system, or their equivalents, although more complex and efficient procedures are available.

In the case of the DTF model, the same basic PED approach to ML optimization can be used, but with $\mathbf{H}_t$ and $\mathbf{P}_t$ in equations (7) to (10) replaced by $\hat{\mathbf{H}}_t$ and $\hat{\mathbf{P}}_t$, respectively. The equations formed in this manner can then be considered as an IV version of PED and as an approximation to the standard TVP regression version. This approximation is justified by the link between optimal IV estimation and the classical ML approach to TF model estimation (Young and Jakeman, 1979; Young, 1984) arising from the theoretical results of Pierce (1972). Its practical efficacy is demonstrated in the following simulation example.

Simulation Example 1

As a simple example of DTF modelling, consider the estimation of the parameters in the following first order TVP model:

$$x_t = \frac{b_0}{1 + a_{1,t} z^{-1}} \, \dot{u}_{t-2}, \qquad u_t = N(0, 6.25),$$

$$y_t = x_t + \xi_t, \qquad \xi_t = N(0, 2.56),$$

or, written in the form of equation (2a),

$$y_t = \frac{b_0}{1 + a_{1,t}} u_{t-2} + \xi_t$$

where $b_0 = 0.5$ is constant and $a_{1,t}$ varies sinusoidally, as $0.9 \sin(0.02t)$. Estimation is based on the measurements of y_t and u_t, $t = 1, 2, \ldots, 2000$, shown in the upper panels of Figure 1; the lower panel shows the variation of $a_{1,t}$. It will be noted that the overall noise/signal ratio on the output measurement y_t is high (0.71 by variance; 0.84 by standard deviation).

It is assumed that no information is available on the variation of the parameters and so RW models are chosen for both of the two unknown parameters. ML optimization of the NVRs, as described in the previous subsection, then yields $\mathbf{Q}_r = \mathrm{diag}[0.00105 \quad 1.905 \times 10^{-20}]$, where it will be noted that the NVR for the b_0 parameter is insignificantly different from zero, indicating that the parameter is identified as being time invariant. This shows how, quite objectively, the ML optimization is able to identify the relative temporal variability of the model parameters from the input-output data, without any other *a priori* information. The FIS estimated TVP $\hat{a}_{1,t|N}$ is shown in Figures 2 and 3 (upper panels), where it is compared with the DARX estimates (lower panels). The superiority of the DTF estimates is particularly clear in Figure 2: not only are the DTF estimates much better than the equivalent DARX estimates, but the estimated standard errors (shown dashed) are more realistic. As in the case of similar situations with constant parameter models, the least squares DARX standard errors are too optimistic and, in contrast to the DTF standard errors, they do not encompass the true variation of the parameters. The DTF model with these estimated parameters explains the data well: the coefficient of determination (COD) based on the simulated model output compared with the noise free output is $R_T^2 = 0.93$ (93% of the data explained by the TVP model); whilst for the DARX model, this is reduced to $R_T^2 = 0.85$. The model residuals (innovations) for the DTF model are also superior: they have an approximately normal amplitude distribution; and, as required, both the autocorrelation function (acf) and the cross correlation function (ccf) between the residuals and the input u_t are insignificant at all lags. In contrast, the ccf for the DARX model residuals shows significant correlation with u_t at some lags.

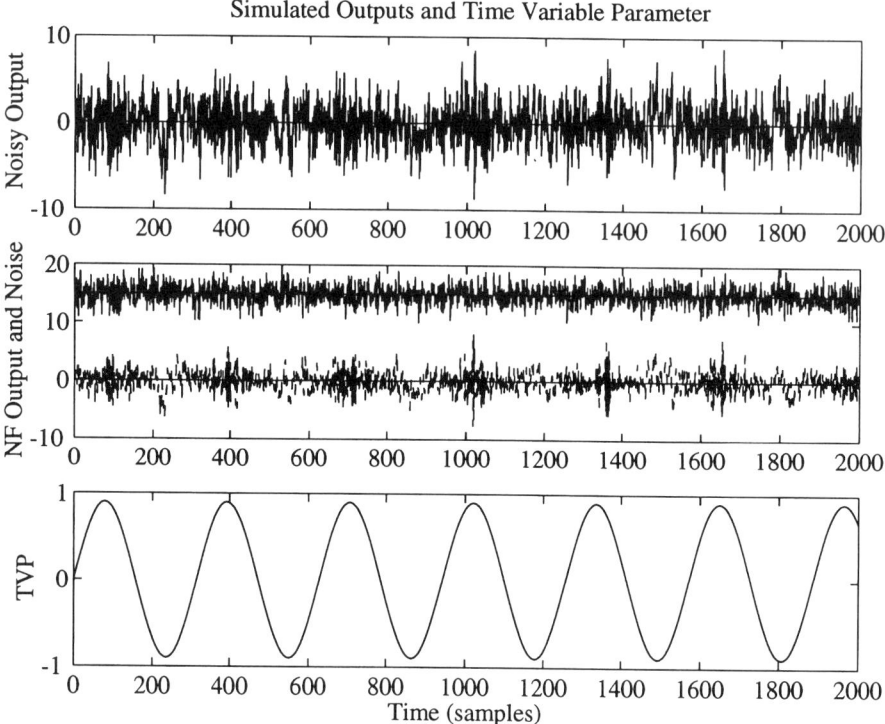

Figure 1. Simulated DTF model. Top panel: noisy output. Middle panel: noise free output and additive noise (+15). Lower panel: variation of $a_{1,t}$ parameter

The advantage of the complete, off-line FIS estimation in relation to forward-pass (filtering) estimation is illustrated in the upper panel of Figure 3, which compares a 250 sample section of the FIS estimates for the DTF model (full line) with both the actual variation (dashed line) and the forward-pass (IV, filtered) estimates generated by algorithm (5a) at the final iteration. In the lower panel, similar results are shown for the DARX estimates. It is clear that the FIS estimates in the DTF case are very close to the actual values and that, as expected, the associated forward-pass (IV, filtered) DTF estimates are much less smooth and have a pronounced lag. Similar characteristics can be observed for the DARX smoothed and filtered estimates but, in addition, they also show clear evidence of the deleterious least squares biasing effects. A measure of the overall estimation accuracy is the COD for the error between the estimated and actual TVPs: in the case of the DTF model, this yields an $R_T^2 = 0.96$ (i.e. 96% of the actual variation in the parameter explained by the TVP estimate) for the FIS estimates, and a much lower $R_T^2 = 0.85$ for the associated IV filtered estimates. For the DARX model, the values are lower still at $R_T^2 = 0.80$ and $R_T^2 = 0.70$, respectively.

Note finally, in this example, that the b_0 parameter has been kept constant

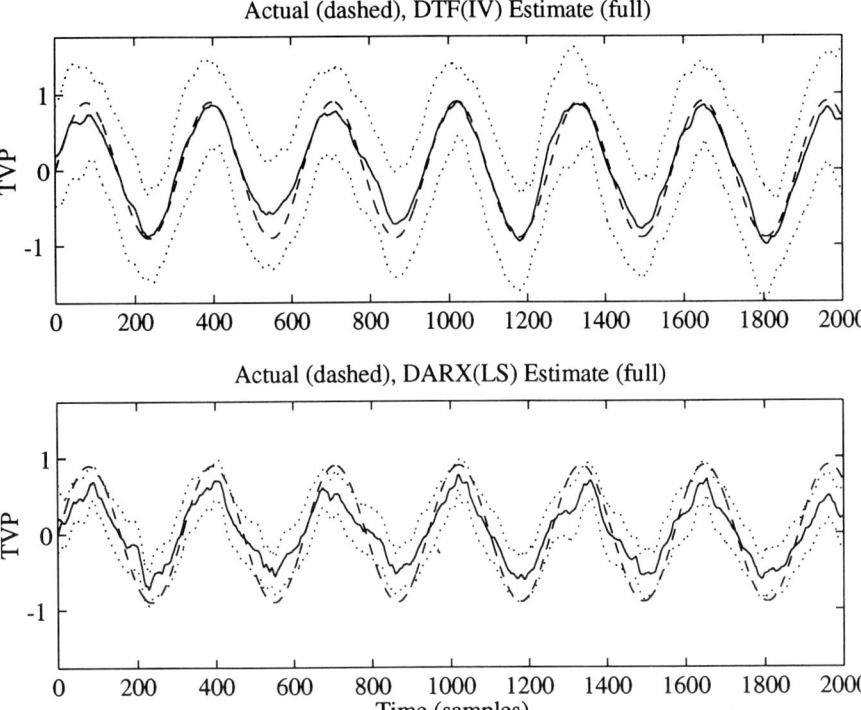

Figure 2. Simulated DTF model. Top panel: FISIV estimate of $a_{1,t}$ (full line) compared with actual variation (dashed), with standard error bound shown dotted. Lower panel: biased least squares FIS estimate of the TVP under the assumption of that the model is DARX (lines as for upper panel)

to show how the ML hyper-parameter optimization is able to detect this fact (without the provision or use of any *a priori* information) and so inform the DTF recursive algorithm (5) that b_0 is time-invariant. However, if both parameters are allowed to vary in a similar manner (e.g. $a_{1,t}$ sinusoidal, as here, and $b_{0,t}$ with a similar frequency, cosine variation between 0.9 and -0.9), then the associated NVRs are optimized as $\mathbf{Q}_r = \mathrm{diag}\,[0.00101\ \ 0.00232]$, showing that the ML optimization has found strong evidence of temporal changes in both parameters. As a result, they are both estimated well (although, as might be expected, there is some deterioration in the $\hat{a}_{1,t}$ estimate, when compared with the above results: see Young and McKenna, 1999).

3   SDP Transfer Function Models

As we have seen above, the approach to TVP estimation discussed in the last section works very well in situations where the parameters are slowly varying

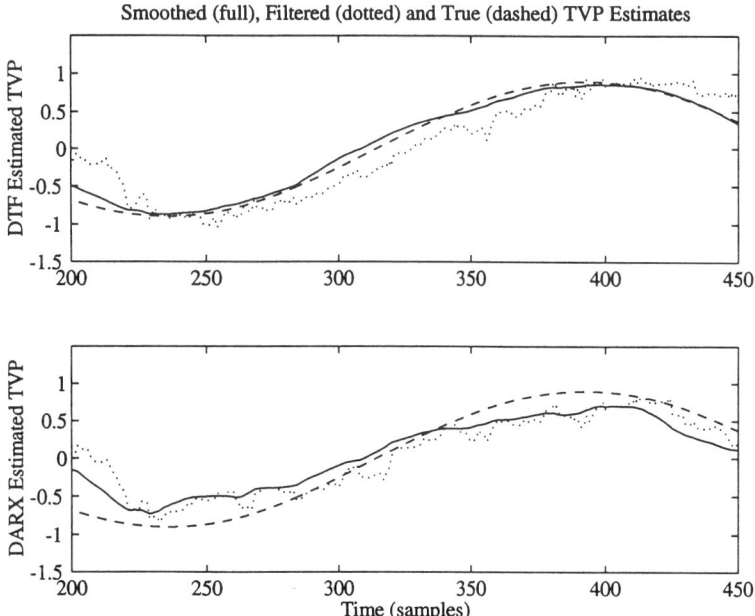

Figure 3. Simulated DTF model. Top panel: FISIV (full line) and forward-pass
filtered (dotted line) estimates of $a_{1,t}$ compared with the actual variation (dashed).
Lower panel: biased least squares estimates of $a_{1,t}$ under the assumption of that the
model is DARX (lines as for upper panel)

when compared to the observed temporal variation in the measured system
inputs and outputs. Although such DARX and DTF models are nonlinear
systems, since the same inputs, injected at different times, will elicit quite
different output responses, the resultant nonlinearity is fairly mild. It is only
when the parameters are varying at a rate commensurate with that of the
system variables themselves that the model behaves in a heavily nonlinear
or even chaotic manner. We will refer to *State Dependent Parameter* (SDP)
models of this type as *State Dependent parameter ARX* (SDARX) and *State
Dependent parameter TF* (SDTF) models respectively. The rest of the chapter
will show how these types of model can be used to represent a wide variety of
nonlinear stochastic systems and time series.

3.1 SDARX Estimation

The SDARX model equation takes the following form,

$$y_t = \mathbf{z}_t^T \mathbf{p}_t + e_t, \qquad e_t = \mathrm{N}(0, \sigma^2), \qquad (11\mathrm{a})$$

where now

$$\begin{aligned}
\mathbf{z}_t^T &= [-y_{t-1} \ -y_{t-2} \ \cdots \ -y_{t-n} \ u_{t-\delta} \ \cdots \ u_{t-\delta-m}], \\
\mathbf{p}_t &= [a_1(\boldsymbol{\chi}_t)a_2(\boldsymbol{\chi}_t)\cdots a_n(\boldsymbol{\chi}_t)b_0(\boldsymbol{\chi}_t)\cdots b_m(\boldsymbol{\chi}_t)]^T,
\end{aligned} \quad (11b)$$

while $a_i(\boldsymbol{\chi}_t)$, $i = 1, 2, \ldots, n$, and $b_j(\boldsymbol{\chi}_t)$, $j = 0, 2, \ldots, m$, are the state dependent parameters, which are assumed to be functions of a non-minimal state vector $\boldsymbol{\chi}_t^T = \begin{bmatrix} \mathbf{z}_t^T & \mathbf{U}_t^T \end{bmatrix}$. Here $\mathbf{U}_t = [U_{1,t} \ U_{2,t} \ \cdots \ U_{r,t}]^T$ is a vector of other variables, not necessarily direct functions of y_t and u_t, that may affect the relationship between these two primary variables (see Young, 1993a; Young and Beven, 1994).

The simpler signal topology for this 'affine' model (see Co, 1996) means that FIS estimation can be based on the standard recursive least squares (RLS) filtering/smoothing equations (e.g. Young, 1984, 1999a): i.e. the following RLS form of the algorithm (5a)–(5e):

1. Forward-Pass Recursive LS Equations

Prediction:

$$\begin{aligned}
\hat{\mathbf{x}}_{t|t-1} &= \mathbf{F}\hat{\mathbf{x}}_{t-1}, \\
\hat{\mathbf{P}}_{t|t-1} &= \mathbf{F}\hat{\mathbf{P}}_{t-1}\mathbf{F}^T + \mathbf{G}\mathbf{Q}_r\mathbf{G}^T.
\end{aligned}$$

Correction:

$$\begin{aligned}
\hat{\mathbf{x}}_t &= \hat{\mathbf{x}}_{t|t-1} + \mathbf{P}_{t|t-1}\mathbf{H}_t^T \left[1 + \mathbf{H}_t\mathbf{P}_{t|t-1}\mathbf{H}_t^T\right]^{-1} \left\{ y_t - \mathbf{H}_t\hat{\mathbf{x}}_{t|t-1} \right\}, \\
\mathbf{P}_t &= \mathbf{P}_{t|t-1} - \mathbf{P}_{t|t-1}\mathbf{H}_t^T \left[1 = \mathbf{H}_t\mathbf{P}_{t|t-1}\mathbf{H}_t^T\right]^{1} \mathbf{H}_t\mathbf{P}_{t|t-1}.
\end{aligned}$$

2. Backward-Pass IV Smoothing Equations

$$\begin{aligned}
\hat{\mathbf{x}}_{t|N} &= \mathbf{F}^{-1}\left[\hat{\mathbf{x}}_{t+1|N} + \mathbf{G}\mathbf{Q}_r\mathbf{G}^T\mathbf{L}_t\right], \\
\mathbf{L}_t &= \left[\mathbf{I} - \mathbf{P}_{t+1}\mathbf{H}_{t+1}^T\mathbf{H}_{t+1}\right]^T \left[\mathbf{F}^T\mathbf{L}_{t+1} - \mathbf{H}_{t+1}^T\left\{y_{t+1} - \mathbf{H}_{t+1}\hat{\mathbf{x}}_{t+1}\right\}\right], \\
\mathbf{P}_{t|N} &= \mathbf{P}_t + \mathbf{P}_t\mathbf{F}^T\mathbf{P}_{t+1|N}^{-1}\left[\mathbf{P}_{t+1|N} - \mathbf{P}_{t=1|t}\right]\mathbf{P}_{t+1|t}^{-1}\mathbf{F}\mathbf{P}_t, \\
\mathbf{L}_N &= 0.
\end{aligned}$$

Since the parameter vector $\mathbf{p}_t$ is potentially state dependent, it may vary at a rate commensurate with the temporal variations in y_t, u_t and $U_{i,t}$, and so it cannot be assumed that the simple GRW model (2e) is appropriate to describe the parametric variation over time. At first sight, it would appear that the stochastic state space model should include prior information on the nature of the parameter variation, if the TVP estimation methodology discussed in previous sections is to work satisfactorily. Fortunately, it is possible to remove this requirement if we resort to the rather unusual procedure, at least within a time series context, of sorting the data in a *non-temporal order*. Then, if the ordering is chosen so that the SDP variations associated with the *sorted* series are smoother and less rapid, it is more likely that a simple GRW process can be utilized to describe their evolution.

For example, if the time series are sorted in some common 'ascending order of magnitude' manner (i.e. the *sort* operation in MATLAB), then the rapid natural variations in y_t and u_t are effectively eliminated from the data and replaced, in the sorted data space, by much smoother and less rapid variations. And if the SDPs are, indeed, related to these variables, then they will be similarly affected by the sorting. Following FIS estimation, however, these SDP estimates can be 'unsorted' (a trivial *unsort* operation to reverse MATLAB's *sort*) and their true, rapid variation will become apparent. Of course, the nature of the sorting will affect the estimation and it seems likely that there will be an optimum sorting which results in minimum variance estimates. However, such optimum sorting will naturally depend upon the nature of the state dependency and its definition would require some sort of iterative estimation procedure. In practical terms, therefore, the common ascending order sorting and unsorting operations seem the most straightforward and will be utilized here.

One obvious requirement with this new approach to SDP estimation is that the sorting of data, prior to FIS estimation, must be *common to all of the variables in the relationship* (11a). If an ascending order strategy is selected, therefore, it is necessary to decide upon which variable in the model the sorting should be based. The simplest strategy is to sort according to the ascending order of the 'dependent' variable y_t. Depending upon the nature of each SDP in the vector $\mathbf{p}_t$, however, a single variable sorting strategy, such of this, may not produce satisfactory results. If this is the case, then a more complicated, but still straightforward, 'backfitting' procedure can be exploited. Here, each parameter is estimated *in turn*, based on the 'modified dependent variable' (mdv) series obtained by subtracting all the other terms on the right hand side of (11a) from y_t. At each such backfitting iteration, the sorting can then be based on the single variable associated with the current SDP being estimated.

Since the SDP estimates resulting from this backfitting algorithm are themselves time series, it will be noted that the algorithm constitutes a special form of non-parametric estimation and, as such, can be compared with other non-parametric methods, such as the Generalized Additive Modelling (GAM) approach of Hastie and Tibshirani (1996). However, in both conceptual and algorithmic terms, the SDP approach described here is significantly different from this earlier approach and seems more appropriate to the estimation of nonlinear, stochastic, dynamic models. Moreover, the recursive methodology, on which SDP estimation is based, is couched in optimal maximum likelihood terms that seem more elegant and flexible than the 'scatter-plot smoothing' procedures used by Hastie, Tibshirani and others.

As before, let $\mathbf{p}_t = [p_{1,t} \; p_{2,t} \; \cdots \; p_{n+m+1,t}]^T$, with $p_{i,t}$, $i = 1, 2, \ldots, n + m + 1$, relating to the TF model parameters $a_{i,t}$ and $b_{j,t}$ through (11b). The backfitting algorithm for the SDP model (11) then takes the following form:

Backfitting Algorithm for SDP Models

(1) Assume that FIS estimation has yielded prior TVP estimates $\hat{p}^0_{i,t|N}$, $i = 1, 2, \ldots, m + n + 1$ of the SDPs.[6]

(2) Iterate: $i = 1, 2, \ldots, m + n + 1; k = 1, 2, \ldots, k_c$

    (i) form the mdv $y^i_t = y_t - \sum_{j \neq i} z_{j,t} \cdot \hat{p}^k_{j,t}$

    (ii) sort[7] both y^i_t and $z_{i,t}$ according to the ascending order of $z_{i,t}$

    (iii) obtain an FIS estimate $\hat{p}^k_{i,t}$ of $p_{i,t}$ in the mdv relationship $y^i_t = p_{i,t} \cdot z_{i,t}$

(3) Continue (2) (each time forming the mdv and then sorting according to the current right hand side variable $z_{i,t}$, prior to FIS estimation), until iteration k_c, when the individual SDPs (which are each time series of length N) have not changed significantly according to some chosen criterion. The smoothing hyper-parameters required for FIS estimation at each stage are optimized by Maximum Likelihood (ML), as explained earlier in section 2.1 and discussed further below.

Note that the ML optimization can be carried out in various ways: after every complete iteration (each involving $m + n + 1$ FIS operations) until convergence is achieved; only at the initial complete iteration, with the hyper-parameters maintained at these values for the rest of the backfitting; or just on the first two iterations. The last choice seems most satisfactory in practice, since very little improvement in convergence occurs if optimization is continued after this stage. Normally, convergence is completed after only a few iterations, although it can be more lengthy in some circumstances (see later discussion in the Conclusions, section 5).

Simulation Example 2

As a simulation example of SDARX modelling, consider the following forced logistic growth equation,

$$y_t = 2.0y_{t-1} - 2.0y^2_{t-1} + u_t + e_t, \qquad u_t = \mathrm{N}(0, 0.08), \quad e_t = \mathrm{N}(0, 0.008) \quad (12a)$$

---

[6]As a default, these can be simply the constant least squares parameter estimates, since the convergence of the backfitting procedure is not too sensitive to the prior estimates, provided they are reasonable: see simulation example 2 below

[7]Depending on the nature of the state dependency, sorting may need to be with respect to another variable in χ_t

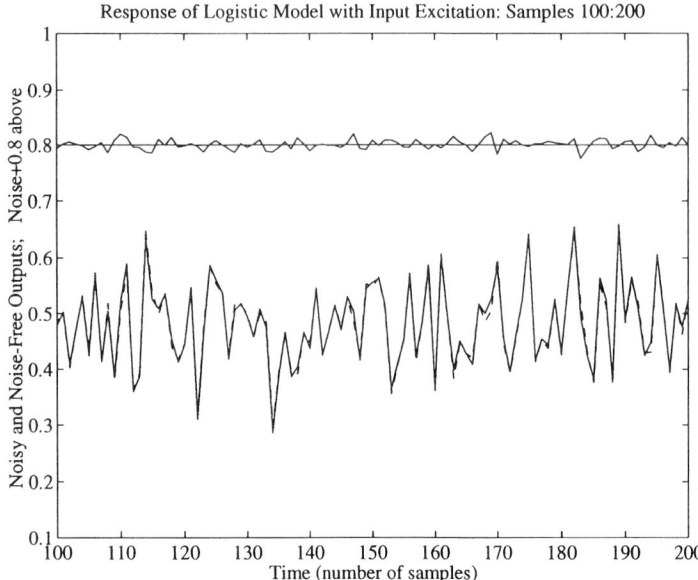

Figure 4. Simulated forced logistic growth model. Noise-free output (full line); noisy output (dashed line); and noise (+0.8) above

or,

$$y_t = a_1(y_{t-1}) . y_{t-1} + b_0(u_t) . u_t + e_t \qquad (12b)$$

where

$$a_1(y_{t-1}) = 2.0 - 2.0y_{t-1}, \qquad b_0(u_t) = 1.0 \quad \forall t, \qquad (12c)$$

Here, u_t is a measured input and e_t is white 'system' noise. The unforced equivalent of this model (i.e. $u_t = 0$ for all t: see simulation example 3) is, in fact, the example used by the author in the first publication on SDP modelling (Young, 1978) although, at that time, the more powerful FIS algorithms described in this chapter had not been developed and simple recursive (filtering) estimation was utilized.

The estimation is based on 1000 samples of u_t and y_t, a 100 sample segment of which (for clarity) is shown in Figure 4. This is a fairly low noise situation with about 10% noise level (by standard deviation). ML optimization at the second iteration yields an NVR matrix $\mathbf{Q}_r = \text{diag}[0.000357 \; 2.5 \times 10^{-17}]$, showing how the optimization has, once again, identified that the parameter $b_0(u_t)$ associated with the input u_t is constant and that only the 'lag' parameter $a_1(y_{t-1})$ is varying. This value of $\mathbf{Q}_r$ is then utilized for the subsequent three iterations of the backfitting procedure that leads to convergence of the SDP estimates.

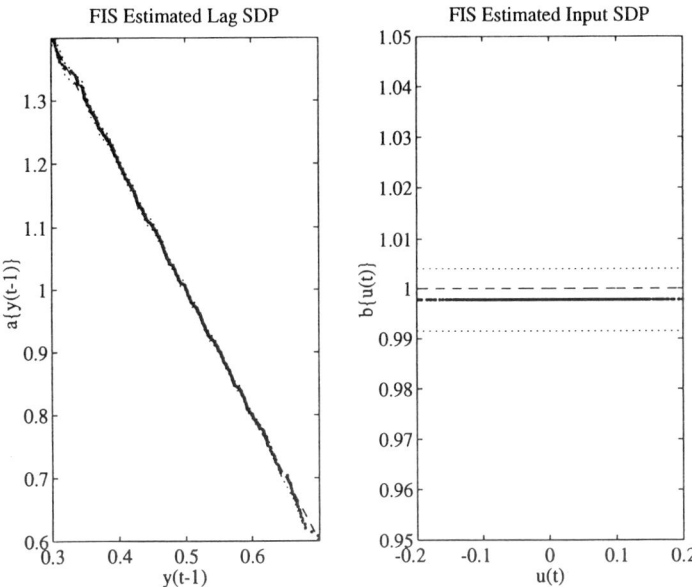

Figure 5. Simulated forced logistic growth model. Left panel: FIS estimate $\hat{a}_1(y_{t-1}|N)$ vs y_{t-1}. Right panel: FIS estimate $\hat{b}_o(u_t|N)$ vs u_t. True values shown as dashed lines and standard error bands shown as dotted lines.

The results obtained in this manner are shown in Figures 5 and 6. The former provides plots of the estimated parameter variation against the associated non-minimal state variable, in each case. The left hand panel shows the estimated relationship between $\hat{a}_{1,t|N} = \hat{a}_1(y_{t-1} \mid N)$ and y_{t-1}; while the right hand panel shows the estimated $\hat{b}_{0,t|N} = \hat{b}_0(u_t \mid N)$ as a function of u_t. As expected, these correspond to those used in the simulated model: $\hat{a}_1(y_{t-1} \mid N)$ is a clear linear function of y_{t-1} with slope and intercept close to -2 and 2, respectively, in correspondence with equation (12c); while $\hat{b}_0(u_t \mid N)$ is constant for all t, at an estimated value close to unity. The actual values of the parameters are shown as dashed lines and the standard error bounds are shown dotted. Figure 6 presents the associated estimates of the nonlinear functions $\hat{f}_1(y_{t-1}) = \hat{a}_1(y_{t-1} \mid N).y_{t-1}$ and $\hat{f}_2(u_t) = \hat{b}_0(u_t \mid N).u_t$, again with actual values shown dashed and the standard error bounds show dotted.

Figure 7 shows results similar to those in Figure 5, again based on a 1000 sample data set, but with a much higher noise level of 69% by standard deviation (48% by variance). Although the uncertainty on the estimates is greater, as would be expected, the general nature of the state dependent relationships is still clear and the identification of a forced logistic growth equation is quite unambiguous. In this noisy situation, the NVR matrix $\mathbf{Q}_r = \text{diag}[9.89 \times 10^{-7}\ 0.0]$

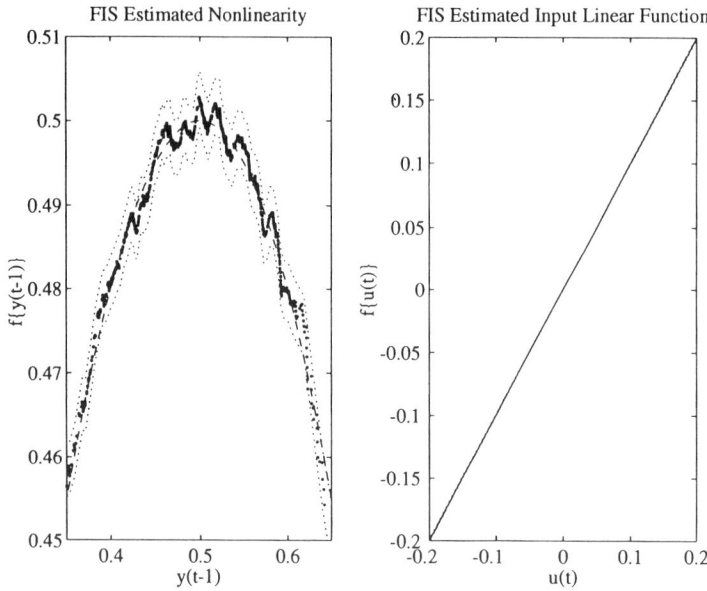

Figure 6. Simulated forced logistic growth model. Left panel: FIS estimate of feedback nonlinearity $\hat{f}_1(y_{t-1})$. Right panel: FIS estimate of input nonlinearity $\hat{f}_2(u_t)$. True values shown as dashed lines and standard error bands shown as dotted lines.

3.2 SDTF Estimation

In the SDTF model, it is assumed that noise can enter as either system or measurement noise, or both. For example, in the case of the forced logistic equation (12), it could take the form:

$$x_t = 2.0x_{t-1} - 2.0x_{t-1}^2 + u_t + e_t, \quad u_t = \mathrm{N}(0, 0.08), \quad e_t = \mathrm{N}(0, 0.008)$$
$$y_t = x_t + \xi_t, \qquad\qquad\qquad\qquad \xi_t = \mathrm{N}(0, 0.08)$$

$$(13a)$$

or,

$$y_t = a_1(y_{t-1}). \, y_{t-1} + b_0(u_t). \, u_t + \zeta_t \qquad (13b)$$

where

$$a_1(y_{t-1}) = 2.0 - 2.0y_{t-1}, \qquad b_0(u_t) = 1.0 \quad \forall t, \qquad (13c)$$

and the noise ζ_t is a complex nonlinear function of e_t, ξ_t and y_t. In this situation, estimates obtained under the assumption that the model is of the simpler SDARX form are nominally biased to a level dependent on the noise/signal ratio. Fortunately, however, this bias is often fairly small, even for quite high noise levels and, in consequence, it does not interfere substantially with the

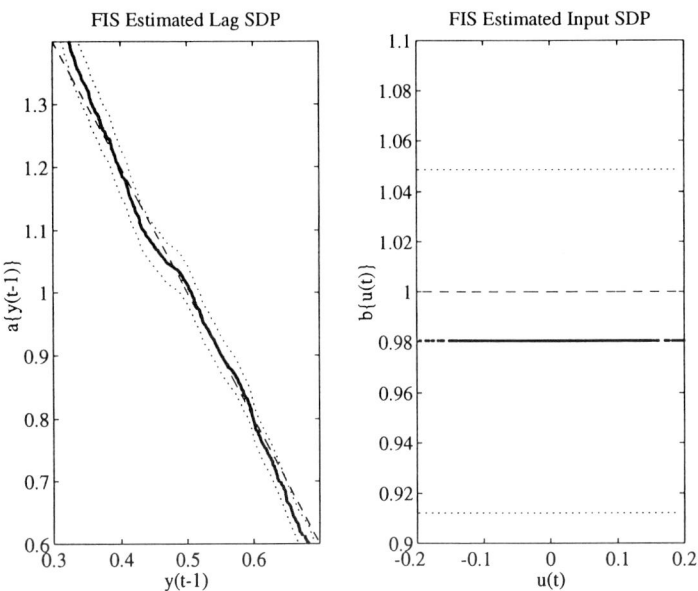

Figure 7. Simulated forced logistic growth model. As for Figure 5 but with higher level of noise on the data

identification of any state dependent relationships (note that the analysis is aimed at *identifying* the form of the nonlinearity and more efficient statistical estimation follows this identification step). For example, in the above example, the estimation results for a measurement noise level similar to that of the system noise level used to obtain the results in Figure 7 (69% by standard deviation; 48% by variance) are quite acceptable and only a little worse than those shown in Figure 7.

Nevertheless, it would be advantageous if a truly bias-free estimation method were available in the SDTF model case and research is continuing on the development of an IV backfitting algorithm which exploits the methodology discussed above. The main problem with such an approach is in maintaining the stability of the auxiliary model (see equation (5c)) that generates the instrumental variables: in the case of nonlinear models with chaotic properties, for example, only small uncertainties can lead to wide differences in response and possible instability. Consequently, other approaches that extend the model to include (SDP) noise terms are also being investigated.

Final Parametric Estimation

If any underlying parametric state dependency can be identified in non-parametric form using the recursive estimation methods discussed in previous sections, then it is often possible to continue further and parameterize the identified nonlinear relationships in terms of a finite set of constant parameters. For example, given the non-parametric estimation results shown in Figures 5, 6 and 7, it is clear that the forced logistic growth equation, with three unknown parameters, provides a parsimonious representation of the data. It is straightforward, therefore, to obtain estimates of these parameters by simply fitting appropriate parameterized functions to the estimated state dependent relationships using weighted or ordinary least squares (WLS: see Young, 1993a, 1998a; Young and Beven, 1994). In this case, the WLS estimates of the two parameters of the linear relationship in (12c) obtained from the low noise results (Figure 5) are 2.002(0.0013) and −2.004(0.0025). The corresponding estimates in the high noise case (Figure 7) are 2.039(0.008) and −2.08(0.14); and similar results are obtained in the SDTF case. However, it is more satisfactory to use ML estimation of the parameters in (12c) directly from the data. In the low noise case, this yields 2.0332(0.008), −2.0679(0.017) and 0.9813(0.008); while in the high noise case, the estimates are 2.0375(0.036), −2.1485(0.075) and 0.9834(0.033). So, in all cases, the estimates are very good (although with signs of very small positive bias) and the finally estimated model produces responses that are insignificantly different from noise-free output of the actual system, even in the high noise case. This is shown in Figure 8, where the error shown above (+0.8) is between the model output and the noise-free output. The noisy output on which the estimation is based is shown with circle points.

3.3   Estimation of Purely Stochastic Systems

A special example of the SDP model (11) is the following *State Dependent parameter Auto-Regressive* (SDAR) model:

$$y_t = -a_{1,t}y_{t-1} - a_{2,t}y_{t-2} - \cdots - a_{n,t}y_{t-n} + e_t \tag{14a}$$

or

$$y_t = \mathbf{z}_t^T \mathbf{p}_t + e_t, \tag{14b}$$

where

$$\begin{aligned} \mathbf{z}_t^T &= [-y_{t-1} \ -y_{t-2} \ \cdots \ -y_{t-n}], \\ \mathbf{p}_t &= [a_1(\boldsymbol{\chi}_t) \ a_2(\boldsymbol{\chi}_t) \ \cdots \ a_n(\boldsymbol{\chi}_t)]^T, \end{aligned} \tag{14c}$$

Clearly, the same SDARX estimation methods discussed previously can be applied to this model, a simple example of which is the chaotic version of the logistic equation. Typical simulation results for this model are discussed below.

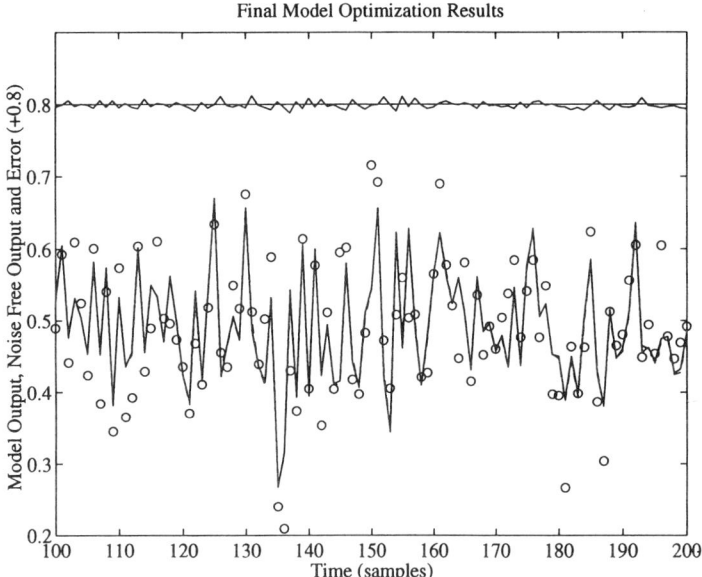

Figure 8. Simulated forced logistic growth model: final parametric estimation results. Comparison of estimated model output (full line) and noise free output (dashed), with the error (+0.8) shown above. The circle points show the noisy measured output used in the estimation.

Simulation Example 3

In order to consider the effects of measurement noise, the following Noisy SDAR (NSDAR) version of the chaotic logistic growth model (cf. equations (12)) will be used in this example:

$$
\begin{aligned}
x_t &= 4.0x_{t-1} - 4.0x_{t-1}^2 + e_t, & e_t &= N(0, 0.0064), \\
y_t &= x_t + \xi_t, & \xi_t &= N(0, 0.0012)
\end{aligned} \tag{15a}
$$

or

$$
y_t = a_1(y_{t-1}) \cdot y_{t-1} + \zeta_t, \qquad a_1(y_{t-1}) = 4.0 - 4.0y_{t-1}. \tag{15b}
$$

As in equation (13b), the noise ζ_t is a complex nonlinear function of e_t, ξ_t and y_t, and the noise level is about 10% (by standard deviation). As expected, without this measurement noise present, SDP estimation is straightforward with excellent, low variance SDP estimates that identify the nature of the nonlinearity without any difficulty.

Even with the measurement noise, this is a simpler estimation problem than in the forced logistic growth model since there is only one SDP, $a_1(y_{t-1})$, and ML optimization of the associated scalar NVR quickly yields $\mathbf{Q}_r = 0.000016$. The subsequent SDP estimation results are illustrated in Figure 9, where the

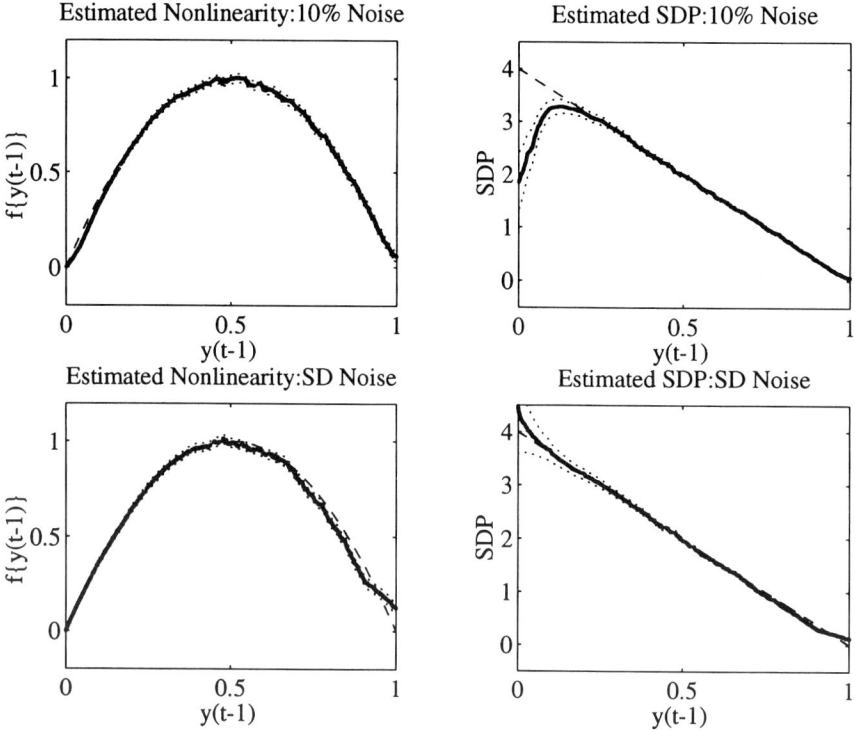

Figure 9. Simulated chaotic logistic model. Left panels: estimated nonlinear functions with ordinary (upper) and state dependent (lower) noise. Right panels: estimated state dependent parameters with ordinary (upper) and state dependent (lower) noise

top left hand panel shows the FIS estimated nonlinear function with its characteristic quadratic shape; while the top right hand panel shows the estimated state dependency of the SDP estimate $\hat{a}_1(y_{t-1} \mid N)$. In both plots, the true relationship is shown as a dashed line, while the estimated standard error bounds are shown dotted.

The most notable feature of the results in Figure 9 is the larger errors in the SDP estimate at low values of y_{t-1}. Since the measurement noise variance is constant, this is the region where the measurement noise is having its most deleterious biasing effect. Even with these errors, however, the WLS estimates of the parameters (3.884(0.111) and $-3.857(0.135)$, respectively) are quite close to the true values (4 and -4). The results are better still, however, if the noise is itself made state dependent, in the sense that it is set proportional to the signal level (i.e. $\xi_t^{\text{sd}} = [x_t/\max(x_t)].\xi_t$). This is a common situation with real data and it significantly reduces the noise effect at low signal levels

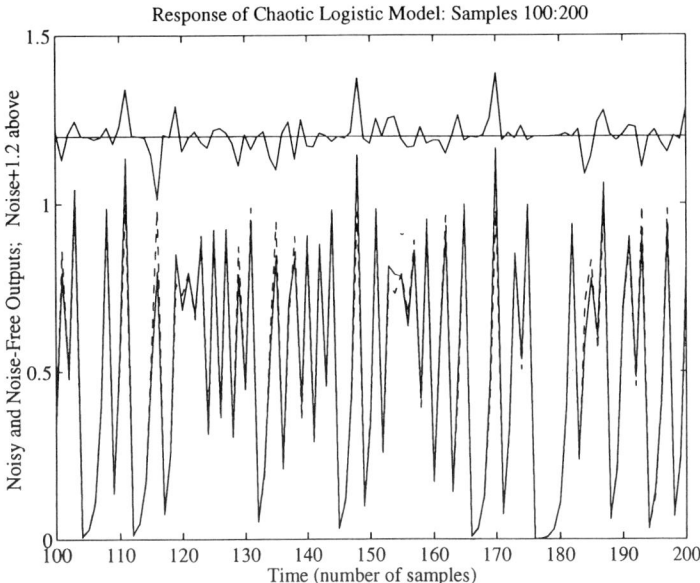

Figure 10. Simulated chaotic logistic model. Noisy (full line) and noise-free (dashed line) output, with the state dependent noise (+1.2) shown above

where the bias is largest. As a result, the WLS estimates are improved to 3.960(0.013) and $-3.946(0.016)$, respectively. Indeed, in this state dependent noise situation, the results are still good even if the variance of ξ_t is doubled, so that the noise level on the data is visibly quite large, as shown in Figure 10. The resulting SDP estimation results are presented in the lower panels of Figure 9, which identify clearly that the data were generated by the logistic model with chaos inducing parameter values.

4 Further Examples

In this section, we present two further examples. The first is another, more difficult simulation example based on the so-called *Cosine Map* model. The second is a brief description of a practical example: namely the analysis of the famous set of time series data on the Australian sheep blowfly *Lucilia cuprina* by Nicholson (1954). This example is described in more detail elsewhere (Young, 1998a; Young and Fawcett, 1999). Other, practical examples, in a variety of application areas from the environment to economics, are described in Young (1993a, 1996, 1998a,b, 1999a; Young and Beven, 1994; Young and Pedregal, 1997, 1999; Young *et al.*, 1996).

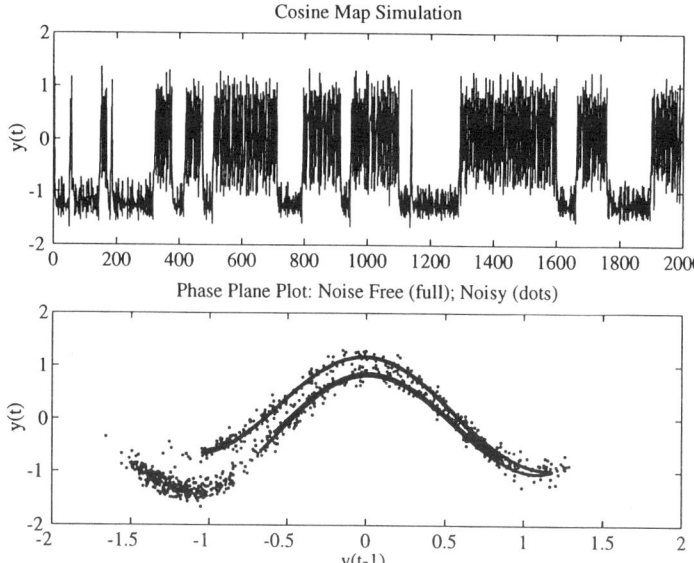

Figure 11. Simulated cosine map model. Upper panel: measured output. Lower panel: phase plane (embedding) plot of y_t vs y_{t-1} with noise-free response shown as a full line

4.1   Identification and Estimation of the Cosine Map Model

The cosine map model (e.g. Zhan-Qian and Smith, 1998) takes the form

$$y_t = \cos(2.8y_{t-1}) + 0.3y_{t-2} + e_t, \qquad e_t = \mathrm{N}(0, 0.01), \tag{16}$$

and a typical 2000 sample simulation of the model is presented in Figure 11, which shows the time response in the upper panel and the phase plane $(y_t \sim y_{t-1})$ plot in the lower panel.

In the latter graph, the noise free response is shown as a full line with the noisy response plotted as dots. This a typical stochastic model that exhibits underlying chaotic response characteristics. It provides a testing example for the SDP approach, however, because it is not in the assumed affine form: in particular, the SDP term $a_1(y_{t-1}) \cdot y_{t-1}$ in the most appropriate SDAR model,

$$y_t = a_1(y_{t-1}) \cdot y_{t-1} + a_2(y_{t-2}) \cdot y_{t-2} + e_t, \qquad e_t = \mathrm{N}(0, \sigma^2), \tag{17}$$

is not able to represent the equivalent term $\cos(2.8y_{t-1})$ in (16) exactly, since $\cos(2.8y_{t-1})/y_{t-1}$ has a singularity at $y_{t-1} = 0$. Despite this difficulty, SDAR estimation yields excellent results, as illustrated in Figures 12 to 14. These were obtained with the ML optimized NVR matrix $\mathbf{Q}_r = \mathrm{diag}[0.0057\ 1.28 \times 10^{-7}]$ and it is clear, yet again, that the optimization has successfully identified that the potential state dependency resides in the first lag parameter $a_{1,t} = a_1(y_{t-1})$, while the second lag parameter $a_{2,t} = a_2(y_{t-2})$ is effectively time-invariant.

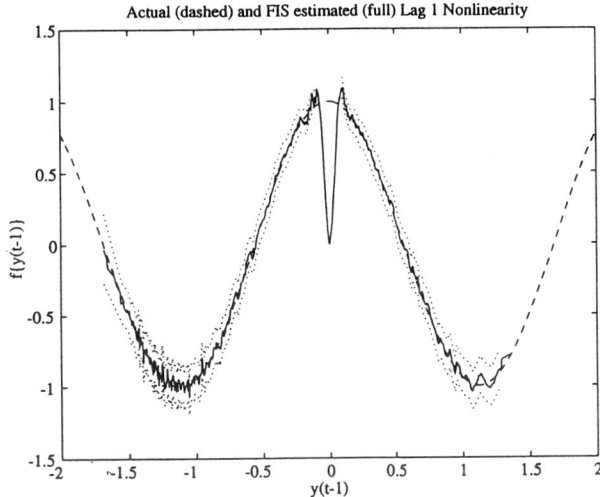

Figure 12. Simulated cosine map model. Comparison of FIS estimated and actual cosine nonlinearity

Figure 12 shows the FIS estimate of the cosine nonlinearity subsequent to the convergence of the backfitting procedure, which took 6 iterations in this case. Except for the region around the singularity at $y_{t-1} = 0$ (see below), the estimation is very good. The associated $\hat{a}_2(y_{t-2} \mid N) = 0.291(0.0038)$, for all t, is estimated as being time-invariant despite the fact that the NVR, $\mathbf{Q}_r(2,2) = 1.28 \times 10^{-7}$, is not too small in this case. Figure 13 compares the actual, simulated phase plane plot for the data used in the estimation (left hand panel) with a similar plot based on data from a typical random realization of the SDAR model (right hand panel). Similar agreement is found in both the time plots of the two series and the histograms.

Figure 14 compares the FIS estimate $\hat{a}_1(y_{t-1} \mid N)$, plotted as a function of y_{t-1}, with the theoretical function given by

$$a_1(y_{t-1}) = \cos(2.8y_{t-1})/y_{t-1}$$

(which migrates to $\pm\infty$ at the point $y_{t-1} = 0$). Over the most important region $-0.1 > y_{t-1} > 0.1$, the estimate is very accurate. Not surprisingly, it becomes inaccurate close to the singularity, but the algorithm is robust enough to handle this well, without impairing the SDP estimates elsewhere. And the overall cosine shape of the nonlinear function is clearly estimated accurately in Figure 12.

Finally, on the basis of the above SDP results, the form of the nonlinear equation is identified correctly as

$$y_t = \alpha \cos(\beta y_{t-1}) + \gamma y_{t-2} + e_t, \tag{18}$$

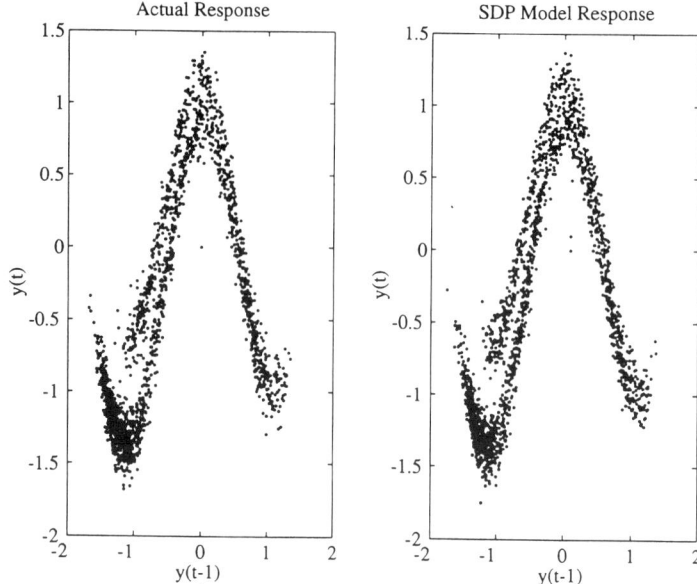

Figure 13. Simulated cosine map model. Comparison of the phase plane plot for the data used in the estimation (left panel) with the phase plane obtained from a random realization of the estimated SDAR model

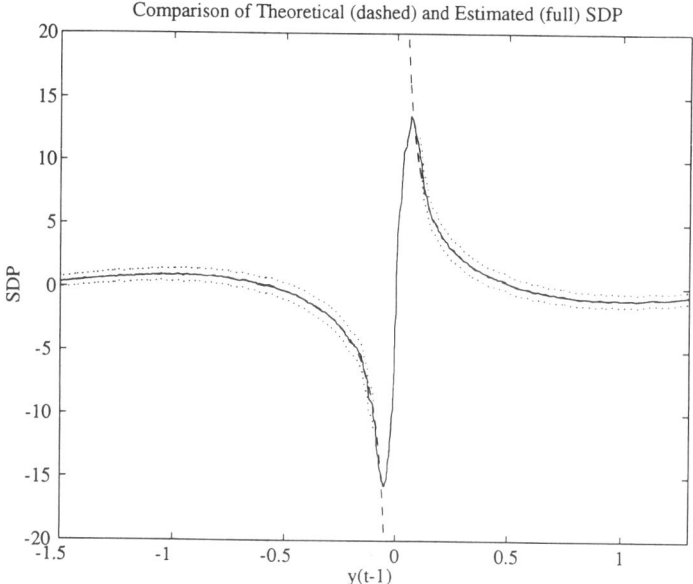

Figure 14. Simulated cosine map model. Comparison of the FIS estimate $\hat{a}_1(y_{t-1} \mid N)$ as a function of y_{t-1}, with the theoretical function given by $a_1(y_{t-1}) = \cos(2.8y_{t-1})/y_{t-1}$ (which migrates to $\pm\infty$ at the point $y_{t-1} = 0$).

and the optimized ML estimates of the, now constant, parameters α, β and γ in this model are $\hat{\alpha} = 0.998(0.004)$, $\hat{\beta} = 2.797(0.004)$ and $\hat{\gamma} = 0.303(0.003)$.

4.2   The Nicholson Blowfly Data Revisited

Figure 15 is the most well-known example of the data collected so laboriously by Nicholson (1954) in his investigation of the Australian sheep blowfly *Lucilia cuprina*. It is clear that, in this particular experiment where the food (liver) supplied to the blowflies was limited to 0.5g per day, the adult blowfly population y_t (upper graph) and the eggs laid per day by the blowflies u_t (lower graph) vary in an apparently systematic fashion that is redolent of nonlinear, limit cycle behaviour. Not surprisingly, these data have received much attention in the scientific literature (e.g. May, 1973, 1976; Banks, 1994; Gurney *et al.*, 1980; and the references therein).

In contrast to the analysis described in the preceding references, however, the SDP modelling approach used here makes no *a priori* assumptions about the nature of the blowfly system but starts with a relatively non-prejudicial analysis of the data, recognizing only that the eggs and blowfly series are causally related in some manner. The data in Figure 15 are sampled daily[8] and the most obvious relationship between u_t and y_t, namely the blowfly response to the egg production rate, seems quite linear and can be described well by the following first order, *constant* parameter, discrete-time TF:

$$y_t = \frac{b_0}{1 + a_1 z^{-1}} u_{t-\delta} + \xi_t \qquad (19a)$$

or

$$y_t = a_1 y_{t-1} + b_0 u_{t-\delta} + \eta_t, \qquad (19b)$$

where the identified time delay of $\delta = 15$ days accounts for the development of the eggs through a larval stage, prior to the emergence of the adult blowflies. The estimates of the parameters and their standard errors are obtained by constant parameter SRIV estimation (e.g. Young, 1984) as $\hat{b}_0 = 0.865(0.031)$ and $\hat{a}_1 = 0.759(0.01)$. This simple linear model explains the experimental data very well with $R_T^2 = 0.86$ and the model residuals, with ξ_t modelled as a constant parameter, AR(3) process, satisfy the usual statistical diagnostic tests.

But the modelling story does not end with this linear model of the 'forward-path' dynamics: of much more interest is the nature of the 'feedback-path' dynamics, namely the mechanism by which the blowflies produce their eggs. In contrast to the forward-path, this mechanism is obviously nonlinear, with the blowflies producing eggs only when their numbers are low and the food

---

[8]The data shown in Figure 15 have been digitized at a daily sampling interval from the graphical plots in Nicholson's paper. The original data, which are no longer available (McNeil, 1996), were collected every two days.

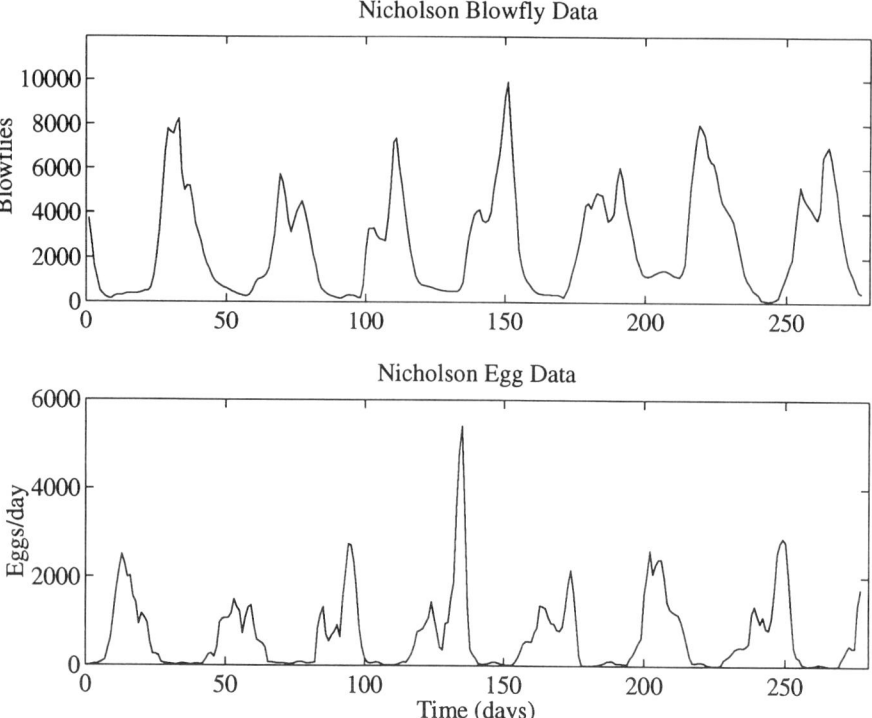

Figure 15. The Nicholson blowfly data example. Upper panel: daily numbers of
adult blowflies. Lower panel: numbers of eggs laid per day.

supply per blowfly is plentiful. There are two obvious ways in which the data
can be analysed to infer the nature of these nonlinear feedback dynamics. Most
straightforwardly, the model (19) can be modified directly to acknowledge that
the egg production rate u_t is a nonlinear function of the blowfly population
y_t: this then yields an SDAR-type model with lag terms in y_t and $y_{t-\delta}$.

Alternatively, *since the time delay in the system is large*, it is possible to
investigate the feedback-path dynamics separately by considering directly the
relationship between y_t, now considered as the 'input' in the feedback path,
and u_t, as the output, i.e.,

$$u_t = b_0(y_t) \cdot y_t + e_t. \tag{20}$$

In other words, we are able to investigate the presence of a simple *static*
nonlinearity in the feedback path that has direct ecological significance. SDP
estimation yields the estimate $\hat{b}_0(y_t \mid N)$ shown in Figure 16 as a function of y_t,
with the standard error bounds shown dotted. Also shown as a dash-dot line is
the WLS estimate of the nonlinearity based on the following exponential-type

Estimated Nonlinearities

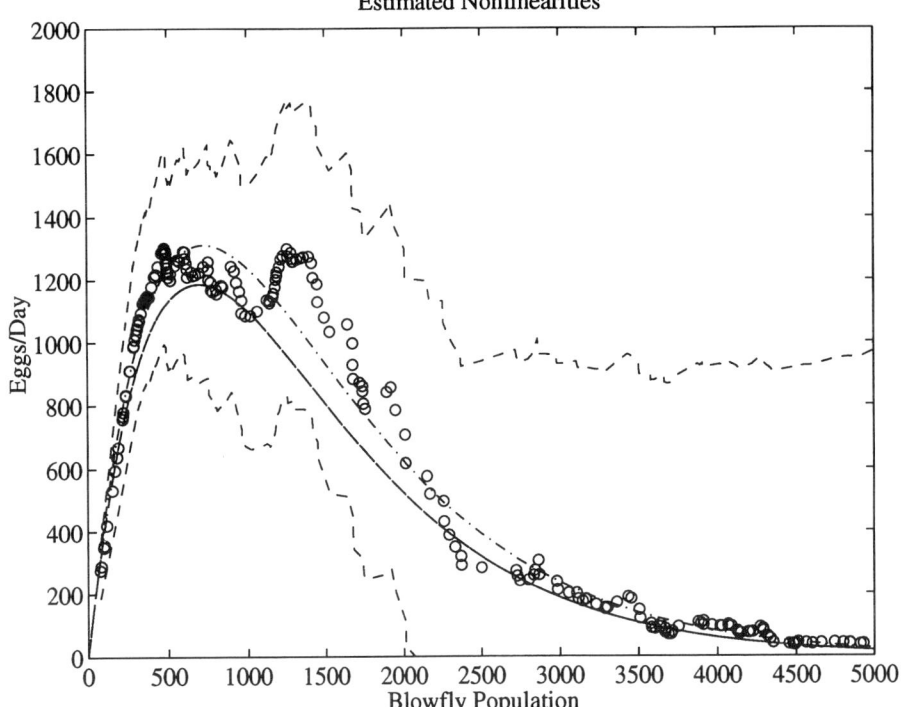

Figure 16. The Nicholson blowfly data example. FIS estimated feedback
nonlinearity (circle points) compared with the weighted least squares estimate
(dash-dot line) and final ML estimate (dashed line) of the parameterized nonlinear
function. The standard error bands on the FIS estimate are shown as dashed lines

parameterization:

$$u_t = g \cdot y_t \cdot f(y_t) + e_t, \qquad f(y_t) = \exp\left(\frac{-1}{N_0 \frac{fd}{y_t}}\right), \qquad e_t = \mathrm{N}(0, \sigma^2), \quad (21)$$

where fd is the food supplied per day to the blowfly colony (here 0.5g); while
g and N_0 are unknown parameters with WLS estimates of $\hat{g} = 4.916(0.26)$
and $N_0 = 1451(94)$, respectively. The results in Figure 16 were obtained with
an optimized NVR $= 8.16 \times 10^{-9}$.

Other parameterizations than (21) are clearly possible (e.g. Young, 1998a;
Young and Fawcett, 1999) but this particular one was chosen here because
it conforms to the prior analysis of Gurney *et al.* (1980). In particular, the
present model, i.e.,

$$y_t = a_1 y_{t-1} + g \cdot y_{t-\delta} \cdot \exp\left(\frac{-1}{N_0 \frac{fd}{y_t}}\right) + \mu_t, \qquad (22)$$

where μ_t represents the overall residual coloured noise, is closely related to the *deterministic, continuous-time differential equation* suggested by Gurney *et al.* However, their analysis is deterministic and only semi-quantitative, in the sense that they do not use statistical identification and estimation at all. Rather, they simply speculate on the form of this nonlinear system and show that, for a range of parameter values, its *general*, deterministic dynamic behaviour conforms reasonably with that exhibited by the Nicholson data.

Having identified a suitable nonlinear model for the blowfly data in (22), it is now possible to move on to the final ML estimation stage in the analysis. In this case, the nature of the coloured noise on the data requires that the model (22) is enhanced to include stochastic elements (Young and Fawcett, 1999). ML optimization of the resulting stochastic model yields the following result:

$$
\begin{aligned}
x_t &= 0.818x_{t-1} + u_{t-15} + \xi_t, \\[4pt]
u_t &= 4.63x_t . \exp\left(\frac{-1}{1392\frac{0.5}{x_t}}\right), \\[4pt]
\xi_t &= 1.137\xi_{t-1} - 0.491\xi_{t-2} + \varepsilon_t, \\[4pt]
y_t &= x_t + e_t, \qquad e_t = N(0, \sigma^2),
\end{aligned}
\tag{23}
$$

where $\sigma^2 = 1.52 \times 10^5$ and $\varepsilon_t = N(0, 44.5\sigma^2)$. In this ML optimization stage, the model (23) is formulated as the associated 17th order, discrete-time, stochastic state equation (15 orders account for the pure 'larval' time delay and 2 represent the noise dynamics). The optimization is then performed using prediction error decomposition based on a state dependent parameter implementation of the Kalman filter *that does not require linearization* (Young and Fawcett, 1999). The nonlinearity associated with this model is shown as the dashed line in Figure 16 and we see that it is similar to the WLS estimate from the non-parametric estimation stage in the identification analysis, lying well within the standard error bounds.

The residuals of the model (23) have satisfactory correlation properties but they are very heteroscedastic (clearly dependent on the blowfly population). Also, the Hessian associated with the parameter and hyper-parameter estimates suggests that some of the parameters have quite large standard errors. Despite this, the model performs well in forecasting and validation tests. Figure 17, for example, provides typical forecasting results, with one-step-ahead forecasts up to the 200th day and true, multi-step *ex ante* forecasts thereafter.

The explicit inclusion of fd in (22) and (23) is useful because it allows us to evaluate the ecological realism of the model still further by examining its prediction of what should happen if the food supply is modified. In particular, Nicholson (1954) showed experimentally that, when the food supply was reduced from 0.5g per day to 0.1g per day, the average adult population dropped from 2520 to 527. In the case of the model (23) the average populations in these two same situations are 2657 and 548 respectively; a remarkable level

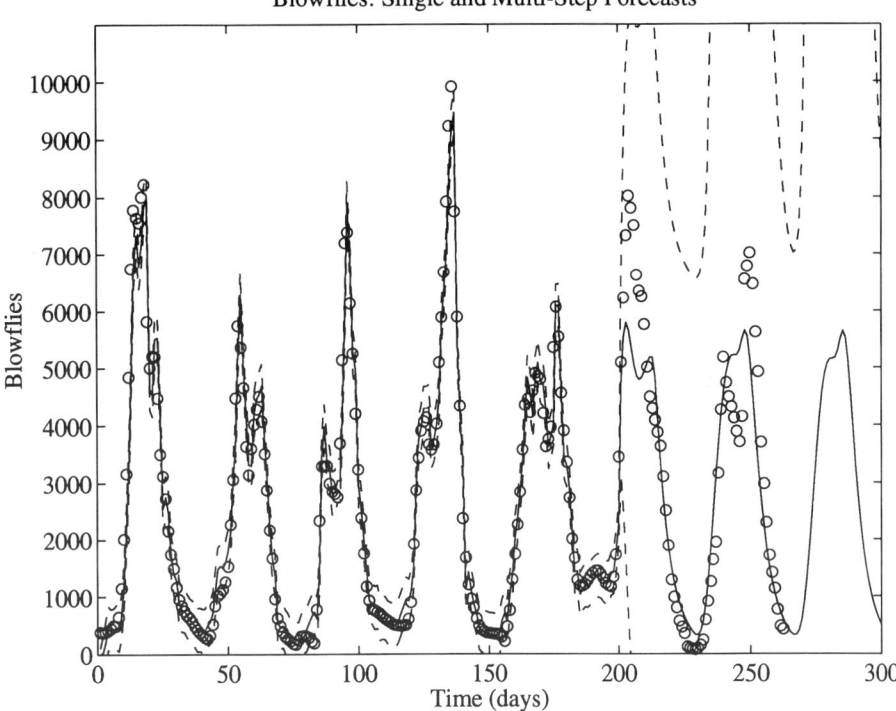

Figure 17. The Nicholson blowfly data example. Kalman filter forecasting results: forecast (full line) compared with the data (circle points), with the standard error band shown as dashed lines. Up to day 200, the forecasts are one-step-ahead; after this, the forecast is a true *ex ante* multi-step-ahead forecast, based only on the first 200 days

of agreement in the circumstances (since the average values here are based on the deterministic limit cycle data produced by (23) when the stochastic inputs are removed, rather than the actual data). Certainly this result provides a suitable initial validation of the model.

These results tend to confirm and further quantify the deterministic analysis of Gurney *et al.* However, the statistical diagnostics and associated stochastic simulations suggest that the stochastic model (23) requires a little further work, taking into account the heteroscedasticity and the poor definition of the optimized parameters, before a fully satisfactory stochastic model is confirmed for the blowfly data. Such future research could be based, for instance, on Markov Chain Monte Carlo (MCMC) methods (e.g. Ruanaidh and Fitzgerald, 1996; Gamerman, 1997) or on a simpler stochastic approach, such as that suggested recently by Durbin and Koopman (1999), where the assumption of Gaussian disturbances is not necessary. Nevertheless, as it stands, the model

(23) is clearly a reasonable one in predictive terms, as shown by the results in Figure 17. And the fact that this forecasting performance is good, despite the limitations of the stochastic model, is testament to the well known robustness of the Kalman filter when it is used as a basis for forecasting.

5   Conclusions

Recursive estimation has a long and rich history: from its beginnings in Gauss's original derivation of recursive least squares (Gauss, 1823; see Appendix 2 of Young, 1984), through its re-discovery by Plackett (1950) and Kalman's seminal work on stochastic state estimation (Kalman, 1960), to the burgeoning of research on recursive estimation in a whole range of different academic disciplines between 1960 and the present. In the last ten years, however, the advent of fast computers and the desire of theorists to extend the boundaries of time series analysis has led to an explosion of research on Monte Carlo based numerical methods, from either classical (e.g. Durbin and Koopman, 1999) or Bayesian (e.g. Ruanaidh and Fitzgerald, 1996; Gamerman, 1997) perspectives.

The motivation of this more recent research is clearly to extend the 'Gaussian' methods of standard recursive estimation to non-Gaussian and nonlinear time series, using models in which the *stochastic inputs* are non-Gaussian. But models of non-Gaussian and nonlinear processes do not necessarily require the assumption of non-Gaussian inputs. As we have seen in this chapter, a fairly wide class of non-Gaussian and nonlinear time series can be represented by time variable (TVP) and state dependent (SDP) parameter, nonlinear, stochastic models with *Gaussian* inputs. When it is possible (and the methods do seem quite widely applicable), this is clearly advantageous, since it allows for the use of well tried and robust algorithms that are computationally much less demanding than even the 'classical' non-Gaussian methods (see Young, 1999c; Durbin and Koopman, 1999).

As far as the author is aware, the idea of SDP modelling originated in his 1978 paper on the modelling of badly defined dynamic systems (Young, 1978), and was then taken up by Priestley in a series of papers and a book on the subject (Priestley, 1988). These earlier publications do not, however, exploit the power of recursive fixed interval smoothing (FIS), which provides the main engine for the developments described in this chapter. The combined Kalman filter (KF) and FIS (or FISIV: see section 2) algorithms, as used here, clearly have a more powerful potential than their more conventional usage would suggest. Moreover, the fact that these same algorithms can function with system matrices characterized by time, or even state dependent parameters, extends their range of applicability to a considerable extent. Thus, the non-parametric models described in this chapter can provide the basis for rather novel non-parametric or state dependent parameter KF-FIS design, with implications for both modelling and nonlinear optimization based on prediction error decomposition.

The fact that the FIS algorithm can function well as a non-parametric estimator means that it provides a powerful, recursive alternative to other, more conventional, methods of smoothing, such as regularization, smoothing splines, kernel smoothers and locally weighted kernel regression (see Young and Pedregal, 1998). It also provides a non-parametric method for transforming random variables (e.g. Gaussian to non-Gaussian), or identifying the nature of a parametric transform between random variables.

Finally, it is clear that the simulated and real examples presented in the chapter, combined with those discussed in other cited references, demonstrate the efficacy of the proposed SDP approach to modelling for a fairly wide and practically useful class of nonlinear stochastic systems. However, the proposed technique is relatively new and it raises a variety of interesting theoretical questions and possibilities for extending the approach to an even richer class of nonlinear stochastic systems. For example:

- How can the approach be extended to handle multivariable state dependencies, where the SDPs may be functions of several state variables?

- What is the best method of handling the *errors-in-variables* problem and the estimation bias that occurs when the proposed SDP modelling approach is applied to *errors-in-variables* TF models? An Instrumental Variable (IV) method, such as that used successfully in the case of DTF models (section 2), has been devised to handle this problem in the case of 'well-behaved' nonlinear models. But alternative approaches will be required in the case of sensitive chaotic models.

- Although no convergence problems have been encountered so far in the evaluation of the proposed SDP estimation procedure, what conditions are required for convergence of the backfitting procedure? Hastie and Tibshirani (1996) use a similar backfitting procedure for estimation of their Generalized Additive Model (GAM). It needs to be established whether their conclusions as regards convergence (which are not entirely persuasive, in any case) are applicable to the models and backfitting procedure described in this chapter. Unlike the GAM, for instance, the nonlinear functions in the SDP models are factorized into the product of the SDP and the model variable; and the SDP is estimated by optimal FIS smoothing (rather than the more conventional scatter-plot smoothing used by Hastie and Tibshirani).

- The backfitting procedure does not provide complete covariance information on the SDP estimates. Could this be distorting the standard errors on the estimates (e.g. the standard errors in Figure 5 seem very small)? In more general terms, what are the full *theoretical* statistical properties of the SDP estimates obtained by backfitting?

- Finally, what are the identifiability conditions on the SDP models? It is clear that problems analogous to collinearity in constant parameter model estimation can occur and that backfitting convergence will be affected by such problems. Also, in the case of input-output models, the nature of the input signals will affect the identifiability of the model parameters. It is necessary to explore these factors further and establish what other factors may affect the indentifiability of the model.

Regardless of the answers to these questions, however, the SDP approach to the identification of nonlinearities in stochastic systems appears to hold great promise. In contrast to other approaches, such as neural networks and NARMAX models, for example, it attempts to identify the type of nonlinearity and, therefore, the form of the nonlinear model, prior to the estimation of the parameters in the finally identified model. This helps to ensure that the final nonlinear model is efficiently parameterized (parsimonious) and it should avoid the over-parameterization that normally accompanies neural network and, to a lesser extent, the 'black-box' NARMAX models. Indeed, the SDP approach has been developed as a primary tool in *Data-Based Mechanistic* (DBM) modelling (e.g. Young, 1993a,b, 1998a,b, 1999b; Young and Beven, 1994; Young and Pedregal, 1997; and the prior references therein), where its ability to obtain parametrically efficient and physically meaningful models is essential.

SDP estimation also provides a non-parametric model that can be useful in its own right. As we have seen, the SDP model can be simulated easily in programs such as Simulink, thus removing the need for the final parametric estimation in some applications, such as simulation, forecasting and automatic control. In the latter case, for instance, it is clearly possible to develop state estimation and control system design methods based on this new class of nonlinear models, in parametric or non-parametric form (the latter providing a completely new way of considering control and estimation system design). Research on such developments is continuing and has so far led to encouraging initial results.

Acknowledgments

Part of this research was supported by the EPSRC under grant GR/K77884, and the BBSRC under grant EO6813. The author is also grateful to the EPSRC and the Newton Institute for Mathematical Sciences in Cambridge for support during his stay at the Newton Institute, where most of this chapter was written.

References

Banks, R.B. (1994) *Growth and Diffusion Phenomena*, Springer.

Box, G.E.P., and Jenkins, G.M. (1970) *Time Series Analysis: Forecasting and Control*, (revised edition 1976), Holden-Day.

Brown, R.L., Durbin, J. and Evans, J.M. (1975) 'Techniques for testing the constancy of regression relationships over time', *Jnl. Royal Stat. Soc., Series B* **37** 141–192.

Bryson, A.E. and Ho, Y.C. (1969) *Applied Optimal Control*, Blaisdell.

Co, T. (1996) 'Parameter estimation of nonlinear systems using modulating functions methods', in *Identification in Engineering Systems*, M.I. Friswell and J.E. Mottershead (eds.), University of Wales, Swansea, 87–96.

Daubechies, I. (1988) 'Orthonormal bases of compactly supported wavelets', *Commun. Pure and Appl. Math.* **41** 906–966.

Dempster, A.P., Laird, N.M. and Rubin, D.B. (1977) 'Maximum likelihood from incomplete data via the EM algorithm', *Jnl. Royal Stat. Soc., Series B* **39** 1–38.

Durbin, J. and Koopman, S.J. (1999) 'Time series analysis of non-Gaussian observations based on state space models from both classical and Bayesian perspectives', *Jnl. Royal Stat. Soc., Series B* **62**, in press.

Gamerman, D. (1997) *Markov Chain Monte Carlo*, Chapman and Hall.

Gauss, K.F. (1821, 1823, 1826) 'Theoria combinationis observationum minimis obnoxiae', Parts 1, 2 and supplement, *Werke* **4** 1–108.

Gurney, W.S.C., Blythe, P.C. and Nisbet, R.M. (1980). 'Nicholson's blowflies revisited', *Nature* **287** 17–21.

Harvey, A.C. (1981) *Time Series Models*, Phillip Allen.

Harvey, A.C. (1984) 'A unified view of statistical forecasting procedures' (with comments), *Jnl. Forecast.* **3** 245–283.

Harvey, A.C. (1989) *Forecasting Structural Time Series Models and the Kalman Filter*, Cambridge University Press.

Harvey, A.C. and Peters, S. (1990) 'Estimation procedures for structural time series models', *Jnl. Forecast.* **9** 173–204.

Hastie, T.J. and Tibshirani, R.J. (1996) *Generalized Additive Models*, Chapman and Hall.

Holst, U., Hössjer, O., Björklund, C., Ragnarsson, P. and Edner, H. (1996) 'Locally weighted least squares kernel regression and statistical evaluation of LIDAR measurements', *Environmetrics* **7** 410–416.

Jakeman, A.J. and Young, P.C. (1979) 'Recursive filtering and the inversion of ill-posed causal problems', CRES Report No. AS/R28/1979, Centre for Resource and Environmental Studies, Australian National University.

Jakeman, A.J. and Young, P.C. (1984) 'Recursive filtering and the inversion of ill-posed causal problems', *Utilitas Math.* **35** 351–376.

Kalman, R.E. (1960) 'A new approach to linear filtering and prediction problems', *ASME Trans. J. Basic Eng.* **82D** 35–45.

Kitagawa, G. (1981) 'A non-stationary time series model and its fitting by a recursive filter', *Jnl. Time Series Anal.* **2** 103–116.

May, R.M. (1973) *Stability and Complexity in Model Ecosystems*, Princeton University Press.

May, R.M. (ed.) (1976) *Theoretical Ecology*, Blackwell Science.

McNeil, D. (1996) *Personal Communication.*

Ng, C.N. and Young, P.C. (1990) 'Recursive estimation and forecasting of nonstationary time-series', *Jnl. Forecast.* **9** 173–204.

Nicholson, A.J. (1954) 'An outline of the dynamics of animal populations', *Austral. Zool. Jnl.* **2** 9–65.

Norton, J.P. (1975) 'Optimal smoothing in the identification of linear time-varying systems', *Proce. Inst.Elec. Eng.* **122** 663–668.

Norton, J.P. (1986) *An Introduction to Identification*, Academic Press.

Pedregal, D.J. and Young, P.C. (1996) 'Modulated cycles, a new approach to modelling seasonal/cyclical behaviour in unobserved component models', Centre for Research on Environmental Systems and Statistics (CRES), Tech. Note No. TR/145.

Pedregal, D.J. and Young, P.C. (1998) 'Extensions of trend models in unobserved component models', Centre for Research on Environmental Systems and Statistics (CRES), Tech. Note No. TR/156.

Pierce, D.A. (1972) 'Least squares estimation in dynamic disturbance time-series models' *Biometrika* **59** 73–78

Plackett, R.L. (1950) 'Some theorems in least squares', *Biometrika* **37** 149–157.

Priestley, M.B. (1988) *Nonlinear and Nonstationary Time Series Analysis*, Academic Press.

Ruanaidh, J.J. and Fitzgerald, W.J. (1996) *Numerical Bayesian Methods Applied to Signal Processing*, Springer.

Schweppe, F. (1965) 'Evaluation of likelihood function for Gaussian signals', *IEEE Trans. Inf. Th.* **11** 61–70.

Young, P.C. (1969) 'Applying parameter estimation to dynamic systems: Part 1, Theory', *Cont. Eng.* **16** (10, 119–125; 'Part ll Applications', **16** (11), 118–124.

Young, P.C. (1970a) 'An instrumental variable method for real-time identification of a noisy process', *Automatica* **6** 271–287.

Young, P.C. (1970b) *Differential equation error method of real-time process identification*, PhD Thesis, Cambridge University.

Young, P.C. (1978) 'A general theory of modeling for badly defined dynamic systems', in *Modeling, Identification and Control in Environmental Systems*, G.C. Vansteenkiste (ed.), North-Holland, 103–135.

Young, P.C. (1983) 'The validity and credibility of models for badly defined systems', in *Uncertainty and Forecasting of Water Quality*, M.B. Beck and G. Van Straten (eds.), Springer, 69–100.

Young, P.C. (1984) *Recursive Estimation and Time-Series Analysis*, Springer.

Young, P.C. (1988) 'Recursive extrapolation, interpolation and smoothing of non-stationary time series', in *Identification and System Parameter Estimation*, C.F. Chen (ed.), Pergamon Press, 33–44.

Young, P.C. (1993a) 'Time variable and state dependent modelling of nonstationary and nonlinear time series', in *Developments in Time Series Analysis*, T. Subba Rao (ed.), Chapman and Hall, 374–413.

Young, P.C. (1993b) *Concise Encyclopedia of Environmental Systems*, Pergamon Press.

Young, P.C. (1994) 'Time-variable parameter and trend estimation in non-stationary economic time series', *Jnl. Forecast.* **13** 179–210.

Young, P.C. (1996) 'A general approach to identification, estimation and control for a class of nonlinear dynamic systems', in *Identification in Engineering Systems*, M.I. Friswell and J.E. Mottershead (eds.), University of Wales, Swansea, 436–445.

Young, P.C. (1998a) 'Data-based mechanistic modelling of environmental, ecological, economic and engineering systems', *Env. Modell. Software* **13** 105–122.

Young, P.C. (1998b) 'Data-based mechanistic modelling of engineering systems', *J. Vibr. Cont.* **4** 5–28.

Young, P.C. (1999a) 'Nonstationary time series analysis and forecasting', *Prog. Env. Sci.* **1** 3–48.

Young, P.C. (1999b) 'Data-based mechanistic modelling , generalised sensitivity and dominant mode analysis', *Comp. Phys. Comm.* **115** 1–17.

Young (1999c) 'Comments on time series analysis of non-Gaussian observations based on state space models from both classical and Bayesian perspectives', *Jnl. Royal Stat. Soc., Series B* **62** 33–34..

Young, P.C. and Beven, K.J. (1994) 'Data-based mechanistic modelling and the rainfall-flow nonlinearity', *Environmetrics* **5** 335–363.

Young, P.C. and Fawcett, C. (1999) 'Data-based mechanistic modelling and the Nicholson blowfly data', Centre for Research on Environmental Systems and Statistics (CRES), Tech. Note No. TR140.

Young, P.C. and Jakeman, A.J. (1979) 'Refined instrumental variable methods of recursive time-series analysis: part I, single input single output systems' *Int. Jnl. Contr.* **29** 1–30.

Young, P.C. and Lees, M.J. (1993) 'The Active Mixing Volume (AMV): a new concept in modelling environmental systems', Chapter 1 in *Statistics for the Environment*, V. Barnett and K.F. Turkman (eds.), Wiley, 3–44 .

Young, P.C. and McKenna, P. (1999) 'An instrumental variable approach to recursive fixed interval smoothing and the estimation of time variable parameter transfer function models', Centre for Research on Environmental Systems and Statistics (CRES), Tech. Note No. TR141.

Young, P.C. and Minchin, P. (1991) 'Environmetric time-series analysis: modelling natural systems from experimental time-series data', *Int. Jnl. Biol. Macromol.* **13** 190–201.

Young, P.C. and Ng, C.N. (1989) 'Variance intervention', *Jnl. Forecast.* **8** 399–416.

Young, P.C. and Pedregal, D.J. (1996) 'Recursive fixed interval smoothing and the evaluation of LIDAR measurements', *Environmetrics* **7** 417–427.

Young, P.C. and Pedregal, D.J. (1997) 'Data-based mechanistic modelling', in *System Dynamics in Economic and Financial Models*, C. Heij, B. Hanzon and K. Paggman (eds.), Wiley, 169–213.

Young, P.C. and Pedregal, D.J. (1998) 'Recursive and en-bloc approaches to signal extraction', *J. Appl. Stat.* **26** 103–128.

Young, P.C. and Pedregal, D.J. (1999) 'Macro-economic relativity: government spending, private investment and unemployment in the USA 1948–1998', *Jnl. Struct. Change Econ. Dyn.* **10** 359–380.

Young, P.C. and Runkle, D.E. (1989) 'Recursive estimation and modelling of nonstationary and nonlinear time series', *Adaptive Systems in Control and Signal Processing* **1**, IFAC/Inst. Measurement and Control, London, 49–64.

Young, P.C., Ng, C.N. and Armitage, P. (1989) 'A systems approach to economic forecasting and seasonal adjustment', *Int. J. Comp. Math. Applic.* **18** 481–501.

Young, P., Parkinson, S. and Lees, M. (1996) 'Simplicity out of complexity: Occam's Razor revisited', *J. Appl. Stat.* **23** 165–210.

Young, P.C., Pedregal, D.J. and Tych, W. (1998) 'Dynamic harmonic regression', Centre for Research on Environmental Systems and Statistics (CRES), Tech. Note No. TR96; *Jnl. Forecastg* **18** 369–394.

Young, P.C., Lane, K., Ng, C.N. and Parker, D. (1991) 'Recursive forecasting, smoothing and seasonal adjustment of non-stationary environmental data', *Jnl. Forecast.* **10** 57–89.

Zhan-Qian, J.L. and Smith, R.L. (1998) 'Estimating local Lyapunov exponents by local polynomial regression', University of North Carolina, Department of Statistics, http://www.stat.unc.edu./postcript/rs/chaos.

The Use of Generalised Likelihood Measures for Uncertainty Estimation in High-Order Models of Environmental Systems

Keith Beven, Jim Freer, Barry Hankin and Karsten Schulz

Introduction

The available mechanistic models for most environmental systems often do not give very good predictions in comparisons with measurements. There are many good reasons for this, including lack of knowledge of the controlling processes, lack of appropriate descriptive theories at the scales at which predictions are required, errors in initial and boundary conditions, errors in input forcing data and calibration data, the difficulties of measuring the complex characteristics of the system of interest at appropriate scales, and the limitations of numerical algorithms and computational power. One result is that in many areas of environmental modelling there are no commonly agreed mechanistic modelling strategies but rather many competing models. Thus, although environmental systems are complex and cannot or cannot easily be controlled for experimental purposes, the problem is not usually that there is a lack of available models since, even in data scarce situations, models are still commonly constructed for quantitative prediction. The problem is rather that there are too many models available, all of which work (at least if some calibration or flux adjustments are deemed to be allowable) and all of which are known to be wrong. As modellers we tend not to talk about that too much, but see the analysis of Morton (1993).

Of course, as both knowledge and modelling in different fields develop and become more sophisticated, and as each generation of computers becomes more powerful allowing finer model solution grids and more extensive sensitivity analysis, it would be expected that the variations between modelling structures of a particular system might be reduced. Indeed, in some fields, such as numerical weather forecasting, this seems to have been the case. However, another effect also takes place that can offset this effect.

Common to most mechanistic modelling studies of environmental systems is the desire to incorporate more understanding into the model structures used, as knowledge grows based on experiment, field study and analysis. The result is generally to make models more complex, with two effects. One is that they demand more computer time, but let us assume that this is not a constraint in what follows. More importantly, incorporating understanding is normally concomitant with increasing the number of parameters in the model. These parameters may be required to represent the physical and chemical characteristics of the system under study and may have excellent physical and chemical justification, but in many systems there may be no way of measuring or specifying values of those parameters with any precision at the space or time scales required in the model. The measurement techniques to obtain such values often do not exist, or even if they exist, are too time consuming or expensive to allow sufficient measurements of the values required. There are many examples of such parameters, including soil hydraulic conductivity; canopy resistance and active rooting density of a vegetation cover; or, in geochemistry, effective surface areas and field reaction coefficients. These will then add to the number of parameters requiring calibration, but will also (generally) add to the degrees of freedom of the model in fitting the calibration data (as expressed in terms of some performance measure or objective function).

Available calibration data tends to be limited and will, in itself, be subject to errors of measurement and also perhaps scale problems relative to the scale at which the model predictions are being made. Very often, therefore, it will be found that mechanistic models of environmental systems are overparameterised with respect to the data available for calibration. This problem has long been recognised in systems analysis where a variety of criteria have been developed to try to identify the point of complexity or model order at which models become overparameterised (see for example, Young, 1984). Environmental modellers also recognise the problem of overparameterisation, but do not often appear to also recognise that the process of increasing the complexity of models by incorporating more understanding (and parameters) will tend to make the problem worse rather than better, especially where simple models can fit the data equally well (see for example Parkinson and Young, 1998). There are, of course, dangers that fitting simpler models to a data series may not result in the correct dynamics for prediction under changed conditions (Beck, 1987) but if more complex models are indeed overparameterised then there can be no guarantee that the uncertainties in prediction might be any less.

Paradoxically, therefore, incorporating more mechanistic understanding into models might lead to more uncertain model predictions if model parameters cannot be easily estimated. Only if incorporating understanding into more complex models leads to parameters that can be estimated *a priori* or by simple measurements, would there appear to be an easy resolution of the paradox. This frequently stated aim has been a driving force behind much of the development of 'physically based' and geochemical models of environmental systems. Unfortunately, the aim has rarely been achieved, primarily because the measurement techniques that are available do not commonly provide parameter values at the scales required in the model, even if that model uses a fine spatial discretisation. The parameters of the model may be given the same name as the quantities that can be measured but there is generally no theory that allows the estimation of the effective values within different parts of a heterogeneous flow domain from a limited number of small scale or laboratory measurements. The quantities at the different scales may effectively be incommensurable (see discussions in Beven, 1995, 1996).

The problem of incommensurability or scale dependence of effective parameters also implies that the appropriate descriptive equations may not be the same at different scales due to different sub-grid scale averaging requirements in heterogeneous domains. Few environmental models attempt to take such considerations into account; but rather they directly apply the theory developed at the experimental scale to the larger scales at which predictions are required. This is generally true in process modelling in hydrological, hydraulic, and geochemical modelling (for a discussion of hydrological examples see Beven, 1989, 1996). There will then be an inevitable effect on the values of parameters derived by calibration which, even if they have names that reflect their origin as measurable quantities at a certain scale, cannot be considered as independent of the model structure and scale of discretisation for which they were calibrated.

Thus, for many environmental systems, it may be necessary to accept that, even if a model appears to have a rigorous theoretical base, scale and heterogeneity considerations may force some form of calibration of parameter values. Traditionally, this has required the definition of an appropriate objective function, followed by the optimisation of that objective function to find the parameter set that best fits the observations available. From the 1960s onwards, as digital computers became more widely available and increasingly powerful, a variety of optimisation algorithms have been developed, including gradient based algorithms, non-gradient based algorithms, algorithms with random search components etc. (see for example Press *et al.* 1992). Problems in optimisation, such as finding the gradients of the parameter response surface for complex models, multiple local optima in the response surface, insensitive parameter values leading to 'flat' areas in the (hyperdimensional) response surface, were recognised early on, and one of the aims in developing optimisation techniques has been to find an algorithm that is robust to such problems (e.g. Duan *et al.*, 1992). The idea of the optimum parameter set has, however, persisted throughout. In what follows we will question the applicability of the optimum parameter set in environmental modelling and propose that it be replaced by a concept of equifinality of acceptable model structures and parameter sets within model structures. This name has been chosen deliberately (rather than indeterminacy, non-uniqueness or model uncertainty) to focus attention more on the multiple possibilities for producing simulations that are, in some sense, acceptable rather than on the problem of identification of an optimum.

The Problem of Model Calibration

Model calibration has been the saving grace of most mechanistic environmental modelling. It allows a demonstration of success in modelling capability and it allows some degree of faith in model predictions. There is a large literature on methods and problems of environmental model calibration including approaches based on classical likelihood estimation and Bayesian concepts (see for hydrological examples, Bates and Townley, 1988; Duan *et al.*, 1992; Gan and Biftu, 1996; McLaughlin and Townley, 1996; and Kuczera, 1997). Such approaches do, however, make certain assumptions about the nature of the modelling errors, which then controls the shape of the likelihood surface around the optimum parameter set. In principle, these assumptions can be checked against the calculated model residual series and likelihood surface but more often these are treated as implicit assumptions of the calibration procedure.

Recent work in statistics has started to recognise that likelihood surfaces in the parameter space may not take ideal forms, particularly for higher dimensional spaces. Techniques have been developed to identify the likelihood surface efficiently, including Gibbs sampling, the Metropolis algorithm and, more generally, Monte Carlo Markov Chain techniques (e.g. Gelfand and Smith, 1990; Sen and Stoffa, 1995). In all these cases an initial kernel shape is gradually refined by iterative sampling of the surface until, asymptotically, a stable representation of the surface is reached. It is still the case, however, that an initial estimate of the form of the kernel is expected to be known from the nature of the estimation problem as defined by the error model and likelihood function.

Other, more direct techniques for parameter calibration have also been continuously improved. Techniques such as simulated annealing (Tarantola, 1987; Sen and Stoffa, 1995) and the shuffled complex algorithm (Duan *et al.*, 1992; Sorooshian and Gupta, 1993; and Gan and Biftu, 1996) have taken advantage of the greater computing power available now to make more extensive searches of the parameter space and try to ensure that some global optimum parameter set is found rather than one of perhaps many local optima.

As noted above, common to all these techniques is the idea that there should be an optimum parameter set. This has been central to nearly all approaches to model calibration, with the exception of some Monte Carlo set theoretic approaches (e.g. Spear and Hornberger, 1980; Hornberger and Spear, 1981; Keesman and van Straten, 1990; van Straten and Keesman, 1991; Spear *et al.*, 1994) where some set of models that are acceptable simulators of the system of interest is identified, an approach that may also be generalised to make use of fuzzy measures of acceptability (e.g. Franks *et al.*, 1997; Aronica *et al.*, 1998). Our experience suggests that the concept of an optimum parameter set is severely compromised in most mechanistic models of environmental systems. Most such models are sufficiently complex that there may be many different sets of parameter values within a model structure that may be compatible with the data available for calibration (see, for example, Figure 1). There may, indeed, be many different model structures that may be compatible with the data (see also Draper, 1995, within a more traditional statistical framework). Certainly, one of those models/parameter sets will be the 'optimum' according to some measure of goodness of fit, but that optimum may not survive application to a different data set or different measure of goodness of fit. Parameter sets that give almost equally good fits may also be scattered throughout the parameter space. The implication is that these modelling problems, while of great practical import, are not well posed (and may even be pathological) in terms of the available statistical techniques.

It is perhaps worth noting that the set of acceptable models is not always necessarily large. In one of the earliest examples of Monte Carlo evaluation of a geophysical model, Press (1968, also discussed in Tarantola, 1987) reported finding only 6 acceptable models out of 5 million random models tested. The 'worst' case (their words!) reported by Spear *et al.* (1994) was 20 acceptable models out of 2.6 million random trials, but they report a general problem of finding more than 5% of all simulations as acceptable. That has not been our experience with hydrological models. There are generally many models that are not significantly different from the parameter set with the optimal goodness of fit for a particular calibration data set (as in Figure 1). Clearly the criterion of acceptability is important here. If it is relaxed, many more models become 'acceptable'; when it is too tight there may be no models left that can be considered acceptable. The problems of defining an acceptable model and finding a suitable compromise measure of acceptability will be considered in more detail below.

Living with equifinality: the GLUE methodology

What if the concept of the correct or optimal model is rejected in favour of an expectation of equifinality of models? There is then an implication that predictions should be made using at least a sample of the set of acceptable models. Each model will result in a different set of predictions and a range of uncertainty in model predictions will necessarily follow. In assessing that uncertainty it seems reasonable that those predictions should be weighted according to how well a given model has performed in the past. Those that have performed well should be given the greatest weight in prediction, those that have performed badly enough to be considered non-behavioural should be rejected and given zero weight. This simple idea is the basis of the Generalised Likelihood Uncertainty Estimation (GLUE) methodology first outlined by Beven and Binley (1992) and applied to a wide range of different environmental models since (see Table 1).

In the GLUE methodology a large sample of model realisations is generated by Monte Carlo simulation. This is normally a sample of randomly chosen independent parameter sets within a single model structure, but the concept is easily extended to the evaluation of multiple model structures. At this stage, any prior expectations about the parameter sets can be taken into account by associating each realisation with a prior likelihood weight. Each simulation is then run and evaluated with respect to any observed variables or other information that may be available. This evaluation is quantified in terms of a likelihood measure chosen to reflect the performance of individual models in reproducing the behaviour of the system under st udy. The values of the likelihood measure should increase with increasing levels of performance. Application of Bayes' equation then allows a posterior likelihood to be calculated for each model realisation from

$$L[M(\underline{\Theta}_i)] = \frac{L_0[M(\underline{\Theta}_i)]L_T[M(\underline{\Theta}_i \mid \underline{Y}]}{C} \tag{1}$$

where $L[M(\underline{\Theta}_i)]$ is a posterior likelihood for the ith model $M(\underline{\Theta}_i)$ with parameter set $\underline{\Theta}_i$, $L_0[M(\underline{\Theta}_i)]$ is the prior likelihood of the ith model, $L_T[M(\underline{\Theta}_i)]$ is the likelihood measure calculated for a period of T time steps with dataset $\underline{Y}$, and C is a scaling constant in general equal to $\sum_{i=1}^{i=N} L[M(\underline{\Theta}_i)]$ over N parameter sets or model realisations, such that the cumulative posterior likelihood is unity. Essentially, the N realisations are being used here to sample the response surface of the scaled likelihood measure.

The application of Bayes' equation in this form requires a certain orthogonality of the samples which is why the parameter sets should be chosen to be independent samples from the parameter space. That sample is used to define the response surface of the chosen likelihood measure within the parameter space. To get a good definition in a parameter space of high dimensions will therefore require a large number of samples (and consequently a large amount of computer time). Simple random sampling may not be the most efficient way of defining the response surface where that surface has a simple form. However, where it has a complex form, the simplicity inherent in this methodology may be considered an advantage. The same equation may be applied to multiple model structures provided that the likelihood measure used in the evaluation of each model can be identical. In all cases, the model (structure and parameter set) is treated as an entity. While it may be desired to extract sensitivity information on individual models or parameters by calculating appropriate marginal distributions of the likelihood measure (see below), it is not necessary to do so. It is the model as a whole that gives a good or bad performance in simulating the observations and marginal distributions on individual components may not always be very informative in applications where there is a variety of models that give a good fit to the available data.

Given the posterior likelihood values, the predictions from each realisation can then be weighted by the associated likelihood value to calculate prediction quantiles as

$$P\left(\widehat{Z}_t < z\right) = \sum_{i=1}^{i=n} L\left[M(\underline{\Theta}_i) \mid \widehat{Z}_{t,i} < z\right] \qquad (2)$$

where $\widehat{Z}_{t,i}$ is the value of variable $\overset{.}{Z}$ at time t simulated by model $M(\underline{\Theta}_i)$. Within this framework, accuracy in estimating such prediction quantiles will depend on having an adequate sample of models to represent the behavioural part of the model space.

Choosing a likelihood measure

In defining an appropriate likelihood measure, it is well known that simple least squares methods of parameter estimation can lead to biased optimal estimates of parameter values in cases where there is significant autocorrelation in the residuals. This is commonly the case in applications of environmental models for both spatial and temporal predictions. A suitable likelihood measure can be defined for cases where the structure of the error series can be specified, such that the final residual series is iid. This is the approach commonly taken in maximum likelihood optimisation. For example, assuming an error series for a single simulated variable that is Gaussian, of stationary variance and autocorrelated in time, a likelihood function is defined as

$$L_T[M(|\underline{\Theta}_i \mid \underline{Y})] = \left(2\pi\sigma_i^2\right)^{-T/2}\left(1 - \alpha_i^2\right)^{1/2}\exp\left\{-\frac{1}{2\sigma_1^2}\Psi(\mu_i, \sigma_i, \alpha_i)\right\} \qquad (3)$$

where

$$
\begin{aligned}
\Psi(\mu_i, \sigma_1, \alpha_i) = \ & (1 - \alpha_i^2)(\hat{y}_1 - y_1 - \mu_i)^2 \\
& + \sum[(\hat{y}_t - y_t - \mu_i) \\
& - \alpha_i(\hat{y}_{t-1} - y_{t-1} - \mu_i)]^2
\end{aligned}
$$

and where $\hat{y}_t$ is the simulated value of variable y at time t, y_t is the observed value, μ_i is the mean of the error series for the ith model, σ_i^2 is the variance of the error series and α_i is the lag 1 autocorrelation coefficient of the error series (see for example the application of Romanowicz *et al.*, 1994). The expression may be extended to higher-order correlations as necessary. However, even where the autocorrelation of the error series is high, for a large number of time steps, the dominant term in (3) is normally $(2\pi\sigma_{2i}^2)^{-T/2}$. The large power on this term will often make the likelihood response surface very peaked around the optimum parameter set. This is clearly an advantage if it is the intention to find an optimum parameter set but if there are many other models in the model space that perform almost equally well but which are some distance from the optimum model, the effect may be to limit the apparent predictive uncertainty too greatly, with the result that the observed data frequently falls outside the estimated prediction bounds.

In the current context of multiple behavioural models, it may not always be possible to assume that the error series from each model will have the same structure. There may also be the problem that if an error model is added to a mechanistic model in this way, the error model could compensate for deficiencies (such as a consistent bias) in the mechanistic model. This is true also, of course, in the case of maximum likelihood optimisation, though in that case the possibility of variations in such a compensation as one moves away from the optimum parameter set is not normally considered. Even if this might lead to improved predictions overall, it is effectively adding an additional stochastic model component, with additional parameter values and degrees of freedom in fitting the data. This is not always what is required of the conditioning process, where the aim is frequently to learn about the performance of the mechanistic model itself.

One possibility that has been used in a number of studies is to use a power function of the error variance with shaping parameter N, where N is ideally chosen to reflect the effective information content of the observations (Box and Tiao, 1973). Thus, the likelihood measure is defined as

$$L_T\left[M(\underline{\Theta}_i)\mid \underline{Y}\right] = \frac{\left(\sigma_i^2\right)^{-N}}{C} \tag{4}$$

where C is a scaling constant. In practice the value of N may be chosen empirically to control how far the model prediction uncertainty brackets the observations. Studies using this type of likelihood measure in rainfall-runoff modelling applications (see Table 1) have shown that, even with time series with large numbers of time steps, the appropriate value of N is not high, reflecting the lack of constraint provided by measurements in the calibration of many environmental models. Another form of performance measure based on the error variance, commonly used in hydrological studies, is the proportion of the observed variance explained by the model (sometimes called the efficiency measure)

$$L_T\left[M(\underline{\Theta}_i)\mid \underline{Y}\right] = \frac{\left(1 - \frac{\sigma_i^2}{\sigma_0^2}\right)^N}{C} \tag{5}$$

where σ_0^2 is the variance of the observed data and C is again a scaling constant. Again a shaping parameter has been added in the form of a power N (see Table 3 below). For N equal to 1, (5) is analogous to a coefficient of determination. Clearly equation (5) gives a different likelihood scaling for any given σ_i^2 than (4).

Where a likelihood function based on assumptions about the structure of the error series can be defined in a way that leads to iid residuals for all models, then the likelihood values for sample models would be a direct estimate of a probability of that (joint mechanistic + stochastic) model simulating the available observations. For any other likelihood measure, treating the measure as an estimate of the probability of that model as a simulator of the observed data will not be theoretically rigorous. Indeed, since such conditions will rarely be satisfied it might be better to treat the likelihood measure as a subjective measure of the probability of a model being an acceptable representation of the system. Prior estimates of the probabilities associated with any model are, before any conditioning, subjective in any case. There would appear to be no reason why subjective likelihood measures should not be precluded from use in the conditioning process in cases where the theoretical rigour of a truly objective likelihood function may be difficult to achieve for all behavioural models.

Combining multiple model evaluations

There will often be more than one possibility of model evaluation, in terms of both quantitative and qualitative performance. These may range from the calculation of different likelihood measures based on different simulated variables to rejection criteria based on whether the way in which the model is simulating a certain process properly reflects some qualitative understanding of the system responses. Each criterion or additional information should allow some refinement of the likelihood distribution associated with each model, and in particular the rejection of some models that had previously been considered as behavioural.

The addition of different types of information in this context may be handled by repeated use of Bayes equation in the form of (1) above. At each stage, a posterior likelihood is calculated conditioned on the additional data. The multiplicative nature of (1) will mean that if at any stage a model is given a likelihood of zero, it will be rejected from further consideration as non-behavioural. In cases where a model cannot predict adequately all available observables, this may lead to the rejection of all the model realisations (see for example the application of Zak and Beven, 1999, and discussion of next section).

The multiplicative nature of (1) also means that the order of application of (1) might be important. This can be seen when the additional information is derived from conditioning based on observations of the same variable but for different time periods. If a likelihood measure is calculated as a function of the inverse error variance for each period then the repeated application of (1) will produce different posterior likelihood values than using the inverse error variance of the whole period treated as a single series. This may, of course, be a desirable property in cases where there is an expectation that the system (and therefore the parameter values) may be changing over time. It may be avoided by using an alternative likelihood measure, such as a negative exponential function of the error variance, such that repeated applications of

(1) will be equivalent for multiple (equal) subdivisions of the whole period, i.e.

$$L_T\left[M(\underline{\Theta}_i) \mid \underline{Y}_1, \underline{Y}_2, \underline{Y}_3, \ldots\right] = \frac{\exp(-\sigma_1^2)\exp(-\sigma_2^2)\exp(-\sigma_3^2)\ldots}{C}$$
$$= \frac{\exp(-\sigma_1^2 - \sigma_2^2 - \sigma_3^2 - \cdots)}{C} \tag{6}$$

In the most general case, defining the error itself may be difficult where the observations are imprecise or fuzzy in nature. In this case, it may be more appropriate to use a fuzzy measure in the model evaluation or conditioning. The values of the fuzzy measure can still be used to weight the predictions of a sample model to create prediction quantiles, as in equation (2). They essentially therefore serve in the same way as the likelihood measures within the GLUE methodology. The use of fuzzy measures, however, also expands the ways in which measures might be combined by utilising operators from fuzzy set theory. Given m different fuzzy measures of performance, $L_1, L_2, \ldots, L_m$, the two simplest ways of combining two fuzzy measures are by fuzzy union and fuzzy intersection as

$$L_T\left[M(\underline{\Theta}_i) \mid L_1, L_2, \ldots\right] = L_1 \cup L_2 \cup \cdots \cup L_m = \max[L_1, L_2, \ldots, L_m]$$
$$\text{and} \tag{7}$$
$$L_T\left[M(\underline{\Theta}_i) \mid L_1, L_2, \ldots\right] = L_1 \cap L_2 \cap \cdots \cap L_m = \min[L_1, L_2, \ldots, L_m]$$

Other combinations are also possible (see for example Klir and Folger, 1988, and the application of Aronica *et al.*, 1998).

There are thus a wide variety of ways in which to approach the evaluation and conditioning of models within the GLUE framework. The choice of a measure will be generally a subjective, but argued and reasonable for the model purpose, choice. The resulting prediction quantiles will therefore also be dependent on this choice. The wide variety of applications summarised in Table 1 suggests that it is possible to obtain useful prediction quantiles from this methodology.

Are many mechanistic environmental models simply poor models?

It has been noted above that mechanistic environmental models may often fail to predict one or more observables adequately. Within a Bayesian GLUE framework, using a multiplicative combined likelihood measure, then failure to predict any single observable will lead to a combined likelihood value of zero. Consistent failure to predict one or more observables could lead to the rejection of all the model parameter set trials. An example is given in Zak and Beven (1999). We have encountered other cases of this in a variety of applications that have not been published since rejection of all models implies that the mechanistic model is poor and has no predictive capability.

Table 1. A summary of applications of the GLUE methodology

Reference	Application	Type of Likelihood Measure
Binley & Beven, 1991	Rainfall-Runoff Modelling	Eqn.(4), $N = 1$
Beven & Binley, 1992	Rainfall-Runoff Modelling	Eqn.(4), $N = 1$
Beven, 1993	Rainfall-Runoff Modelling	Eqn.(5), $N = 1$
Romanowicz *et al.*, 1994	Rainfall-Runoff Modelling	Eqn.(3)
Buckley *et al.*, 1995	Groundwater Transport	Eqn. (4), $N = 1$
Freer *et al.*, 1996	Rainfall-Runoff Modelling	Eqn.(5), $N = 1, 30$, Eqn.(6), Sum abs. errors
Romanowicz *et al.*, 1996	Flood Inundation Modelling	Eqn.(3)
Franks *et al.*, 1997a	Soil-Vegetation-Atmosphere Modelling	Eqn.(5), $N = 1$
Franks & Beven, 1997	Soil-Vegetation-Atmosphere Modelling	Eqn.(4), $N = 1, 2.5, 5$
Piñol *et al.*, 1997	Rainfall-Runoff Modelling	Eqn.(5), $N = 1$
Zak *et al.*, 1997	Critical Loads Modelling	Average fuzzy measure
Aronica *et al.*, 1998	Flood Inundation Modelling	Fuzzy rules
Franks *et al.*, 1998	Rainfall-Runoff Modelling	Eqn.(5), $N = 1+$ Fuzzy measures
Hankin & Beven, 1998	River Dispersion Modelling	Eqn.(5), $N = 1$
Lamb *et al.*, 1998	Rainfall-Runoff Modelling	Eqn. (6)
Romanowiz & Beven, 1998	Flood Inundation Modelling	Eqn. (3)
Blazkova & Beven, 1999	Flood Frequency Modelling	Fuzzy rules and centroid defuzzification
Cameron *et al.*, 1999	Flood Frequency Modelling	Various
Schulz *et al.*, 1999	Soil Nitrate Modelling	Average fuzzy measure
Zak *et al.*, 1997; Zak & Beven 1999	Critical Loads Modelling	Average fuzzy measure

This is, of course, not necessarily true. A model may predict some variables of interest to a level of uncertainty that is entirely acceptable, but it may not predict all observables adequately. Thus, a model could have utility but would be rejected within this Bayesian framework. The model may be based on the best knowledge that is currently available and if some predictions are required total rejection may not be an acceptable response. In the case of Global Circulation Models, for example, it is well known that local regional predictions of these models may be poor, but predictions of the nature and impacts of possible future climate change are still required. It is expected that these predictions will be improved in the future as computer power increases and the sub-grid scale parameterisations improve. Simple rejection is not acceptable.

A number of implications follow. One is to accept that not all observables can be predicted adequately and to therefore change the criterion for evaluation such that some models at least are retained. Additive (rather than multiplicative) combinations of likelihood measures, for example, will average over any zero values on some criteria and will take account of good performance on others. Fuzzy union will emphasise the best performance of any model over all criteria. This does not obviate the poor prediction of some variables but does recognise that some model predictions may still be useful.

However, the rejection of models on some criteria also suggests that both the model structure and the observed variables should be critically reviewed. The rejection may be due to a failing of the concepts on which the model is based. It might also be due to a mismatch in meaning of the measured and predicted variables, for example as a result of differences in spatial scale of available measurement techniques and the scale of the model predictions (see discussion in Beven, 1996). Variables and parameter values are often compared without taking proper account of such incommensurability. In short, it is clear that for many environmental systems the goal of a truly realistic process description remains a long way off. It may still therefore be necessary to recognise multiple competing descriptions that may yet have some predictive utility.

The GLUE framework provides one way of doing that but does require an explicit definition of what is accepted as a behavioural model when there may be a continuum of responses from good to poor using different sets of parameters or model structures (see, for example, Figure 1). There will not always be a clear distinction between behavioural and non-behavioural models. Defining some threshold in the likelihood measure, below which models will be considered non-behavioural, is one solution (see rainfall-runoff modelling case study below). The use of high values of the shaping parameter N in equation (4) or (5) might avoid the use of an arbitrary threshold by reducing the effective likelihood for poor models, as long as the resulting prediction bounds are not too narrow. An approach based on fuzzy measures provides an alternative way of avoiding the choice of a particular threshold (see the study of Blazkova and Beven, 1999, which uses fuzzy rules to combine multiple performance measures, with a final weight determined by centroid defuzzification).

The specification of prior distributions of parameters

A Bayesian type of analysis allows for the specification of prior distributions for the parameter values. In the GLUE methodology it is the prior likelihood weights for the different parameter sets (and possibly different model structures) that are required. In many areas of mechanistic environmental modelling it may be very difficult to decide on what these prior distributions should be for individual parameters and especially for the covariation of parameter values. In most applications of the GLUE methodology to date an assumption of minimum prior information has been made, coupled to the uniform independent random sampling of the model space. Each parameter set is then given an equal prior likelihood value. This is not a necessary prior assumption and there is no reason in principle why the individual random parameter set samples should not be associated with different prior likelihood values. In practice, there may be very little information on which to base the prior estimates, even in terms of defining the model space to be sampled in terms of the ranges of feasible parameter values. It is commonly found that when comparing simulated and observed model values good model fits are found up to the limits of the specified ranges (see for example, Figure 1). Beyond those sampling limits, however, models have been rejected *a priori*. Such a sudden threshold in likelihood values can be avoided by specifying a prior distribution that falls off more gradually at the edges of the range considered so that the posterior likelihoods will also fall off more gradually (at least in a multiplicative Bayesian combination of likelihoods). However, this perhaps says more about a modeller's desire to avoid such a sharp boundary than about either the available knowledge of effective parameter values or the performance of the model.

Computational efficiency in sampling the model space

The GLUE methodology samples the model space in terms of uniform independent sampling. In high dimensional model spaces a very large number of sample parameter sets may be required in order to characterise adequately the likelihood response surface in the model space. For response surfaces in which the high likelihood parameter sets are found in only a contiguous small region of the model space, this will be a highly inefficient way of characterising the response surface. Many different techniques have been developed to try to improve the efficiency of the sampling in such cases, including Monte Carlo Markov Chain (MCMC) processes of different types. All such methods are aimed at concentrating the sampling in areas of the response surface where the likelihoods are high. MCMC methods aim to make the sampling density proportional to the likelihood values, so that in prediction each sample is effectively given equal weight. The GLUE methodology is different in that uniform independent sampling is used and the likelihood value associated with any particular model is used to weight the predictions, as in (2). Where good (high likelihood) models are found scattered throughout the model space, this remains a useful strategy. It has the advantage of ease of implementation at the expense of increased computational cost. At Lancaster University, the GLUE calculations are generally carried out on a 20 processor parallel Pentium PC system running PVM under Linux, which allows a large number of Monte Carlo runs to be made even for models of some complexity. Efficiency can still be gained by trying to avoid sampling in areas of very low or zero likelihood (for example by using the interpolative resampling procedure of Beven and Binley (1992) or adapting the tree structured random search with locally uniform sampling of Spear *et al.*, (1994)).

Case Studies

In what follows we illustrate the application of the GLUE methodology in three different applications in environmental modelling. The first is a rainfall-runoff modelling study of the Maimai catchment in New Zealand. This study compares a number of possible likelihood measures in model evaluation and prediction uncertainty estimation. The second looks at the problem of modelling latent and sensible heat fluxes from the land surface atmosphere. This is an area where modelling advances are introducing more and more complexity and more and more parameters into process descriptions without any consideration of predictive uncertainty. The land surface parameterisations are essential components of Global Circulation Models of the atmosphere, whose predictions (for example of the impacts of greenhouse gases) have been shown to be sensitive to the land surface components. Consequently, predictions of future change will also be sensitive to uncertainty in the land surface parameterisations. Here three different, relatively simple, model parameterisations are compared. Finally, a third example illustrates the equifinality of different model structures and the model evaluation using fuzzy measures in an application to predicting contaminant dispersion in open channel flows. Reference to other published applications may be found in Table 1.

Case study: rainfall-runoff modelling of the Maimai catchment

The Maimai M8 catchment is located in the Tawhai State Forest, North Westland, South Island, New Zealand. Mean annual gross rainfall in this area is approximately 2600mm, producing some 1550mm of runoff from 1950mm of net rainfall (Rowe, 1979), with rapid responses to rainfall throughout the year. M8 is one of eight small adjoining catchments that have been studied since

1974 as part of a land use change study. Much of this work has focused on understanding the contributions of subsurface stormflow at different scales. The research has generated a number of diverse and interesting data sets with which to assess conceptual hydrological models; however, in this study only the rainfall-runoff data for a 2.3 year period have been considered.

The rainfall-runoff model TOPMODEL was chosen to simulate the rainfall-runoff processes of the Maimai catchment. This model makes a number of simplifying assumptions about the runoff generation processes that are thought to be reasonably valid in this wet, humid temperate catchment with short slopes and shallow soils (see Beven *et al.*, 1995; Beven, 1997). An initial assessment of the form of the master discharge recession curve at Maimai suggested that the preferred formulation of TOPMODEL was the original exponential transmissivity version (Beven and Kirkby, 1979; Ambroise *et al.*, 1996). Previously TOPMODEL has been applied to Maimai using classical optimisation techniques which suggested this form of the model was able to produce good simulations of the observed discharge data. Here the model is applied within the GLUE framework. An assessment is made of the predictive uncertainties resulting from a number of different likelihoood measures and behavioural model thresholds, similar to the work presented by Freer *et al.* (1996) for the Ringelbach research catchment.

There are six parameters to be calibrated in this version of the model (Table 2). Each has a physical interpretation. The model can make predictions that are distributed in space but all the parameters have been assumed to be homogeneous effective values for the catchment as a whole. Uniform sampling strategies over specified ranges have been used except for parameters which are known, through field data, to vary over orders of magnitude (T_0 and K_0, where a uniform sampling strategy for log values was used, see Table 2). From previous experiences with GLUE, where behavioural simulations were found throughout the total range of multiple sampled parameters (i.e. Freer *et al.*, 1996; and other studies in Table 1), the limits were set wider than indicated by the available field data.

As noted previously the choice of the likelihood measure has been shown to affect the calculation of model uncertainty (see Freer *et al.*, 1996). To demonstrate the sensitivity of the GLUE approach to the choice of the likelihood measure using the original form of TOPMODEL six different combinations of likelihood measures and thresholds were assessed and are defined in Table 3.

The model was run for 20,000 Monte Carlo simulations with different parameter sets. Rainfalls and other forcing data were assumed to be known and were kept the same for all realisations. Likelihood weights were calculated separately for each year of discharge data and for each of the likelihood measures shown in Table 3. Bayesian updating of the likelihood weights for each year of record was calculated using equation (1).

Figure 1 presents scatter plots for all six likelihood measures defined in

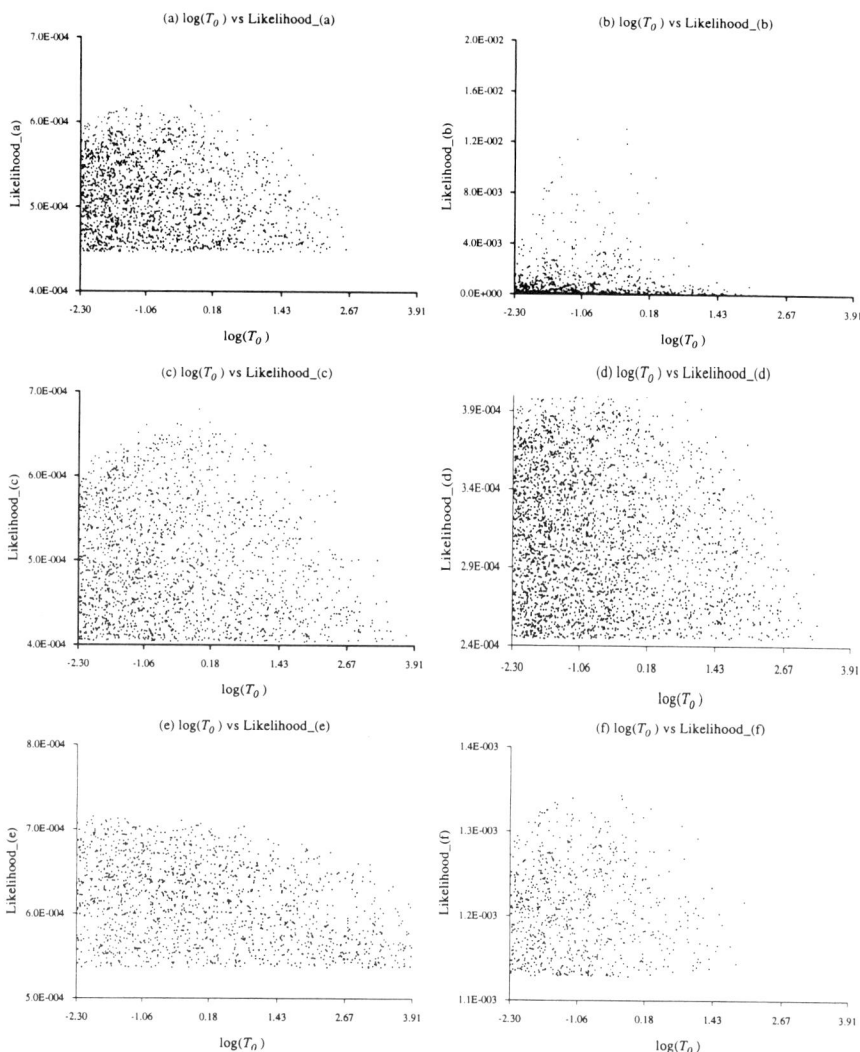

Figure 1. Scatter plot of likelihood values for different definitions of the likelihood measure and rejection criteria from Monte Carlo simulations for Maimai catchment, likelihoods conditioned on the 1985 discharge data (1 hour timesteps).

Table 2. Parameter ranges used in Monte Carlo simulations for Maimai catchment. The parameter f controls the rate of decline of transmissivity with depth into the soil, S_{rmax} is a maximum root zone storage, $\Delta\theta_1$ is the effective storage coefficient for gravity drainage in the soil, K_0 is the hydraulic conductivity at the soil surface, T_0 is the transmissivity of the soil when the profile is saturated to the surface, and P_{mac} is the fraction of the rainfall inputs that goes directly to recharge the water table bypassing the root zone storage.

Parameter	Minimum Value	Maximum Value	Sampling	Mean Field Estimates*
f [m]	2.00	12.00	Uniform	9.425
S_{rmax} [m]	0.01	0.30	Uniform	0.086
$\Delta\theta_1$ [fraction]	0.01	0.35	Uniform	0.070
K_0 [m hr^{-1}]	0.10 **(−2.33)	50.00 (3.91)	Uniform log values	5.026
T_0 [m^2 hr^{-1}]	0.10 **(−2.33)	50.00 (3.91)	Uniform log values	0.833
P_{mac} [fraction]	0.00	0.50	Uniform	0.195

* Estimates are shown for comparison; the details of this analysis have been given in Freer (1998)
** Ranges for T_0 and K_0 shown also in log to relate to the graph scales

Table 3. Definition of the likelihood measures and behavioural model thresholds used in the GLUE procedure for the Maimai catchment. σ^2 is the variance of the modelling residuals, σ_0^2 is the variance of the observations, $\hat{y}_t$ is the predicted variable at time step t, Y_t is the observed variable at time step t, and the summation is over all time steps.

Likelihood Measure	Behavioural Threshold	Equation	Shaping Parameter N	Observed Series
Likelihood_(a)	0.60	Eqn.(5)	1	Data
Likelihood_(b)	0.60	Eqn.(5)	30	Data
Likelihood_(c)	Best 10%	$\propto \sum \left\| \hat{Y}_t - Y_i \right\|^N$	1	Data
Likelihood_(d)	0.50	Eqn.(5)	1	Data
Likelihood_(e)	0.60	Eqn.(5)	1	log(Data)
Likelihood_(f)	0.70	Eqn.(5)	1	Data

Table 3 for the parameter T_0, after conditioning on the first 1985 discharge dataset. These results suggest that the distribution of behavioural parameter sets, in this case for T_0, does not seem to be greatly affected by the choice of likelihood measure, except that for Likelihood_(e), and to some degree Likelihood_(c), the range of behavioural simulations includes parameters sets having higher values of T_0. Scatter plots for the other individual parameters (not shown) suggest that behavioural parameter sets are available throughout most of the ranges specified in Table 2.

Figure 2 examines the effect that the different likelihood measures have on the calculation of 90% uncertainty bounds, determined from equation (2) using the likelihood weights associated with each behavioural parameter set. The changes to the 90% uncertainty bounds using different likelihood measures are comparable with previous studies with GLUE (Freer *et al.*, 1996). A high value of the shaping parameter N causes very narrow prediction bounds to be estimated, similar but less pronounced to having a higher threshold value (Likelihood_(f)). Likelihood_(e), based on the error variance of the log transformed discharges, will take more account of errors in predicting the lower discharges and recession periods (i.e. see small storm on the 11th August). Likelihood_(c), based on the absolute prediction errors, seems to be a compromise in this respect, having recession characteristics which are between those of Likelihood_(e) and the results obtained from error variance based likelihoodmeasures. These results suggest that the (subjective) choice of likelihood measure is not so critical to the GLUE approach as has previously been suggested (e.g. by Melching, 1995 and Clarke, 1994) and also suggests that different likelihood measures can be used to some extent to highlight the model performance in simulating different aspects of the discharge record.

The sensitivity of individual parameters can also be assessed using the GLUE approach by calculating the marginal distributions of the likelihood weights over all the behavioural parameter sets. Figure 3 shows the cumulative posterior Likelihood_(a) weighted distributions for the behavioural parameter sets in comparison with the prior (uniform) distribution. With the exception of the S_{rmax} and P_{mac} parameters, four of the parameters show a marked deviation between the prior and posterior distributions as a result of sensitivity to the conditioning process

It is important within the GLUE procedure to ensure that there is sufficient sampling of the n-dimensional parameter space so that a representative range of behavioural model simulations are obtained. Figure 4 demonstrates the effect of the sample size on the distributions of a variety of model output responses for simulations of the 1987 data set. Each graph has been divided into 20 equal intervals across the range of the associated model output response. To ensure that the evaluation of these distributions was relevant to the efficiency of the Monte Carlo sampling, only the retained behavioural parameter sets for Likelihood_(d) conditioned on all three periods of data were included. The data presented in Figure 4 represent the predictive uncertainty in these model output variables after conditioning on the discharge data (for this likelihood measure). The results show that there is little change to the distributions between 15,000 and 20,000 runs, although ideally more simulations would be

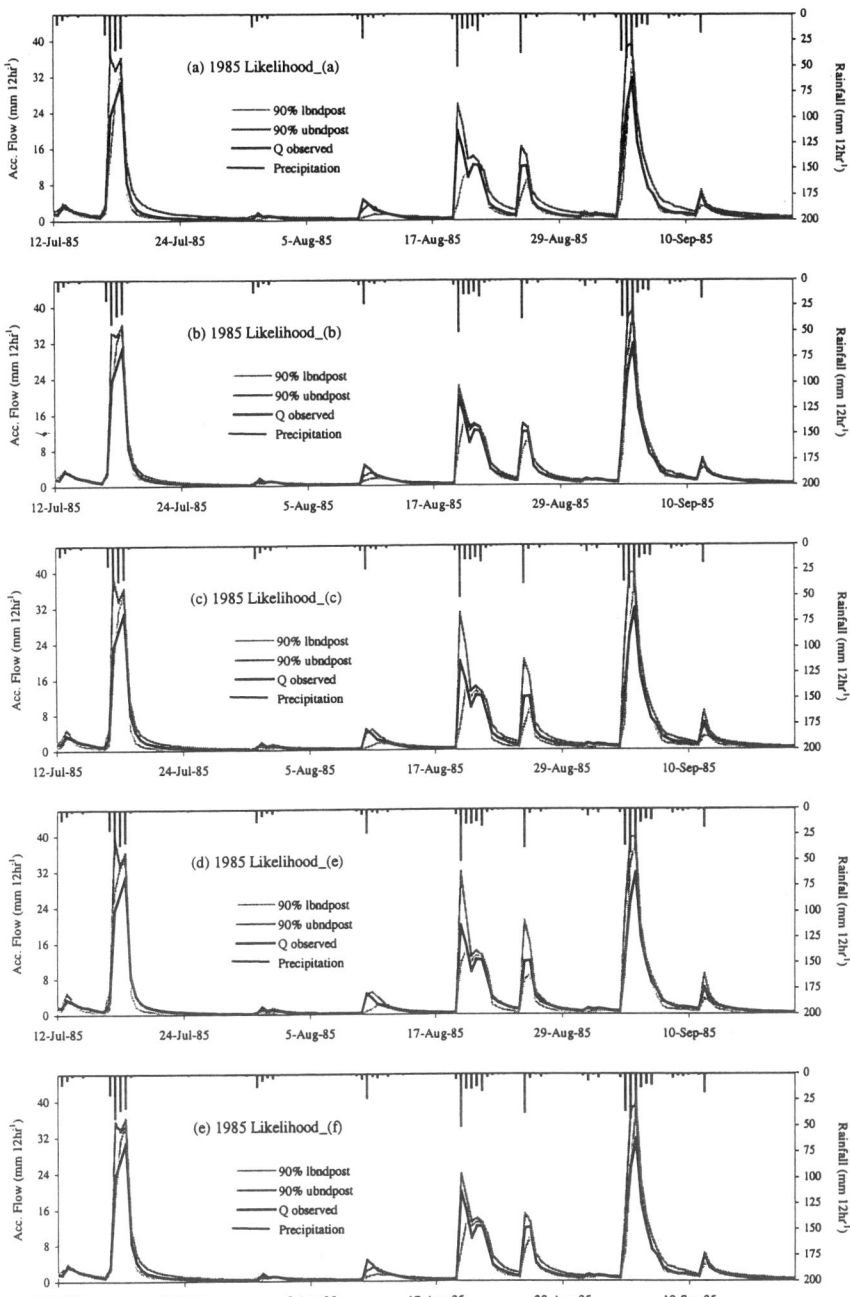

Figure 2. Uncertainty bounds for 1985 discharge predictions for 5 likelihood measures, Maimai catchment (note Likelihood$_{(d)}$ has been left out for clarity).

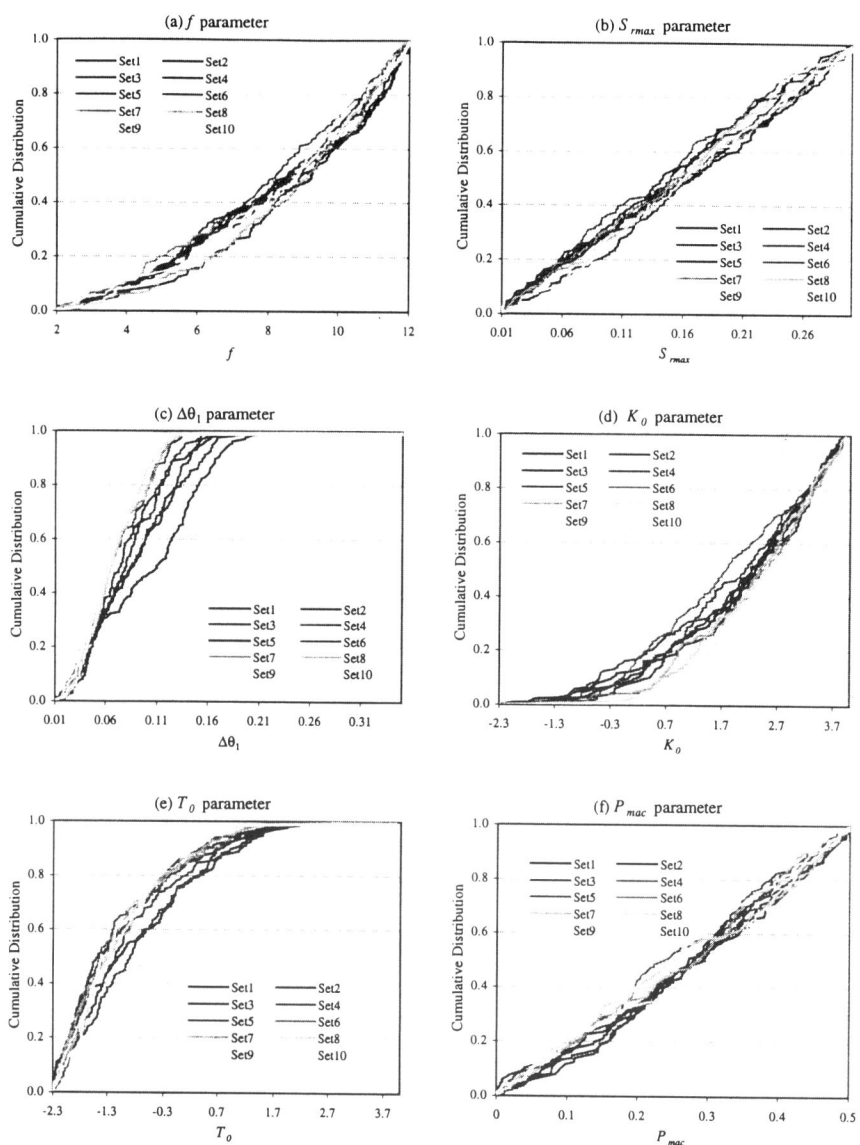

Figure 3. Sensitivity of individual parameters expressed as cumulative distributions of values in 10 equal sets (by number) of the retained behavioural simulations conditioned on the 1985–7 discharge data for Maimai catchment, likelihood measure is equal Likelihood$_{(d)}$ differentiated by the likelihood values (Set 10, highest values; Set 1, lowest values).

needed to ensure stability of these resulting distributions. The number of simulations required will clearly be both model and data dependent. In this case, the efficiency of sampling could be improved to take account of the parameter sensitivities apparent in Figure 3. Figure 4 also reflects the considerable uncertainty regarding the variability in the model characterisation of the dynamics of the runoff response when conditioned on discharge data alone. This is especially apparent for the dynamics of the saturated zone, characterised directly in Figure 4 by the extent of the maximum saturation (Figure 4e) and the minimum depth to water table (Z_{bar}, Figure 4c). It may be unrealistic to classify simulations as behavioural which produce a saturated area that covers 96% of the total catchment area for Maimai. In this case the parameter sets could be conditioned using data other than discharge (i.e. water table information, see for example, Lamb *et al.*, 1998) or by some more qualitative rejection criterion.

Case study: modelling land surface to atmosphere fluxes

The proper description of land surface to atmosphere fluxes plays a crucial role not only for numerical weather forecasting but also in the prediction of climatic changes due to global warming within General Circulation Models (GCM). This importance has led to the development of a large number of different soil-vegetation-atmosphere transfer (SVAT) schemes. They largely vary in complexity and range from simple bucket type models (Manabe, 1969) to vertically very complex multilayer models such as BATS (Dickinson *et al.*, 1986), or SiB2 (Sellers *et al.*, 1996). Recent trends in SVAT modelling have resulted in the incorporation of ever more complex process structures (for example the addition of a carbon component in SiB2) with the underlying rationale that improved process representations will result in parameters that are easier to measure or to estimate on the basis of physical or biological principles.

In this case study, GLUE is used to demonstrate the utility of simpler, rather than more complex, process formulations to describe and predict evaporative fluxes. Three SVAT schemes, differing in complexity, are compared. Two are based upon the Penman–Monteith equation (Monteith, 1981,1995a,b); the third uses a simple evaporation fraction formulation. The models are compared in terms of parameter identification and predictive uncertainty using meteorological forcing data from a variety of different climate and vegetation conditions.

The meteorological data used in this analysis (consisting of net radiation, temperature, humidity, windspeed, rainfall, and latent and sensible heat fluxes as the variables to be predicted) are from three different sites. Two periods, IFC3 (August 6–21, 1987) and IFC4 (October 5–16, 1987), are available from the FIFE site in Kansas, USA, where data from five meteorological stations located above tall grass were averaged according to Famiglietti and Wood

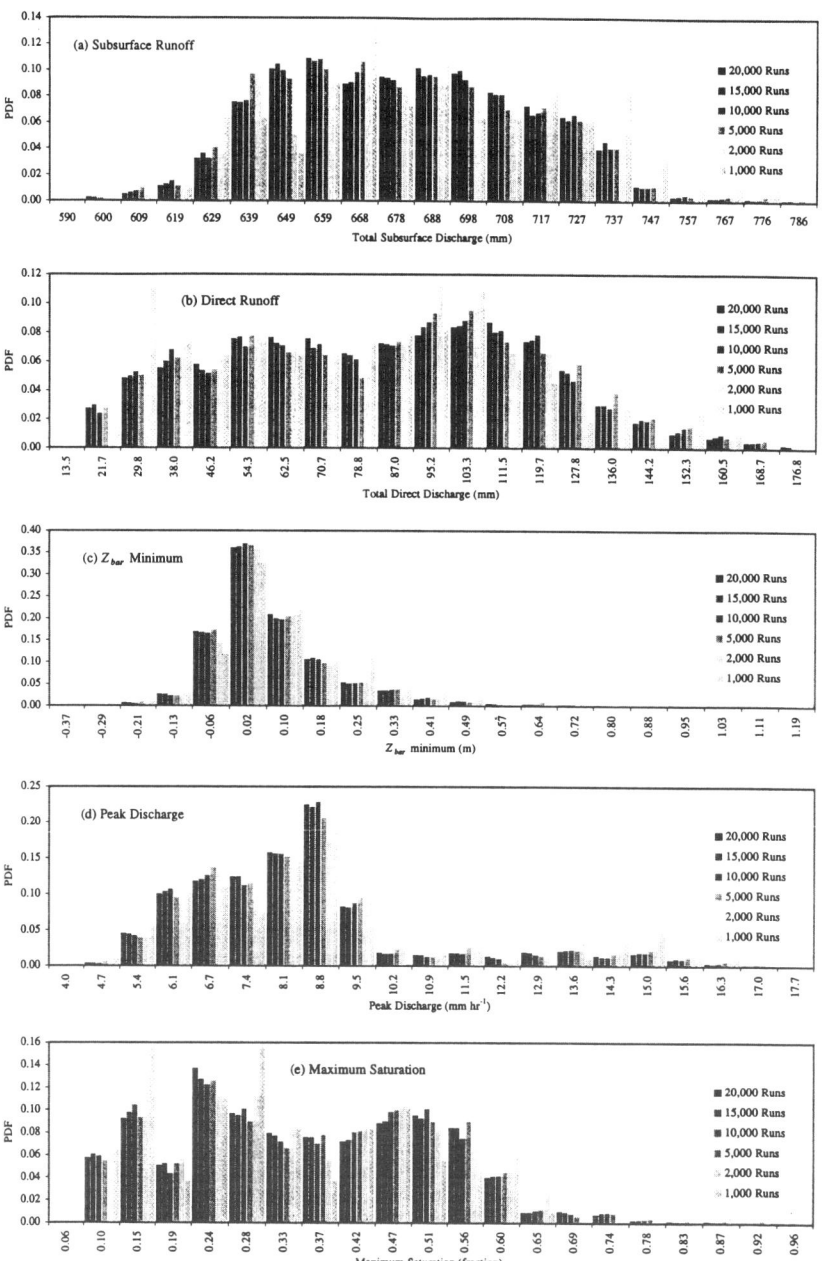

Figure 4. Probability distribution of output variables from behavioural parameter sets for different sample sizes of Monte carlo simulations. 1987 discharge data set, Maimai catchment.

(1994). The same kind of meteorological forcing data and evapotranspiration measurements were employed for an Amazonian, post-deforestation pasture site, collected as part of the ABRACOS UK–Brazilian collaboration (Shuttleworth *et al.*, 1991). Two data sets were available, AMAZ90 (October 16 to November 2, 1990) and AMAZ91 (June 29 to September 10, 1991), where the 1991 period was significantly wetter than that in 1990. Details of the instrumentation and measurements are given by Wright *et al.* (1992). A fifth meteorological dataset (TISBY95) was available from a boreal, agricultural fieldsite located west of Uppsala, Sweden. Measurements were taken above spring wheat crops over a two week period (June 29 to July 13, 1995) within the framework of the NOrthern hemisphere climate Processes land surface EXperiment (NOPEX) (Halldin *et al.*, 1998; Soegaard and Thorgeirsson, 1998; and Soegaard, 1999).

The Soil-Vegetation to Atmosphere Transfer (SVAT) Models

MODEL1 is similar to the TOPUP SVAT used by Beven and Quinn (1994) and Franks *et al.* (1997). It is a variable surface resistance bucket-type model based on the Penman-Monteith equation (Monteith, 1981). Evapotranspiration is calculated from three sources: a canopy-topsoil interception store, a root zone store and a variable saturated zone beneath the water table. A given rainfall event will first be routed to the interception store. When exceeded, water will overflow to fill the root zone and excess moisture from the root zone is then routed to the water table with a time delay parameter. A surface resistance r_S is calculated each time step according to the water availability of the individual stores, and is scaled between a minimum and maximum value which will depend on the plant characteristics. More details on the process description can be found in Franks *et al.* (1997). *MODEL1* has six parameters.

MODEL2 is similar to the first, but also includes a functional description for the dependence of stomatal response on transpiration rate suggested by Monteith (1995a,b). The response of stomata to solar irradiance and the dependence of evapotranspiration rate on the available soil water are also parameterised as suggested in Monteith (1995a). *MODEL2* has seven parameters.

These two models already require significantly fewer parameters than most contemporary SVATs (see Table 4), the description of soil vegetation to atmosphere transfer processes is further simplified in *MODEL3*. This model uses the same subsurface hydrological process descriptions to account for the available soil moisture but calculates latent heat flux on the basis of a simple evaporative fraction *EF*, defined as the ratio between latent heat flux and available energy. This evaporative fraction is known to vary both diurnally and over longer periods but has often been shown from field observations to be remarkably steady and conservative during daytime periods (Nichols and Cuenca, 1993; Bastiaanssen, 1995; Crago, 1996; Crago and Brutsaert, 1996). *MODEL3* has only two parameters.

The specification of the sampling parameter ranges of *MODEL1* follows the previous study of the TOPUP SVAT model behaviour by Franks *et al.* (1997) and Franks and Beven (1997a). For *MODEL2* and *MODEL3* we used the same ranges for all parameters describing subsurface hydrological processes. Table 4 summarises the selected ranges for the plant characteristics and meteorological parameter of all three SVAT schemes and for the different catchment and periods employed.

For each simulation period 20,000 random sets of parameters were generated from uniform distributions across the specified ranges and the model was run for each of these sets. No attempt was made to account for possible covariation between the parameter values in these prior choices. As noted earlier, since each parameter set is treated as a set, the conditioning process within GLUE will implicitly account for covariation in the posterior distribution. The performance of the models was compared in terms of the measure of equation (5) with $N = 1$. The 'optimal' model performances over all realisations for each model and each data set are given in Table 5. While the more complex *MODEL1* and *MODEL2* (having more degrees of freedom in terms of number of model parameters) are performing 'better' for the IFC3 and TISBY95 data sets, the simpler *MODEL3* produces comparable or even higher efficiencies for the AMAZ90 and AMAZ91 data. All the models perform relatively poorly for the IFC4 period (which was after the start of senescence of the grass vegetation).

Dotty plots for the likelihood measure given of (5) derived for the IFC3 dataset are shown in Figure 5 for the parameters of *MODEL1* and *MODEL3*. These plots represent the projection of the multidimensional parameter response surface, as sampled by the Monte Carlo simulations, onto single parameter axes. For the *MODEL1* aerodynamic parameters d, the zero displacement height, $\log(z_0/z_h)$, the log of the ratio of roughness length for momentum flux, and z_0, the roughness length for momentum flux, very good but also very poor simulations are available throughout the whole of the chosen parameter ranges. This also holds for the parameter RSMIN, the minimum surface resistance. For the parameter RSMAX, the minimum surface resistance, there is at least a slight tendency for the better simulations to be concentrated towards one part of the chosen parameter range, but good fits are found right up to the *a priori* lower limit of the range considered. There is no single clearly differentiated optimum parameter set for this model, given this choice of performance measure.

In contrast, the EF_{cor} parameter of the simpler *MODEL3* does show a well defined optimum parameter range (Figure 5), suggesting that the more complex models are overparameterised with respect to the data available for use in conditioning. Qualitatively the same results are obtained for all the other available data sets (figures not shown here). The scatter plots for the parameters of *MODEL2* (see Table 4) in general also show equifinality over

Table 4: Plant characteristics and meteorological parameters for the three different SVAT schemes and *a priori* selected parameter ranges for (corresponding model and data set in parentheses)

Parameter	Description	Range of values
MAXINT	Interception storage capacity [m], (*MODEL1, MODEL2, MODEL3*)	0.0005–0.005
z_0	Roughness length for momentum flux [m], (*MODEL1, MODEL2*)	0.02–0.12
$\log(z_0/z_h)$	Log of the ratio of roughness lengths for momentum and heat flux [-] (*MODEL1, MODEL2*)	1.0–3.0
d	Zero displacement height [m], (*MODEL1, MODEL2*) (IFC3, IFC4, AMAZ90, AMAZ91) (TISBY95)	0.15–0.35 0.5–1.2
EF_{cor}	Effective evaporative fraction [-], (*MODEL3*)	0.0–1.5
RSMIN	Minimum surface resistance [s m^{-1}], (*MODEL1*)	50–150
RSMAX	Maximum surface resistance [s m^{-1}], (*MODEL1*)	300–1000
$E_{m,max}$	Upper level of the max. evapotranspiration rate [m d^{-1}], (*MODEL2*)	0.01–0.025
f_{gm}	Ratio of max. stomatal conduct. and radiation [m^3 J^{-1}], (*MODEL2*)	2×10^{-5}–10^{-6}
S_m	Threshold of solar radiation for a g_m increase [W m^{-2}], (*MODEL2*)	300–600

Table 5. Performance of the best 20,000 model runs, based on the measure of equation (5) ($N = 1$) expressed as a percentage, for each of the three models and each of the five data sets.

Model / Data	IFC3	IFC4	AMAZ90	AMAZ91	TISBY95
MODEL1	92.0	−38.5	83.6	91.3	84.7
MODEL2	90.6	33.3	84.9	91.3	85.5
MODEL3	79.8	25.5	87.8	92.3	74.7

the whole range of the parameters space and are not shown here. The same holds for the subsurface parameters of all three SVAT models.

Following the GLUE procedure as described above, prediction bounds for

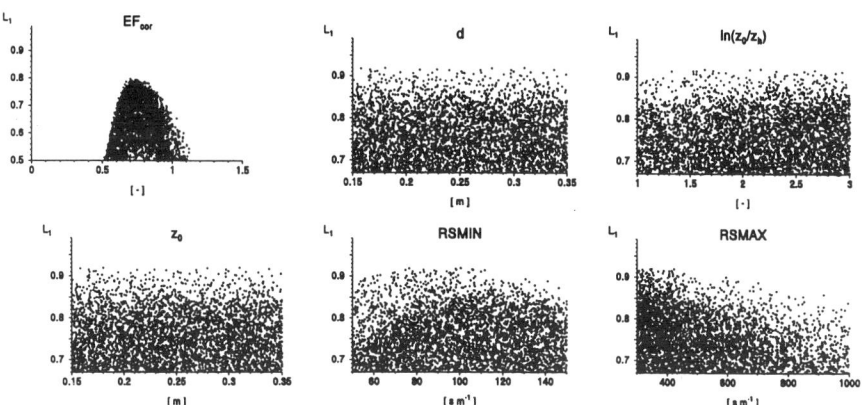

Figure 5. Model efficiencies for the parameter EF_{cor} of the BUCKEF model and for the five parameters of the Penman–Monteith formulation of the BUCKUP model using the IFC3 data set.)

the latent heat fluxes were calculated for each model and data set. Figure 6 shows a comparison of the resulting prediction quantiles (5% and 95%) of *MODEL1* and *MODEL3* for a three day period of AMAZ90, when behavioural model parameterisations were defined by a likelihood measure > 0.7. The widths of the bounds for all three models show that, in general, large predictive uncertainties result from equifinality of model parameters with maximum values of around $100\mathrm{Wm}^{-2}$ during midday. All three models produce approximately the same width of uncertainty bounds; however, there is an important difference in that the prediction bounds of *MODEL1* and, to a lesser extent, *MODEL2* (figure not shown) are not fully encompassing the measured data during daytime.

In general, for the three models a significant uncertainty was also observed for all other data sets with similar widths of the predicted bounds (dependent on the absolute latent heat fluxes) as presented for the AMAZ90 data in Figure 6. When compared for all data sets, none of the models used showed a significant advantage in terms of having narrower uncertainty bounds or a better ability to envelope the measured data, and therefore no justification for the more complex process descriptions used here can be made.

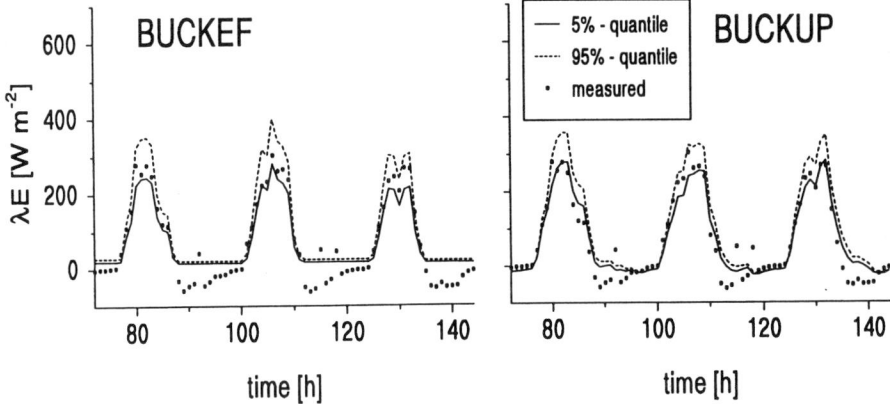

Figure 6. Comparison of the predicted uncertainties bounds (5% and 95% quantiles) of two of the SVAT models for the AMAZ90 data set.

Case study: modelling dispersion in open channel flows

In this case study, detailed measurements of the velocity distributions in an overbank flow in the Flood Channel Facility at HR Wallingford (SERC, 1992) were used in conjunction with tracer test data (Guymer *et al.*, 1989) to assess the effectiveness of dispersion models based on a random particle tracking methodology (for more detail see Hankin and Beven, 1998). In this case, different model structures are being considered within the GLUE framework, as well as different parameter sets within each model structure. This is handled easily provided the evaluation of each model structure can be made on the basis of a consistent likelihood measure.

Random particle tracking (RPT) models simulate the diffusive process via the successive application of random perturbations to the trajectory of computer generated particles representing the tracer. The particles are subjected to random displacements in the cross-sectional plane, and at the same time they are advected in the downstream direction according to the pattern of velocities within a given cross-section. The random motions result in dispersion of the particles and such models have been used elsewhere to simulate longitudinal dispersion in inhomogeneous, turbulent river flows (Allen, 1982, 1992; Heslop and Allen, 1993) and in laboratory scale channel flows (Davis and Guymer, 1994).

A variety of different random particle tracking (RPT) models were investigated which were based around various physically based, but ultimately heuristic assumptions about the nature of the Lagrangian turbulence field in overbank flow conditions in the Flood Channel Facility (see Knight and Shiono, 1989). For each model, 20,000 computer generated particles were advected in accordance with an interpolation of the mean downstream Eulerian velocity field and were subjected to random perturbations, to simulate the fluctuating part of the Lagrangian turbulent velocity field. The precise distribution which the random perturbations should take in the highly anisotropic, inhomogeneous flow is not known, nor is the relationship between the Eulerian and the Lagrangian velocity fields. For this reason, ten RPT models based on different physical assumptions were investigated. The most complex is based on a Markov chain perturbation sequence and incorporates velocities scaled using the detailed measurements of turbulence fields in the Flood Channel Facility (based on models of Zannetti, 1990).

The models were calibrated by varying multiplicative factors (effective diffusivity scaling parameters, f_v and f_w) to estimates of the average transverse and vertical velocity perturbations. In an application of the GLUE methodology, 900 different combinations of these parameters, spread uniformly in the parameter space, were simulated. The predicted particle concentration distribution for each model was compared with the measured concentration distribution at the first measurement site downstream (4m) using a coefficient of determination based on the predicted and measured concentrations at each of 201 measurement points. The particles were then allowed to disperse downstream and the predicted particle distributions were compared once again, as a validation test, with the measured concentration distributions at 16m from the source (258 measurement points).

It was found that the more complex RPT models, which used more information about the turbulent velocities, did not provide very good fits to the observations. However, four of the simpler models investigated produced acceptable simulations. The four models (which are called RPT5, RPT6, RPT9 and RPT10) are based on quite different representations of the Lagrangian velocity statistics, and are briefly described here to emphasise their structural differences. RPT5 is the most complex model, and simulates the particle Lagrangian velocity statistics at each step based upon statistical distributions of Lagrangian particle motions, observed in a moving camera experiment, by Sullivan (1974). A simpler model, RPT6, accounts for the spatial inhomogeneity by relating the time step between perturbations to the local mean downstream velocity. Model RPT9 is simpler still, treating all of the mixing processes as a diffusive process represented by random perturbations drawn from a zero mean Gaussian distribution. It is similar to models such as that of van Dam (1992) and Allen (1982), using a constant velocity scale and constant time step between perturbations everywhere in the flow. Finally RPT10 is an even more simplified version of RPT9, for which only the orientations of

Table 6. Summary statistics of particle cloud distributions for best fit RPT model structures

Model	Centroid(m) $Z = 0.159$m $x = 4$m	Spread (m) $Z = 0.159$m $x = 4$m	*Calibration* Coefficient of Determination	Centroid(m) $Z = 0.159$m $x = 16$m	spread(m) $Z = 0.159$m $x = 16$m	*Prediction* Coefficient of Determination
DATA	1.07 ± 0.05	0.17 ± 0.02	—	1.07 ± 0.05	0.31 ± 0.03	—
RPT5	1.12	0.19	0.899	1.19	0.35	0.871
RPT6	1.10	0.17	0.932	1.13	0.33	0.878
RPT9	1.11	0.17	0.932	1.14	0.33	0.875
RPT10	1.10	0.17	0.947	1.11	0.32	0.853

the random steps are random, the magnitude of the step displacements being constant everywhere in the flow for each set of parameter values. It is similar to the model used by Davis and Guymer (1994). Table 6 contains summary statistics and the optimal values of the objective function for the four models.

There are two possible responses to a situation in which the models that contained more information about the turbulent flow field did not give better simulations of tracer dispersion, but even the very simplest model gave acceptable simulations. The first is to demand even more information about the flow field, either by detailed CFD simulation of the turbulent flow field (Younis, 1996), or by even more detailed measurement of the velocities, preferably direct Lagrangian measurements. The second response is to recognise explicitly the generic equifinality of model representations of the dispersive process. Here we have taken the second approach since the first will not be generally practical for many applications. There are clearly still limitations to the performance of the four models which have been considered as more feasible simulators of the system in this study. A recognition of equifinality of the different model structures considered does not imply that there is no room for future improvement of the process description used in an RPT model.

The initial evaluation of the different model structures was made on the basis of the best fit parameter set of the 900 runs made with each model structure. The GLUE methodology is used here to allow for possible equifinality in the parameter sets for each model structure. For all the retained models, a fuzzy measure has been defined to express a relative degree of belief that a model structure is a good simulator of the system. This relative possibility measure (RPM) is used in equation (2) to weight the predictions of different models in an assessment of predictive uncertainty. The RPM is the membership function of the fuzzy set of 'feasible' simulations, taking a value between 1 (the model fits the data exactly) and zero (the model is outside the fuzzy set of 'feasible' simulators of the system), and is summarised in Figure 7.

Let RPM[4] express the relative degree of belief that a model structure (parameter combination) is a good simulator of the system, *based on the fit to*

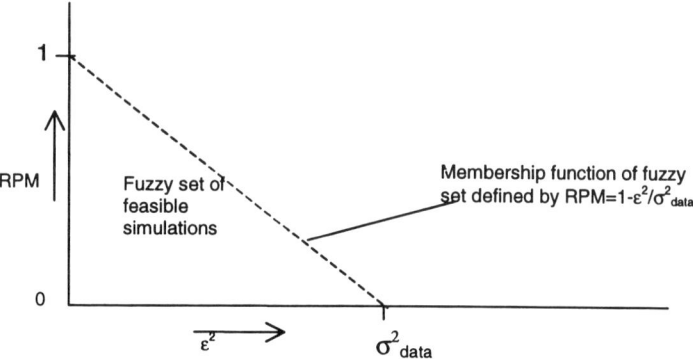

Figure 7. The Relative Posssibility Measure is the membership function of the fuzzy set of 'feasible' simulations, ε^2 is the mean of the sum of the squares of the residuals (difference between predicted and observed non-dimensionalised concentrations) and σ^2_{data} is the observed variance.

the data at the calibration cross-section. RPM[4] will represent *a posterior* distribution of weights after conditioning on the observations at 4m downstream and can then be used in an assessment of predictive uncertainty at 16m downstream. The RPM measures as calculated for each model simulation can be used as weights in the combined predictions of all four models retained as behavioural (RPT5, RPT6, RPT9 and RPT10). Figure 8 shows the prediction limits for the measurement sites 16m downstream based on all four representations of the Lagrangian velocity perturbations. These final prediction limits define the fuzzy set of 'reasonable' simulations, covering all those model structures which exhibit a functionality that captures the principal features of the tracer dispersion, irrespective of the original heuristic representation of the Lagrangian turbulent perturbations. The prediction limits generally show an improved fit over the limits predicted by some of the individual models (especially RPT5) at sites where the observed concentrations were not enclosed, although of course the improvement is limited by the best performing model structure at each site.

Conclusions

The variety of different types of models and methods of evaluation presented above demonstrate the conceptual simplicity and flexibility of the GLUE methodology. In essence GLUE represents an extension of Bayesian or fuzzy averaging procedures to less formal likelihood or fuzzy measures. It can be

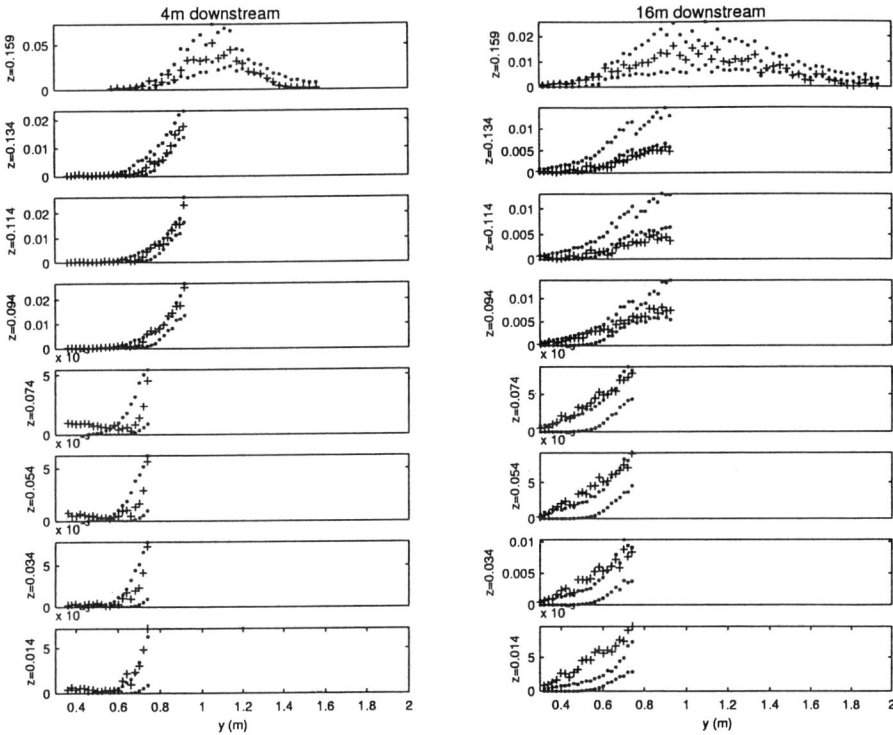

Figure 8. Observed tracer concentrations (+) and 5% and 95% prediction limits (•) assuming equifinality of models RPT5, RPT6, RPT9 and RPT10 for different depths at cross-sections 4m and 16m downstream of tracer source. In both cases, the upper cross-section is taken through the overbank flow, while the other cross-sections are in the main channel. The observations at 4m are used in model conditioning. the variable y is the distance from the centreline of the laboratory channel.

used in cases where the assumptions of more rigorous likelihood functions may be difficult to justify and where it cannot be shown that there is a true model or clear optimal model of the system under study.

The main criticism of GLUE is that the subjective choice of a likelihood measure means that significance of the resulting prediction bounds may be difficult to interpret (e.g. Melching, 1995; Clarke, 1994). Certainly the likelihood weights cannot be considered to be estimates of the true probability of predicting a variable given the model unless very strong constraints on the likelihood function can be met. It is worth noting that GLUE is a generalised methodology in that traditional statistical likelihood functions can be used in model evaluation where their assumptions can be shown to be valid (although in such a case an MCMC approach is likely to prove more computationally efficient). It is the argument here that in many applications of environmental models these constraints are unlikely to be met because of deficiencies in the model structures available, the difficulty of estimating effective parameter values for a particular model structure, and errors in the forcing data for the model. As demonstrated above, in these situations it is possible within GLUE to produce prediction bounds that appear reasonable in bracketing any observations that are available to verify the predictions. The estimated uncertainties will be conditional: conditional on the model structures considered, on the ranges of parameter values sampled, on the period of forcing data used, and on the particular likelihood measure chosen to evaluate the different models. It is important, however, that all these choices must be made explicit, so that the analysis can be critically assessed, reformulated and revisited if necessary. In any such study it should be expected that the choices made will reflect the nature of the problem under study, so that the estimated prediction quantiles will have utility.

Acknowledgements

One of the most important features of the GLUE methodology is that it focuses attention on the value of data for model evaluation. These studies would not be possible without the effort of all the researchers who collect the observations. Thanks are due particularly to Jeff McDonnell and the staff of the Maimai catchments; to Ian Wright, John Gash, Henrik Soegaard and all others involved in the collection of the Amazon and Tisby evapotranspiration data; and to Ian Guymer, Donald Knight, Koshi Shiono and Nick Brockie for the velocity and dispersion data from the Wallingford flood channel facility. There are many other colleagues, former research staff and graduate students who have contributed to the development of the GLUE concepts. Special thanks are due to George Hornberger who instigated some original latent thoughts along these lines back in 1980. Andy Binley, Jonathan Tawn, Peter Young, Robin Clarke, Bruno Ambroise, Renata Romanowicz, Stewart Franks, Rob Lamb, James Fisher and Susan Zak have all helped refine the ideas in fruitful discussions, criticisms and applications.

References

Allen, C.M., (1982) 'Numerical simulation of contaminant dispersion in estuary flows', *Proc. Roy. Soc. Lond.* **A381** 179–194

Allen, C.M., (1992) 'Particle tracking models for pollutant dispersion', in *Computer Modelling in the Environmental Sciences*, D.G. Farmer and M.J. Rycroft (eds.), Clarendon Press, 65–74.

Ambroise, B., Freer, J. and Beven, K.J. (1996) 'Application of a generalised TOP-MODEL to the small Ringelbach catchment, Vosges, France', *Wat. Resour. Res.* **32**(7) 2147–2159.

Aronica, G, Hankin, B.G. and Beven, K.J., (1998) 'Uncertainty and equifinality in calibrating distributed roughness coefficients in a flood propagation model with limited data', *Adv. in Wat. Resour.* **22**(4) 349–365.

Bastiaanssen, W.G.M., (1995) *Regionalisation of surface flux densities and moisture indicators in composite terrain*, PhD Thesis, University of Wageningen, Wageningen, Netherlands.

Bates, B.C. and Townley, L.R., (1988) 'Nonlinear discrete flood event models. 1. Bayesian estimation of parameters', *J. Hydrol.* **99** 61–76.

Beck, M.B., (1987) 'Water quality modelling: a review of the analysis of uncertainty', *Wat. Resour. Res.* **23**(8) 1393–1442.

Beven, K.J., (1989) 'Changing ideas in hydrology: the case of physically-based models', *J. Hydrol.* **105** 157–172.

Beven, K.J., (1993) 'Prophecy, reality and uncertainty in distributed hydrological modelling', *Adv. in Wat. Resour.* **16** 41–51.

Beven, K.J., (1995) 'Linking parameters across scales: subgrid parameterisations and scale dependent hydrological models', in *Scale Issues in Hydrological Modelling*, J.D. Kalma and M. Sivapalan (eds.), Wiley, 263–281.

Beven, K.J., (1996) 'A discussion of distributed hydrological modelling', in *Distributed Hydrological Modelling*, M.B. Abbott and J.C. Refsgaard (eds.), Kluwer, 255–278.

Beven, K.J., (1997) 'TOPMODEL: a critique', *Hydrol. Process.* **11**(3) 1069–1085.

Beven, K.J. and Binley, A.M., (1992) 'The future of distributed models: model calibration and uncertainty prediction', *Hydrol. Process.* **6** 279–298.

Beven, K.J. and Franks, S.W., (1999) 'Functional similarity in landscape scale SVAT modelling', *Hydrol. Earth Syst. Sci.* **3**(1) 85–94.

Beven, K.J. and Kirkby, M.J., 1979) 'A physically-based variable contributing area model of basin hydrology', *Hydrol. Sci. Bull.* **24**(1) 43–69.

Beven, K.J., Lamb, R., Quinn, P. Romanowicz, R. and Freer, J. (1995) 'TOPMODEL, in *Computer Models of Watershed Hydrology*, V.P. Singh (ed.), Water Resource Publications, Highlands Ranch, CO, 627–668.

Beven, K.J., and Quinn, P.F., (1994) 'Similarity and scale effects in the water balance of heterogeneous areas', in *Proc. AGMET Conference on The Balance of Water – Present and Future*, AGMET, Dublin.

Binley, A.M. and Beven, K.J., (1991) 'Physically-based modelling of catchment hydrology: a likelihood approach to reducing predictive uncertainty', *Computer Modelling in the Environmental Sciences*, D.G. Farmer and M.J. Rycroft (eds.), Clarendon Press, 75–88.

Blazkova, S. and Beven, K.J., (1999) 'Flood frequency estimation by continuous simulation for an ungauged catchment with fuzzy possibility uncertainty estimation', submitted to *Wat.Resour. Res.*

Box, G.E.P. and Tiao, G.C., (1973) *Bayesian Inference in Statistical Analysis*, Addison-Wesley.

Buckley, K.M., Binley, A.M. and Beven, K.J., (1995) 'Calibration and predictive uncertainty estimation of groundwater quality models: application to Twin Lake Tracer Test', in *Proc. of Groundwater Quality Models 93*, Tallinn, Estonia, IAHS Pubn. **220** 205–214.

Cameron, D., Beven, K.J., Tawn, J., Blazkova, S. and Naden, P., (1999) 'Flood frequency estimation by continuous simulation for a gauged upland catchment (with uncertainty)', *J. Hydrol.* **219** 169–187.

Clarke, R.T., (1994) *Statistical Modelling in Hydrology*, Wiley.

Crago, R.D., (1996) 'Comparison of the evaporative fraction and the Priestley-Taylor α for parameterizing daytime evaporation', *Wat. Resour. Res.* **32** 1403–1409.

Crago, R.D. and Brutsaert, W., (1996) 'Daytime evaporation and the self-preservation of the evaporation fraction and the Bowen ratio', *J. Hydrol.* **178** 241–255.

Davis, P.M. and Guymer, (1994) 'Evaluation of a random walk model for solute mixing in open channel flow', in *2nd International Conference on Hydraulic Modelling*, A.J. Saul (ed.), BHR Group Conference Series publication (11), Mechanical Engineering Publications Ltd., 451–460.

Dickinson, R.E., Henderson-Sellers, A., Kennedy, P.J. and Wilson, M.F. (1986) 'Biosphere-Atmosphere Transfer Scheme (BATS) for the NCAR community climate model', NCAR Tech. Note NCAR/TN-275+STR, National Center for Atmospheric Research, Boulder, CO.

Draper, D., (1995) 'Assessment and propagation of model uncertainty', *J. Roy. Statist. Soc.* **B37** 45–98.

Duan, Q., Sorooshian, S. and Gupta, V.K., (1992) 'Effective and efficient global optimisation for conceptual rainfall-runoff models', *Wat. Resour. Res.* **28**(4), 1015–1031.

Famiglietti, J.S. and Wood, E.F., (1994) 'Application of multiscale water and energy balance models on a tallgrass prairie', *Wat. Resour. Res.* **30** 3079–3093.

Fisher, J.I. and Beven, K.J., (1996) 'Modelling of streamflow at Slapton Wood using TOPMODEL within an uncertainty estimation framework', *Field Studies* **8** 577–584.

Franks, S.W. and Beven, K.J. (1997a) 'Bayesian estimation of uncertainty in land surface-atmosphere flux predictions', *J. Geophys. Res.* **102** (D20), 23991–23999.

Franks, S. and Beven, K.J., (1997b) 'Estimation of evapotranspiration at the landscape scale: a fuzzy disaggregation approach', *Wat. Resour. Res.* **33**(12), 2929–2938.

Franks, S.W. and Beven, K.J., (1999) 'Conditioning a multiple patch SVAT model using uncertain time-space estimates of latent heat fluxes as inferred from remotely-sensed data', *Wat. Resour. Res.*, in press.

Franks, S.W., Beven, K.J., Quinn, P.F. and Wright, I.R. (1997) 'On the sensitivity of soil-vegetation-atmosphere transfer (SVAT) schemes: equifinality and the problem of robust calibration', *Agric. For. Meteorol.* **86** 63–75.

Franks, S.W., Gineste, Ph., Beven, K.J. and Merot, Ph., (1998) 'On constraining the predictions of a distributed model: the incorporation of fuzzy estimates of saturated areas into the calibration process', *Wat. Resour. Res.* **34** 787–797.

Freer, J., (1998), *Uncertainty and Calibration of Conceptual Rainfall Runoff Models*, unpublished PhD thesis, Lancaster University.

Freer, J., Beven K.J. and Ambroise, B. (1996) 'Bayesian estimation of uncertainty in runoff prediction and the value of data: an application of the GLUE approach', *Wat. Resour. Res.* **32**(7), 2161–2173.

Gan, T.Y. and Biftu, G.F., (1996) 'Automatic calibration of conceptual rainfall-runoff models: optimisation algorithms, catchment conditions and model structure', *Water Resour. Res.* **32**(12), 3513–3524.

Gelfand, A.E. and Smith, A.F.M., (1990) 'Sampling-based approaches to calculating marginal densities', *J. Amer. Statist. Assoc.* **85** 398–409.

Guymer, I., Brockie, N., Allen, C.M. and Beven, K., (1989) 'Hydraulics Research – SERC flood channel facility. Straight flood channel dispersion tests', Dept. Civil. Eng. , Heriot-Watt University, Edinburgh .

Halldin, S., Gottschalk, L. van de Griend, A.A., Gryning, S.E., Heikinheimo, M., Högström, U., Jochum, A. and Lundin, L.C. (1998) 'NOPEX – a northern hemisphere climate processes land surface experiment', *J. Hydrol.* **212–213** 172–187.

Hankin, B. and Beven, K.J., (1998) 'Modelling dispersion in complex open channel flows: 2. Fuzzy calibration', *Stoch. Hydrol. Hydraul.* **12**(6), 397–412.

Heslop, S. and Allen, C.M., (1993) 'Modelling contaminant dispersion in the River Severn using a random walk model', *J. Hydraul. Res.* **31**(3), 323–331.

Hornberger, G.M. and Spear, R.C., (1981) 'An approach to the preliminary analysis of environmental systems', *J. Env. Manage.* **12** 7–18.

Keesman, K. and van Straten, G., (1990) 'Set membership approach to identification and prediction of lake eutrophication', *Wat. Resour. Res.* **26**(11), 2643–2652.

Klir, G.J. and Folger, T.A., (1988) *Fuzzy Sets, Uncertainty and Information*, Prentice-Hall.

Knight, D.W. and Shiono, K., (1989) 'Turbulence measurements in a compound overbank flow', *J. Hydraul. Res.* **28**(2), 175–196.

Kuczera, G., (1997) 'Efficient subspace probabilistic parameter optimisation for catchment models', *Wat. Resour. Res.* **33**(1), 177–185.

Lamb, R., Beven, K.J. and Myrabo, S., (1998) 'Use of spatially distributed water table observations to constrain uncertainty in a rainfall-runoff model', *Adv. Wat. Resour.* **22**(4), 305–317.

Manabe, S., (1969) 'Climate and the ocean circulation. 1. The atmospheric circulation and the hydrology of the earth's surface', *Mon. Weather Rev.* **97** 739–774.

McLaughlin and Townley, L.R., (1996) 'A reassessment of the groundwater inverse problem', *Wat. Resour. Res.* **32**(5), 1131–1161.

Melching, C., (1995) 'Reliability estimation'. Chapter 3 in *Computer Models of Watershed Hydrology*, V.P. Singh (ed.), Water Resource Publications, Highlands Ranch, CO.

Monteith, J.L., (1981) 'Evaporation and surface temperature', *Q. J. R. Meteorol. Soc.* **107** 1–27.

Monteith, J.L., (1995a) 'Accomodation between transpiring vegetation and the convective boundary layer', *J. Hydrol.* **166** 251–263.

Monteith, J.L., (1995b) 'A reinterpretation of stomatal responses to humidity', *Plant Cell Env.* **18** 357–364.

Morton, A., (1993) 'Mathematical models: questions of trustworthiness', *Brit. J. Phil. Sci.* **44** 659–674.

Nichols, W.E., and Cuenca, R.H. (1993) 'Evaluation of the evaporative fraction for parameterization of the surface energy balance', *Wat. Resour. Res.* **29** 3681–3690.

Parkinson, S. and Young, P.C., (1998) 'Uncertainty and sensitivity in global carbon cycle modelling', *Clim. Res.* **9** 157–174.

Piñol, J., Beven, K.J. and Freer, J., (1997) 'Modelling the hydrological response of mediterranean catchments, Prades, Catalonia – the use of distributed models as aids to hypothesis formulation', *Hydrol. Process.* **11**(9) 1287–1306

Press, F., (1968) 'Earth models obtained by Monte Carlo inversion', *J. Geophys. Res.* **73** 5223–5234.

Press, W.H., Flannery, B.P., Teukolsky, S.A. and Vetterling, W.T., (1992) *Numerical Recipes: the Art of Scientific Computing (2nd edition)*, Cambridge University Press.

Romanowicz, R. and Beven, K.J., (1998) 'Dynamic real-time prediction of flood inundation probabilities', *Hydrol. Sci. J.* **43**(2), 181–196.

Romanowicz, R., Beven, K.J. and Tawn, J., (1994) 'Evaluation of predictive uncertainty in non-linear hydrological models using a Bayesian approach', in *Statistics for the Environment II. Water Related Issues*, V. Barnett and K.F. Turkman (eds.), Wiley, 297–317.

Romanowicz, R., Beven, K.J. and Tawn, J., (1996) 'Bayesian calibration of flood inundation models, in *Flood Plain Processes*, M.G. Anderson, D.E. Walling and P.D. Bates (eds.), Wiley.

Rowe, L.K., (1979) 'Rainfall interception by a beech-pedocarp-hardwood forest near Reefton, New Zealand', *J. Hydrol. (N.Z.)* **18** 63–72.

Schulz, K., Beven, K. and Huwe, B., (1999) 'Equifinality and the problem of robust calibration in nitrogen budget simulations', *Soil Sci. Soc. Amer. J.*, in press.

Schulz, K., Franks S.W. and Beven, K. (1988) 'TOPUP – a TOPMODEL-based SVAT model to calculate evaporative fluxes between the landsurface and the atmosphere, Version 1.1', Program Documentation, IENS, Dept. Environ. Sci., Lancaster University, Lancaster, UK.

Sellers, P.J., Randall, D.A., Collatz, G.J., Berry, A., Field, C.B., Dazlich, D.A., Zhang, C., Collelo, G.D. and Bounoua, L. (1996) 'A revised land surface parameterization (SiB2) for atmospheric GCMs. Part I: Model Formulation', *J. Clim.* **9** 676–705.

Sen, M.K. and Stoffa, P.L., (1995) *Global Optimisation Methods in Geophysical Inversion*, Elsevier.

SERC, (1992) 'Experimental data phase A', School of Civil Eng., University of Birmingham vols. 1 and 14, series 10, report SR314.

Shuttleworth, W.J., Gash J.H.C., Roberts, J.M., Nobre, C.A., Molion, L.C.B. and Ribeiro, M.N.G., (1991) 'Post-deforestation Amazonian climate: Anglo-Brazilian research to improve prediction', *J. Hydrol.* **129** 71–85.

Soegaard, H., (1999) 'Fluxes of carbon dioxide, water vapour and sensible heat in a boreal agricultural area of Sweden – scaled from canopy to landscape level', *Agric. For. Meteorol.*, in press.

Sorooshian, S. and Gupta, V.K., (1995) 'Model calibration'. In *Computer Models of Watershed Hydrology*, V.P. Singh (ed.), Water Resource Publications, Highlands Ranch CO, 23–68..

Soegaard, H., and Thorgeirsson, H., (1998) 'Carbon dioxide exchange at leaf and canopy scale for agricultural crops in the boreal environment', *J. Hydrol.* **212–213** 51–61.

Spear, R.C. and Hornberger, G.M., (1980) 'Eutrophication in Peel Inlet, II. Identification of critical uncertainties via generalised sensitivity analysis', *Wat. Research* **14** 43–49.

Spear, R.C., Grieb, T.M. and Shang, N., (1994) 'Parameter uncertainty and interaction in complex environmental models', *Wat. Resour. Res.* **30**(11) 3159–3169.

Sullivan, P.J , 1974) 'Instantaneous velocity and length scales in a turbulent shear flow', *Adv. Geophys.* **18A** 213–223.

Tarantola, A., (1987) *Inverse Problems Theory, Methods for Data Fitting and Model Parameter Estimation*, Elsevier.

van Dam, G.C., (1992) 'The study of shear dispersion in tidal waters by applying discrete particle techniques', technical paper, Rijkswaterstaat, Tidal Waters Division, The Hague, Netherlands.

van Straten, G. and Keesman, K.J., (1991) 'Uncertainty propagation and and speculation in projective forecasts of environmental change: a lake eutrophication example', *J. Forecast.* **10** 163–190.

Wright, I.R., Gash, J.H.C., Da Rocha, H.R., Shuttleworth, W.J., Nobre, C.A., Maitelli, G.T., Zamperoni, C.A.G.P. and Carvalho, P.R.A., (1992) 'Dry season micrometeorology of central Amazonian ranchland', *Q. J. R. Meteorol. Soc.* **118** 1083–1099.

Young, P.C., (1984) *Recursive Estimation and Time Series Analysis*, Springer-Verlag.

Younis, B.A., (1996) 'Progress in turbulence modelling for open channel flows', in *Flood Plain Processes*, M.G. Anderson, D.E. Walling and P.D. Bates (eds.), Wiley.

Zak, S. and Beven, K.J., (1999) 'Equifinality, sensitivity and uncertainty in the estimation of critical loads', *Sci. of Total Env.* **236** 191–214.

Zak, S.K., Beven, K.J. and Reynolds, B., (1997) 'Uncertainty in the estimation of critical loads: a practical methodology', *Soil, Wat. Air Poll.* **98** 297–316.

Zannetti, P., (1990) *Air Pollution Modeling*, Computational Mechanics Publications, Winchester, UK.

Spatial Statistics in Environmental Science

Richard L. Smith

1 Introduction

Spatial statistics is the natural generalization of signal processing to higher dimensions. In traditional signal processing, one has a signal $X(t)$ dependent on a scalar variable t, which may belong to a discrete set or which may be continuous (e.g. the whole real line). Spatial statistics is concerned with cases in which t is a multidimensional index of dimension $d > 1$. In most practical examples, $d = 2$, though much of the basic theory and methodology is the same whatever the dimension. Although the models and methods of spatial statistics have not developed as rapidly as those for one-dimensional signal processing, there have nevertheless been substantial new developments in recent years. Standard references on spatial statistics include the books of Ripley (1981, 1988) and Cressie (1993).

Applications of spatial statistics cover many areas. Much of the original impetus for the subject was driven by geostatistics, e.g. given measurements of the concentration of a mineral at a finite set of positions within a geological field, the aim was to determine the overall volume of the mineral over the whole field. It was in this context that the technique of 'kriging' – optimal least squares interpolation over a random spatial field – was originally developed. In recent years, the applications of spatial statistics have increased enormously, with particularly rich applications in the environmental and ecological sciences. A typical problem is the sampling of a pollution field, such as ozone in the atmosphere or toxic chemicals in rivers and lakes. Another example is the use of meteorological measurements in studies of global climate change. In these fields, as in geostatistics, the objective may be to interpolate spatially between measurements, but there are also otherobjectives which may be quite different. For example, in the context of global climate change, a natural question is to what extent the data support hypotheses of increasing temperature or rainfall, and how the resulting trends, if they exist, vary over the earth's surface. This kind of problem is discussed in section 7. Spatial statistics has also found applications in such diverse fields as sociology – for example, social networks theory – and financial economics: an example of the latter is the term structure of interest rates, where there are two distinct 'time' parameters, one the time at which a loan is taken out, the other the term of the loan. However, in the present chapter we shall concentrate on environmental applications, which is where the most rapid growth has occurred in recent years.

Sections 2–6 are a review of various concepts and methods in spatial statistics. Section 7 discusses recent developments in spatial trend estimation, motivated by the problem of characterizing the spatial pattern of climate change, but also having applications in a number of other fields.

2 Geostatistics and kriging

A spatial process will be represented by $Z(s)$, where s varies over a domain $\mathcal{D}$ contained in d-dimensional Euclidean space $\mathbf{R}^d$ for some $d > 1$. Typically, but not necessarily, $d = 2$. As an example, $Z(s)$ might be the concentration of atmospheric ozone taken at a specific place s on the earth's surface (where, for the purpose of the present discussion, we regard the earth's surface as two-dimensional). Suppose also we have a finite number of observations, $z_i = Z(s_i)$ for $i = 1, \ldots, n$, $s_1, \ldots, s_n$ denoting the positions of the monitoring sites. Typically in environmental applications (and a contrast with traditional time series analysis), there is no fixed lattice of measuring sites and we regard the variable s as varying over a continuous set $\mathcal{D}$.

The classical 'kriging' problem is as follows: given z_i, $i = 1, \ldots, n$, predict $z_0 = Z(s_0)$ for some new location s_0 which is not one of the given sites $s_1, \ldots, s_n$. Once this problem is solved, it is easily extended to other problems such as jointly estimating the values of $Z(s)$ at several unmonitored sites s, or estimating a quantity such as $\int_A Z(s)ds$ for some set $A \subseteq \mathcal{D}$.

Models for z_i are traditionally of two types:

$$z_i = \mu + \eta_i, \tag{1}$$

$$z_i = x_i^T \beta + \eta_i, \tag{2}$$

where in each case $\{\eta_i\}$ represent some spatially correlated zero-mean 'noise' process. In (1), μ is a single (usually unknown) parameter assumed to be constant at all points on the surface, while in (2), the mean at a specific point is assumed to depend on a given set of covariates x_i through a linear regression model with unknown parameters β. Traditionally (1) is described as the 'ordinary kriging' problem and (2) as 'universal kriging'.

Covariance assumptions on $\eta(\cdot)$ may be represented through

$$\text{Cov}\{\eta(s), \eta(s')\} = C(s, s'), \tag{3}$$

for some covariance function $C(\cdot, \cdot)$.

Various homogeneity assumptions may be made on the covariance function. For example, if C is of the form

$$C(s, s') = C_0(s - s') \tag{4}$$

for some C_0, then the process is described as *stationary*. It captures the property (the obvious generalization of stationarity in time series analysis) that the dependence between two sites s and s' depends only on the (vector) distance between the sites, $s - s'$.

Instead of the covariance function, it is common in spatial statistics to work instead with the *semivariogram function* $\gamma(\cdot)$, defined by

$$\text{Var}\{\eta(s) - \eta(s')\} = 2\gamma(s - s').$$ (5)

For somewhat odd historical reasons, the left hand side of (5) is called the *variogram*, and the function γ the semivariogram. One motivation for considering the (semi)variogram rather than the covariance function is that the assumption (5), also known as *intrinsic stationarity*, is actually somewhat weaker than (4), in that there are models for which the variogram exists when the covariance function does not. For the applications considered in this chapter, however, it will be good enough to assume that the covariance function exists, and then it does not really matter whether the covariance function or the variogram is used to characterize the process.

Denoting the argument of C_0 or γ as h, if either $C_0(h)$ or $\gamma(h)$ depends only on $||h||$, i.e. the length of h, usually measured via the usual Euclidean metric, the process is said to be *isotropic*. Sometimes the word *homogeneous* is used to describe a process which is both stationary and isotropic.

2.1   Kriging

Rewrite the model (2) in the form

$$\mathbf{Z} = X\beta + \eta,$$ (6)

where $\mathbf{Z}^T = (\, z_1 \quad z_2 \quad \ldots \quad z_n \,)$, X is the $n \times p$ matrix of covariates, β is a p-dimensional vector of unknown regression coefficients, and η is a vector of random errors with mean 0 and covariance matrix of the form $C = \alpha V$, where $\alpha > 0$ is allowed to be an unknown positive scalar but the matrix V is assumed known. In most cases, we also assume that η has a multivariate normal distribution. As stated, with the regression term $X\beta$, this is the universal kriging problem, but ordinary kriging is a special case in which $X\beta$ is replaced by $\mathbf{1}\mu$, $\mathbf{1}$ being an n-dimensional vector of ones and μ a fixed unknown constant. Suppose we wish to predict a value

$$z_0 = x_0^T \beta + \eta_0,$$ (7)

in which the covariates x_0 at a new site are given, and η_0 is a random variable with mean 0, variance αv_0 and covariance with η given by $\text{E}\{\eta^T \eta_0\} = \alpha w$ with the scalar v_0 and the vector w both known. Consider a linear predictor of the form $\hat{z}_0 = \lambda^T \mathbf{Z}$ where λ satisfies the constraint

$$X^T \lambda = x_0.$$ (8)

The rationale for the constraint (8) is that it justifies the reduction

$$\hat{z}_0 - z_0 = \lambda^T(X\beta + \eta) - x_0^T\beta - \eta_0 = \lambda^T\eta - \eta_0,$$

or in other words, the prediction error does not depend on the unknown quantity β. There are now several ways to go about the solution:

1. Find λ to minimize $E\{(\hat{z}_0 - z_0)^2\}$ subject to the constraint (8). This does not involve any normality assumption, since the formulation of the problem depends solely on first- and second-order moments. The most direct solution is via the method of Lagrange multipliers, leading to

$$\lambda = \{(x_0 - X^TV^{-1}w)^T(X^TV^{-1}X)^{-1}X^T + w^T\}V^{-1}, \tag{9}$$

with an accompanying (complicated) expression for the variance of $\hat{z}_0$ given **Z**. See, for example, Ripley (1981) or Cressie (1993).

2. A Bayesian solution (assuming normality) is to fix the improper prior distribution $\pi(\beta) \equiv 1$, calculate the posterior density $\pi(\beta|\mathbf{Z})$, and then

$$\hat{z}_0 = \int E\{z_0|\mathbf{Z}, \beta)\}\pi(\beta|\mathbf{Z})d\beta.$$

3. Suppose $\mathbf{Z}_1 = A\mathbf{Z}$ is an $(n - p)$-dimensional vector of linearly independent contrasts, i.e. linear functions of $\mathbf{Z}$ whose distributions do not depend on β. General vector space theory implies that such a matrix A must exist, though it may not be so easy to calculate explicitly. Also let $\hat{\beta} = (X^TV^{-1}X)^{-1}X^TV^{-1}\mathbf{Z}$ denote the generalized least squares estimator of β. It is easily checked that the distribution of $z_0 - x_0^T\hat{\beta}$ does not depend on the true value of β. Then $\hat{z}_0 - x_0^T\hat{\beta}$ is the conditional expectation of $z_0 - x_0^T\hat{\beta}$ given $\mathbf{Z}_1$.

The equivalence of formulations 2 and 3 is an interesting connection between Bayesian and conditional inference which has been noted in other contexts – see the discussion of REML estimation in the next subsection.

These approaches to kriging all make full allowance for the fact that the regression coefficient β are *a priori* unknown, but they make no allowance at all for the fact that V is also typically unknown, which, arguably, is the more important problem! So let us now turn to consideration of this feature.

2.2 Estimation of V

The most common estimation approaches assume that the process is stationary and isotropic. For methods getting away from this assumption, see section 3.

The first step is very often a plot to determine the shape of either the covariance or the semivariogram function. Assuming the process $Z(s)$ has common mean, a common estimator for the semivariogram γ based on a finite number of observations $s = s_1, \ldots, s_n$ is of the form

$$2\hat{\gamma}(h) = \frac{1}{|N(h)|} \sum_{(s_i, s_j) \in N(h)} \{Z(s_i) - Z(s_j)\}^2, \tag{10}$$

where $N(h)$ denotes a collection of (s_i, s_j) pairs whose Euclidean distance lies within a given neighborhood of h, and $|\cdot|$ denotes cardinality.

The estimate (10) is sometimes criticized for being overly sensitive to outliers and there are various 'robust' alternatives, for example, the Cressie–Hawkins estimator (Cressie, 1993)

$$2\tilde{\gamma}(h) = \frac{1}{0.457 + \frac{0.494}{n}} \left\{ \frac{1}{|N(h)|} \sum_{N(h)} |Z(s_i) - Z(s_j)|^{1/2} \right\}^4, \tag{11}$$

which is also an approximately unbiased estimator when the data are normally distributed, but which is less affected by outliers than (10).

The second step in the estimation procedure is to choose from a family of positive-definite covariance functions (or, equivalently, negative-definite semi-variogram functions). Simple estimators such as (10) and (11), when viewed as a function of h, do not have the negative-definiteness property which is a necessary condition for a legitimate semivariogram function. Therefore, in most cases, we choose from a parametric family which does have this property.

Formulating the problem in terms of covariances rather than semivariograms, a general condition for a positive-definite covariance in a stationary process is that it be expressible in the form

$$C_0(h) = \int \cos(\omega^T h) G(d\omega), \tag{12}$$

where G is a non-negative measure on $\mathbb{R}^d$. In isotropic cases, where the vector argument h is replaced by a scalar argument t, the formula reduces to

$$C_0(t) = \int_{(0,\infty)} Y_d(\omega t) \Phi(d\omega), \quad 0 < t < \infty, \tag{13}$$

with Φ a non-negative measure on $(0, \infty)$. In (13), Y_d is given (depending on dimension d) by

$$Y_d(t) = \left(\frac{1}{t}\right)^{(d-2)/2} \Gamma\left(\frac{d}{2}\right) J_{(d-2)/2}(t),$$

$\Gamma(\cdot)$ being the usual gamma function and $J_{(d-2)/2}(\cdot)$ a Bessel function of the first kind of order $(d-2)/2$. See, for example, Ripley (1981) or Cressie (1993).

In practice, there are a number of standard families of covariance functions (or, equivalently, semivariograms) which are consistent with (13). Examples include:

Spherical model:

$$\gamma_0(t) = \begin{cases} 0, & t = 0, \\ c_0 + c_1 \left\{ \frac{3}{2}\frac{t}{R} - \frac{1}{2}\left(\frac{t}{R}\right)^3 \right\}, & 0 < t \le R, \\ c_0 + c_1, & t \ge R, \end{cases} \qquad (14)$$

Exponential model:

$$\gamma_0(t) = \begin{cases} 0, & t = 0, \\ c_0 + c_1 \left(1 - e^{-t/R} \right), & t > 0, \end{cases} \qquad (15)$$

Gaussian model:

$$\gamma_0(t) = \begin{cases} 0, & t = 0, \\ c_0 + c_1 \left(1 - e^{-(t/R)^2} \right), & t > 0, \end{cases} \qquad (16)$$

Matérn model: Best defined in terms of the covariance function C_0 as

$$C_0(t) = \frac{1}{2^{\nu-1}\Gamma(\nu)} \left(\frac{2\sqrt{\nu}t}{R} \right)^\nu \mathcal{K}_\nu \left(\frac{2\sqrt{\nu}t}{R} \right), \qquad (17)$$

where $\nu > 0$ is a shape parameter and $\mathcal{K}_\nu$ is the modified Bessel function of the third kind of order ν (Abramowitz and Stegun, 1964, Chapter 9). The special cases $\nu = \frac{1}{2}$ and $\nu \to \infty$ correspond respectively to the exponential and Gaussian models, (15) and (16).

In each of (14)–(17), $R > 0$ is a scale parameter known as the *range*, and in (14)–(16), if $c_0 \ne 0$ it is known as the *nugget*. This reflects the commonly observed property that, even at very small distances, observed variograms are non-negligible, which is often interpreted to reflect measurement errors in the observations rather than real discontinuities in the surface being measured.

Discrimination among parametric models is often carried out by visual assessment based on the sample semivariogram $\hat{\gamma}(\cdot)$ or $\tilde{\gamma}(\cdot)$, but may also be carried out (after fitting the model) by more formal criteria such as likelihood ratio tests, AIC, BIC, etc.

The third step in the estimation procedure is the estimation of parameters of the assumed covariance (or variogram) model. One relatively simple, and reasonably efficient, technique for this is Cressie's weighted least squares procedure (Cressie, 1993): given a sample variogram $\hat{\gamma}(h)$ evaluated at a finite number of values of h, say $h_1, h_2, \ldots$, and a model $\gamma(h; \theta)$ depending on unknown parameters θ, choose θ to minimize

$$\sum_j |N(h_j)| \left\{ \frac{\hat{\gamma}(h_j)}{\gamma(h_j; \theta)} - 1 \right\}^2. \qquad (18)$$

The method is not dependent on a particular sample estimator; for example, $\hat{\gamma}(\cdot)$ from (10) may be replaced by $\tilde{\gamma}(\cdot)$ from (11). This method has the advantage of being relatively straightforward to calculate, requiring a nonlinear optimization but no complicated likelihood evaluation. A practical disadvantage is that there is no easy way to obtain standard errors for the estimators, or tests of hypotheses about the parameters.

Most of the alternatives to this method are based on some form of likelihood procedure, assuming a Gaussian process:

• *Maximum likelihood estimation* (Kitanidis, 1983; Mardia and Marshall, 1984). This is more complicated than Cressie's method because the evaluation of the exact likelihood is appreciably harder than (18), but it is computationally feasible for reasonably sized problems. If there are n data points, likelihood evaluation requires storage and inversion of an $n \times n$ covariance matrix; the author has successfully applied this for n up to 500, but there would clearly be problems if n were of the order of several thousand.

• *Restricted maximum likelihood (REML) estimation* (Cressie, 1993). This is an alternative to maximum likelihood, especially well adapted to models of the form of (6), for which it uses a likelihood function for θ based on a set of contrasts of $\mathbf{Z}$, orthogonal to the design matrix X, whose distribution is unaffected by β. Although asymptotically equivalent to maximum likelihood, the method is generally believed to have superior properties in small samples, especially when the dimension of β is large. As shown originally by Harville (1974), the method is also equivalent to a simple form of Bayesian analysis in which the parameter β is given a uniform prior density, though the treatment of the θ parameter is not Bayesian.

• *Bayesian methods* (Le and Zidek 1992, Handcock and Stein 1993, Brown *et al.* 1994), in which all the unknown parameters are given prior distributions and a joint posterior distribution is calculated, have become much more popular in recent years, though they are more complicated computationally than ML or REML estimation.

An example of these estimation methods if shown in Fig. 1. This is based on the Texas aquifer data set of Cressie (1993), in which the underground water level was measured at various places in Texas. Cressie's original analysis used an intrinsically stationary model without directly adjusting for any deterministic spatial trend. In the present analysis, a linear trend was incorporated through the regression function in (2), and the residuals modeled as a stationary, isotropic process. The standard (10) and robust (11) variograms of the residuals are shown, along with an exponential semivariogram model fitted by each of the weighted least squares, maximum likelihood and REML methods. The results appear to show good consistency among the different methods.

The fourth and final step in the estimation procedure is the application to kriging. In traditional geostatistical applications, once the covariance matrix V was estimated by any of the methods we have described, this was simply treated as a known matrix for the calculation of kriging formulae and their prediction variances. However, as already noted, it is somewhat illogical to

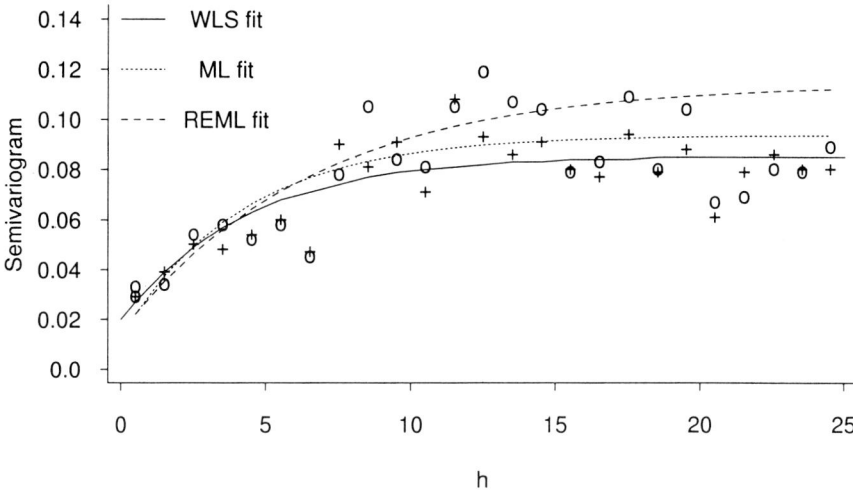

Fig. 1. Standard $(+)$ and robust (o) estimates of the sample semi-variogram with fitted exponential semivariogram function by the weighted least squares, maximum likelihood and restricted maximum likelihood methods.

adapt the kriging procedure for the fact that β is unknown without making a similar allowance for the unknown parameters in V. Two possible approaches are:

• *Corrections based on the delta method.* These usually use the standard kriging formulae for point predictions but correct the prediction variances, using Taylor expansions to represent the effect of parameter uncertainty. See Zimmerman and Cressie (1992), Harville and Jeske (1992).

• *Fully Bayesian approach:* It is straightforward to formulate the problem from a Bayesian viewpoint, since it treats (β, θ, z_0) (where β are the regression parameters, θ the unknown parameters of the covariance matrix, and $z_0 = Z(s_0)$ the unknown quantity being predicted) as a random vector, and calculates the conditional distribution of z_0 given observed data Z after integrating out the effect of β and θ. As already noted, in the special case that θ is known and β has a uniform (improper) prior distribution, this is equivalent to the standard kriging formulae.

3 Nonstationary models

The approach described in section 2 is based on an assumption that the spatial random field is stationary and isotropic. In the original geophysical applications which motivated the development of the field, this assumption was often justified by the fact that with sparse data, there was no reasonable alternative. A further point is that many geostatistical applications involved only one measurement at each site (or equivalently, only one replication of the random field) so there was no way of determining the complete spatial covariance function without some kind of stationarity assumption. In modern environmental applications, however, there are very often enough monitoring stations to go beyond such assumptions, and with multiple observations per site, it is also possible to estimate the covariance between any pair of sites without assuming stationarity across the field. Another consideration is that very often, simple topography makes a stationarity assumption implausible. Therefore, there are by now many reasons to go beyond a stationary model.

In spite of this obvious need for nonstationary models, however, there is not as yet a wide variety of approaches to the problem. In the present section we concentrate on one particular approach, pioneered by Sampson and Guttorp (1992) and also developed by, among others, Mardia and Goodall (1993), Guttorp *et al.* (1994) and Smith (1996).

The idea is a 'deformation approach' to nonstationarity: we assume the observed process is nonstationary, but that it can be deformed into a stationary (and, in most applications, isotropic) process by some nonlinear map. For a spatial process $Z(s)$ with constant mean, defined for sites s, t within some domain $\mathcal{D}$, define the *dispersion function*

$$D(s,t) = \mathrm{E}\left[\{Z(s) - Z(t)\}^2\right]$$

for each pair of sites (s, t). We look for models of the form

$$D(s,t) = 2\gamma_0(\|f(s) - f(t)\|) \tag{19}$$

where f is some nonlinear function on $\mathcal{D}$ and γ_0 is the semivariogram of a stationary, isotropic process. In the terminology developed by Sampson and Guttorp, the original 'geographic space', or G-space, in which the observations are located, is transformed by the function f into a 'dispersion space', or D-space, and in that space the process is stationary and isotropic.

To estimate such a model, it is not sufficient to have only one observation per location, because without any stationarity assumption, it is not possible to measure the covariance between any pair of points. However, virtually all the practical applications of this methodology have been in contexts for which there is no shortage of available data to estimate the covariances required.

Guttorp and Sampson, and their co-authors, have developed a variety of ingenious but somewhat *ad hoc* fitting techniques for these models. For example, Guttorp *et al.* (1994) proposed an estimation scheme in which f and γ_0 were chosen to minimize

$$\sum_i \sum_j \left(\frac{d_{ij} - D_{ij}}{d_{ij}}\right)^2 + \lambda\{J(f^{(1)}) + J(f^{(2)})\}, \tag{20}$$

where $f^{(1)}$ and $f^{(2)}$ are the two coordinate functions of $f : \mathbf{R}^2 \to \mathbf{R}^2$, and

$$J(f^{(j)}) = \int\int \left\{\left(\frac{\partial^2 f^{(j)}}{\partial x^2}\right)^2 + 2\left(\frac{\partial^2 f^{(j)}}{\partial x \partial y}\right)^2 + \left(\frac{\partial^2 f^{(j)}}{\partial y^2}\right)^2\right\} dx\, dy. \tag{21}$$

Note that we are representing a generic point of $\mathbf{R}^2$ by (x, y) rather than s as previously. In (20), D_{ij} is the observed dispersion between sites i and j as determined empirically from the data, d_{ij} is the model-based dispersion determined by the functions f and γ_0, $\lambda > 0$ is a smoothing parameter, and $J(f^{(j)})$, $j = 1, 2$, is a 'bending energy' functional whose presence ensures that the function f is not allowed to become too irregular. This kindof functional is often used as a penalty function in spline-based approaches to smoothing and interpolation.

An alternative approach (Mardia and Goodall 1993, Smith 1996) is to choose f and γ_0 to minimize the profile negative log likelihood function

$$L = \frac{N}{2}\log|\Sigma| + \frac{N-1}{2}\,\mathrm{tr}\left(\Sigma^{-1}\hat{\Sigma}\right), \tag{22}$$

where we assume there are N replications of the field, $\hat{\Sigma}$ is the sample covariance matrix and Σ is the model-based covariance matrix.

To make this approach work effectively, we would really like to represent both γ_0 and f as parametric functions. Parametric models for γ_0 have already been discussed, but an alternative approach, motivated by the representation (13), is to represent the covariance function C_0 in the form

$$C_0(h) = \sum_{c=1}^{C} \phi_c J_0(\omega_c h) \tag{23}$$

for a fixed C and positive constants $\phi_1, \ldots, \phi_C$, $\omega_1, \ldots, \omega_C$. The idea underlying (23) is that it forms a discrete approximation to the integral model (13). Note that the function Y_d in (13) reduces to J_0 when $d = 2$, as we assume throughout this section.

The other issue involved in parametrizing the model is to obtain a finite-parameter representation for f. Assuming again that we are working in dimension $d = 2$, f can be represented as $(f^{(1)}, f^{(2)})$, where $f^{(1)}$ and $f^{(2)}$ are scalar functions of location $s = (x, y)$. There are by now many approaches to nonlinear function reconstruction that use an expansion in terms of basis functions: for example, thin-plate splines, radial basis functions (RBFs) and wavelets all use representations of this form. In dimension 2, the thin-plate

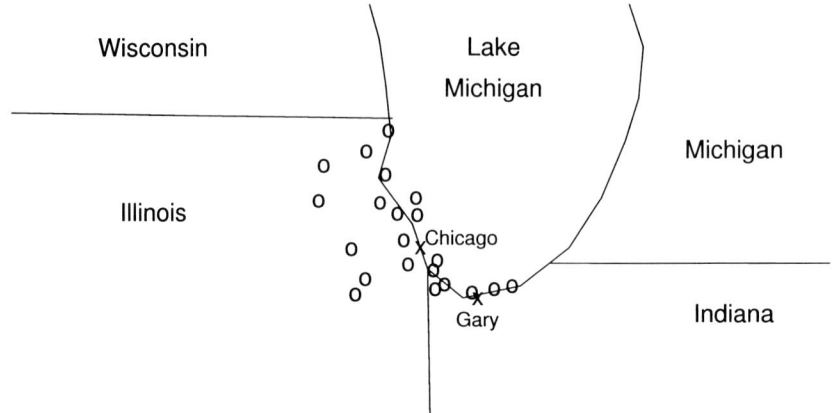

Fig. 2. Map of ozone stations: G-space.

spline and RBF approaches coincide, and lead to $f^{(1)}$ and $f^{(2)}$ being represented as linear combinations of basis functions of the form

$$\eta_i(x, y) = r^2 \log r,$$

where $r = \{(x - x_i)^2 + (y - y_i)^2\}^{1/2}$, the distance between the current location (x, y) and the ith 'center' (x_i, y_i). The model is completely determined
once we specify the number and positions of the centers. We then write

$$f^{(j)}(x, y) = \sum_{i=1}^{I} \delta_i^{(j)} \eta_i(x, y), \quad j = 1, 2,$$

where I is the number of centers and $\{\delta_i^{(j)}, \ i = 1, \ldots, I, \ j = 1, 2\}$ are coefficients to be determined.

As an example of this approach (taken from Smith, 1996), Fig. 2 shows a
map of ozone stations in the neighborhood of Chicago in the original 'G-space'.
The bulk of the stations shown are in the city of Chicago or surburban Illinois,
but the three lower right stations are near the city of Gary, Indiana, a heavily
industrialized region. Sample correlations among ozone measurements show
that these three stations have much lower correlations (with each other, as well
as with the remaining stations in the figure) than the rest of the stations, so
we would not expect a stationary model to hold. Throughout the examples in
this section, we work with spatial correlations rather than spatial covariances
to avoid having to deal with the fact that the variances may also vary from

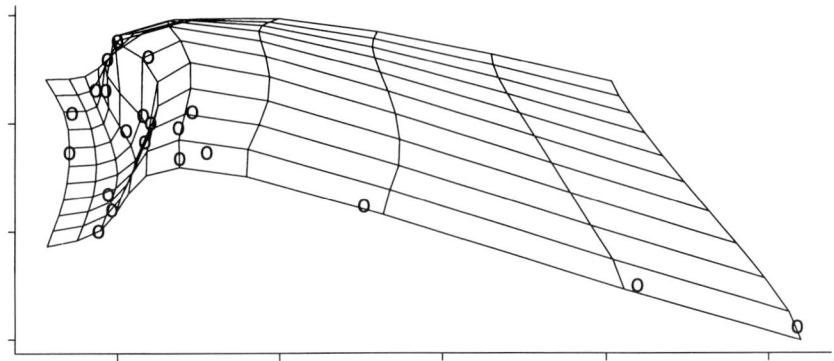

Fig. 3. Map of ozone stations: D-space.

site to site. Fig. 3 indeed shows that (one reconstruction of) the 'D-space' in this example is highly distorted, with the three outlying stations drawn out into positions very distant from the remainder.

As an example of the effect of the transformation on the sample dispersion, Fig. 4 shows the dispersions in the original G-space, with dispersions involving one of the three outlying stations distinguished by a separate plotting symbol. It is evident that the plot cannot easily be represented by a smooth curve. The transformed picture (Fig. 5), together with the fitted Matérn variogram function, is much cleaner.

A second example is based on temperature averages for stations in a subset of the United States 'Historical Climatological Network'. The top plot in Fig. 6 shows the locations of the stations superimposed on a map of the country. The bottom plot shows the transformed D-space. In this example, the striking feature of the plot is that stations in the southwestern states (California, Nevada, Arizona) have been pulled far away from the rest of the plot, while stations in the northwest (Washington, Oregon) are drawn closer to the rest of the country. However, the corresponding 'before and after' variogram plots (Fig. 7) do not have nearly so clear-cut an interpretation as Fig. 5. In this case, the fitted Matérn curve in D-space still does not come close to fitting the whole dispersion function, and one is led to suspect that the transformed model still does not fit the data very well. Further investigation based on the covariance function (23) improves the fit (Smith, 1996), but still without adequately fitting the whole of the data.

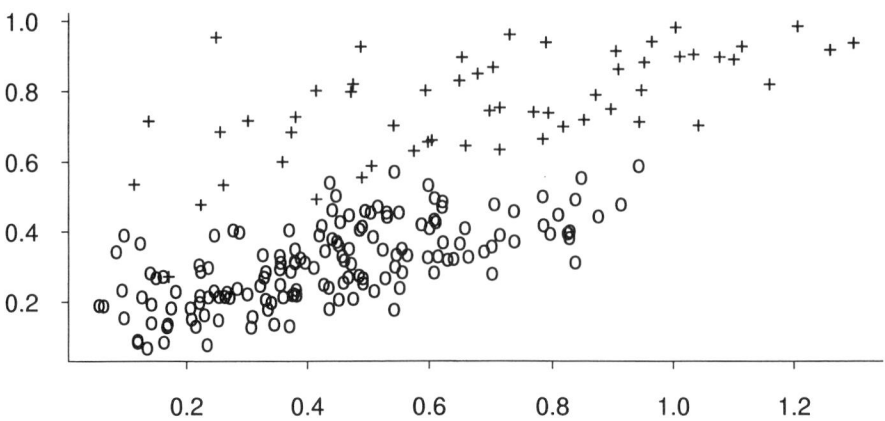

Fig. 4. Dispersion plots in G-space for the ozone example. For each pair of stations, the distance between the stations is plotted along the horizontal axis and the dispersion is plotted along the vertical axis. For any dispersion calculation involving one of the three outlying stations, the plotted point is +; for the remainder, it is o.

3.1   Alternative approaches to nonstationary covariances

Although the deformation approach to nonstationary spatial processes is the one which has been most extensively developed, it is far from being the only approach to this problem. One fairly straightforward approach is the 'moving window' method due to Haas (1990, 1995). According to this, kriging at a particular location is based only on a subset of the monitoring stations within a given distance of the location for which a prediction is needed. In a space-time context, the window is defined in time as well as space, thus allowing for temporal nonstationarity as well. The actual size of the window is chosen by some form of cross-validation scheme. This approach is relatively easy to apply, in part because it avoids the complications inherent is specifying a full model for the nonstationary case. On the other hand, the lack of a fully specified model is a disadvantage in some contexts.

Another approach is the *orthogonal expansion approach*, discussed by Nychka and Saltzman (1998), Holland *et al.* (1999). This is based on

$$C(s,t) = \sigma(s)\sigma(t)\left[\rho e^{-\|s-t\|/R} + \sum_{\nu=1}^{m} \lambda_\nu \psi_\nu(s)\psi_\nu(t)\right], \qquad (24)$$

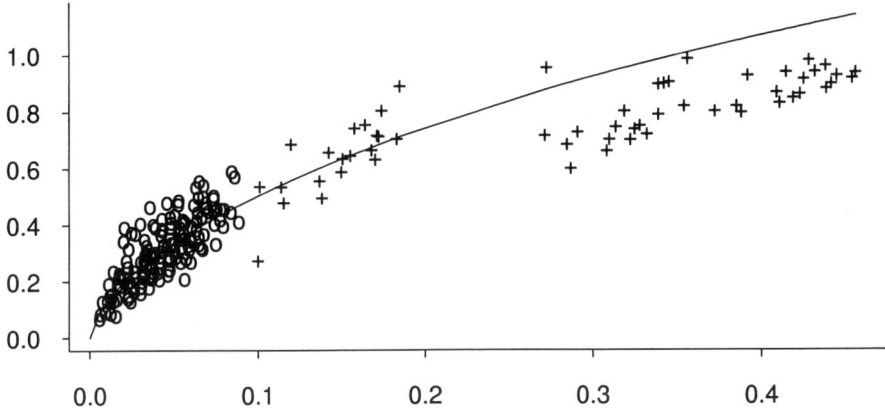

Fig. 5. Figure 4 redrawn in D-space, with fitted Matérn variogram curve.

where $\sigma(\cdot)$ is a spatially varying standard deviation, $0 \le \rho \le 1$, λ_ν is a non-negative weight, and ψ_ν is the νth eigenfunction in some orthogonal function expansion. As Nychka and Saltzman point out, the representation (24) is equivalent to a representation of the random field as

$$Z(s) = \sigma(s)\left\{\rho Z_0(s) + \sum_{\nu=1}^{m} a_\nu \lambda_\nu^{1/2}\psi_\nu(s)\right\}, \qquad (25)$$

where Z_0 is a stationary isotropic random field and $a_1, \ldots, a_m$ are independent (of each other and of Z_0) standard normal random variables. This is therefore a hybrid between the geostatistical approach based on stationary random fields, and an approach known as empirical orthogonal functions, which is popular in atmospheric sciences. Presumably, the approach is not restricted to the exponential covariance function in (24) but other forms of stationary isotropic covariance function could also be used. At the present time, the whole approach based on expansions of the form of (24) has not been developed very extensively.

4 Models defined by conditional probabilities

An entirely different approach to modeling spatial fields is through families of conditional distributions for the observation at a site given its neighbors. Such models are most naturally defined for a discrete set of locations, though they

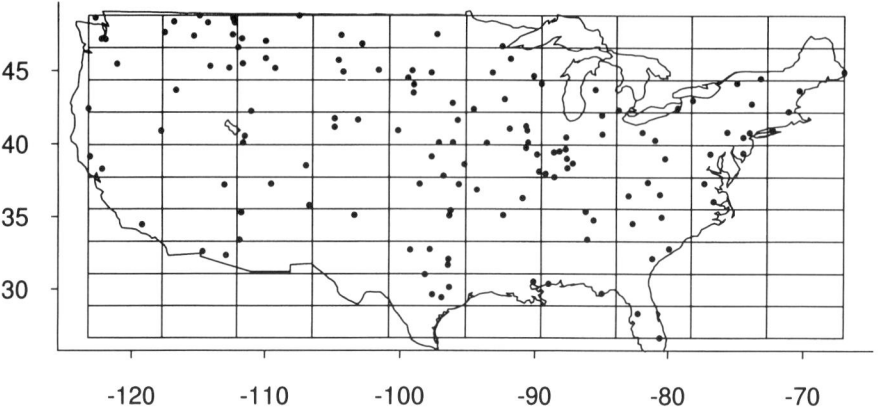

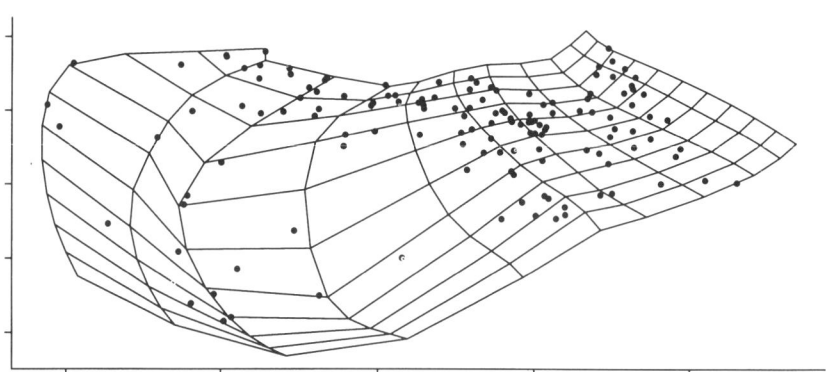

Fig. 6. G-space and D-space for climatological example.

are often applied in situations where the underlying random field is defined continuously in space. The whole approach stems from Besag (1974), though there have been many extensions and variations on the approach in recent years. A small sample of recent papers are Besag *et al.* (1995), Waller *et al.* (1997), Best *et al.* (1999), Diggle *et al.* (1998), Wolpert and Ickstadt

Raw Dispersions

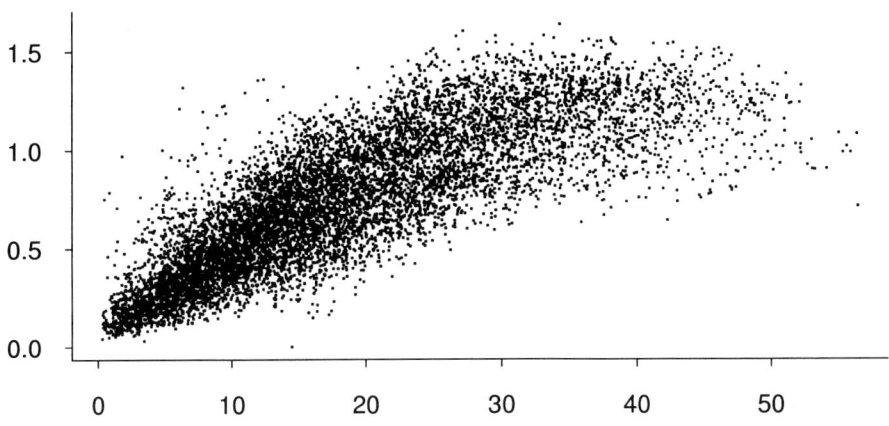

Transformed Dispersions and Matern fit

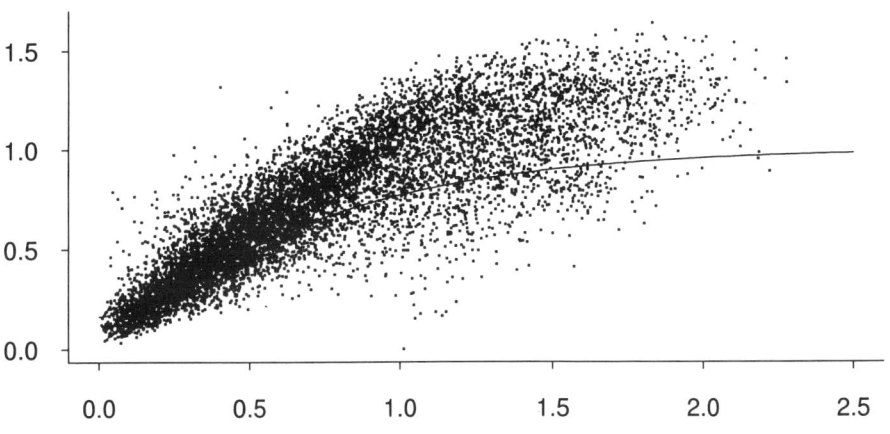

Fig. 7. Top plot: sample dispersion function for climatological data in G-space. Bottom plot: same in D-space, with fitted Matérn curve.

(1998), Besag and Higdon (1999). In the present discussion, we do not attempt anything like a complete survey of recent developments, but will concentrate on outlining the fundamental ideas, stemming from Besag (1974).

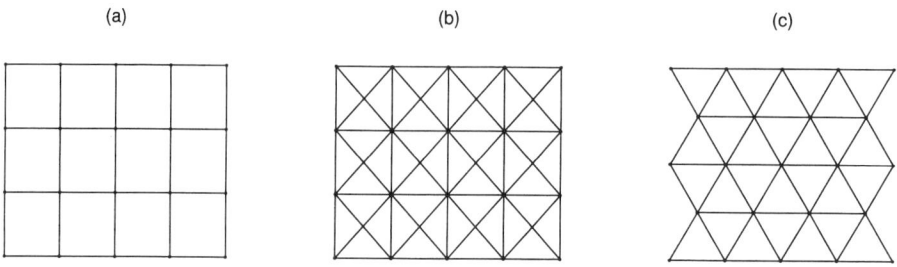

(a)   (b)   (c)

Fig. 8. Examples of regular lattices. The vertices of the graph represent spatial locations, and two vertices are said to be neighbors if there is edge of the graph joining them.

The starting point is that one considers data defined on a discrete lattice with an underlying neighborhood structure. The lattice may have some regular shape as in Fig. 8, though often a lattice is defined simply as a graph with no regular structure. Where the true field is distributed continuously in space, the spatial locations are usually aggregated into discrete cells so that such a model can be applied.

A model is defined by specifying the conditional distribution at a particular site given its neighbors. For example, in the *autologistic model*, each site value X_i is either 0 or 1, and satisfies

$$
\begin{aligned}
\Pr\{X_i = 1 \mid X_j = x_j,\ j \neq i\} &= \Pr\{X_i = 1 \mid X_j = x_j,\ j \in N_i\} \\
&= \frac{\exp\left(\alpha_i + \sum_{j \in N_i} \beta_{ij} x_j\right)}{1 + \exp\left(\alpha_i + \sum_{j \in N_i} \beta_{ij} x_j\right)}
\end{aligned}
\tag{26}
$$

where N_i is the set of neighbors of the site i, α_i is a coefficient for each site, and β_{ij} is an interaction coefficient between neighboring sites i and j.

A corresponding model for normally distributed systems is the *autonormal model*, defined as follows: the conditional distribution of X_i given X_j, $j \neq i$, is normal with variance σ^2 and mean

$$
\mu_i + \sum_{j \in N_i} \beta_{ij}(X_j - \mu_j).
$$

Thus μ_i is the mean at site i and β_{ij} is again a pairwise interaction component defined for neighboring sites i and j.

It should be noted that the autonormal model just defined is not the same as

$$
X_i = \mu_i + \sum_{j \in N_i} \beta_{ij}(X_j - \mu_j) + \epsilon_i, \quad \epsilon_i \sim N(0, \sigma^2) \text{ (independent)},
$$

which is called the *simultaneous equation model* and corresponds to a quite different joint distribution of the $\{X_i\}$. Besag (1974) discussed this distinction in detail.

A key question with models of this form is whether the family of conditional probabilities is consistent with some set of joint probabilities on all the random variables. If it is not, then clearly the model is not well defined. Both the autologistic and autonormal models are examples of a *Markov random field* (MRF), and the question of whether they are well-defined models is answered through a very general result for MRFs known as the Hammersley–Clifford theorem (Besag 1974, Clifford 1990). For pairwise interaction models such as logistic and autonormal, the answer essentially reduces to the statement that both the neighborhood structure and the interaction coefficients must be symmetric: $j \in N_i$ if and only if $i \in N_j$, and $\beta_{ij} = \beta_{ji}$. For example, in the autologistic case, the joint density of $\mathbf{X} = \{X_i\}$, evaluated at $\mathbf{x} = \{x_i\}$, is of the form

$$p(\mathbf{x}) \propto \exp \left(\sum_k \alpha_k x_k + \frac{1}{2} \sum_j \sum_{k \in N_j} \beta_{jk} x_j x_k \right), \qquad (27)$$

while in the autonormal case, the corresponding joint density is

$$p(\mathbf{x}) = (2\pi\sigma^2)^{-1/2} |B|^{1/2} \exp \left\{ -\frac{1}{2\sigma^2} \sum_j \sum_k (x_j - \mu_j) b_{jk} (x_k - \mu_k) \right\}, \qquad (28)$$

where

$$b_{jk} = \begin{cases} 1 & \text{if } j = k, \\ -\beta_{jk} & \text{if } j \in N_k, \\ 0 & \text{otherwise.} \end{cases}$$

It is readily checked that (27) implies (26), and that (28) implies the autonormal model. Note that (28) is equivalent to the statement that the $\{X_i\}$ are multivariate normally distributed with covariance matrix $\Sigma = B^{-1}$.

4.1 Estimation of MRF models

(i) *Exact MLE*

Exact maximum likelihood is usually feasible only for Gaussian processes. To see why, note that $p(\mathbf{x})$ in (27) is defined only up to an unknown constant of proportionality, and direct maximum likelihood is not possible without evaluating that constant. However, exact calculation of the constant can only be performed by summing (27) over all possible states of the system, which is impossible unless the system is extremely small. On the other hand, in (28), we are able to evaluate the constant analytically, so in this case it is possible to calculate the likelihood function exactly. Virtually all non-Gaussian cases are like (27), in that the model is specified up to an unknown constant of proportionality, but there is no direct method of evaluating the constant.

(ii) *Maximum pseudo-likelihood*

Besag (1975) proposed estimating the unknown parameters of the model, θ say, by maximizing the quantity

$$PL(\theta) = \prod_i p(X_i \mid X_j, \ j \in N_i; \ \theta), \tag{29}$$

which he called the pseudo-likelihood. This has the advantage of being easy to calculate, and behaving in many respects like a likelihood function, though it is not equivalent to the likelihood function, even asymptotically, and in some contexts the maximum pseudo-likelihood estimates are much less efficient than the maximum likelihood estimates. In Gaussian cases, it is possible to compare the two methods directly (Besag and Moran 1975, Besag 1977). The method has fallen under something of a cloud in recent years, partly because of the growing popularity of simulation-based estimation methods such as the Gibbs sampler, though it remains of interest as a theoretical technique (see e.g. Comets, 1992).

(iii) *Simulated maximum likelihood estimators*

The idea behind this was proposed by Penttinen (1984), and extended by Geyer and Thompson (1992). Suppose we have a model of the form

$$p(\mathbf{x}; \theta) = C(\theta) F(\mathbf{x}; \theta),$$

where $F(\mathbf{x}; \theta)$ is a specified function of data $\mathbf{x}$ and unknown parameter θ, and $C(\theta)$ is a normalizing constant, in principle computable by summing or integrating over all possible values of $\mathbf{x}$, but in practice not computable. Our objective is to calculate a simulation-based approximation to the maximum likelihood estimate for a particular realization of the Markov random field, $\mathbf{X}$ say.

Suppose we generate a Monte Carlo sample $\mathbf{X}^{(1)}, \ldots, \mathbf{X}^{(M)}$ of realizations of the random field $\mathbf{X}$ for some particular value of θ, say θ_0. It is not essential that $\mathbf{X}^{(1)}, \ldots, \mathbf{X}^{(M)}$ be independent. Then for $1 \le m \le M$,

$$
\begin{aligned}
\mathrm{E}_{\theta_0} \left\{ \frac{F(\mathbf{X}^{(m)}; \theta)}{F(\mathbf{X}^{(m)}; \theta_0)} \right\} &= \sum_{\mathbf{x}} \frac{F(\mathbf{x}; \theta)}{F(\mathbf{x}; \theta_0)} \cdot C(\theta_0) F(\mathbf{x}; \theta_0) \\
&= C(\theta_0) \sum_{\mathbf{x}} F(\mathbf{x}; \theta) \\
&= \frac{C(\theta_0)}{C(\theta)}.
\end{aligned}
\tag{30}
$$

If the distribution of $\mathbf{x}$ is continuous, then the sum in (30) is replaced by an integral. Based on (30), therefore,

$$\frac{1}{M} \sum_{m=1}^{M} \frac{F(\mathbf{X}^{(m)}; \theta)}{F(\mathbf{X}^{(m)}; \theta_0)} \cdot \frac{F(\mathbf{X}; \theta_0)}{F(\mathbf{X}; \theta)} \tag{31}$$

is an unbiased estimate of the likelihood ratio of θ_0 to θ, in other words

$$\frac{C(\theta_0)F(\mathbf{X};\theta_0)}{C(\theta)F(\mathbf{X};\theta)}.$$

An estimator of θ defined so as to minimize (31), then, may be regarded as a simulated maximum likelihood estimator, and provided M is sufficiently large, may be expected to be a good approximation to the true MLE.

Note that as the method is commonly applied, the sample $\mathbf{X}^{(1)}, \ldots, \mathbf{X}^{(M)}$ is generated only once, with a fixed θ_0, and then (31) is treated as a deterministic function of θ. However, partly as a check on the simulation-sensitivity of the method, and also in some cases to speed up the convergence to the MLE, it is possible to update θ_0 during the procedure, for example, minimizing (31) to get an initial estimate $\hat{\theta}^{(1)}$, then taking $\theta_0 = \hat{\theta}^{(1)}$ and repeating the process to get an estimate $\hat{\theta}^{(2)}$, and so on.

Many practical issues are raised by this method – for example, the choice of M, and the method of simulation used to generate the individual $\mathbf{X}^{(m)}$ fields. Many modern ideas such as the Gibbs sampler (Geman and Geman, 1984), auxiliary variable methods (Swendsen and Wang 1987), Metropolis-Hastings algorithms (Besag *et al.* (1995) reviewed all of these methods) and perfect simulation (Propp and Wilson, 1996) have been developed in recent years for simulation from random fields, but we shall not attempt to review these here, beyond mentioning that they are all relevant for the kinds of models that have been discussed.

4.2   Comparisons between geostatistical and MRF approaches

One aspect of spatial models which has not been very much explored is the connection between the geostatistical approaches of sections 2 and 3, and the MRF models of the present section.

In multivariate normal cases, one way to characterize the difference is that geostatistical approaches work by specifying the covariance matrix of the observations, Σ say, in terms of unknown parameters θ. The MRF approach, through (28), amounts to a parametric model for the inverse autocovariance matrix, $B = \Sigma^{-1}$. There is no obvious reason for preferring one to the other.

One issue is that of *marginalization* (the author thanks Julian Besag for drawing his attention to this issue). Suppose we have a random field defined for a continuous space variable. In order to fit it in with a MRF approach, we must first restrict the data to some form of lattice. For example, one approach which has been adopted in some agricultural or epidemiological contexts is to aggregate data by county, treating counties with common boundaries as neighbors in the MRF specification. But the question then arises: how much are the resulting joint distributions invariant to the arbitrary specifications of county boundaries?

As a concrete example of this problem, we consider a model with data $x_1, \ldots, x_{n+1}$ corresponding to averages of some spatially distributed quantity over $n + 1$ counties, and suppose a joint density $p_1(x_1, \ldots, x_{n+1})$ is specified. Now suppose some administrative authority decides to amalgamate the nth and $(n+1)$st counties, defining $x'_n = (x_n + x_{n+1})/2$. Let $p_2(x_1, \ldots, x_{n-1}, x'_n)$ denote the joint distribution under the new model. Logically, p_2 should be derived from p_1 through a marginalization condition of the form

$$p_2(x_1, \ldots, x_{n-1}, x'_n) = \int p_1(x_1, \ldots, x_{n-1}, x_n, 2x'_n - x_n) dx_n. \qquad (32)$$

The question then arises: do natural specifications of the models p_1 and p_2 satisfy relationships such as (32) ?

Although this is a very specific question it is meant to illustrate a general point, that there are natural consistency relations among probability distributions, and it may require some care to ensure that there are satisfied.

In the case of geostatistical models, the difficulty just described does not arise, because such models work by specifying the covariance between any pair of sites, and the covariance between two county averages is computed by an obvious integral of the pointwise covariance function over the two counties. Such an operation will always satisfy consistency relations such as (32) . However, for a typical MRF model, (32) will not be satisfied unless unusual care is taken in specifying the model. This appears to be an argument against the use of MRF models in cases where the lattice structure is not defined by the natural geometry of the system.

On the other hand, from other points of view, MRF models are more flexible — for example, we have already seen that they may be defined for binary data (through the autologistic model) and there are by now many models for count data with either marginal or conditional Poisson distributions, whereas all of our geostatistical discussion has been (implicity or explicitly) for normally distributed data. For this reason, the MRF models potentially lead to a much richer class of models.

4.3   Use of MRFs in hierarchical models

One of the most rapid developments in recent years has been the use of MRFs as a component model in some hierarchical structure, this greatly increasing the scope of applications for such models. As an example, Best *et al.* (1999) considered extensions of the Clayton–Kaldor (1987) model of the form

$$Y_i \mid \mu_i \sim \text{Pois}\{E_i \exp(\mu_i)\},$$

$$\mu_i = x_i^T \beta + u_i + v_i, \qquad (33)$$

in which Y_i is the count (of disease incidences, say) in a particular county or region i, E_i is the susceptible population and μ_i is a random intensity function;

μ_i is specified through a regression model $x_i^T \beta$ and additional random errors u_i and v_i. In the model of Best *et al.*, v_i are independent random errors but u_i are spatially dependent and specified through a conditional autoregressive (or autonormal) model.

Diggle *et al.* (1998) proposed something similar but based on geostatistical models for μ_i. Wolpert and Ickstadt (1998) also considered a model with Poisson counts, but for them, the underlying field was assumed to have a special structure with gamma marginal distributions. Evidently, these are only a few examples of what is possible with this kind of structure, and there are many possibilities for extensions to other kinds of marginal or conditional distributions.

4.4 Environmental applications of MRF models

So far, the applications of MRF models in physical environmental modeling have been very limited, but an exception is the recent paper by Cressie *et al.* (1999). That paper applied both geostatistical and MRF approaches to the prediction of a particulate matter field based on 27 monitoring stations in the area of Pittsburgh, Pennsylvania. They also gave a more detailed discussion than we have here of the relative merits of the two approaches.

5   Spatial design of experiments

In this and the following section, we deal much more briefly with two other topics which are of major importance, but which space does not permit us to develop in more detail in the present review.

The issue of *spatial design of experiments* arises most commonly in developing a monitoring network. To take one example where the problem arises, the US Environmental Protection Agency is responsible for monitoring a large number of airborne and water-borne pollutants. There are legal requirements and political considerations in deciding where to place monitors but, in most cases, the Agency still has discretion over how many monitors to place in a particular city or region, and over the precise location of these monitors. For example, in a city such as Chicago, should there be a higher concentration of monitors near Lake Michigan (where, experience shows, there is a higher variability in atmospheric conditions due to local meteorological and lake-based effects), or should the monitors be evenly distributed over the suburban regions near the city? Another version of the problem (Oehlert, 1996) is when there is an existing network of monitors but, for cost-saving reasons, it is desired to close down a certain fract ion of the network.

One formulation of the problem developed by Le and Zidek (1992) is to assume a relatively large but discrete set of potential sites of interest, divided into 'gauged' sites (the ones where monitors are actually located) and 'ungauged' sites. In broad terms, the problem then becomes to select the set of gauged sites so that the predictions at the ungauged sites are as accurate as possible. There are then two issues: (i) how to specify a suitable criterion for accuracy at a large set of ungauged sites, (ii) how to find designs which perform well under such a criterion.

Problem (i) is essentially the classical problem of optimal design of experiments, for which there are classical criteria such as D-optimality, E-optimality and so forth (see, e.g., Fedorov (1972) or Atkinson and Donev (1992)), or Bayesian approaches which are often formulated in terms of information-theoretic criteria, following Bernardo (1979).

Even when the criterion is well defined, the problem of selecting, for example, the best 99 out of 249 potential sites (Oehlert, 1996) is a formidable combinatorial problem which defies exact solution. In practice, *ad hoc* addition-deletion rules have been developed. Other references include Brown *et al.* (1994), Oehlert (1993, 1995), Le *et al.* (1997) and Nychka and Saltzman (1998).

6   Spatial-temporal data

Spatial-temporal analysis is concerned with random fields of the form $X(s,t)$, where s is a location or site variable and t is time. (In contrast to previous sections, here we use t specifically to denote time.) A model for the random field then requires that we specify the joint distribution of $\{X(s_1, t_1), \ldots, X(s_n, t_n)\}$ for any combination of space-time pairs $(s_1, t_1), \ldots, (s_n, t_n)$. In the simplest case where we assume the field is Gaussian, this means specifying the covariance between $X(s,t)$ and $X(s', t')$ for any s, t, s', t'.

The simplest models are the *separable models*, for which the spatial-temporal covariance function factorizes as

$$\text{Cov}\left\{X(s,t), X(s', t')\right\} = C(s, s')\gamma(t - t') \tag{34}$$

where $C(\cdot, \cdot)$ is a spatial covariance function and $\gamma(\cdot)$ is the covariance function of a stationary time series.

An early example of the application of (34) was the paper by Haslett and Raftery (1989), who used it to model the joint distribution of wind speeds at 12 stations in Ireland. In their model, spatial covariance was represented by a stationary, isotropic model with an exponential variogram (15), while the temporal covariance function they adopted was the fractional ARIMA process (Beran, 1994), which incorporates long-range dependence in time.

Apart from the convenience of the mathematical representation, another advantage of separable models is their computational tractability. For example, Mardia and Goodall (1993) considered the case of an $m \times n$ data matrix $\mathbf{X} = (X_{ik})$, where X_{ik} is the value at the ith spatial location and the kth time point, with a model of form

$$\mathbf{X} \sim N(\mu, \Lambda \otimes \Gamma)$$

where $\Lambda = (\lambda_{ij})$ is a spatial covariance matrix and $\Gamma = (\gamma_{kl})$ is a temporal covariance matrix. (The $\otimes$ notation is interpreted to mean $\mathrm{Cov}(X_{ik}, X_{jl}) = \lambda_{ij}\gamma_{kl}$.) Assuming multivariate normality, the joint density of $\mathbf{X}$ may be written as

$$(2\pi)^{-mn/2} |\Lambda|^{-m/2} |\Gamma|^{-n/2} \exp\left[-\frac{1}{2} \, \mathrm{tr}\left\{\Lambda^{-1}(X-\mu)\Gamma^{-1}(X-\mu)^T\right\}\right]. \quad (35)$$

The importance of (35) is that it shows that one only needs to compute the determinant and inverse for the matrices Λ and Γ, and not for the $(mn) \times (mn)$ matrix $\Lambda \otimes \Gamma$. To do the latter directly would, of course, both be much slower and consume far more computer storage space.

Nevertheless, despite the pragmatic advantages of separable models, it is now increasingly recognized that they are not a realistic assumption for much practical data. It is likely that much work over the next few years will be devoted to the development of new non-separable models for spatial-temporal processes. At the present time, the literature is scattered, with few coherent themes. A few recent developments are:

(i) Carroll *et al.* (1997) proposed a model for the spatial-temporal distribution of atmospheric ozone in the Houston area. By examining correlation vs. distance plots at different time lags, they proposed a specific parametric form for the spatial-temporal covariance function and fitted it through a cross-validation-type method, full maximum likelihood being infeasible for the data sizes they were considering. The model appeared to have good practical properties for the specific data to which it was applied, but the general form of the covariance does not appear to be positive definite (Cressie, 1997), and without this, the model is of limited general utility.

(ii) Jones and Zhang (1997) have proposed a class of continuous-parameter models derived from stochastic partial differential equations.

(iii) The most promising approach at the present time is based on generalizations of the Kalman filter, in which the set of spatial variables at each time is viewed as a random vector with dynamic equations for the temporal evolution. The approach is highly computationally intensive, but it is feasible within the structure of Bayesian hierarchical models, and has the additional advantage (e.g., in an atmospheric science context) that the dynamical equations can sometimes be suggested by the physics of the process being observed. Some representative papers on this approach are those by Wikle *et al.* (1998), Wikle and Cressie (1999).

7   Hierarchical models for spatial trends

In the remainder of this chapter, we outline some recent ideas for modeling of a spatially varying temporal trend. The canonical problem is suggested by global warming: empirical data show an increasing temperature trend at many points of the earth's surface, which some scientists interpret as evidence of an anthropogenically induced greenhouse effect. However, the trend is not the same at all places. For example, within the continental United States, the increasing temperature trend is greatest in the northern midwest states, while in other parts of the country, such as the south east, there has been little or no observed trend. This effect is clearly seen in individual time series at different spatial locations, but there is wide variability in estimated trends from one location to another which may simply be due to the statistical error in estimating the trends. Therefore, the problem arises of 'smoothing' the trends available from individual time series, to obtain an overall picture of how the trend varies with space. A similar problem has been studied in connection with trends in atmospheric SO_2 levels across the United States (Holland *et al.* 2000), and in a rather different context, for the variability of particulate matter-based mortality across different cities (Dominici *et al.* 2000).

One plausible model is as follows. Suppose there is a linear spatially dependent trend, denoted $Z_1(s)$ for location s, for which

$$\mathrm{E}\{Z_1(s)\} = x(s)^T \beta, \quad \mathrm{Cov}\{Z_1(s), Z_1(s')\} = C(s, s'; \theta), \qquad (36)$$

in which $x(s)$ is a known spatially dependent covariate vector, β an unknown vector of regression coefficients, and C a spatial covariance function depending on parameters θ. We assume that we cannot observe $Z_1(s)$ directly, but instead, for a fixed set of spatial locations $s = s_1, \ldots, s_m$, we observe a time series $\{Y(s, t_1), \ldots, Y(s, t_n)\}$, whose distribution depends on $Z_1(s)$ as well as other unknown parameters which we shall denote by ϕ. This suggests a natural 'hierarchical model' structure in which there is a top level of the hierarchy represented by the unknown parameters (β, θ, ϕ), a middle level represented by the unobserved process $Z_1(\cdot)$, and a bottom level represented by the observed data $\{Y(s_i, t_j), \ i = 1, \ldots, m, \ j = 1, \ldots, n\}$. The (Bayesian) specification of the model is completed by a prior distribution on (β, θ, ϕ), and the whole structure would then be analyzable by modern methods of hierarchical models analysis (see, e.g., Gilks *et al.*, 1996). However, for typical data sets with large numbers of spatial locations as well as many time points, such an approach, directly implemented, would be very time consuming.

We therefore propose an alternative approach which avoids the full complications of a hierarchical models analysis. Suppose, for each observed spatial location s_i, we calculate an estimate of $Z_1(s_i)$, which we denote $\tilde{Z}_1(s_i)$, based just on the time series $Y(s_i, t_j), \ j = 1, \ldots, n$. This may be based on any model appropriate for that time series. Since most statistical methods lead to approximately normal distributions of estimators in large samples, we may assume

$$\tilde{Z}_1(s_i) = Z_1(s_i) + \xi(s_i), \qquad (37)$$

where $\{\xi(s_1), \ldots, \xi(s_m)\}$ is a zero-mean vector of errors such that

$$\mathrm{Cov}\{\xi(s_i), \xi(s_j)\} = w_{ij}, \qquad (38)$$

$W = (w_{ij})$ being the error covariance matrix. Moreover, these errors, corresponding to measurement errors at individual stations, may be assumed independent of the true trend surface $Z_1(\cdot)$.

By combining (36)–(38), we have a model

$$\mathrm{E}\{\tilde{Z}_1(s_i)\} = x(s_i)^T\beta, \quad \mathrm{Cov}\{\tilde{Z}_1(s_i), \tilde{Z}_1(s_j)\} = C(s_i, s_j; \theta) + w_{ij}, \qquad (39)$$

together with (approximate) joint normality.

We make one final simplifying assumption, which is to assume that the $\{w_{ij}\}$ in (38) are known. This is justified by the fact that, since these estimates arise from the statistical errors in individual time series, they can be characterized from standard error calculations in the individual time series estimations. At any rate, we should be able to approximate the w_{ij}s much more accurately than we could initially guess the covariances of the true Z_1 process, in other words, the C function in (39).

Thus we are led to a model of the form (39) in which $C(\cdot, \cdot; \theta)$ is a parametric covariance function depending on parameter vector θ, and $W = (w_{ij})$ is a known error covariance function.

We now give two examples of this approach. The first is again based on the Historical Climatological Network (see section 3), from which we have calculated time trends in winter mean temperatures, over the period 1965–1996, for each of 184 stations, representing the data as the sum of a linear trend and an AR(p) staionary time series and estimating the slope of the linear trend by maximum likelihood. This leads to an estimate $\tilde{Z}_1(s_i)$ of the temperature trend at each station, with an accompanying standard error. The squares of the standard errors are taken as values of the diagonal entries w_{ii}, while the off-diagonal entries, w_{ij} for $i \neq j$, are taken as 0. A homogeneous spatial model with Gaussian semivariogram function (16) is assumed for Z_1, and a regression model $x(s)^T\beta$ corresponding to a cubic polynomial function of s, after testing various alternative models both for the spatial covariance function and for the polynomial regression model. Kriging is then used to construct an estimate of the surface $Z_1(s)$. The resulting estimate (contour plot at the top, perspective plot at the bottom) is shown in Fig. 9. The result shows how the estimated trend varies across the country, with the largest trend around the great lakes region, consistent with the earlier description. Although we do not give error estimates here, it should be pointed out that the methodology does allow us to obtain approximate error bounds of the reconstructed surface.

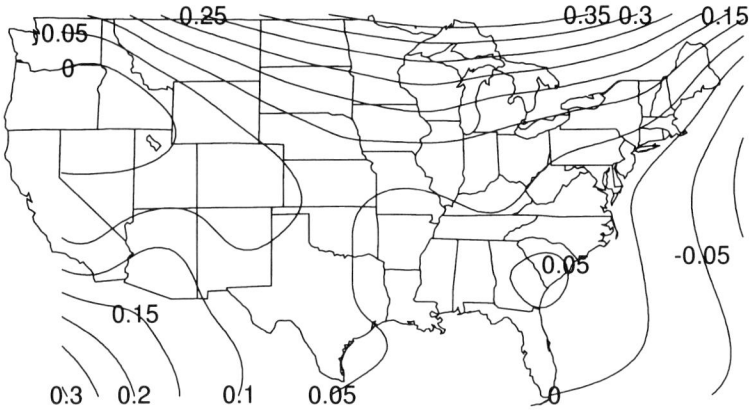

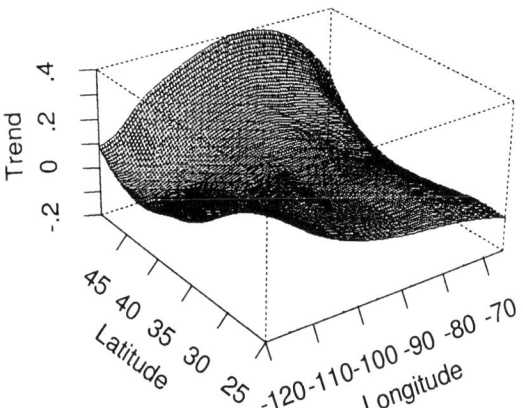

Fig. 9. Reconstructed trend surface for US temperatures. Top picture: contour plot. Bottom picture: perspective plot. In the perspective plot, increasing values of Y correspond to increasing latitude °N, and increasing X to decreasing longitude (°W).

The second example is based on Holland *et al.* (2000). In this paper, sulfur dioxide measurements at 35 locations in the eastern US over the time period 1989–1995 were characterized as functions of seasonal trends, meteorology, and an overall additive linear trend. A generalized additive model (Hastie and Tibshirani, 1990), applied to the logarithms of weekly sulfur dioxide totals, was used to estimate the trend at each station, after adjusting for the seasonal and meteorological terms. In this example, instead of assuming the $W = (w_{ij})$ matrix in (39) is diagonal, w_{ij} is estimated for each (i, j) pair by a jackknife procedure. The model (39) is then fitted, again with a Gaussian semivariogram kernel (16). The resulting estimate of the trend surface is shown in Fig. 10.

The importance of this analysis, in the context of evaluating improved regional-scale air quality resulting from electric utility emission reductions, is that it allows the characterization of estimated trends in sulfur dioxide, not only at the monitoring stations themselves, but also on a regional basis. The trends can be compared to corresponding changes in sulfur dioxide emissions to evaluate the impact of reduced emissions. For the period, 1989–1995, reduced emissions levels from large electric utilities are similar to the estimates of regional trends.

8   Conclusions

In this chapter, I have attempted to give a broad overview of current themes in spatial statistics, though concentrating on the geostatistical approach, which remains the most widely applied method in environmental statistics. Modern methods of estimation, such as REML or Bayesian estimation, allow these processes to be estimated without some of the *ad hoc* features of earlier proposals, and the Bayesian procedures in particular have the advantage that when applied to the spatial prediction or kriging problem, they automatically allow for the uncertainty of the estimated model parameters, a deficiency of classical kriging. Extensions to nonstationary processes and spatial-temporal models are major themes of current research and may be expected to remain so for some time to come.

The final substantive part of the chapter discussed a particular application of these techniques, to the estimation of spatially dependent trends. The method described in section 7 is intended to be fairly straightforward to apply as an extension of classical kriging, but here also there are possibilities for more general approaches, including fully Bayesian approaches.

Acknowledgements

This work was supported in part by N.S.F. grant DMS-9705166, and by E.P.S.R.C. Visiting Fellowship GR-K99015 at the Isaac Newton Institute.

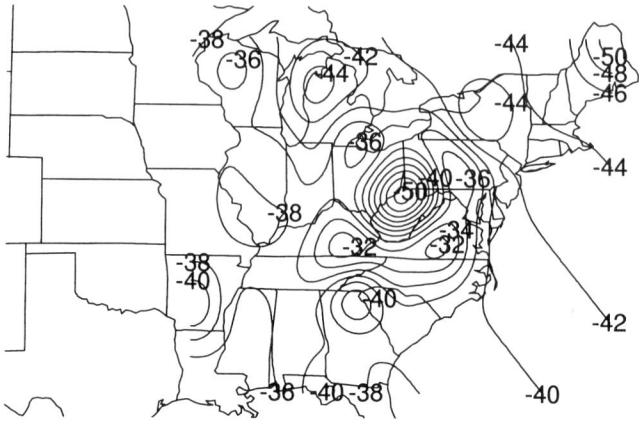

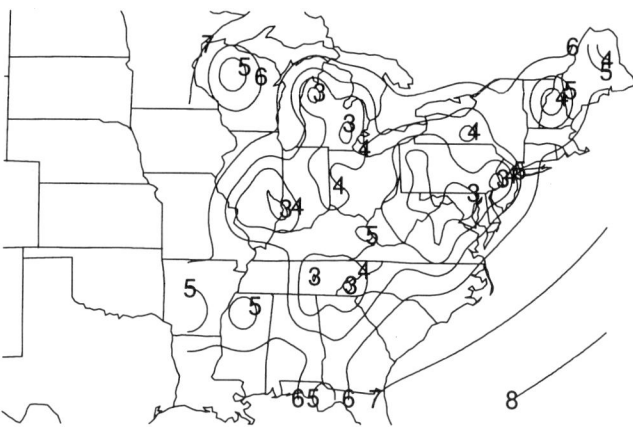

Fig. 10. Reconstructed trend surface for sulfur dioxide data (top plot) and prediction error variances (bottom plot).

References

Abramowitz, M. and Stegun, I.A. (1964), *Handbook of Mathematical Functions*. National Bureau of Standards, Washington D.C., reprinted by Dover.

Atkinson, A.C. and Donev, A.N. (1992), *Optimum Experimental Designs*. Oxford University Press.

Beran, J. (1994), *Statistics for Long-Memory Processes*. Chapman and Hall.

Bernardo, J.M. (1979), Expected information as expected utility. *Ann. Statist.* **7**, 686-690.

Besag, J. (1974), Spatial interaction and the statistical analysis of lattice systems. *J.R. Statist. Soc. B* **36**, 192–225.

Besag, J. (1975), Statistical analysis of non-lattice data. *Statistician* **24**, 179–195.

Besag, J. (1977), Efficiency of pseudolikelihood estimation for simple Gaussian field. *Biometrika* **64**, 616–618.

Besag, J., Green, P.J., Higdon, D. and Mengersen, K. (1995), Bayesian computations and stochastic systems. *Statist. Sci.* **10**, 1–66.

Besag, J. and Higdon, D. (1999), Bayesian analysis of agricultural field experiments (with discussion). *J. R. Statist. Soc. B* **61**, 691–746.

Besag, J. and Moran, P.A.P. (1975), On the estimation and testing of spatial interaction in Gaussian lattice processes. *Biometrika* **62**, 555–562.

Best, N.G., Arnold, R.A., Thomas, A., Waller, L.A. and Conlon, E.M. (1999), Bayesian models for spatially correlated disease and exposure data (with discussion). In *Bayesian Statistics 6*, edited by J.M. Bernardo, J.O. Berger, A.P. Dawid and A.F.M. Smith, Oxford University Press, 131–156.

Brown, P.J., Le, N.D. and Zidek, J.V. (1994), Multivariate spatial interpolation and exposure to air pollutants. *Canad. J. Statist.* **22**, 489–509.

Carroll, R.J., Chen R., George, E.I., Li, T.H., Newton, H.J., Schmiediche, H. and Wang, N. (1997), Ozone exposure and population density in Harris County, Texas. *J. Amer. Statist. Assoc.* **92**, 392–404.

Clayton, D.G. and Kaldor, J. (1987), Empirical Bayes estimates of age-standardized relative risks for use in disease mapping. *Biometrika* **43**, 671–681.

Clifford, P. (1990), Markov random fields in statistics. In *Disorder in Physical Systems: a Volume in Honour of John M. Hammersley*, edied by G.R. Grimmett and D.J.A. Welsh, , Oxford University Press.

Comets, F. (1992), On consistency of a class of estimators for exponential families of Markov random fields on a lattice. *Ann. Statist.* **20**, 455–468.

Cressie, N. (1993), *Statistics for Spatial Data*, second edition. John Wiley.

Cressie, N. (1997), Comment on Carroll *et al.* (1997). *J. Amer. Statist. Assoc.* **92**, 411–413.

Cressie, N., Kaiser, M.S., Daniels, M.J., Aldworth, J.W., Lee, J., Lahiri, S.N. and Cox, L.H. (1999), Spatial analysis of particulate matter in an urban environment. Preprint, Iowa State University.

Diggle, P.J., Tawn, J.A. and Moyeed, R.A. (1998), Model-based geostatistics (with discussion). *Appl. Statist.* **47**, 299–350.

Dominici, F., Samet, J.M. and Zeger, S.L. (2000), Combining evidence on air pollution and daily mortality from the 20 largest US cities: a hierarchical modelling strategy (with discussion). *J. R. Statist. Soc. A* **163**, to appear.

Fedorov, V.V. (1972), *Theory of Optimal Experiments.* Academic Press.

Geman, S. and Geman, D. (1984), Stochastic relaxation, Gibbs distributions and the Bayesian restoration of images. *IEEE Trans. Patt. Anal. Mach. Intell.*, **6**, 721–741.

Geyer, C.J. and Thompson, E.A. (1992), Constrained Monte Carlo maximum likelihood for dependent data (with discussion). *J. R. Statist. Soc. B 54*, 657–699.

Gilks, W.R., Richardson, S. and Spiegelhalter, D.J. (eds.) (1996), *Markov Chain Monte Carlo in Practice.* Chapman and Hall.

Guttorp, P., Meiring, W. and Sampson, P. (1994), A space-time analysis of ground level ozone data. *Environmetrics* **5**, 241–254.

Haas, T.C. (1990), Lognormal and moving-window methods of estimating acid deposition. *J. Amer. Statist. Assoc.* **85**, 950–963.

Haas, T.C. (1995), Local prediction of a spatio-temporal process with an application to wet sulfate deposition. *J. Amer. Statist. Assoc.* **90**, 1189–1199.

Handcock, M.S. and Stein, M. (1993), A Bayesian analysis of kriging. *Technometrics* **35**, 403–410.

Harville, D.A. (1974), Bayesian inference for variance components using only error contrasts. *Biometrika* **61**, 383–385.

Harville, D.A. and Jeske, D.R. (1992), Mean squared error of estimation or prediction under a general linear model. *J. Amer. Statist. Assoc.* **87**, 724–731.

Haslett, J. and Raftery, A.E. (1989), Space-time modelling with long-memory dependence: assessing Ireland's wind power resource. *Appl. Statist.* **38**, 1–21.

Hastie, T.J. and Tibshirani, R.J. (1990), *Generalized Additive Models.* Chapman and Hall.

Holland, D.M., De Oliveira, V., Cox, L.H. and Smith, R.L. (2000), Estimation of regional trends in sulfur dioxide over the eastern United States. *Environmetrics*, to appear.

Holland, D., Saltzman, N., Cox, L.H. and Nychka, D. (1999), Spatial prediction of sulfur dioxide in the eastern United States. In *geoENV II — Geostatistics for Environmental Applications*, edited by J. Gómez-Hernández, A. Soares and R. Froidevaux, Kluwer, 65–76.

Jones, R.H. and Zhang, Y. (1997), Models for continuous stationary spatial-temporal processes. In *Modelling Longitudinal and Spatially Correlated Data: Methods, Applications and Future Directions*, edited by T.G. Gregoire *et al.*, *Lecture Notes in Statistics* **122**, Springer Verlag, 289–298.

Kitanidis, P.K. (1983), Statistical estimation of polynomial generalized covariance functions and hydrologic applications. *Wat. Resour.s Res.* **19**, 909–921.

Le, N.D. and Zidek, J.V. (1992), Interpolation with uncertain spatial covariances: a Bayesian alternative to kriging. *J. Multivar. Anal.* **43**, 351–374.

Le, N.D., Sun, W. and Zidek, J.V. (1997), Bayesian multivariate spatial interpolation with data missing by design. *J. R. Statist. Soc. B* **59**, 501–510.

Mardia, K.V. and Goodall, C.R. (1993), Spatial-temporal analysis of multivariate environmental monitoring data. In *Multivariate Environmental Statistics*, edited by G.P. Patil and C.R. Rao, Elsevier, 347–386.

Mardia, K.V. and Marshall, R.J. (1984), Maximum likelihood estimation of models for residual covariance in spatial regression. *Biometrika* **71**, 135-146.

Nychka, D. and Saltzman, N. (1998), Design of air quality monitoring networks. Chapter 4 of *Case Studies in Environmental Statistics*, edited by D. Nychka, W. Piegorsch and L.H. Cox, Lecture Notes in Statistics, **132**, Springer Verlag, 51–76.

Oehlert, G.W. (1993), Regional trends in sulfate wet deposition. *J. Amer. Statist. Assoc.* **88**, 390-399.

Oehlert, G.W. (1995), The ability of wet decomposition networks to detect temporal trends. *Environmetrics* **6**, 327–339.

Oehlert, G.W. (1996), Optimal shrinking of a wet decomposition network. *Atmos. Env.* **30**, 1347–1357.

Penttinen, A. (1984), Modelling interaction in spatial point patterns: parametric estimation by the maximum likelihood method. *J. Stud. Comput. Sci. Econ. Statist.* **7**.

Propp, J.G. and Wilson, D.B. (1996), Exact sampling with coupled Markov chains and applications to statistical mechanics. *Rand. Struct. Algor.* **9**, 223–252.

Ripley, B.D. (1981), *Spatial Statistics.* Wiley.

Ripley, B.D. (1988), *Statistical Inference for Spatial Processes.* Cambridge University Press,

Sampson, P.D. and Guttorp, P. (1992), Nonparametric estimation of nonstationary spatial covariance structure. *J. Amer. Statist. Assoc.* **87**, 108–119.

Smith, R.L. (1996), Estimating nonstationary spatial correlations. Unpublished; University of North Carolina, Chapel Hill. Available at www.unc.edu/depts/statistics/postscript/rs/nonstationary.ps.

Swends en, R.H. and Wang, J.-S. (1987), Nonuniversal critical dynamics in Monte Carlo simulations. *Phys. Rev. Lett.* **58**, 86–88.

Waller, L.A., Carlin, B.P., Xia, H. and Gelfand, A.E. (1997), Hierarchical spatio-temproal mapping of disease rates. *J. Amer. Statist. Assoc.* **92**, 607–617.

Wikle, C., Berliner, L.M. and Cressie, N. (1998), Hierarchical Bayesian space-time analysis. *J. Env. Ecol. Statist.* **5**, 117–154.

Wikle, C. and Cressie, N. (1999), A dimension reduction approach to space time Kalman filtering. *Biometrika* **86**, 815–829.

Wolpert, R.L. and Ickstadt, K. (1998), Poisson-gamma random field models for spatial statistics. *Biometrika* **85**, 251–267.

Zimmerman, D.L. and Cressie, N. (1992), Mean squared prediction error in the spatial linear model with estimated covariance parameters. *Ann. Inst. Statist. Math.* **44**, 27–43.

Useful Lies: Dynamics from Data

Alistair Mees

1   Introduction

Interpretation of observations from an experiment or a natural or artificial process is the foundation of science. When the process or experiment is a dynamical system (as it nearly always is), the traditional approach in the physical sciences has been to try to model it from first principles: that is, to work upwards from known physical laws to make a model of the system. The types of models used have been limited by the principle that useful models should be expressible in equations that are at most a few lines long. The fact that this is entirely a pragmatic decision is often overlooked, and occasionally scientists and engineers seem to believe that the world is in error if it delivers observations that do not fit simple equations. In fact, any model is a simplification, indeed, a lie. A model is a *useful lie* if the simplification leads to insights about the real-world process being observed.[1] A model is a *useless lie* or a *useless truth* if it is oversimplified to the point where it says nothing useful about the system it claims to model, or if it is so complex as to be little more than a listing of the data, without powers of generalization.

Johannes Kepler's great feat was to take a useless truth, the large data set collected by Tycho Brahe,[2] and to approximate it by a useful lie, the statement that each planet orbits the sun in an ellipse with the sun at one focus.[3] This is so much a part of our culture that we forget it is a lie: the planets do not orbit the sun, and their orbits are not ellipses. Nevertheless, it is an astonishingly useful lie. It is also the best kind of lie, a scientific lie, in that it makes falsifiable predictions, something that comes near to the very definition of a scientific theory.

In this article we are interested in useful lies, with the idea of 'most useful lie' quantified in terms of the compactness of a description of the data. We will study an extreme case: we assume we have some time series data for which it is not possible to write down any useful first-principles model, and we try to *reconstruct* the dynamics directly from the data. Here, 'dynamics' refers to a causal relationship: given observations y_t, $t = 1, \ldots, T$, we believe that there is some way in which each y_t depends on previous observations $y_{t-1}, \ldots, y_1$,

---

[1] The expression 'useful lies' appears to have originated with Richard Bandler, though in the context of psychological rather than physical modeling.

[2] 'Useless truth' refers to the data set thought of as a long list of numbers, and not to the intention in collecting the data.

[3] Actually, Kepler only had access to the data from Mars, which fortunately has a very eccentric orbit.

and we are trying to find this relationship. Thus we may be trying to estimate the function F in

$$y_{t+1} = F(y_t, y_{t-1}, \ldots, y_1)$$

or perhaps the conditional probability distribution P in

$$P(y_{t+1}|y_t, y_{t-1}, \ldots, y_1).$$

This estimation will be done typically by choosing some sort of basis in which to approximate F or P, or by selecting the best out of some parametrized model class. The reconstruction practiced in this article assumes that any prior understanding of the dynamics is taken into account in our choice of model class or basis; the question of how to construct models that are based partly on first principles and partly on data does not seem to have any very satisfactory answer at present.

To continue the solar system analogy, the methods of this article try to emulate Kepler's feat of reducing data to geometry. At the same time, the models described here do have predictive power, which is typically available numerically rather than analytically. Such geometric-numeric models may open the way for a Newtonian understanding, in which the geometry is explained by small closed-form equations, or it may not. Most systems cannot be represented by small closed-form equations (Rapp et al., 1999), and one of the lessons that has been learned from the study of nonlinear dynamics is that it is often not possible to model real-world systems by closed-form models, essentially because of sensitivity to initial conditions: the well-known butterfly effect. Regardless of the possibility of a Newtonian understanding, however, geometric-numeric models already have many useful features: through their statements about geometry, they may be able to help in discovering mechanism; they can make quantitative predictions, typically with better accuracy than conventional models; they can calculate dynamical invariants, which assist in classifying systems; they can provide an experimental test-bed which may suggest other experiments or observations that might elucidate the real-world system; and indeed, they can be used in nearly all of the ways that a traditional closed-form system can be used, including identification of steady states, study of bifurcations, and so on.

We begin with a description of the sort of questions that need to be answered if we are interested in nonlinear dynamics, and we point out the need for dynamical models to have both stochastic and dynamic aspects,' and the problems this raises. In the case of systems that operate over a finite number of symbols, we are able to borrow methods from information theory to build models with good properties; this is the theme of Section 2. More traditional dynamical models are discussed in Section 3, which deals with real variables and continuous (indeed, ideally, smooth) dynamical equations. In this case the role of geometry is apparent. For both types of system we discuss reconstruction methods, and show how reconstructions can be used to answer interesting dynamical questions.

1.1 Dynamical questions and answers

Since we are interested in understanding dynamics, we require additional abilities of our time series models over those traditionally required in time series modeling. We do not merely want a good fit, in some suitable sense: we also want some other properties, and we want to be able to calculate certain quantities. We usually need to understand the operation of the models beyond the black box level, at least to the extent that we can identify interesting states or regions of state space and the way they interact.

prediction As with all time series modeling, we need the ability to predict at least a step or two into the future given previously unobserved data. That is, given a state x_t and a model $\hat{F}$ that predicts $\hat{F}(x_t)$ as the next state, we need to ensure that some error measure such as

$$\mathbb{E}|y_t - \hat{F}(x_t)|^2$$

is small, where $\mathbb{E}$ is expected value. This is the familiar generalization problem, and is addressed by using appropriate model types, or at least a large class of models and a high-quality method of selecting a good one.

reproducing dynamics We need more than just predictability: if the model has captured the dynamics, it must be able to behave like the system it is modeling. That is, if we start the system in some initial state x_0, the trajectory $\{x_t\}_{t=0}^{T}$ defined by

$$\hat{x}_0 = x_0, \qquad \hat{x}_t = \hat{F}(\hat{x}_{t-1}) \tag{1.1}$$

should be a plausible trajectory from the system. A significant difficulty here is defining 'plausible'. As it happens, any reasonable definition makes most popular reconstruction methods fail this test miserably; we shall see an example later.

embedding space The embedding theorem, discussed in Section 3.1, defines a space of observations, delayed observations, and possibly other quantities. The coordinates of this space are highly significant, and already tell us a great deal about the dynamics. For example, they must be able to describe essentially all qualitatively different patterns that appear in the time series plot, so they provide a measure of dynamical complexity: a low dimension indicates that the dynamics is relatively simple, even if the time series does not appear to be so.

significant states Often in a dynamical system there will be some interesting parts of state space, such as locations containing saddle equilibria and homoclinic orbits. Part of the process of understanding geometry is to identify such regions, and the connections between them.

prediction horizon In many systems, useful prediction beyond some time horizon is not possible, because of noise effects combined with sensitivity to small errors (the butterfly effect). If the model system predicts much better than expected, either we have discovered something new, or we have made an error in the modeling and testing procedure. A good test of a model is to estimate its prediction horizon against real data and then to check its self-consistency: if we perturb $\hat{x}_0$ in (1.1) by an amount comparable to the noise level, then generate a time series $\{\tilde{x}_t\}$, at what value of t does $|\hat{x}_t - \tilde{x}_t|$ become comparable to the size of the attractor in the system?

Lyapunov exponents Lyapunov exponents are one of the most important dynamical invariants; they measure average sensitivity to small perturbations. Conventional methods of estimating Lyapunov exponents (even those which work from supposedly known system equations) are usually equivalent to local linear or local polynomial reconstruction. The top Lyapunov exponent, together with the level of dynamic noise or of measurement uncertainty, is the main factor in determining prediction horizon.

entropies For a system with finitely many states, with transitions possible between only some of them, it is natural to quantify the degree of uncertainty about the evolution of the system in terms of entropy. There are several different variants and ways to extend this idea to continuous systems. For example, topological entropy only counts connections between states; metric entropy requires that distance between states be defined; and measure-theoretic entropy is closest to the original Shannon entropy. We would expect there to be relationships between entropies and Lyapunov exponents, since they are both measures of uncertainty or surprise. The review paper by Young (1983) is a concise source for the definitions of these and other invariants, and the book by Pollicott and Yuri (1998) gives a more detailed exposition.

dimensions Correlation and other dimensions (Grassberger and Procaccia, 1983) attempt to measure the fractal structure of strange attractors. There are significant difficulties in estimating them from data, but with a high-quality algorithm (Judd, 1992), it is possible to perform the estimation with a good degree of accuracy even in the presence of a moderate amount of noise.

mutual information Fraser (1989) introduced the idea of using mutual information between a signal and a lagged version of the same signal as a way of finding good embedding lags and other information. Estimating mutual information accurately appears to involve estimating the probability measure of the embedded data first.

invariant measures If the invariant measure of an attractor is known, all the dynamical invariants can be calculated by spatially averaging suitable quantities over this measure. Estimating invariant measures is not easy unless one knows a special partition of the state space (a Markov partition, in fact). Methods of estimating invariant measures from data have only recently begun to appear (Hunt and Miller, 1992; Froyland et al., 1995b; Froyland, 1996; Hunt, 1996).

cycles Locating periodic orbits has long been taken to be one of the main parts of the analysis of any dynamical system. Chaotic attractors may be thought of as the closure of an infinity of unstable periodic orbits. Their dynamics can be modeled conceptually as low order cycles with stochastic transitions. Finding low order cycles is useful in many ways; the work of the Cvitanovic school (Artuso et al., 1990a,b; Cvitanovic, 1995) shows how to use such knowledge to estimate dynamical invariants.

bifurcations The way in which dynamics changes as system parameters change is often important; there can be many—even infinitely many— changes in the dynamics as a parameter varies through a small range. Since any model is only approximate, it is extremely useful to know the behavior of other models that are near by in parameter space.

It is rare that all of these tests and calculations are carried out on any one data set, but they serve to indicate that the flavor of dynamical reconstruction differs from traditional time series analysis in many ways. For example, a notable omission is any mention of autocorrelations or spectra; it is not that one would never use these at all, rather that spectra are seldom useful because interesting nonlinear dynamics is usually associated with broad spectra, and linear correlations are less useful than mutual information.

1.2   Reconstruction needs nonlinearity and stochasticity

Even the solar system, the quintessential Newtonian clockwork mechanism, does not perfectly fit the equations it is supposed to. This is because we must neglect most interactions: at best, we might be able to take into account the effects of the sun and the major planets on the orbit of an asteroid, but this means that in the long term our predictions will be wrong. (For a non-technical discussion of the difficulties of long-term prediction in the solar system, and some of the surprising results that have been obtained, such as the chaotic behavior of Mars, see Peterson (1993).) For normal use, this does not matter: we simply agree that there will be small errors in our predictions. In other words, we accept that our model is stochastic rather than purely deterministic. This is, of course, all the more necessary if we are modeling a system less amenable to simple description, since we expect to have errors in our equations as well as disturbances in the dynamics and inaccuracies in the measurements.

It should therefore be uncontroversial to state that all interesting models of dynamics need to account for noise. On the other hand, the difficulty of handling stochastic differential equations, to take only one example, means that stochastic models tend to be greatly oversimplified in their dynamical aspects: most often, they are at best linear. A rough and ready way to distinguish the approaches of nonlinear dynamicists and statisticians is to say that the dynamicists concentrate on the deterministic structure and ignore the noise, while the statisticians concentrate on the noise and ignore the structure. Misunderstandings between dynamicists and statisticians sometimes stem from this caricature, though the caricature is becoming less and less true as bridges like this book are built.

Philosophy aside, the need for models to account for both aspects is apparent with most interesting data, since non-trivial dynamics cannot arise from linear systems, and non-trivial systems cannot be modeled without error. Dynamical models are at risk of over-fitting, where the model has fitted aspects of the data that cannot be handled by that model class, either because the class is too small or because they are inherently unpredictable. In this article we will ensure that our models are capable of handling both nonlinearity and noise, and the method used to tackle the over-fitting problem will also, in principle, choose the balance between determinism and stochasticity.

1.3   Model code length

A string of symbols may be able to be described concisely: for example, the digits of π form an infinite symbol string but one that can be described by a short algorithm. The *Kolmogorov complexity* is the minimal length program in some agreed language that can reproduce the string (Li and Vitanyi, 1997; Cover and Thomas, 1991). The program must include any data such as parameters or fitting residuals so that the string can be reproduced exactly.[4] If the program is shorter than the string, we say the string is compressible.

The program can be thought of as a model for the string; for example '9325 copies of 10' describes a finite periodic string which can naturally be thought of as deterministic, and '9325 copies of 10 except that the 68th symbol is a 1 instead of a 0' is a description of a periodic string with a noise glitch.

---

[4]It is important to realise that this requirement of lossless compression (the string can be reproduced exactly) is not the same things as saying there is an exact model; both the model and its fitting errors are transmitted, with the fitting errors represented in a way that is optimally compressed given the model.

Kolmogorov complexity is not computable, but Rissanen (1989), and, independently, Wallace and Boulton (1968), introduced approximations which are computable. Let us call such approximations *model code lengths*. The claim is that if a class of models is available for a set of data, the model that should be chosen is the one that describes the data in the shortest code length. This is known as the minimum description length principle, or by various equivalent names. It is a very successful way of defeating the over-fitting problem.

The model code length can be thought of as the amount of information in a string that implements a program that contains the dynamical model, as well as the model's parameters, initial conditions and residuals. The string has to be able to be decoded uniquely so that the original data can be recovered exactly. One way to represent such a string is as a two-part code, the first part consisting of parameters and the second part consisting of residuals, with the residuals optimally encoded according to the model's error probability distribution. For example, if our model is

$$P(y_t|y_{t-1},\ldots,y_{t-k}) = \psi(a_1,\ldots,a_m,y_t,y_{t-1},\ldots,y_{t-k}) \qquad (1.2)$$

where ψ is some given function, then we encode the string $y_1,\ldots,y_T$ by storing (or transmitting) the parameters $a_1,\ldots,a_m$ and also the information needed to specify y_t for each t: the code length of the time series $y_1,\ldots,y_T$ is

$$
L(y_1,\ldots,y_T) = \sum_{j=1}^{m} \Lambda(a_j) + \sum_{t=1}^{k} \Lambda(y_t)
$$
$$
- \sum_{t=k+1}^{T} \log_2 \psi(a_1,\ldots,a_m,y_t,y_{t-1},\ldots,y_{t-k}), \qquad (1.3)
$$

where $\Lambda(\alpha)$ is the number of bits to encode α using any agreed code, and we have used the fact that if y is a realisation from a distribution P then $-\log_2 P(y)$ is the optimal-on-average code length of y; see, for example, Cover and Thomas (1991).

Rissanen (1989) has shown that the widely used model selection criterion due to Schwarz (1978) can be thought of as an approximation to such a two-part code.[5] A different way to compute code length is to use a predictive code: we envisage a transmitter and a receiver, each with the same program (in effect, a description of a class or classes of models) with the transmitter sending the compressed data as soon each symbol is read, and the receiver updating its model so that it is always just one symbol behind the transmitter. In this case, the relationship to time series prediction is very strong: we are simply describing an adaptive modeling process, and the data that will have to be transmitted is just the prediction errors. The predictive code appears not to

---

[5]Schwarz's criterion is also called the Bayesian information criterion (BIC).

include parameters, but the parameter cost is still there because in the early stages of transmission, the model is not as accurate as it is later.

Throughout this article we will assume, whenever necessary, that some computable approximation to model code length is available: in Section 2 it will be a predictive code, while in Section 3 it will be a two-part code.

The main theme of this article is modeling dynamics of nonlinear processes. This requires us to make use of statistical tools, and as with more than one of the foundational aspects of statistical analysis, there is some debate about the philosophy; in this case, about using code length as against Bayesian analysis (Rissanen, 1999). It is not the purpose of this article to take sides on this issue: applied nonlinear dynamics is above all a pragmatic field, where the problems are so difficult that anything that works with real data is welcomed, and there is no doubt that description length methods work in dynamical reconstruction. The fundamental questions for us are those of geometry and dynamics, but the intimate relationship between nonlinear dynamics and symbol shifts (Lind and Marcus, 1995) (among other reasons) makes information-theoretic tools such as description length a natural choice for dynamicists when it comes to relating the geometry or the dynamics to the data.

2 Dynamics of Symbol Strings

To understand complex temporal behavior, a fruitful idealization is to assume that observations can take on only a finite number of values, and that the system that produces them can be modeled as a finite stochastic automaton. This is natural in information theory, but is relatively unusual in the hard sciences. The payoff is that we can easily build models that answer many of the dynamicists' questions; the main disadvantage is that we lose geometry.

There is, as was already mentioned, a close connection between nonlinear dynamics and symbols, and most work on symbols and dynamics assumes that symbols have been allocated in a special way that matches the dynamics. We make no such assumption here: the dynamics is to be discovered, not assumed. Even if we happen to believe that the underlying system is described by continuous variables, we accept that measurements will be to finite accuracy (some relatively small number of bits coming from a digitizer), so it is rather natural to to think in terms of a finite alphabet. Typically we will use a very coarse discretization; because of the relationship between dynamics and shifts we expect there to be a tradeoff between coarseness of discretization and the length of history that is required to build a model, but the discretization issue is not well understood and will be glossed over here.

The requirement is to model a symbol string

$$s_1, s_2, \ldots, s_T \qquad (2.1)$$

where each symbol s_t is drawn from a finite alphabet A of m symbols, say $0, 1, \ldots, m-1$. In the symbol string case the relationship with description length is particularly clear, since we explicitly adopt algorithms designed for efficient data compression, emphasizing their modeling aspects and thereby their ability to reconstruct dynamics.

We assume that the states of the system that produced the string are not available, and we attempt to find a causal model: that is, we estimate

$$P(s_{t+1}|s_t, s_{t-1}, \ldots, s_0), \tag{2.2}$$

the next-symbol probability given the entire past. We hope that given enough data, the estimates will converge to something useful. It is usual to suppose the string can be reproduced with a system having finite memory,

$$P(s_{t+1}|s_t, s_{t-1}, \ldots, s_0) = P(s_{t+1}|s_t, s_{t-1}, \ldots, s_{t-k+1}), \tag{2.3}$$

so that the next-symbol probabilities only depend on at most the most recent k symbols. This is an approximation, though it might be a good one if we are observing a hidden Markov process.[6] In this case the system would be an unknown Markov process with states x^j, $j = 1, \ldots, n$, in which we observe symbol i in state x^j, where as above, $i \in [0, m]$ and usually $m < n$. The observations might also be probabilistic, with

$$P(s = i|x) = \psi(i, x)$$

where ψ is also unknown. It should be noted that finite memory is only an approximation in such models, and it is generally undesirable to have to assume a value of k in advance.

2.1   Context trees

Call $c = (s_{t-r}, s_{t-r+1}, \ldots, s_t)$ a *context* of length r at time t. More general contexts are possible and desirable, where there is not a simple unit difference between successive members, but for this presentation we take a context to be just a suffix of the data string. That is, at time t we know part of (2.1), namely $s_1, \ldots, s_t$, and we take as a context the last r symbols in this substring for some r to be determined (and with r not necessarily fixed). The causality assumption in (2.3), especially in its stronger finite memory form, suggests that contexts might be used as states, so we should look for a suitable set of contexts and make a model where the next-symbol probabilities depends only on the matching context.

If the system has maximum memory k then we might consider all contexts of length k as states of the system and try to fit next-symbol probabilities

---

[6]Contrary to the usual hidden Markov modeling methods, however, we are not assuming that any structure is known in advance.

to them. Even if k is known, this is not usually a good approach: with an alphabet of size only 2, there are 2^k parameters to fit, and many of them will have no matching data available. An adaptive approach is required.

A *context tree* (Rissanen, 1983) is a way of representing contexts as leaves of a tree: the path from the root of the tree is the context, read backwards from the most recent symbol. For example, if the alphabet is $\{0, 1\}$, then Figure 1 shows a possible context tree where the contexts are 0, 01 and 11. It is assumed that probabilities are independent when conditioned on contexts: in Figure 1 there is probability 1 that the next symbol will be a 1 and probability 0 that the next symbol will be a 0, regardless of previous symbols before the two represented in the context.

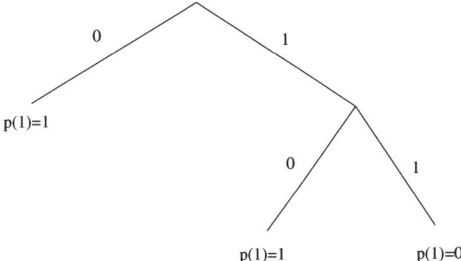

Figure 1: A simple context tree which generates the periodic sequence 011011011.... Each leaf is a *context;* that is, it represents the most recent symbols seen, read backwards in time from the present and downwards from the root of the tree. Probabilities at the leaves are conditional probabilities for the next output symbol, given that context.

It is not essential to use a tree, but doing so provides a convenient way of comparing different contexts, which we must do in deciding what set of contexts to use in the final model. We suppose first that the tree structure (that is, the set of contexts) is known, and determine how to estimate the conditional probabilities at the leaves.

2.2 Estimating transition probabilities when the contexts are known

Given a context tree structure, and assuming it has generated our symbol string, we can estimate the conditional probabilities at the leaves by recording the number of times each symbol is seen at each leaf, then estimating the multinomial distribution of symbols there. A suitable estimator in the binary case is

$$P(0|n_0, n_1) = \frac{n_0 + 1/2}{n_0 + n_1 + 1} \tag{2.4}$$

where n_0 and n_1 are the numbers of observations of 0 and 1; that is, we divide a 'ballast' of one observation equally between the binomial bins before the start of observations. This is the *Krichevsky–Trofimov* (KT) estimator and Willems et al. (1995) show that it has good asymptotic properties; this is in contrast to other estimators that one might consider, including the cases of zero ballast and ballast of one observation per bin. If we have prior knowledge, or prejudices, about the system, we may want to use a different amount of ballast: more if we believe we have a good estimate of the distribution, less if we do not have such an estimate but nevertheless believe the system to be nearly deterministic. It happens that the KT estimator results from a Dirichlet $(1/2, 1/2)$ prior.

In the following we use a fixed ballast of $1/m$ in each bin, where m is the size of the alphabet. We have found this to work well in practice over a wide range of problems, though any serious investigation of a paricular problem ought to consider carefully what would be an appropriate estimator.

Throughout this section, the reader is being invited to think in terms of data compression: a transmitter is sending the compressed symbols as soon as they come in, and a receiver is decoding them as soon as they arrive. Given the current estimate of the distribution just before any time instant, the transmitter can find a minimal length encoding for the symbol that appears at that time, and given the same estimate, the receiver will decode it correctly. Thus they both have to estimate the distribution predictively. The probability they ascribe to the symbol is the probability ascribed by the KT estimator based only on what has been seen in the past. In the binary case, the probability ascribed to an input sequence containing n_0 zeros and n_1 ones is easily seen to be

$$P_{KT}(n_0, n_1) = \frac{\frac{1}{2} \times \frac{3}{2} \times \cdots \times (n_0 - \frac{1}{2}) \times \frac{1}{2} \times \frac{3}{2} \times \cdots \times (n_1 - \frac{1}{2})}{1 \times 2 \times \cdots \times (n_0 + n_1)}. \quad (2.5)$$

As we saw earlier, the code length, which is $-\log_2 P_{KT}(n_0, n_1)$, automatically accounts for parameter estimation, because at each stage both transmitter and receiver have enough information to estimate the distribution to be used to encode the next symbol.

2.3 Selecting the best subtree

We do not initially know the tree structure, or set of contexts. To avoid having to specify a maximum tree depth, we record the entire past for every symbol. This means the maximal depth of the tree is equal to the length T of the input, but with efficient implementation (Willems, 1998) the storage required is linear in T.

Each leaf now represents just one observation, and the KT estimator is trivially found. We can also find KT estimators for the internal nodes, since the counts at these nodes are easily obtained by merging the counts at their children in the tree.

We would now like to identify a good set of contexts. Equivalently, we want to prune the maximal tree just constructed to produce a good model. There is insufficient space here to go into the various methods, and the discussion here will be necessarily oversimplified, and restricted to one heuristic; we discuss the issues in more detail elsewhere (Kennel and Mees, 1998, 2000). To justify the heuristic we are going to describe, we mention an alternative proposed by Willems et al. (1995) which does not prune at all, but instead weights all possible subtrees to produce a model that compresses better than any of them. It can be shown that this is asymptotically optimal and that it has optimality properties for finite strings as well. More detail can be found in the literature (Weingberger and Lempel, 1992; Willems et al., 1995).

The weighted tree estimator at a node s is

$$P_s(x) = \begin{cases} P_{KT}(x) & \text{if } s \text{ is a leaf;} \\ \frac{1}{2}\left(P_{KT}(x) + \prod_{i=1}^{k} P_{i,s}(x)\right) & \text{otherwise.} \end{cases} \tag{2.6}$$

Here $P_{i,s}(x)$ is the probability ascribed to child i of s by the weighted estimator. Recall that k is the alphabet size. These quantities can be easily found by recording counts at the leaves, then computing P_s recursively at nodes.

The code length at the root is the code length of the model in the weighted tree representation. This code can be created explicitly, but the implementation is non-trivial and it is not easy to see how to derive a model from it, so we use a pruning heuristic. We construct the tree for all of the data and compute P_s for nodes by a depth-first scan of the tree, pruning whenever

$$-\log_2 P_s(x) > -\log_2 P_{KT}(x).$$

We declare this tree to be our final model.

Remark As indicated above, it is not actually necessary to prune the tree: the weighted probabilities at the root can be used directly, and this is preferable in many cases. We have pruned both because doing so results in a single model, something more familiar to most scientists than the idea of a blend of many models, and also to allow the Markov chain conversion described next. (It should be noted that the pruning method described above is heuristic: the methods used in practice (Rissanen, 1984) are much more complex but have provable asymptotic properties.) In addition, the contexts that are selected may provide insight: even although context trees do not directly give us information about geometry, they do define regions of 'state space' which are interesting. For example, if the symbols were obtained by discretizing real variables, each context can be thought of as a cluster in embedding space.

2.4 Context trees as Markov processes

Although based on conditional probabilities, context trees are not immediately equivalent to a Markov chain with states corresponding to the leaves of the tree. A closure property must be added. The tree represented by solid lines in Figure 2 is not a Markov process because it requires additional memory outside the tree. If at leaf A a 1 is emitted, the context is now 01, corresponding to internal node E. But node E is not a leaf and so is not part of the machine as such: we need to know what came before the 0 so that we can choose between nodes G and H to determine the distribution for the next output symbol.

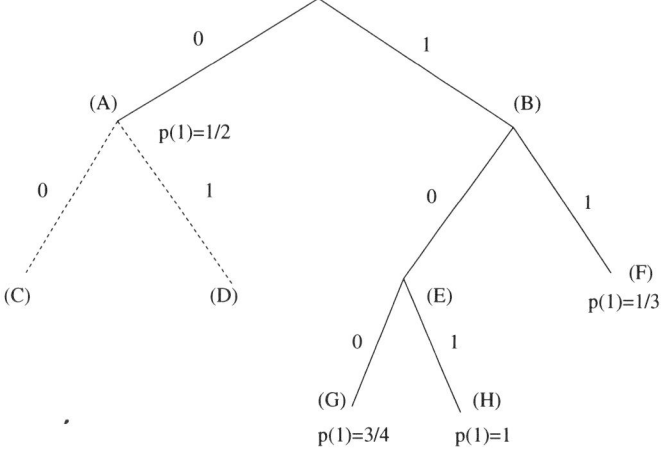

Figure 2: The solid lines show a tree machine which is not a Markov chain: that is, the contexts (leaves of the tree) cannot be used as states because they are incomplete. By extending the tree as shown by the dashed lines, we do obtain a Markov chain. This can be done in general.

By adding children C and D to node A as shown by the dashed lines, we do obtain a Markov chain. In general, we need to add children until the tree has the property that every subtree is also a subtree at the root. The probability distributions at added child nodes such as C and D are defined to be the same as at their parent A.

Clearly, a simulation output from such a Markov chain will be identical to a simulation produced by the original tree machine. More usefully, we can easily compute the equilibrium probability distribution of the Markov chain and hence obtain an invariant probability distribution for the contexts. For a system which really is discretized, this immediately allows us to calculate any of the standard dynamical invariants, which are averages over the equilibrium distribution. For a system which is assumed to have continuous states, which are merely approximated by symbols, there is further work to be done in proving convergence of the equilibrium probabilities to one of the invariant measures, most likely the SBR measure (Froyland et al., 1995b).

The *topological transition matrix* T can be taken to be the same as the transition matrix of the Markov chain, except with all non-zero elements replaced by 1. Then topological entropy in bits per iteration is $\log_2 \lambda$ where λ is the maximal eigenvalue of T.

For example, the tree machines in Figures 1 and 2 have topological transition matrices

$$
\begin{pmatrix} 0 & 1 & 0 \\ 0 & 0 & 1 \\ 1 & 0 & 0 \end{pmatrix} \quad \text{and} \quad \begin{pmatrix} 1 & 0 & 1 & 0 & 0 \\ 1 & 0 & 0 & 1 & 0 \\ 0 & 1 & 0 & 0 & 1 \\ 0 & 0 & 0 & 0 & 1 \\ 0 & 1 & 0 & 0 & 1 \end{pmatrix},
$$

where in each case the states are the leaves of the (possibly augmented) tree read from left to right. The corresponding topological entropies are 0 and 0.91 bits per iteration. The 0 for the first case reflects the fact that the machine is deterministic, so there is no uncertainty at all in the prediction.

A significant problem with this approach is that a very sparse context tree could conceivably produce a Markov chain which is equivalent to a rather dense context tree; this would mean an exponential increase in the number of states and so make it impractical to use context trees to compute dynamical invariants. Current research by the author and David Ridout (unpublished) is directed towards computing a suitable analogue of invariant measure directly on the context tree. This allows the invariants to be computed directly from the tree, without an intervening 'Markovisation' step.

2.5 Applications of context models

We have already seen how to compute one invariant, the topological entropy. Our main interest besides computation of invariants is in simulation but it is worth looking briefly at code lengths first.

2.5.1 Compression

Table 1 shows the code lengths obtained by applying the context tree method to symbol strings which are random, persistent, deterministic with short periodicity, and a concatenation of two deterministic series with different periods. In each case the expected results are obtained. The algorithm fails to compress the random data, and the tree is pruned to zero depth (i.e., it is memoryless). The persistent time series is compressed about as much as its entropy would indicate, and the corresponding pruned tree has depth 1 with essentially the correct transition probabilities. The simple deterministic strings are compressed considerably, and the pruned context trees are easily seen to be the simplest possible. The code length for the concatenated deterministic sequences is slightly longer than the sum of the code lengths when the two are compressed separately, giving a weak indication of non-stationarity of this sequence. A more entertaining test, involving compression of Shakespearean English text, is discussed elsewhere (Mees and Kennel, 1999).

	Bits (uncomp)	Bits (comp)
Fair coin tosses	10,000	10,008
Base 5 random	232,193	232,224
Persistent binary	10,000	4,608
Period 3 binary	57	12.9
Period 4 binary	92	17.8
Above 2 concatenated	149	31.1

Table 1: Compression performance of algorithm on random and simple deterministic inputs. The 'persistent' input is 10,000 binary symbols, where each symbol is followed by itself with probability 0.9 and its complement with probability 0.1. The entropy of this source is 0.47 bits per symbol, which should be compared with the attained compression factor of 0.46.

2.5.2   Simulation

A context tree acts as a model symbol source, and hence as a model for simulation, if it is given a starting context and allowed to emit a symbol which is a realization of the estimated probability distribution for that context. The emitted symbol and the current context together define another context, and the process can be iterated. If we believe we have a good sample of the process, it makes sense to estimate the conditional probabilities *for simulation purposes only* via the maximum likelihood estimator; that is, the emission probability for the ith symbol is taken to be $n_i / \sum n_j$ rather than the KT estimate. This ensures that, for example, a deterministic system is not randomly noised-up.

We already saw an example in Figure 1; it is clear that the tree there emits the sequence claimed. Non-trivial applications of the context tree method include stationarity testing of data from a fluidized bed reactor (Kennel and Mees, 1999) and a model of the electric field amplitude measured in a laboratory laser experiment (Mees and Kennel, 1999). Here we describe an application to biological data.

Figure 3 shows a segment of a time series of 500 measurements of firing voltage from a voltage-clamped squid giant axon (Mees et al., 1992a). A crude discretization to 4 bits accuracy was used in a context tree model. The context tree compressed the 2000 bit truncated data to about 830 bits using 27 contexts. The computation took 0.04 seconds on a 266MHz Pentium II computer.

The two longest contexts had four symbols, but these both had a single observation corresponding to what may be a glitch in the experiment. (See also the caption of Figure 7 where a quite different approach suggests the same thing.) Most contexts were of size one or two symbols, confirming the intuition that the system dynamics are 'one and a bit dimensional'. (See the discussion of embedding in Section 3.1.) Because of the way the model is constructed, all

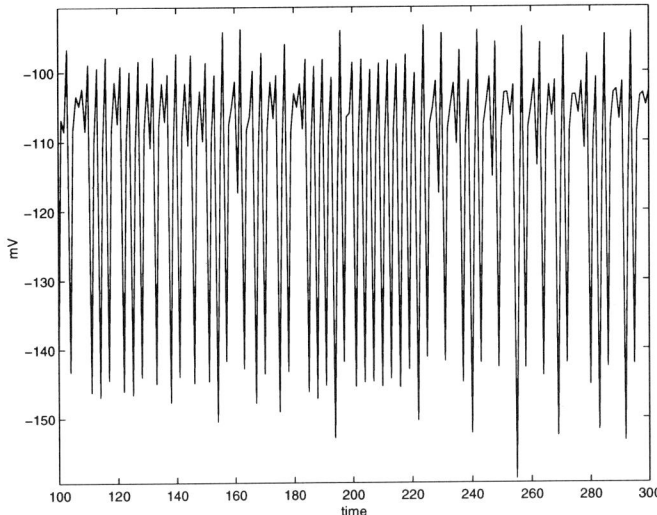

Figure 3: Part of a time series of firing voltage measurements from an in vitro preparation of a squid giant axon.

the simple statistics are captured rather precisely; for example, the one step prediction error is small. Such issues will be described elsewhere.

The context tree produces the simulation shown in Figure 4. Allowing for the discretization to an alphabet of size 16, the simulation performance is remarkably good.

3 Dynamics of Real-valued Time Series

Having seen what can be achieved using the finite alphabet idealization, we now try the more familiar idealization, where data is assumed to be real-valued; although more familiar, and in spite of the fact that enormously more effort has been put into it, this case is in some ways less satisfactorily developed.

The finite memory assumption in Section 2 is now replaced by the very similar idea of time delay embedding. That is, the system's state is taken to be represented by at most k previous observations $(y_{t-\tau_1}, \ldots, y_{t-\tau_k})$. That the state can be represented in this way is known to be true in the case of no noise and smooth dynamics (Takens, 1981; Sauer et al., 1992), but the case with noise is somewhat more difficult (Stark et al., 1997). We will discuss embedding very briefly, and then consider issues of model-building under the assumption that satisfactory embeddings can be found. Several different types of models are in current use: given an embedding, modeling can be reduced to function fitting, so the issues are similar to those faced in other data fitting problems, and popular methods include local linear fitting (both continuous

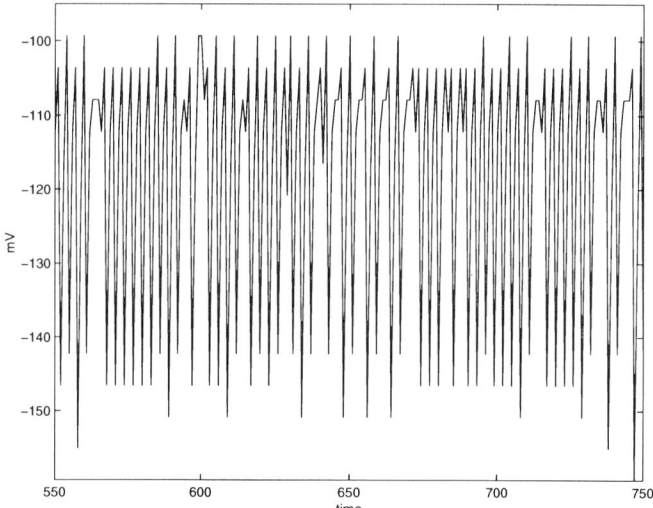

Figure 4: A simulation from the context tree model of the 4 bit truncation of the squid data. The dynamical features appear comparable to those of the experimental data shown in Figure 3.

and discontinuous), radial basis function, neural networks, and others.

The real issue for dynamicists is whether the dynamical behavior has been captured. There is insufficient space for a full discussion of either the ways we might measure the quality of a model's dynamics, or the ways in which we might improve this quality; the brief discussion later can be summarized by saying that none of the methods here is completely satisfactory in this respect, but that recent research indicates the possibility of extensions and modifications that may result in large improvements.

3.1   Embedding

Assume the observations come from a discrete-time dynamical system, with unknown state x_t belonging to some unknown smooth manifold M^n of unknown dimension n and observed real value y_t such that

$$x_{t+1} = f(x_t, \xi_t) \tag{3.1}$$

and

$$y_t = c(x_t, \nu_t) \tag{3.2}$$

where the *dynamic noise* ξ_t and the *observational noise* ν_t are unknown disturbances.

The state x and the readout map c are unknown. The embedding theorem (Packard et al., 1980; Takens, 1981; Noakes, 1991; Sauer et al., 1992; Stark et al., 1997; Stark, 2000) allows states to be reconstructed from observations: in the present case, we can generate a vector time series $\{z_t\}$ which is equivalent to $\{x_t\}$ in a strong sense. The standard method uses the simple time-delay embedding given by

$$z_t = (y_t, y_{t-\tau}, \dots, y_{t-(k-1)\tau}) \tag{3.3}$$

for some *embedding dimension* k and *lag* τ. Let

$$h(x) = (c(x), c(b(x)), \dots, c(b^{k-1}(x)))$$

where $b(x) = f^{-\tau}(x)$ and we are assuming (entirely for notational reasons) that f is invertible. Then h is a map from M^n to $\mathbb{R}^k$, such that

$$z_t = h(x_t), \tag{3.4}$$

and the embedding theorem states that, generically, if k is sufficiently large and in the absence of noise, h is an *embedding*, i.e., a proper injective immersion (Guillemin and Pollack, 1974).

There is no reason to restrict attention only to time-delay embeddings. For example, other information such as the time since the last local maximum or minimum of the time series may be useful, and in experiments there may be several time series to work with. Kilminster (1999) discusses the advantages of complicated embeddings, and Cao et al. (1998) discuss how to find good embeddings when multivariate data is available.

There is a great deal of advice on choice of embedding dimension and related matters in the literature (Albano et al., 1988; Broomhead et al., 1986; Cheng and Tong, 1991; Abarbanel and Kennel, 1992). It can be argued that the question of what is a good embedding cannot be answered separately from the question of what kind of model one is trying to build, and later we will touch on how to treat embedding as part of the modeling problem.

Since h is an embedding, the following diagram commutes—

$$
\begin{array}{ccc}
x_t & \xrightarrow{\;f\;} & x_{t+1} \\
{\scriptstyle h}\downarrow & & {\scriptstyle h}\downarrow \\
z_t & \xrightarrow{\;F\;} & z_{t+1}
\end{array}
$$

—which implies there is a smooth map F acting as the dynamics on the embedded series:

$$z_{t+1} = F(z_t).$$

We hope to find an approximation $\hat{F}$ for F. All of the observables can be predicted from F and therefore can be approximately predicted by $\hat{F}$.

Usually it is not necessary to estimate the vector-valued function F explicitly, since we are only interested in the behavior of the observable y, rather than the reconstructed states z. It is sufficient to estimate the map $\rho : \mathbb{R}^k \to \mathbb{R}$ such that

$$y_{t+1} = \rho(z_t). \tag{3.5}$$

In the simple time-delay embedding, ρ is the first component of F.

The usual modeling process re-introduces the disturbances on the observations, but makes the simplifying assumption that the time series embedding coordinates z_t are known exactly, although the observed variables y_t may be subject to disturbances. This is incorrect since the embedding coordinates are derived from the observables, so that any disturbances on $\{y_t\}$ affect $\{z_t\}$. This assumption is implicit in virtually all nonlinear reconstruction. One price paid for neglecting errors in the z variables is that the amplitudes of peaks and troughs are poorly fit and poorly predicted.

We therefore have the following problem: given a real-valued time series $\{y_t\}$ and a k-vector-valued time series $\{z_t\}$ where it is assumed that

$$y_t = \rho(z_t) + \nu_t, \tag{3.6}$$

find an estimate $\hat{\rho}$ of ρ.

3.2   Local linear reconstruction: triangulation maps

Reconstruction by triangulation uses locally linear approximations to the dynamics, but ensures that the approximations are continuous in the (possibly embedded) state variables. An analogy is with a triangulated representation of a landscape or a sea-bed, with vertices of the triangulation at locations where spot heights or depth soundings have been measured and facets interpolate between these known values.

Start with the noiseless case corresponding to (3.5). We will define an approximate map $\hat{\rho}$ that is locally linear and fits exactly at the embedded data points, so (3.5) holds exactly with ρ replaced by $\hat{\rho}$. The approximation we will choose is such that, for all z in the convex hull of the data,

$$|\rho(z) - \hat{\rho}(z)| \leq O(\varepsilon^2) \tag{3.7}$$

where ϵ will be defined shortly. The approximate map $\hat{\rho}$ is therefore accurate to first order in distance from embedded data.

Consider the triangulation shown in Figure 5. The vertices are points z_t on a short trajectory of the Hénon map,

$$f(\alpha, \beta) = (1 - 1.4\alpha^2 + \beta, 0.3\alpha).$$

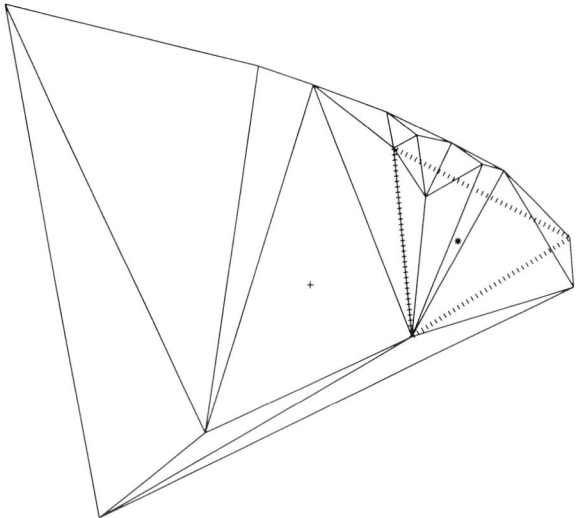

Figure 5: A Delaunay triangulation with vertices on a 15 point orbit of a two dimensional smooth map. The image under the map of vertex z_t is z_{t+1} and a good local linear approximation to the map is one which takes each triangle into the triangle whose vertices are the appropriate images. In the example, the triangle containing the point marked + maps into the triangle shown dotted; the point itself maps into the point * which lies in the same relative position in the image triangle. Note that the image triangle is not a member of the triangulation.

The point z marked + lies in the triangle with vertices z_{10}, z_{11} and z_{14}, and we can write

$$z = \sum_{i=10,11,14} \lambda_i z_i.$$

where $\lambda_i \geq 0$ for all i and $\sum_i \lambda_i = 1$. That is, we write z as a convex combination of the vertices of the triangle it lies in. Assuming the dynamics is linear over this triangle, the image of the triangle is the triangle with vertices z_{11}, z_{12} and z_{15}. A plausible approximation to F is the map $\hat{F}$ defined by

$$\hat{F}(z) = \sum_{i=10,11,14} \lambda_i F(z_i)$$

where $F(z_i) = z_{i+1}$. It is easy to show (Mees, 1991b,a) that this map is continuous and its projections onto individual coordinates have the property (3.7) with ε taken as the diameter of the triangle. A map with triangles of small diameter—that is, as nearly equiangular as possible—will therefore be the most accurate. A triangulation that has this property is the Delaunay triangulation (Mees, 1991a), which minimizes average triangle diameter. The Delaunay triangulation is described elsewhere and there are efficient algorithms for its construction (Watson, 1981). The construction extends easily to higher dimensions.

The triangulation reconstruction is therefore defined by

$$\hat{F}(z) = \sum_{t \in \Delta(z)} \lambda_t z_{t+1} \tag{3.8}$$

where $\Delta(z)$ is the set of indices of the vertices of the triangle containing z. If z is an embedded state, then we may instead write

$$\hat{\rho}(z) = \sum_{t \in \Delta(z)} \lambda_t y_{t+1} \tag{3.9}$$

where as usual, y_t are the observations, but we still require to know the embedded state z since $\hat{\rho}$ is a map from embedding space to the reals.

A less trivial example appears in Figure 6, which shows a Delaunay triangulation of the squid data from Figure 3, embedded in two dimensions with a lag of one. Notice how the data is clustered in three regions. In some problems, the geometry of the triangulation might help in identifying parts of the plane where closer examination and additional experimental measurements might be in order.

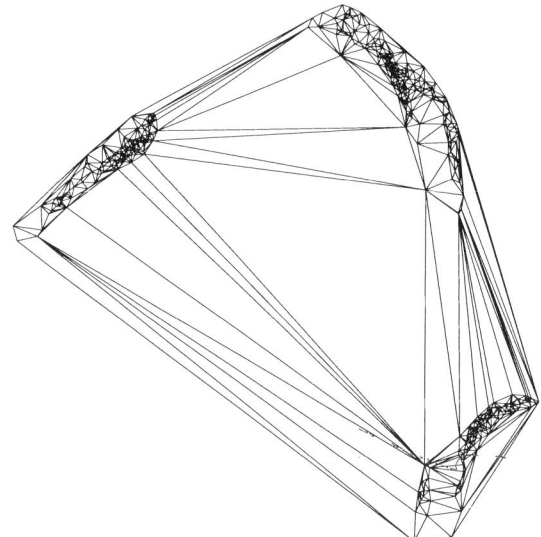

Figure 6: A triangulation of the squid data from Figure 3.

The image of the triangulation under (3.8) is shown in Figure 7. The action of the dynamics is to stretch and fold a region of the plane and map it back onto itself is apparent; this is the famous horseshoe mechanism (Smale, 1967) which gives rise to chaos.

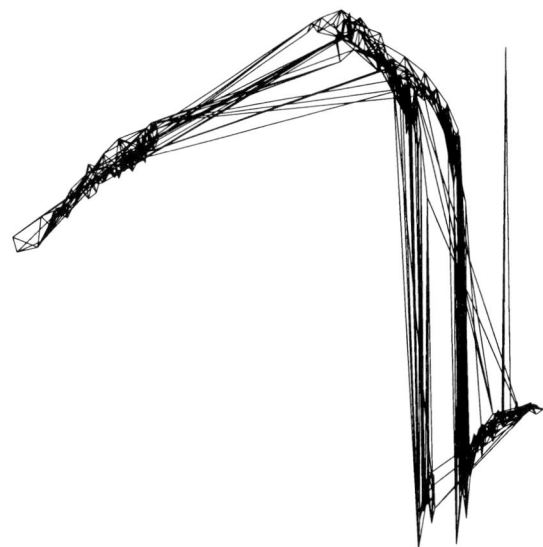

Figure 7: The image of the triangulation in Figure 6. The stretching and folding action of the map is apparent; this geometric information is a very strong indicator of chaos. The spike on the far right is caused by the anomalous last point which may be a measurement error.

Similar ideas have been used to analyze an analytic model of a neuron (Judd et al., 1991).

Reconstruction of dynamics by triangulation is also possible in the presence of noise (Allie and Mees, 1997b). The method is to select vertices by subset selection, and to choose the number of vertices to minimize code length.

Additionally, in the noiseless case there is a dual construction, Dirichlet tessellation (Green and Sibson, 1978; Mees, 1991a), which may give somewhat more accurate approximations.

Triangulation reconstructions turn out to have good dynamical properties (we will examine this in Section 3.5) and so one could consider estimating dynamical invariants by building a model from data then using simulations to provide very long time series to enable calculation of Lyapunov exponents and so on by the standard methods. There are two alternatives, however.

First, it is possible to find invariant densities directly from the triangulation (Froyland et al., 1995b) and hence dynamical invariants such as Lyapunov exponents can be calculated (Froyland et al., 1995a).

Second, triangulations make it easy to locate periodic orbits and to give good estimates of their stability properties (Allie and Mees, 1997a), and this is useful even when the dynamics is known in detail. (Recall that knowing periodic orbits is interesting in itself.) The dynamical invariants of Section 1.1 can be calculated from the periodic orbits instead of from the invariant measure or from simulations.

3.3   Smooth reconstructions: radial basis functions

Keep the assumption of (3.6); as before, it is necessary to approximate ρ : $\mathbb{R}^k \to \mathbb{R}$. For many approximation problems of this kind, radial basis functions have good properties (Powell, 1985; Mees et al., 1992b). Radial basis functions were first used for reconstruction by Casdagli (1989), and have since become a popular and powerful method (Smith, 1992; Mees, 1993; Judd and Mees, 1995). A radial basis function model is of the form

$$\hat{\rho}(z) = \alpha^{\mathrm{T}} z + \beta + \sum_{j=1}^{\ell} \lambda_i \phi(|z - c_i|) \qquad (3.10)$$

where c_i, $i = 1, \ldots, \ell$ are called *centers*, and $\phi : \mathbb{R} \to \mathbb{R}$ is a given function such as $\phi(r) = r^q$ with, say, $q = 1$ or $q = 3$, or $\phi(r) = \exp(-r^2/2\sigma^2)$, with σ chosen somehow from the data. The linear parameters α, β and λ are to be found, generally by a least squares fit to the data. The nonlinear parameters are the centers; there are various ways of choosing these, but the most powerful appears to be by generating many of them in suitable ways, then using a subset selection technique (Mees, 1993; Judd and Mees, 1995) to pick those centers that give the best fit to the data in the sense of minimizing code length. The details of the calculation of code length are in the cited references; in brief, the numbers and precisions of the parameters are optimized, with parameters being allowed to be specified less accurately if this results in a sufficient increase in fitting error. The most accurate of the parameters required about five significant decimal digits of accuracy, and the least accurate required about four. The RMS fitting error was around 7. Similar results were obtained in several different runs with different randomly generated centers.

Such models have been used successfully in bifurcation studies (Glover and Mees, 1992; Judd and Mees, 1996) and in reconstructing the dynamics of difficult data such as speech (Judd and Mees, 1998). An interesting feature of the subset selection process (after some generalization) is that it can be used to find good embeddings, so that the embedding and the modeling problems can be conflated.

Radial basis models give geometric information because the centers are located in the interesting parts of state space, and because they can define local embeddings, which determine the subspaces (locally) or the manifolds (globally) in which the dynamics is taking place. They are generally very good at short term prediction, though often less so at capturing long-term dynamics or in simulations.

Figure 8 shows one-step predictions of the squid data. The model used was a radial basis predictor, with the code length being minimized with 11

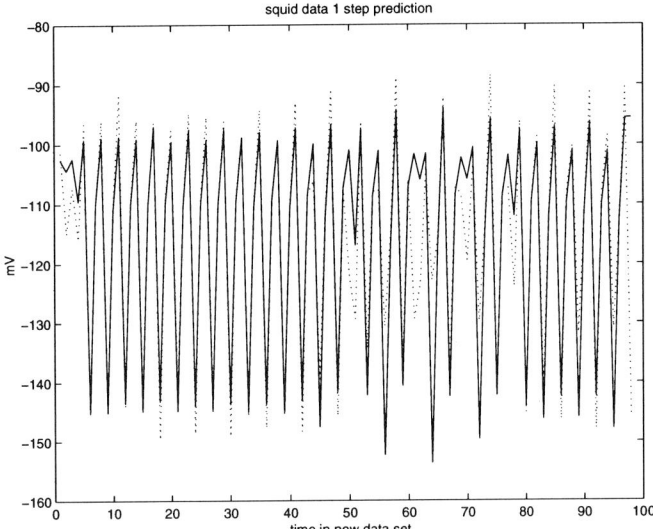

Figure 8: Single step predictions of unseen squid data with a radial basis model. A model with centers chosen to minimize code length was constructed as described in the text, using the first 400 points from the squid time series, embedded in 2 dimensions with a lag of 1. The figure shows single step predictions on the next 100 points (dashed line) and experimental values (solid line).

adaptively chosen centers and no linear or constant terms. The nonlinear basis function ϕ was $\phi(r) = r^2 \log r$. Constructing the model took less than 10 seconds in Matlab 5.3 on a 266MHz Pentium II computer. The one-step prediction performance is fairly good, though it does not do so well on large peaks. We discuss the simulation behavior in Section 3.5.

Since this modeling method and its generalizations and applications are described in great detail elsewhere (Judd and Mees, 1998, 1997; Small et al., 1999; Judd and Mees, 1996; Mees and Judd, 1996; Pilgram et al., 2000) we do not go into any more detail here.

3.4   Other reconstruction methods

Of the several other approaches to reconstruction, perhaps the best-known is neural networks, which often seem to be applied in a black-box manner, without a full awareness of the dynamical or statistical issues involved. For a fully informed discussion, which makes some of the points that have been made in this article, see (Weigend et al., 1990).

One of the first reconstruction methods used, and still one of the most popular, is local linear function fitting, where one selects a neighborhood in embedding space and fits a linear function to the points in the neighborhood and their images using least squares (Farmer and Sidorowich, 1987). A moment's thought will reveal that this produces discontinuous models, so their dynamical behavior is likely to be poor—something that is easily verified—but they are quick to program and do give geometric information. A natural generalization is to include higher-order terms (Abarbanel, 1996) and this can aid in determining the appropriate neighborhood size. Another important method is exponentially weighted local linear models (Sugihara and May, 1990; Dixon and Sugihara, 2000), which share some of the advantages and disadvantages of the unweighted local linear approach, but have advantages of their own.

If one can write down a satisfactory class of parametrized models, such as the radial basis function or neural network models, and if computation time is not a restriction, Monte Carlo methods (Ruanaidh and Fitzgerald, 1996) may be the best way to find good reconstructions. One other possible reconstruction method (Allingham et al., 1999) is the direct estimation of conditional probability estimators: a real-space version of the symbol-space techniques described in Section 2.

3.5   Dynamics of reconstructions

In Section 2 we saw that reconstruction of the dynamics of symbol strings using context trees had many good properties. With some experimental data, such as the laser data discussed in Kennel and Mees (2000), it does not capture the long-term dynamics as well as might be hoped. This is because it concentrates on predicting the next symbol well, rather than, say, predicting the next 10 symbols well. It is likely that improvement would result by modifying the context tree method to more accurately fit the invariant measure—in the symbolic case, the equilibrium probabilities—of the automaton that is assumed to have produced the string.

Much the same is true of most continuous reconstructors, except that they typically fail far more often than the context tree method. Even very good one-step predictors may fail to be good simulators. An extreme case is the radial basis models of the squid data: every attempt to use these for simulation resulted in time series that diverged to infinity in a few cycles. The squid data is a difficult case for radial basis models, because it varies rapidly, and is very tightly clustered with no data in regions outside the clusters. In the author's experience, most other real-space modeling methods behave equally badly on such data.

Contrast this with Figures 9 and 10, which show a part of the simulated time series, and the embedding of the same data produced by the triangulation reconstruction. In this case the obvious defects caused by local linearization, such as the false interpolations along straight lines joining two of the clusters, and the occasional point far from the clusters, are perhaps less striking than the faithfulness with which the simulation captures many of the patterns of the dynamics in the time series, and the geometry in the embedding.

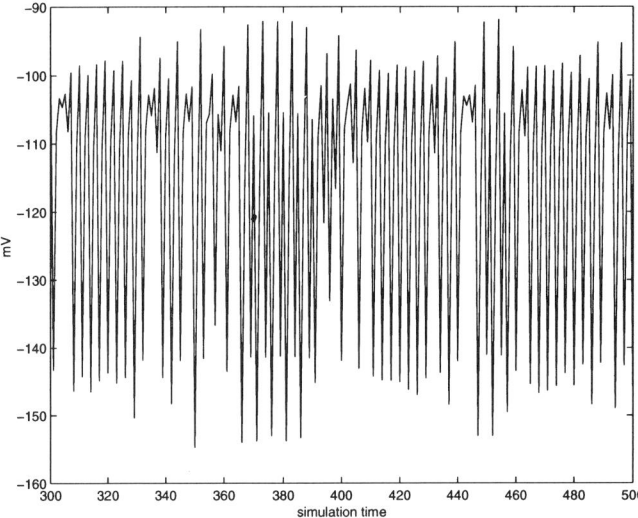

Figure 9: A typical segment of a simulation from the triangulation model of the squid data.

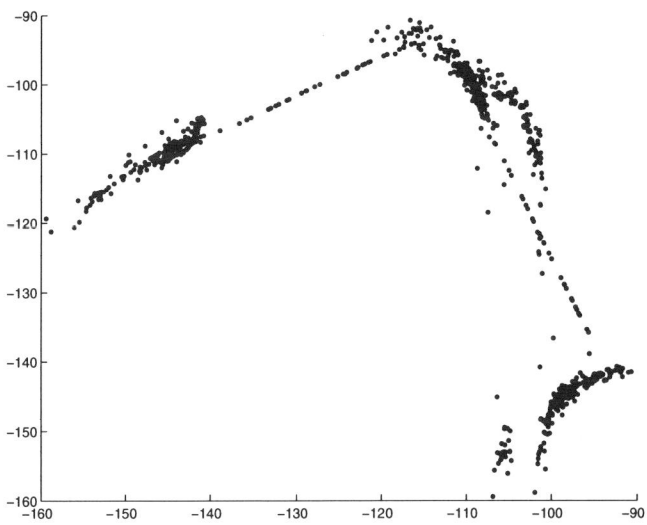

Figure 10: Embedding of the data shown in Figure 9. Black dots: model output; grey dots: original data.

It appears that the triangulation reconstruction is good because not only is it tied to the convex hull of the data, and so cannot diverge, but also it is very closely linked to a method of estimating invariant measures (Froyland et al., 1995b; Mees et al., 1993). In addition it has a fortunate property: the reconstruction will tend to preserve the invariant measure that would be estimated from a histogram on the data. This is a topic of current research, and unfortunately there is insufficient space to discuss it here. Instead, the reader is referred to the forthcoming thesis of Kilminster (2001). We also remark that radial basis function reconstructions can be modified to have simulation and excellent long-term prediction properties, in essence by modeling a function that fits large filtered segments of future and past (Pilgram et al., 2000), and that one can even produce a 'bolt-on' method that will improve the dynamics of models produced in any of the more standard ways (Judd et al., 2000).

4   Concluding remarks

The methods described in this article are only relevant for systems that are stationary and autonomous; the study of nonstationary and/or non-autonomous systems is a vast field, which has not been touched on at all in this article.

Reconstruction of nonlinear dynamics is still partly an art: no one method has been shown to be suitable for all cases. As we have seen, the context tree approach appears to come close to being a satisfactory universal modeling method when data comes from a finite alphabet, but the situation is less clear-cut when we are pretending that the system in question has continuously-valued states and observations. In that case, even although the fundamentals of embedding are fairly well understood, there is no one modeling method that is clearly the best, and function fitting methods abound.

We can draw some tentative conclusions from the work described here. If we have discrete data from a small alphabet, information-theoretic methods are the right choice, with context tree reconstruction being a good candidate. On the other hand, if we have nominally-continuous data, the best way to capture dynamics seems to be to use techniques which are geometrically-based but which also take care of other important issues like continuity; the obvious candidate is triangulation.

But reconstruction that only optimizes something like one-step prediction error, even in its description length generalizations, is not the best way to make models that capture dynamics, and this is an area that needs more research.

Acknowledgments

Support for this research was provided by the Australian Research Council, and by the Isaac Newton Institute as part of the programme 'Nonlinear and Non-stationary Signal Processing'. I also thank The University of Western Australia for leave to visit the Isaac Newton Institute. The squid neuron data was kindly provided by Professor K. Aihara of Tokyo University.

References

H.D.I. Abarbanel. *Analysis of Observed Chaotic Data*. Institute for Nonlinear Science. Springer, New York, 1996.

H.D.I. Abarbanel and M.B. Kennel. Local false nearest neighbors and dynamical dimensions from observed chaotic data. Technical report, Department of Physics, University of California, San Diego, October 18, 1992.

A.M. Albano, J. Muench, C. Schwartz, A.I. Mees, and P.E. Rapp. Singular value decomposition and the Grassberger-Procaccia algorithm. *Phys. Rev. A*, 38A:3017–3026, 1988.

S. Allie and A. Mees. Finding periodic points from short time series. *Phys. Rev. E*, 56(1):346–350, 1997a.

S. Allie and A.I. Mees. Reconstructing noisy dynamical systems by triangulation. *Phys. Rev. E*, 55(1):87–93, 1997b.

D. Allingham, D. Kilminster, and A.I. Mees. Estimating probability distributions using tomographic imaging techniques. In *NOLTA '99*, pages 379–382, Waikoloa, Hawaii, 1999. NOLTA.

R. Artuso, E. Aurell, and P. Cvitanovic. Recycling of strange sets: I: Cycle expansions. *Nonlinearity*, 3:325–359, 1990a.

R. Artuso, E. Aurell, and P. Cvitanovic. Recycling of strange sets: II: Applications. *Nonlinearity*, 3:361–386, 1990b.

D.S. Broomhead, R. Jones, and G.P. King. Topological dimension and local coordinates from time series data. Technical report, Department of Mathematics, Imperial College London, 1986.

L. Cao, A.I. Mees, and K. Judd. Dynamics from multivariate time series. *Physica D*, 121:75–88, 1998.

M. Casdagli. Nonlinear prediction of chaotic time series. *Physica D*, 35(3): 335–356, 1989.

B. Cheng and H. Tong. On consistent non-parametric order determination and chaos. Technical Report 6th draft, Institute of Mathematics and Statistics, University of Kent, May 1991.

T. Cover and J. Thomas. *Elements of Information Theory*. Wiley Interscience, New York, 1991.

P. Cvitanovic. Dynamical averaging in terms of periodic orbits. *Physica D*, 83:109–123, 1995.

P. Dixon and G. Sugihara. Noise and nonlinearity in an ecological system. In *Nonlinear Dynamics and Statistics*, A.I. Mees, editor, Birkhäuser, Boston, 2000.

J.D. Farmer and J.J. Sidorowich. Predicting chaotic time series. *Phys. Rev. Letters*, 59(8):845–848, 1987.

A. Fraser. Information and entropy in strange attractors. *IEEE Trans. Inform. Th.*, 35:245–262, 1989.

G. Froyland. *Estimating Physical Invariant Measures and Space Averages of Dynamical Systems Indicators*. PhD Thesis, The University of Western Australia, 1996.

G. Froyland, K. Judd, and A.I. Mees. Estimation of Lyapunov exponents of dynamical systems using a spatial average. *Phys. Rev. E*, 51(4):2844–2855, 1995a.

G. Froyland, K. Judd, A.I. Mees, K. Murao, and D. Watson. Constructing invariant measures from data. *Int. J. Bifurc. Chaos*, 5(4):1181–1192, 1995b.

J. Glover and A.I. Mees. Reconstructing the dynamics of Chua's circuit. *J. Circ., Syst. Comput.*, 3(1):201–214, 1992.

P. Grassberger and I. Procaccia. Characterization of strange attractors. *Phys. Rev. Lett.*, 50(5):346–349, 1983.

P.J. Green and R. Sibson. Computing Dirichlet tessellations in the plane. *Comput. J.*, 21(2):168–173, 1978.

V. Guillemin and A. Pollack. *Differential Topology*. Prentice-Hall, New Jersey, 1974.

F.Y. Hunt. Approximating the invariant measures of randomly perturbed dissipative maps. *J. Math. Anal. Appl.*, 198:534–554, 1996.

F.Y. Hunt and W.M. Miller. On the approximation of invariant measures. *J. Statist. Phys.*, 66(1/2):535–548, 1992.

K. Judd. An improved estimator of dimension and some comments on providing confidence intervals. *Physica D*, 56:216–228, 1992.

K. Judd and A.I. Mees. On selecting models for nonlinear time series. *Physica D*, 82:426–444, 1995.

K. Judd and A.I. Mees. Modeling chaotic motions of a string from experimental data. *Physica D*, 92:221–236, 1996.

K. Judd and A.I. Mees. Modeling chaos from experimental data. In *Control and Chaos*, K. Judd, A.I. Mees, K. L. Teo, and T. Vincent, editors, pages 25–38. Birkhäuser, Boston, 1997.

K. Judd and A.I. Mees. Embedding as a modeling problem. *Physica D*, 120: 273–286, 1998.

K. Judd, A.I. Mees, K. Aihara, and M. Toyoda. Grid imaging for a two-dimensional map. *Int. J. Bifurc. Chaos Appl. Sci. Eng.*, 1(1):197–210, 1991.

K. Judd, M. Small, and A.I. Mees. Nonlinear modelling: Getting the dynamics right. In *Nonlinear Dynamics and Statistics*, A.I. Mees, editor, Birkhäuser, Boston, 2000.

M. Kennel and A.I. Mees. Stationarity of dynamics from time series. Technical report, INLS, UC San Diego, November 1998.

M.B. Kennel and A.I. Mees. Data compression algorithms for analyzing observed symbolic time series. In *Nonlinear Dynamics and Statistics*, A.I. Mees, editor, Birkhäuser, Boston, 2000.

M.B. Kennel and A.I. Mees. Symbols and dynamics. In *Nonlinear Dynamics and Statistics*, A.I. Mees, editor, Birkhäuser, Boston, 2000.

D. Kilminster. The benefits of complicated embeddings. In *NOLTA '99*, pages 383–386, Waikoloa, Hawaii, 1999. NOLTA.

D. Kilminster. *Invariant Measures, Reconstruction and Estimation in Nonlinear Dynamics*. PhD thesis, Department of Mathematics and Statistics, The University of Western Australia, 2001.

M. Li and P.M.B. Vitanyi. *An Introduction to Kolmogorov Complexity and its Applications*. Springer, New York, 2nd edition, 1997.

D. Lind and B. Marcus. *Symbolic Dynamics and Coding*. Cambridge University Press, 1995.

A.I. Mees. Dynamical systems and tesselations: Detecting determinism in data. *International Journal of Bifurcation and Chaos*, 1(4):777–794, 1991a.

A.I. Mees. Tesselations and dynamical systems. In *Nonlinear Modelling and Forecasting*, M. Casdagli and S. Eubank, editors, volume 12 of *Santa Fe Institute Studies in the Science of Complexity*, pages 3–24. Addison-Wesley, 1991b.

A.I. Mees. Parsimonious dynamical reconstruction. *Int. J. Bifurc. Chaos*, 3 (3):669–675, 1993.

A.I. Mees, K. Aihara, M. Adachi, K. Judd, T. Ikeguchi, and G. Matsumoto. Deterministic prediction and chaos in squid axon response. *Phys. Lett. A*, 169:41–45, 1992a.

A.I. Mees, M.F. Jackson, and L.O. Chua. Device modeling by radial basis functions. *IEEE Trans. CAS/FTA*, 39(1):19–27, 1992b.

A.I. Mees and K. Judd. Parsimony in dynamical modeling. In *Predictability of Complex Dynamical Systems*, Y. Kravtsov and J. Kadtke, editors, pages 123–142. Springer, Berlin, 1996.

A.I. Mees and M.B. Kennel. Context trees and dynamics. In *Computing Anticipatory Systems*, D. Dubois, editor, page 3. CHAOS asbl, Institute of Mathematics, University of Liège, Belgium, 1999.

A.I. Mees, K. Murao, K. Judd, and G. Froyland. Triangulations on tori and density estimation. In *Proceedings of the 1993 International Symposium on Nonlinear Theory and its Applications (NOLTA '93)*, volume 1, pages 275–280. NOLTA, Hawaii, 1993.

L. Noakes. The Takens embedding theorem. *Int. J. Bifurc. Chaos*, 1(4): 867–872, 1991.

N.H. Packard, J.P. Crutchfield, J.D. Farmer, and R.S. Shaw. Geometry from a time series. *Phys. Rev. Lett.*, 45(9):712–716, 1980.

I. Peterson. *Newton's Clock: Chaos in the Solar System*. W.H. Freeman & Co, 1993.

B. Pilgram, K. Judd, and A.I. Mees. Modelling the dynamics of nonlinear time series using canonical variate analysis. *Physica D*, in press, 2000.

M. Pollicott and M. Yuri. *Dynamical Systems and Ergodic Theory*, volume 40 of *London Mathematical Society Student Texts*. Cambridge University Press, 1998.

M.J.D. Powell. Radial basis functions for multivariable interpolation: a review. Technical Report 1985/NA12, Department of Applied Mathematics and Theoretical Physics, Cambridge University, October 1985.

P.E. Rapp, T.I. Schmah, and A.I. Mees. Models of knowing and the investigation of dynamical systems. *Physica D*, 132:133–149, 1999.

J. Rissanen. A universal data compression scheme. *IEEE Trans. Inf. Theory*, 29(5):656–664, 1983.

J. Rissanen. Universal coding, information, prediction and estimation. *IEEE Trans. Inf. Theory*, IT-30(4):629–636, 1984.

J. Rissanen. *Stochastic Complexity in Statistical Inquiry*, volume 15 of *Series in Computer Science*. World Scientific, Singapore, 1989.

J. Rissanen. Hypothesis selection and testing by the mdl principle. *Comput. J.*, 42(4):1–11, 1999.

J.J.K.O Ruanaidh and W.J. Fitzgerald. *Numerical Bayesian Methods Applied to Signal Processing*. Springer, 1996.

T. Sauer, J.A. Yorke, and M. Casdagli. Embedology. *J. Stat. Phys.*, 65(3/4): 579–616, 1992.

G. Schwarz. Estimating the dimension of a model. *Ann. Statist.*, 6(2):461–464, 1978.

S. Smale. Differentiable dynamical systems. *Bull. Amer. Math. Soc.*, 73:747–817, 1967.

M. Small, K. Judd, M. Lowe, and S. Stick. Is breathing in infants chaotic? dimension estimates for respiratory patterns during quiet sleep. *J. Appl. Physiol.*, 86:359–376, 1999.

L.A. Smith. Identification and prediction of low-dimensional dynamics. *Physica D*, 58:50–76, 1992.

J. Stark. Delay reconstruction: Dynamics vs. statistics. In *Nonlinear Dynamics and Statistics*, A.I. Mees, editor, Birkhäuser, Boston, 2000.

J. Stark, D.S. Broomhead, M.E. Davies, and J. Huke. Takens embedding theorems for forced and stochastic systems. *Nonlin. Anal.*, 30:5303–5314, 1997.

G. Sugihara and R.M. May. Nonlinear forecasting as a way of distinguishing chaos from measurement error in time series. *Nature*, 344:734–741, 1990.

F. Takens. Detecting strange attractors in turbulence. In *Dynamical Systems and Turbulence*, D.A. Rand and L.-S. Young, editors, volume 898 of *Lecture Notes in Mathematics*, pages 365–381. Springer, Berlin, 1981.

C.S. Wallace and D.M. Boulton. An information measure for classification. *Comput. J.*, 11:185–195, 1968.

D.F. Watson. Computing the n-dimensional Delaunay tesselation with application to Voronoi polytopes. *Comput. J.*, 24(2):167–172, 1981.

A.S. Weigend, B.A. Huberman, and D.E. Rumelhart. Predicting the future: a connectionist approach. *Int. J. Neur. Syst.*, Vol 1(No 3):193–209, 1990.

M.J. Weingberger and A. Lempel. A sequential algorithm for the universal coding of finite memory sources. *IEEE Trans. Inf. Theor.*, 38(3):1002–1014, 1992.

F.M.J. Willems. The context tree weighting method: Extensions. *IEEE Trans. Inf. Theor.*, 44(2):792–798, 1998.

F.M.J. Willems, Y.M. Shtarkov, and T.J. Tjalkens. The context tree weighting method: Basic properties. *IEEE Trans. Inf. Theor.*, 41(2):653–664, 1995.

L.-S. Young. Entropy, Lyapunov exponents, and Hausdorff dimension in differentiable dynamical systems. *IEEE Trans. Circ. Syst.*, CAS-30(8):599–607, 1983.

A Modelling Framework for the Prices and Times of Trades made on the New York Stock Exchange

Tina Hviid Rydberg and Neil Shephard

1 Introduction

1.1 The data and model

Most modern theoretical and empirical finance is based on continuous time models with continuous sample paths or, in other words, diffusion processes which are driven by Wiener processes. Prominent recent references include (16) and (1), while the most well known example is geometric Brownian motion used in the (7) option pricing model.

In practice almost all the prices at which financial assets transact live on a lattice structure. Figure 1 displays the transaction prices (in US dollars) for the IBM stock traded on the New York Stock Exchange (NYSE) on four randomly selected days in 1995. This shows prices are integer multiples of 1/8 and that transactions are irregularly spaced in time. Although this multiple or 'tick' size varies with the financial market, the lattice structure is always present and is often important. In the case of trades on the NYSE, the tick size of 1/8 of a dollar was determined by the NYSE, and was set in order to avoid unnecessary negotiations between seller and buyer (see, for example (26) and the references therein). In 1997 it was reduced to 1/16 of a dollar.

There are at least three different ways of dealing with this kind of data:

1. Ignore the fact that the state space of the prices is a lattice and use a continuous sample path model. The literature assessing the effect of this type of misspecification is surveyed by (10, pp. 109–128). Their broad conclusion is that methods which use low frequency data (e.g. daily or monthly returns) are not overly influenced by discreteness, but higher frequency analysis can be sensitive.

2. Build a model assuming an underlying unobserved continuous sample path process and a latent continuous exposure cost and then round it to match the discreteness. This method is used by (27) in his model of discrete bid and ask quotes.

3. Directly model what is observed. Such models will have a state space which is a lattice. This approach involves a distinct stochastic model for

217

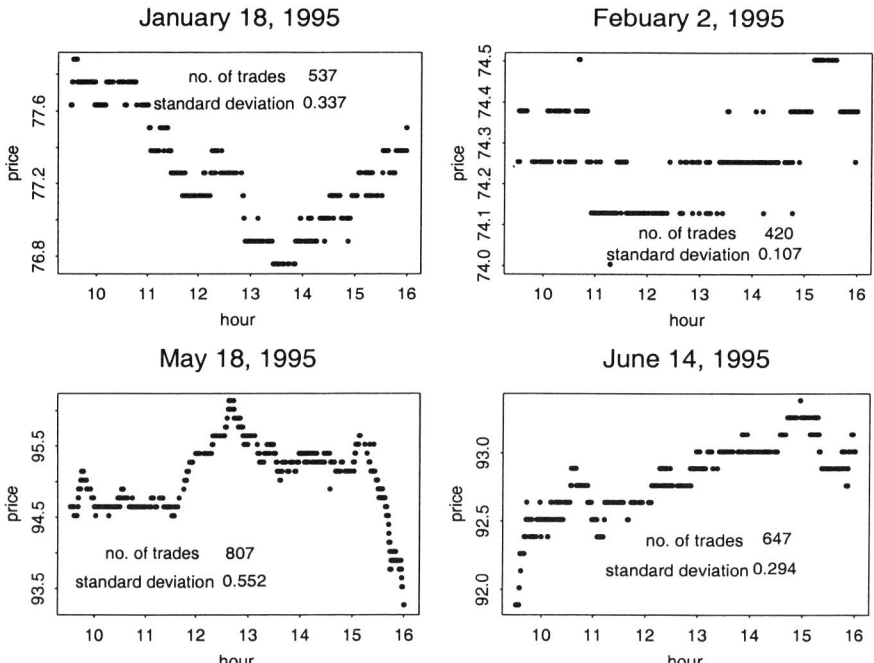

Figure 1: *Plot of all traded IBM prices at the New York stock exchange on four different days in 1995. A trade is represented as a dot (•); i.e. what appears as lines in the graphs are trades at the same price.*

the times of trades as well as a model for the price changes at the times of trades.

In this article we will discuss the third of these approaches. To simplify our exposition we have normed prices so that the tick size is one.

Our basic model structure will be a compound Poisson process in the wide sense of, for example, (23). Let $p(u)$ denote the price of the asset at time u, then we allow the non-stationary and non-linear price process to follow

$$p(u) = p(0) + \sum_{t=1}^{N(u)} Z_t, \qquad u \geq 0, \text{ and } \sum_{t=1}^{0} Z_t \overset{\text{def}}{=} 0,$$

where $\{N(u)\}_{u\geq 0}$ is a counting process[1] which counts the number of trades recorded up until time u, and Z_t is the price movement associated with the

---

[1]There are many equivalent definitions of a counting process. The one which is most helpful in our context states that if $\{N(u)\}_{u\geq 0}$ is a process with state space $\mathbf{Z} \cup \{+\infty\}$ and non-decreasing right continuous paths, then $\{N(u)\}_{u\geq 0}$ is a counting process. Since the paths are non-decreasing and right continuous we automatically get that $\{N(u)\}_{u\geq 0}$ is càdlàg (continu à droit – limite à gauche).

*t*th trade. It is important to note that Z_t can be exactly zero, for many trades take place without moving the price. There is no loss of generality in writing down this representation since both $\{N(u)\}_{u \geq 0}$ and $\{Z_t\}_{t \in \mathbf{N}_+}$ can be either continuous or discrete (for ease of notation we will use the slightly abbreviated expressions $\{N(u)\}$ and $\{Z_t\}$). However, we will tend to be more specific than this. We suggest modelling $\{N(u)\}$ as Cox process,[2] that is a Poisson process with a random intensity.[3] In general, the dynamics of the Cox and price movements processes can be adapted to a wide class of filtrations involving just their own past or more extensive information sets. This is purely an issue of combining both the empirical evidence and *a priori* economic theory, reflecting both the purpose of the modelling exercise and the data generating mechanism. A simple example of this is that the models we specify should prevent the price process from going negative. This is easy to do by careful modelling of the $\{Z_t\}$ process, which must depend upon the level of the price.[4]

1.2 Related work

Independently of our initial draft of this article (36) have suggested a similar type of compound Poisson process for asset prices. This article will focus on the study of $\{N(u)\}$ and only allow the $\{Z_t\}$ to be modelled by simple descriptive Markov chains. Papers which have previously looked at the $\{N(u)\}$ process include (18) and, subsequently, (32), (39), (21) and (28). We will compare our suggestion with this earlier work in Section 5.

Following an initial draft of this article, (38) have studied the dynamics of the $\{Z_t\}$ process within the context of our compound Poisson process framework, while not discussing the specification of the counting process $\{N(u)\}$. Another approach to modelling the dynamics of the $\{Z_t\}$ has also been previously proposed by (37). These two models have rather different features since the models proposed in (38) could potentially include discrete as well as continuous returns, whereas the model in (37) only allows $\{Z_t\}$ to live on a fixed number of points.

Our models are also related to stochastic volatility (SV) or time deformation models, see e.g. (11), (29), (41), (20) and (5). In SV models Brownian motion is deformed, while in the compound Poisson process the Cox process is a deformed Poisson process, the intensity process playing the role of that of the volatility for the Brownian motion. (5) have studied the connection between

---

[2]A convenient example of a Cox process is the influential autoregressive conditional duration (ACD) model advocated by (18), which allows straightforward likelihood based econometric inference.

[3]This is based on the assumption that the counting process is *simple* which means that $N(u)$ increases one unit at its epochs of increase, see (22, p. 34). Such an assumption is fulfilled if the intensity measure is continuous.

[4]In practice trading would probably be halted if information released caused very dramatic falls in the share price. At, e.g., the NYSE there are specific rules about when trading should be paused.

the modelling framework we propose and an SV model in a thickly traded market. Those results have been elaborated by (19).

Compound Poisson processes with discrete innovations resemble binomial models of stock prices (see, for example, (15)), because they also live on a grid. Our framework is more complicated since binomial models typically live in discrete time, with independent and identically distributed price movements only occurring at deterministic points in time.

Jump diffusion models, which allow discontinuous sample paths for prices, have also been used in finance. The discontinuities are usually introduced as a standard Poisson process (see e.g. (33)) and are used to model big events such as interventions by government or monetary authorities. They are not appropriate for the empirical phenomenon we are modelling in this article.

Compound Poisson processes are extensively used in insurance mathematics as a model for the capital of an insurance company. The capital is effected by the known income from premiums and the randomly arriving stream of insurance claims from policy holders. The earliest reference we know of to the use of Cox processes in this context is (2), while a textbook exposition of this literature is given in (22).

The article is organized as follows. In Section 2 we look at some descriptive statistics for the trade-by-trade data we are analysing — studying both the basic features of the counting process $\{N(u)\}$ and the price innovations process $\{Z_t\}$. Section 3 looks at the general properties of counting processes in our context. In Section 4 we discuss using two signal extraction methods for estimating the current level of intensity of trading in the market. Two alternatives are studied in Section 5, while in Section 6 we look at the implication of our model structure for the dynamics of changes in the price level over intervals of length Δ. In Section 7 we look at connecting our theoretical model of $N(u)$ and the price changes with the empirical evidence. We draw our conclusions in Section 8.

2   Basic features of trade-by-trade data

2.1   The data

The trade data used in this article is for the IBM share recorded electronically at the New York Stock Exchange in 1995 (NYSE TAQ–data base). The market itself is not electronic but is open out cry and there is one market maker for each stock. Each market maker can make the market in several stocks, for an excellent exposition of market microstructure, see (34). The precision of the time stamp is one second and it is the duty of the seller to report the trade. We first construct a time series for each day on which the exchange was open, computing the price changes at each trade (rescaling the data to have a tick size of one). We cut out all trades registered after 16.00 as this is the official closing of the exchange and our initial data analysis suggested the data was significantly different when it had a time stamp which was after 16.00. Also all trades occurring with an error mark are discarded and in order to reduce the size of the data set we only consider trades which took place at the NYSE.

2.2   Price movements

The dynamics of the price level in calendar time is determined by the properties of $\{N(u)\}$ and $\{Z_t\}$. In order to formulate models for these processes, we first look at their basic empirical features. We start with the price movements $\{Z_t\}$.

The $\log_{10}$ of the counts of $\{Z_t, Z_{t-1}\}$ is given in Figure 2. This shows the dramatic concentration of the data on $0, 0$ which accounts for 60% of the data and substantial mass along the lines

$$\{Z_t = k, Z_{t-1} = 0\}_k \quad \text{and} \quad \{Z_t = 0, Z_{t-1} = k\}_k .$$

Most importantly there is mass along the diagonal

$$\{Z_t = k, Z_{t-1} = -k\}_k ,$$

which represents a move in the price which is reversed at the next trade. A high proportion of this represents a single tick up (down) which is followed by an immediate reversal of one tick down (up). This is caused by the discreteness of the sample space and the action of bid/ask bounce. This very significant diagonal is not matched by one along

$$\{Z_t = k, Z_{t-1} = k\}_k$$

which has almost no mass.

Figure 2 implies the $\{Z_t\}$ process has extremely significant negative first order autocorrelation. However, Figure 3 shows that by lag two this correlation is close to zero (although lags up to 6 are statistically different from zero). This suggests, up to a very rough Wold representation, the $\{Z_t\}$ are a Markov process.

Figure 3(c) counts the numbers of price movements of the $\{Z_t\}$ equal to particular integers. It demonstrates that most of the trades on the NYSE do not move the price and only rarely does the price move by many ticks. The vast majority of the data is one of $-2, -1, 0, 1, 2$, while the density is seemingly slightly skewed to the left. This is more easily seen by looking at the log-histogram which is again given in Figure 3(d).

Figure 3(b) also gives the correlogram of the $|Z_t|$. This has a large number of lags which are significantly different from zero and take a great deal of time to die down. Indeed there are indications of long-memory type behaviour in

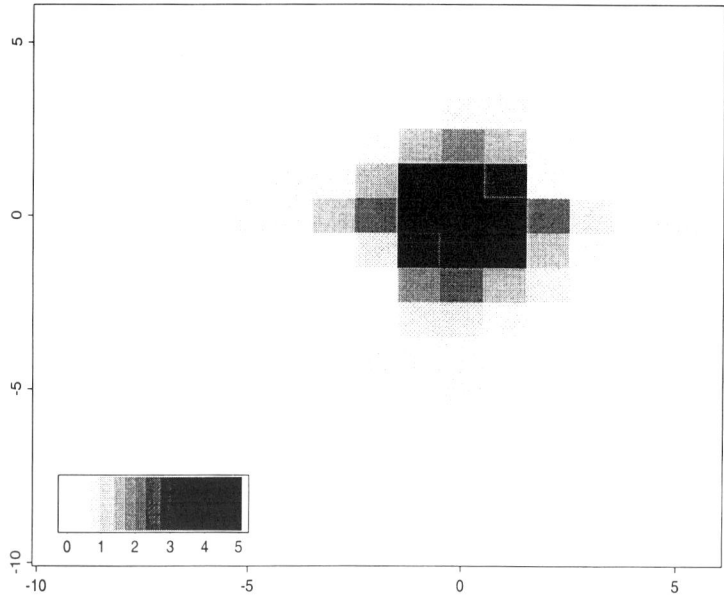

Figure 2: *This figure shows in a* $\log_{10}$ *scale the number of observations in each coordinate. To all cells* 1 *was added before taking* log *in order to avoid problems with* log *of zero.*

this plot. This is not surprising as this is close to the usual volatility cluster-ing that is often observed in financial returns over 5 minute or daily intervals. See, for example, the econometric literature on autoregressive conditional het-eroskedastic (ARCH) and SV models ((9), (20) and (40)). This feature of the data will not be the focus of this article. Instead we look at a broadbrush analysis of the data, for which we maintain the Markov assumption on $\{Z_t\}$. For a thorough analysis of the $\{Z_t\}$ process, see (38).

2.3   Trading times

2.3.1   Stylized intra-day, week and monthly effects

The intensity of trading on the NYSE varies considerably through time. In this subsection we study the basic features of the observed sequence of $\{N(u)\}$ for the IBM stock. We do this via a difference operator

$$N_n = N\left[\{(n+1)\,\Delta\}\,-\right] - N(n\Delta), \quad \Delta > 0, \tag{2.1}$$

which creates a discrete time series from the time continuous counting process by recording the number of trades which occurred in time intervals, or bins, of length Δ. Typically in this article we take Δ to equal one second. Note

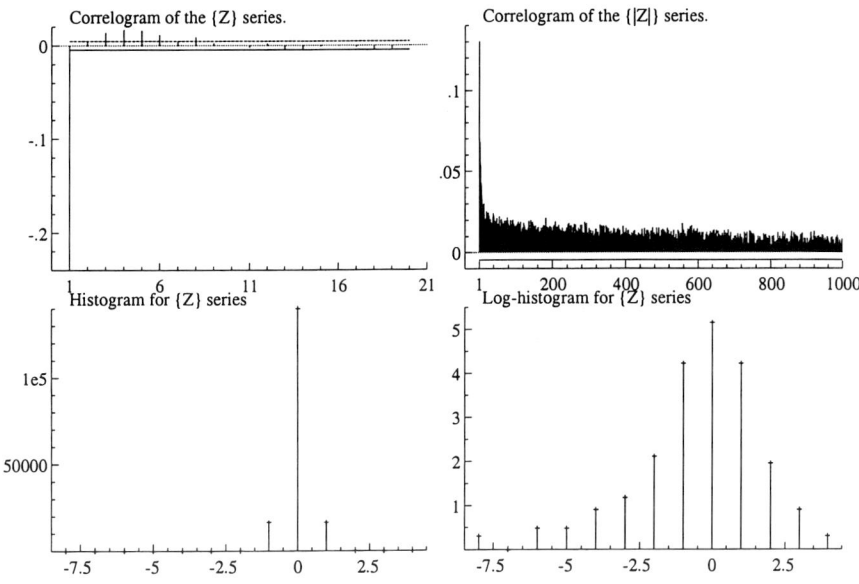

Figure 3: *(a) Shows the correlogram of the price movements, Z_t; (b) shows correlogram for $|Z_t|$; (c) is histogram of Z_t, while (d) is the* log *of the histogram plus one.*

that N_n is well defined, since $\{N(u)\}$ is càdlàg, and counts all arrivals with time stamps τ_t such that $n\Delta \leq \tau_t < (n+1)\Delta$.

The top graph of Figure 4 shows an estimate of the average number of trades which occur at each second for each day of the week. The estimate is generated using a natural cubic spline with a different bandwidth selected by generalized cross-validation for each day of the week (see, for example, (24)). We can see that for each day trading is brisk in the morning hours, slows down around lunch time and picks up again in the afternoon. In addition there are changes in these patterns between the days of the week. In particular Monday mornings and Friday afternoons are comparatively inactive, while the first 30 minutes of Friday mornings are the most active trading period of the week during 1995 for the IBM stock. Finally, we can see that the first ten minutes of each day are unlike most of the rest of the day — for the activity rate changes very dramatically during this time. Patterns of intra-day trading have been studied by several researchers, e.g. (3), (4) and (25).

The bottom graph of Figure 4 shows the number of trades on each day that the NYSE was open. We can see very significant changes in the activity level during the year, with low levels at the beginning of the year and high levels

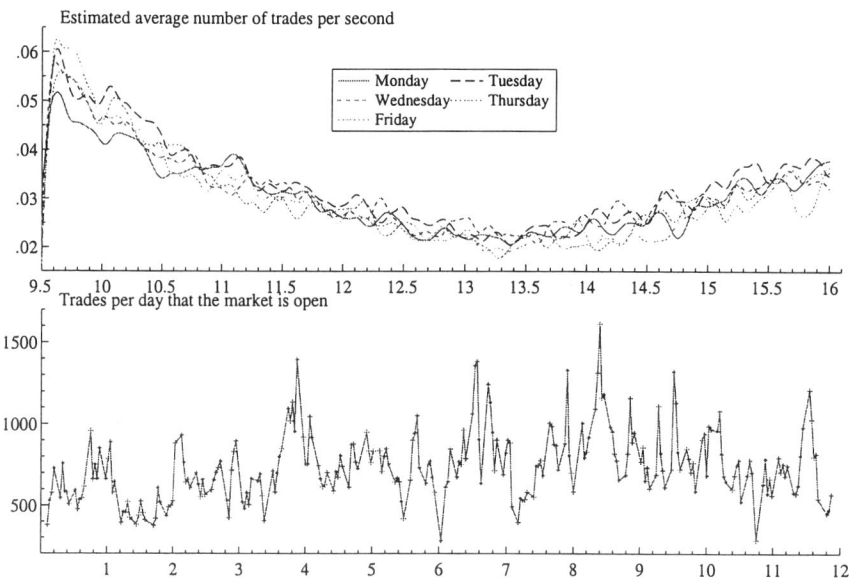

Figure 4: *The top graph is the intensity of the trading per second. Estimated daily curves using a spline with the smoothness penalty selected using generalized cross-validation. The bottom graph is the number of trades per day for each day the market is open during the year. The x-axis indicates the month.*

in September and October. Some of the variation of this series arises due to a seasonal component. However, there is also important serial dependence in the series.

2.4   Dynamics of $\{N_n\}$

Throughout we will study the dynamics of the $\{N_n\}$ process with Δ set to one second, focusing on the stochastic properties rather than the deterministic seasonal features. Our first analysis is to look at the daily time series generated by looking at the difference between the $\{N_n\}$ sequence in each day and the corresponding daily seasonal pattern given in Figure 4. For each day we computed the correlogram and plotted the average of these 251 correlograms in Figure 5. This picture shows a negative correlation at lag 1, followed by very significant (although quite small) correlations at longer lags. These die down quite slowly, but are mostly irrelevant after 5000 seconds.

The negative correlation at lag 1 indicates that there are fewer runs of trades in the series than one would expect if the trades were independently spread

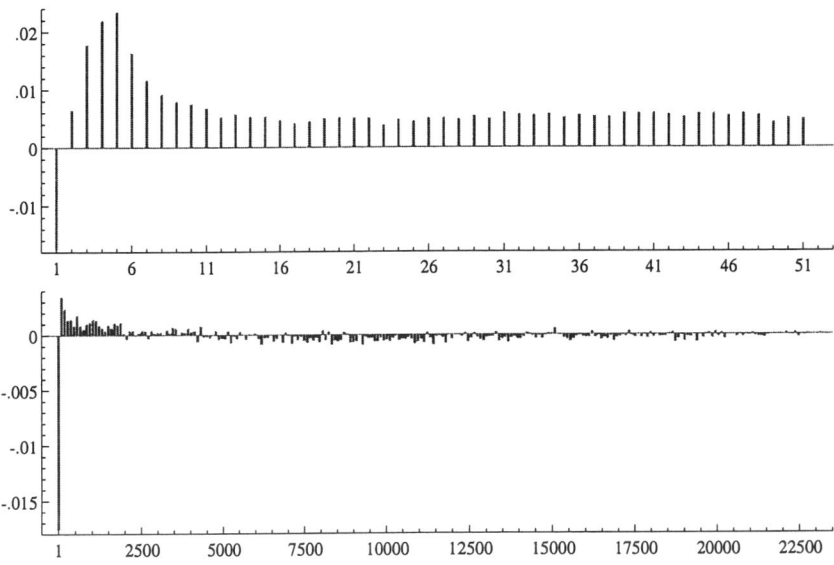

Figure 5: *Averaged correlogram for* 251 *active days. For each day we computed the correlogram for the day using the* 23,400 *second-by-second data. The top correlogram shows first* 50 *lags, the bottom, a thinned version of* 23,400 *lags.*

throughout the day. This is almost certainly due to the inability of the market maker to record trades quickly enough at active times of the market. The positive correlations at other lags are more important to the overall dynamics of the counting process as they are sustained over a large number of lags.

A difficulty with the above analysis is that it may not be picking up very long pieces of memory in the series which wash over from one day onto the next. We now try to measure this. To carry this out we will construct 4 time series each of length 60 working days by simply sticking together 60 days of the difference between $\{N_n\}$ sequence in each day and the corresponding daily seasonal pattern. Each of these series has 1,404,000 observations. Figure 6 gives the correlograms for each of these massive series. On the left we give a thinned version of the correlogram up to 50,000 lags, plotting every 200th correlation to make it easier for the eye to pick up patterns. On the right we plot the first 100 correlations in order to easily see the very short run correlations in the series. These pictures are remarkably similar over the 4 periods of 60 days and show very long levels of persistence in the correlations which are qualitatively different from the averaged correlogram approach that we showed earlier.

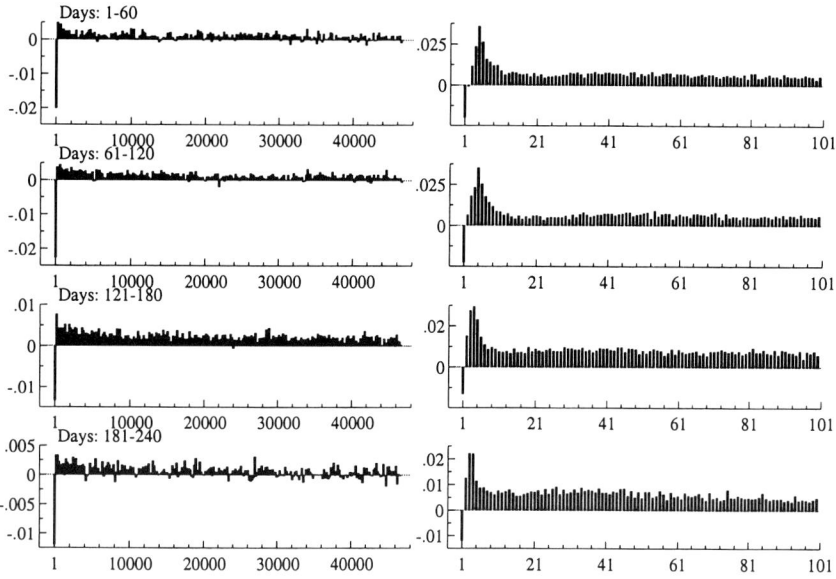

Figure 6: *Correlogram for 60 days of collated data minus seasonal pattern. Length of series is 1, 404, 000 observations.*

We can reconcile the long time series analysis with the average correlogram figure by working with the bottom of Figure 4 which shows the number of trades on each day that the NYSE was open. This series reveals a great deal of memory (or a neglected seasonal pattern) between days which will thus impact on the correlograms of the massive time series but have no impact on the averaged correlograms. In order to adjust for this feature of the $\{N_n\}$ we have constructed an adjusted massive series which multiplies the seasonal term by the ratio of the number of trades in the previous day to the average number of trades in the year.

The resulting correlograms for the new massive series are given in Figure 7. They show moderate correlations after about 15,000 lags which seems much more in line with the averaged correlogram analysis. Further, the analysis looks remarkably stable over time as each of these 4 series looks basically the same. As a result we will focus our analysis on this type of series. However, before we carry out some empirically based modelling we need to improve our understanding of the basic theory of Cox processes in order to model accurately $\{N_n\}$. This is carried out in the next section.

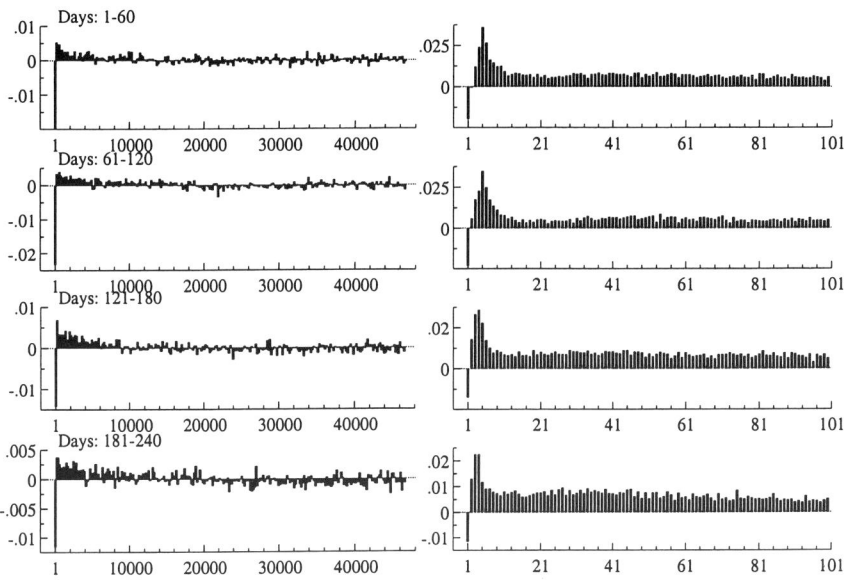

Figure 7: *Correlogram for 60 days of collated data minus seasonal pattern adjusted by dynamics of day-by-day data.*

3 Specifying a framework for Cox processes

3.1 Background

The focus in this section will be on constructing simple models which generate the counting process $\{N(u)\}$, which counts the number of trades up to time u. These will be based upon the framework of Cox or doubly stochastic Poisson processes.

We first recall that a stochastic process $\{\Lambda(u)\}$ with

$$\Lambda(0) = 0, \Lambda(u) < \infty \quad \text{forall} \quad u < \infty$$

and non-decreasing realizations is called a random measure. Then let $\left\{\tilde{N}(u)\right\}$ be a standard Poisson process and further let Λ and $\tilde{N}$ be independent of each other. Then the point process $\{N(u)\} = \left\{\tilde{N}\left\{\Lambda(u)\right\}\right\}$ is called a Cox process (see, for example, (23)). An elegant discussion of Cox processes from the viewpoint of subordination is given in (13, p. 154).

The random measure Λ is linked to the intensity λ via the integral equation

$$\Lambda(u) = \int_0^u \lambda(s)ds,$$

where $\{\lambda(u)\}$ is a càdlàg positive stochastic process, which implies that $\{\Lambda(u)\}$ is continuous.

The previous subsection analysed the number of trades in the nth bin denoted by N_n, which occurred in intervals or bins of length one second. We call this a binned time series. Since

$$\{N(u) - N(v) \mid \Lambda(u) - \Lambda(v)\} \sim \text{Po}\{\Lambda(u) - \Lambda(v)\}$$

where Po denotes the Poisson distribution, we know that

$$N_n \mid \lambda_n \sim \text{Po}(\lambda_n), \qquad \text{where} \qquad \lambda_n = \Lambda\{(n+1)\Delta\} - \Lambda(n\Delta). \qquad (3.1)$$

Note that we do not need $\Lambda[\{(n+1)\Delta\}-]$ since $\{\Lambda(u)\}$ is continuous by construction. Furthermore, since the Cox process conditionally on $\{\Lambda(u)\}$ has independent increments we have that the binned counts are conditionally independent. An interesting special case of this is where Δ is very small, in which case

$$\Pr(N_n = 0 \mid \lambda_n) = 1 - \lambda_n + o(\Delta).$$

3.2   Generic properties of the bins

It is possible to work out the autocorrelation pattern of $\{N_n\}$ simply under the condition that $\{\lambda(u)\}$ is covariance stationary. This work follows closely some related ideas on stochastic volatility due to (5). These general results will be helpful in allowing us to derive empirically realistic and simple models for the counting process.

Let ξ, ω^2 and r denote, respectively, the mean, the variance and the autocorrelation function of the process $\{\lambda(u)\}$. It is useful to define the notation r^* for the cumulative autocorrelation function, i.e.

$$r^*(t) = \int_0^t r(u)\mathrm{d}u \quad \text{and} \quad R^*(t) = \int_0^t r^*(u)\mathrm{d}u.$$

For use below we note that

$$\int_0^t \int_0^t r(u - v)\mathrm{d}u\mathrm{d}v = 2R^*(t)$$

and consequently, assuming that $\lambda(t)$ is square integrable,

$$\text{Var}\{\Lambda(u)\} = 2\omega^2 R^*(u)$$

and

$$\begin{aligned}
\text{Cov}\{\lambda_n, \lambda_{n+s}\} &= \omega^2\left[R^*\{(s+1)\Delta\} - 2R^*(s\Delta) + R^*\{(s-1)\Delta\}\right] \\
&= \omega^2 \Diamond R^*(\Delta s),
\end{aligned} \qquad (3.2)$$

where $\Diamond R^*(s)$ is defined as

$$\Diamond R^*(s) = R^*(s + \Delta) - 2R^*(s) + R^*(s - \Delta).$$

The moments of N_n follow immediately. Notice

$$
\begin{aligned}
\mathrm{E}(N_n) &= \mathrm{E}(\lambda_n) = \Delta\xi, \\
\mathrm{Var}(N_n) &= \mathrm{Var}(\lambda_n) + \mathrm{E}(\lambda_n) = 2\omega^2 R^*(\Delta) + \Delta\xi.
\end{aligned}
$$

Furthermore,

$$
\begin{aligned}
\mathrm{Cov}(N_n, N_{n+s}) &= \mathrm{E}(N_n N_{n+s}) - \mathrm{E}(N_n)^2 && (3.3) \\
&= \mathrm{Cov}(\lambda_n, \lambda_{n+s}) = \omega^2 \Diamond R^*(\Delta s). && (3.4)
\end{aligned}
$$

The implication is that

$$\mathrm{Cor}(N_n, N_{n+s}) = q\Diamond R^*(\Delta s), \qquad \text{where} \qquad q = \frac{\omega^2}{2\omega^2 R^*(\Delta) + \Delta\xi}.$$

Example Suppose that

$$r(s) = e^{-\beta|s|}$$

for some $\beta > 0$. Such autocorrelation functions occur when we use the Ornstein-Uhlenbeck process

$$d\lambda(u) = -\beta\lambda(u)dt + dz(\beta u), \tag{3.5}$$

with $0 < \beta < \infty$. The process z is a homogeneous Lévy process[5] with positive increments (also termed a subordinator). They are studied at length in (5). Exactly the same autocorrelation function results from the 'constant elasticity of variance' process

$$d\lambda(u) = -\beta\left\{\lambda(u) - \overline{\lambda}\right\}dt + \gamma\left\{\lambda(u)\right\}^d dW(u), \quad d \geq 1/2, \tag{3.6}$$

where $W(u)$ is standard Brownian motion. This general structure, which is always covariance (and strictly) stationary if $0 < \beta < \infty$, has been recently highlighted by (31) who strongly argue that it provides a great deal of tractability in terms of studying temporal aggregation of stochastic volatility using different information sets. Then,

$$R^*(\Delta) = \frac{1}{\beta^2}\left(\beta\Delta + e^{-\beta\Delta} - 1\right)$$

and for $s > 0$,

$$\Diamond R^*(\Delta s) = \beta^{-2}(1 - e^{-\beta\Delta})^2 e^{-\beta\Delta(s-1)}$$

---

[5]Note that $\mathrm{Var}\{\lambda(u)\} = \omega^2$ does not depend on β. This is due to the reparametrization used on the definition of Ornstein–Uhlenbeck processes, where the β enters the time of z.

which falls exponentially with s. Hence

$$\mathrm{Var}\,(N_n) = \frac{2\omega^2}{\beta^2}\left(e^{(-\beta\Delta)} - 1\right) + \Delta\left(\xi + \frac{2\omega^2}{\beta}\right)$$

which is linear for large Δ. Furthermore, we get that

$$\mathrm{Cor}\{N_n, N_{n+s}\} = \frac{\omega^2}{2\omega^2 R^*(\Delta) + \Delta\xi}\diamond R^*(\Delta s) = ce^{-\beta\Delta(s-1)}, \tag{3.7}$$

where

$$c = \frac{\omega^2(1 - e^{-\beta\Delta})^2}{2\omega^2\left(\beta\Delta + e^{-\beta\Delta} - 1\right) + \beta^2\Delta\xi}.$$

Note that $0 < c < 1$ and that N_n has a Wold representation which is an ARMA(1,1) process with weak white noise errors.

4 Signal extraction

4.1 Estimation of intensity

We could think of $N_n \mid \lambda_n \sim \mathrm{Po}(\lambda_n)$ as a state space model and then perform signal extraction on the random integrated intensity, λ_n. In this section we will study two ways of performing signal extraction in this context: a best linear method generated by the Kalman filter and an efficient method computed using a particle filter. Both are adapted to our case from the treatment of stochastic volatility developed by (5).

The number of trades in the nth interval of length Δ is

$$N_n = \lambda_n + (N_n - \lambda_n) = \lambda_n + u_n,$$

where $\{u_n\}$ is a Martingale difference sequence. Further, so long as $\{\lambda_n\}$ is covariance stationary, $\{u_n\}$ is a zero mean, white noise process. We will write the variance of $\{u_n\}$ as σ_u^2. If we were to adopt the (5) model for $\lambda(t)$, given in (3.5), then we would have two available approaches: a discrete time approximation or the exact continuous time version. For ease of exposition we have set $\Delta = 1$ throughout and we will only discuss the discrete time case. The extension to the continuous time version follows using results in (5).

4.2 Discrete time model

First we can take an Euler style approximation to λ_n so that

$$\lambda_{n+1} = e^{-\beta}\lambda_n + \left(1 - e^{-\beta}\right)\eta_n, \quad \text{and} \quad \mathrm{E}(\eta_n) = \mathrm{E}(\lambda_n) = \xi. \tag{4.1}$$

Here $\{\eta_n\}$ are independent and identically distributed strictly positive random variables. Then this model is in a linear state space form and so the Kalman filter provides the best linear estimator, written $a_{n+1|n}$, of the unobserved λ_{n+1} given $N_1, \ldots, N_n$. In particular the Kalman filter is given by the

$$a_{n+1|n} = e^{-\beta} \left\{ a_{n|n-1} + \frac{p_{n|n-1}}{p_{n|n-1} + 1} \left(N_n - a_{n|n-1} \right) \right\} + \left(1 - e^{-\beta} \right) \xi,$$

and its associated mean square error $\sigma_u^2 p_{n+1|n}$, where $v_n = \left(1 - e^{-\beta} \right) \left(\eta_n - \xi \right)$,

$$p_{n+1|n} = \frac{e^{-2\beta} p_{n|n-1}}{p_{n|n-1} + 1} + \frac{\sigma_v^2}{\sigma_u^2}, \qquad \text{where} \qquad \sigma_v^2 = \text{Var}(v_n).$$

In many senses the Kalman filter solution is unsatisfactory for it does not give $f(\lambda_{n+1} \mid \mathcal{F}_n)$, nor in particular the fully efficient estimator $\text{E}(\lambda_{n+1} \mid \mathcal{F}_n)$. We employ the auxiliary sampling importance resampling particle filtering of (35) to carry out this task.

We use the notation $f(\lambda_{n+1} \mid \lambda_n)$ to denote the Markov evolution of the unobserved intensity over time of the discrete time model. The particle filter has the following basic structure. The density of $\lambda_n \mid \mathcal{F}_n$ is approximated by a sample $\lambda_{1,n}, \ldots, \lambda_{M,n}$. The particle filter regenerates these points into an approximate sample from $\lambda_{n+1} \mid \mathcal{F}_{n+1}$ by sampling from

$$\widehat{f}(\lambda_{n+1} \mid \mathcal{F}_{n+1}) \propto f(N_{n+1} \mid \lambda_{n+1}) \sum_{k=1}^{M} f(\lambda_{n+1} \mid \lambda_{k,n}). \qquad (4.2)$$

This is carried out by sampling k^j with probability proportional to

$$f\left(N_{n+1} \mid \mu_{n+1}^k \right), \quad \text{where } \mu_{n+1}^k = \text{E}(\lambda_{n+1} \mid \lambda_{k,n}),$$

and then drawing from $\lambda_{j,n+1} \sim \lambda_{n+1} \mid \lambda_{k^j,n}$. This is carried out R times. The resulting population of particles are given weights proportional to

$$w_j = \frac{f(N_{n+1} \mid \lambda_{j,n+1})}{f\left(N_{n+1} \mid \lambda_{k^j,n+1} \right)}, \qquad \pi_j = \frac{w_j}{\sum_{i=1}^{R} w_i}, \qquad j = 1, \ldots, R.$$

We resample this population with probabilities $\{\pi_j\}$ to produce a sample of size M, $\lambda_{1,n+1}, \ldots, \lambda_{M,n+1}$. This sample is approximately from $\lambda_{n+1} \mid \mathcal{F}_{n+1}$. In this way we update the sample at each time step through the entire sample, $n = 1, 2, \ldots, T$. We can estimate $\text{E}(\lambda_{n+1} \mid \mathcal{F}_n)$ by

$$e^{-\beta} \frac{1}{M} \sum_{j=1}^{M} \lambda_{j,n} + \left(1 - e^{-\beta} \right) \xi.$$

In practice when we have applied the ASIR particle filter in this context we have taken in the order of $M = 10,000$ and $R = 3M$.

5   Alternatives models of $N(u)$

5.1   Time between trades

In the preceding section we have studied the stochastic properties of the number of trades up to time u, written $N(u)$. We did this via modelling the number of trades occurring in bins of length Δ, $N_n = N\left[\{(n+1)\,\Delta\} -\right] - N(n\Delta)$. An alternative is to model the time between trades, also termed *durations*. Let τ_t be the time of the tth trade. Then it is given by

$$\tau_t = \min\{u : N(u) = t\}, \qquad t = 1, 2, \ldots, N(S),$$

recalling that $N(S)$ is the number of trades in the period of length S we are studying. Then the length of time between trades is

$$L_t = \tau_t - \tau_{t-1}, \qquad t = 1, 2, \ldots.$$

For our data a small number of these times are exactly zero.

¿From a statistical viewpoint we can think of $\{L_t\}$ as a time series of duration times. There is an enormous literature on the analysis of durations, although most of it does not have a time series interpretation. We refer to (30), (42) and (14) for general discussions of this literature.

In the econometric literature an influential model of the durations is the autoregressive conditional duration (ACD) model of (18). This puts

$$L_t = \varepsilon_t \psi_t, \qquad \varepsilon_t > 0, \qquad \mathrm{E}(\varepsilon_t) = 1$$

and the ε_t's are independent identical distributed (*i.i.d.*), with

$$\psi_t = \alpha + \sum_{j=1}^{p} \gamma_j L_{t-j} + \sum_{j=1}^{q} \beta_j \psi_{t-j}.$$

Here $\psi_t = E(L_t \mid \mathcal{F}_{t-1})$, the conditional expected waiting time, where $\mathcal{F}_t$ is a filtration, potentially containing all information up till time $t - 1$. The mathematical structure of this model is identical to that of the square of a GARCH model associated with the work of (17) and (8). The model has many similarities with earlier work by (43) and (12). In practice (18) have used an exponential or Weibull distribution on the $\{\varepsilon_t\}$. Straightforward alternative structures would be to parameterize the $\log \psi_t$ instead of the ψ_t.

A key feature of this model is that, conditional on ψ_0, the likelihood can be computed via a prediction decomposition. Further the number of terms that need to be evaluated is only $N(S)$, rather than the number of seconds. However, if the focus is on events in calendar time this model has a serious drawback, namely that the intensity λ_n cannot be calculated analytically, this is discussed in more detail in (39).

5.2 Modelling the price level

The ACD model implies the following structure for the price level of the stock. It has the evolution according to the process

$$p\left(\sum_{t=1}^{n} \mathcal{L}_t\right) = p(0) + \sum_{t=1}^{n} Z_t,$$

which tells us the price at n irregularly spaced time points. In continuous time this has a less elegant expression as

$$p(u) = p(0) + \sum_{t=1}^{N(u)} Z_t, \qquad \text{where} \qquad N(u) = \arg\max_{n} \left\{\sum_{t=1}^{n} L_t \le u\right\}.$$

Note that we need $\sum_{t=1}^{n} L_t \le u$ in order for $N(u)$ to be càdlàg. This implies

$$
\begin{aligned}
N_n &= N\left[\{(n+1)\,\Delta\}\,-\right] - N(n\Delta) \\
&= \arg\max_{r}\left\{\sum_{t=1}^{r} L_t < (n+1)\,\Delta\right\} - \arg\max_{r}\left\{\sum_{t=1}^{r} L_t \le n\Delta\right\}.
\end{aligned}
$$

The complexity of the relationship between the $\{L_t\}$ and $\{N_n\}$ implies studying the behaviour of returns in calendar time implied by ACD style models is difficult.

Other difficulties with the ACD style of model include the following.

1. The intra-day seasonal pattern of changing activity has quite a complicated impact on the conditional waiting times, for the index t does not correspond to a physical time. This is quite important in this context as the intra-day pattern is very strong and quite quickly changing.

2. The conditional duration is only changed when there is a trade in the stock. However, in terms of economic theory it might be easier to parameterize the model if we were to allow the conditional waiting time to change with any new information arriving in the market. At the most refined level this would argue that the conditional intensity should change every second.

3. At the end of each trading day, there is a period which does not result in a trade. This has an impact on the likelihood function, although this is easy to compute.

5.3 A BIN model

In a recent paper (39) have suggested a simple direct model for $\{N_n\}$ in discrete time. That is they model, as a time series, the number of counts in the interval of length Δ. In their simplest model they write, with $\mathcal{F}_n^N$ as the natural filtration of the $\{N_n\}$ sequence,

$$N_n \mid \mathcal{F}_{n-1}^N \sim \text{Po}(\lambda_n), \quad \text{where} \quad \lambda_n = \alpha + \gamma N_{n-1} + \delta \lambda_{n-1}.$$

They impose the constraints that $\alpha, \gamma, \delta > 0$. (39) show that for this model the following results hold

1. The conditional likelihood $f(N_1, \ldots, N_T \mid \lambda_1)$ can be computed.

2. The process is covariance stationary if and only if $\gamma + \delta < 1$.

3. If the process is stationary then

$$\text{Cor}(N_n, N_{n+1}) = \frac{\gamma \{1 - \delta (\gamma + \delta)\}}{1 + \delta^2 - 2\delta (\gamma + \delta)},$$

$$\text{Cor}(N_n, N_{n+s}) = \text{Cor}(N_n, N_{n-1})\gamma^{s-1}, \quad s = 2, 3, \ldots.$$

This is the same autocorrelation as that derived from a Cox process for $\{N(u)\}$ when the intensity was an Ornstein–Uhlenbeck process, see formula (3.7).

This model allows us to predict future $\{N_n\}$ or time aggregations of that process. As a result is seems ideally placed if our goal is to model the price process.

6 The properties of returns

6.1 Model structure

In this section we will attempt to put together a model for the intensities and the price movements in order to produce some simple properties of returns over discrete periods of time, which are given by

$$p_n = p\left[\{(n+1)\,\Delta\} -\right] - p(n\Delta).$$

p_n is well defined since $\{p(u)\}$ is càdlàg. We will use the structure

$$p(u) = p(0) + \sum_{t=1}^{N(u)} Z_t,$$

together with three basic assumptions.

(1) The $\{Z_t\}$ are a zero mean, first order moving average process, with autocorrelation $\rho_Z(1)$ and unconditional variance σ_Z^2. Sometimes it will be helpful to write this out explicitly in terms of $Z_t = \varepsilon_t + \theta \varepsilon_{t-1}$, where

$\{\varepsilon_t\}$ is an *i.i.d.* zero mean process. An important consequence of this structure is that $|\rho_Z(1)| \le \frac{1}{2}$ and

$$\text{Cov}(Z_t, Z_{t-1}) = \theta \text{Var}(\varepsilon_t) = \frac{\theta}{1 + \theta^2} \sigma_Z^2.$$

(2) Trades occur irregularly in time according to a Cox process.

(3) The price movements will be stochastically independent of the timing of trades.

Assumptions (1) and (3) are not entirely reasonable; however, to a first approximation, they will be helpful. For a detailed discussion and model of this process, see (38).

6.2 Two moments

We first discuss the linear structure of the model. We immediately get that

$$\text{E}\{p(u)\} = p(0).$$

For the variance, by using that $\mu_N^e(u) = \text{E}\{N(u) \mid N(u) > 0\}$, we get the following structure:

$$
\begin{aligned}
\text{Var}\{p(u)\} &= \text{E}\left[\text{Var}\left\{\sum_{t=1}^{N(u)} Z_t \mid N(u)\right\}\right] && (6.1) \\
&= \sigma_Z^2 \Pr\{N(u) > 0\}\left[\mu_N^e(u) + 2\{\mu_N^e(u) - 1\}\rho_Z(1)\right] && (6.2) \\
&= \sigma_Z^2 \left[\text{E}\{N(u)\}\{1 + 2\rho_Z(1)\} - 2\rho_Z(1)\Pr\{N(u) > 0\}\right]. && (6.3)
\end{aligned}
$$

If u is quite large then for an active stock $N(u)$ must also be large. Thus

$$\text{Var}\{p(u)\} \simeq \sigma_Z^2\{1 + 2\rho_Z(1)\}\text{E}\{N(u)\}, \quad \text{for large } u.$$

In practice we have found $2\rho_Z(1)$ to be around -0.5, so the dependence in the price movements has a very considerable influence on this volatility measure. An important feature of this result is that $\text{Var}\{p(u)\}$ is approximately proportional to $\text{E}\{N(u)\}$ for large u, with the constant being $\sigma_Z^2\{1 + 2\rho_Z(1)\}$. However, is should be noted that u has to be pretty large in practice for this to hold. If $\Delta = 30$ seconds then $\Pr\{N(u) > 0\}$ is around $\frac{1}{2}$. One use of this result is that

$$\text{Var}(p_n) = \text{E}\left[\text{Var}\left\{\sum_{t=1}^{N_n} Z_t \mid N_n\right\}\right] = \text{Var}\{p(\Delta)\},$$

and so for large Δ

$$\text{Var}(p_n) \simeq \sigma_Z^2\{1 + 2\rho_Z(1)\}\text{E}\{N(\Delta)\}. \tag{6.4}$$

When Δ is very small we have that $\Pr\{N(u) > 0\} \simeq \mathrm{E}\{N(\Delta)\}$ and so

$$\mathrm{Var}(p_n) \simeq \sigma_Z^2 \mathrm{E}\{N(\Delta)\}, \text{for small } \Delta. \tag{6.5}$$

These results for returns are not immediately obvious, but they are important. For small Δ it is very rare to get more than a single price movement and so the presence of correlation amongst the price movements is irrelevant (note the variance would be exactly $\sigma_Z^2 \mathrm{E}\{N(\Delta)\}$ if the $\{Z_t\}$ were $i.i.d.$ and had zero mean). For larger Δ it is possible to get more than a single price movement occurring in the interval and hence the correlation reduces the variation in the returns and causes the nonlinearity in the relationship between $\mathrm{Var}(p_n)$ and $\mathrm{E}\{N(\Delta)\}$.

6.3   Conditional independence

In order to study the dependence between p_n and p_{n+s} it is helpful to work conditionally on

$$N_n, N\{(n+s)\Delta\} - N\{(n+1)\Delta\}, N_{n+s}.$$

A crucial feature of our setup is that

$$p_n \perp\!\!\!\perp p_{n+s} \mid N_n, N_{n+s}, N\{(n+s)\Delta\} - N\{(n+1)\Delta\} > 0,$$

that is the returns will be conditionally independent if there are trades in between these time periods. Hence when we look at any dependence structure we will only be interested in situations where $N\{(n+s)\Delta\} = N\{(n+1)\Delta\}$. An implication of this, which we will use twice in our calculations, is that we can write

$$\left\{\begin{array}{rcl} p_n &=& \varepsilon_{N(n\Delta+\Delta)} + x \\ p_{n+s} &=& \theta\varepsilon_{N(n\Delta+\Delta)} + y \end{array}\right\}, \tag{6.6}$$

where

$$\left\{\begin{array}{rcl} x &=& \theta\varepsilon_{N(n\Delta+\Delta)-1} + Z_{N(n\Delta+\Delta)-1} + \ldots + Z_{N(n\Delta)+1} \\ y &=& Z_{N\{(n+s)\Delta+\Delta\}} + \ldots + Z_{N\{(n+s)\Delta\}+2} + \varepsilon_{N(n\Delta+\Delta)+1} \end{array}\right\}.$$

We can think of x as shocks to the price which occur before $\varepsilon_{N(n\Delta+\Delta)}$, while y are the shocks after that period. The only shared shock in this framework is $\varepsilon_{N(n\Delta+\Delta)}$. As a result y, $\varepsilon_{N(n\Delta+\Delta)}$ and x are conditionally (on $N_{n+s}, N_n, N\{(n+s)\Delta\} - N\{(n+1)\Delta\} = 0$) mutually independent.

6.4   Autocorrelation

Let $\gamma_p(s) = \mathrm{Cov}(p_{n+s}, p_n)$. The only way there can be linear dependence is if there is a trade between times $n\Delta, n\Delta + \Delta$, then no trade in the interval

$(n + 1) \Delta, (n + s) \Delta$ and finally a trade between times $(n + s) \Delta, (n + s + 1) \Delta$. Thus the autocorrelation $\gamma_p(s)$ equals

$$\text{Cov} \left\{ \varepsilon_{N(n\Delta + \Delta)}, \theta \varepsilon_{N(n\Delta + \Delta)} \right\}$$
$$\times \Pr[N_{n+s} > 0, N\{(n + s)\Delta\} = N\{(n + 1)\Delta\}, N_n > 0],$$

which in turn equals

$$\sigma_Z^2 \rho_Z(1) \Pr[N_{n+s} > 0, N\{(n + s)\Delta\} = N\{(n + 1)\Delta\} \mid N_n > 0] \Pr(N_n > 0).$$

Hence

$$\begin{aligned} \rho_p(s) &= \text{Cor}(p_{n+s}, p_n) \\ &= \frac{\rho_Z(1) \Pr[N_{n+s} > 0, N\{(n + s)\Delta\} = N\{(n + 1)\Delta\} \mid N_n > 0]}{\text{Var}(p_n)}. \end{aligned}$$

As $\rho_Z(1) < 0$, this correlation will be negative for every $s > 0$. If $s = 1$, then as $\Delta \to 0$ this correlation approaches zero, as there is little chance there will be two contiguous trades and so there cannot be any linear dependence. For larger Δ the dependence becomes apparent with a strong negative correlation, but for large Δ the series will be basically uncorrelated. All of these effects weaken as s increases beyond one due to the requirement that there is no trade in the interval $(n + 1) \Delta, (n + s) \Delta$.

6.5 Volatility clustering

6.5.1 No dependence case

In order to study volatility clustering we will initially assume the $\{Z_t\}$ are independent over time. Then let $s > 0$,

$$\begin{aligned} \gamma_{p^2}(s) &= \text{Cov}(p_{n+s}^2, p_n^2) \\ &= \text{E}\left\{ \text{Cov}\left(p_{n+s}^2, p_n^2 \mid N_{n+s}, N_n\right) \right\} \\ &\quad + \text{Cov}\left\{ \text{E}\left(p_{n+s}^2 \mid N_{n+s}\right), \text{E}\left(p_n^2 \mid N_n\right) \right\} \\ &= \text{Cov}\left\{ \text{Var}\left(p_{n+s} \mid N_{n+s}\right), \text{Var}(p_n \mid N_n) \right\} \\ &= \sigma_Z^4 \text{Cov}\left(N_{n+s}, N_n\right) \\ &= \sigma_Z^4 \omega^2 \Diamond R^*(\Delta s), \end{aligned}$$

using (3.3) and $\text{Cov}\left(p_{n+s}^2, p_n^2 \mid N_{n+s}, N_n\right) = 0$. This is a completely general expression for volatility clustering under the overly strong independence assumption on the $\{Z_t\}$.

6.5.2 Dependence case

We are interested in the problem where there is dependence amongst the $\{Z_t\}$. We start by noting that we still have

$$\begin{aligned} \gamma_{p^2}(s) &= \text{E}\left\{ \text{Cov}\left(p_{n+s}^2, p_n^2 \mid N_{n+s}, N_n\right) \right\} \\ &\quad + \text{Cov}\left\{ \text{Var}\left(p_{n+s} \mid N_{n+s}\right), \text{Var}(p_n \mid N_n) \right\} \end{aligned}$$

and that, using (6.1),

$$\text{Var}(p_n \mid N_n) - \text{E}\{\text{Var}(p_n \mid N_n)\}$$
$$= \sigma_Z^2 \left\langle \begin{array}{c} \{1 + 2\rho_Z(1)\}[N_n - \text{E}\{N(\Delta)\}] \\ -2\rho_Z(1)[I\{N_n > 0\} - \Pr\{N(\Delta) > 0\}] \end{array} \right\rangle .$$

As a result, for $s > 0$

$$\text{Cov}\{\text{Var}(p_{n+s} \mid N_{n+s}), \text{Var}(p_n \mid N_n)\}$$
$$= \sigma_Z^4 \left\langle \begin{array}{c} \{1 + 2\rho_Z(1)\}^2 \text{Cov}(N_{n+s}, N_n) \\ +4\rho_Z(1)^2 \text{Cov}[I\{N_{n+s} > 0\}, I\{N_n > 0\}] \\ -4\{1 + 2\rho_Z(1)\}\rho_Z(1)\text{Cov}[I\{N_{n+s} > 0\}, N_n] \end{array} \right\rangle . \quad (6.7)$$

For large values of Δ, in a thickly traded market $N_{n+s} > 0$ and so

$$\text{Cov}[I\{N_{n+s} > 0\}, N_n] = 0$$

and

$$\text{Cov}[I\{N_{n+s} > 0\}, I\{N_n > 0\}] = 0.$$

The implication will be that

$$\text{Cov}\{\text{Var}(p_{n+s} \mid N_{n+s}), \text{Var}(p_n \mid N_n)\} \simeq \{1 + 2\rho_Z(1)\}^2 \text{Cov}(N_{n+s}, N_n).$$
$$(6.8)$$

For small Δ,

$$\begin{aligned} \text{Cov}(N_{n+s}, N_n) &\simeq \text{Cov}[I\{N_{n+s} > 0\}, N_n] \\ &\simeq \text{Cov}[I\{N_{n+s} > 0\}, I\{N_n > 0\}] \end{aligned}$$

and so

$$\text{Cov}\{\text{Var}(p_{n+s} \mid N_{n+s}), \text{Var}(p_n \mid N_n)\} \simeq \text{Cov}(N_{n+s}, N_n), \quad (6.9)$$

which again reflects the irrelevance of the dependence amongst the Z_t for small Δ.

In order to evaluate $\text{E}\{\text{Cov}(p_{n+s}^2, p_n^2 \mid N_{n+s}, N_n)\}$ it will be convenient to use the structure introduced in (6.6). Then

$$\text{Cov}\left(p_{n+s}^2, p_n^2 \mid N_{n+s} > 0, N_n > 0, N\{(n+s)\Delta\} = N\{(n+1)\Delta\}\right)$$
$$= \text{Cov}\left\{\left(\theta\varepsilon_{N(n\Delta+\Delta)} + y\right)^2, \left(\varepsilon_{N(n\Delta+\Delta)} + x\right)^2 \mid N_{n+s}, N_n\right\}$$
$$= \theta^2 \text{Var}\left(\varepsilon_{N(n\Delta+\Delta)}^2 \mid N_{n+s}, N_n\right)$$
$$= \sigma_Z^4 \rho_Z(1)^2$$

Of course unconditionally

$$E\left\{Cov\left(p_{n+s}^2, p_n^2 \mid N_{n+s}, N_n\right)\right\}$$
$$= \sigma_Z^4 \rho_Z(1)^2 E\left[N_{n+s} > 0, N_n > 0, N\left\{(n+s)\Delta\right\} = N\left\{(n+1)\Delta\right\}\right].$$

If $s = 1$ this term can have a reasonably important contribution when Δ is small. Then

$$\gamma_{p^2}(1) \simeq \left\{1 + 2\rho_Z(1)^2\right\} Cov\left(N_{n+1}, N_n\right),$$

but for larger values of s it is likely to be swamped by (6.7) and is thus irrelevant. For large values of Δ the contribution of this term will be basically irrelevant even at lag 1.

The dominant terms in these expressions are given in (6.9) and (6.8). They suggest the dynamic features of the autocorrelations amongst the squares change with Δ in an important way which is not apparent when there is no dependence in the price movements or in the corresponding SV type model. It is the combination of discreteness and serial dependence which is needed in order to get this feature of the model.

7 Empirical analysis

7.1 The arrivals

Let us briefly study the behaviour of the arrival process. Recall that in the example in Section 3.2 we discussed Ornstein–Uhlenbeck processes as possible models for the intensity process. For such types the autocorrelation function is of the type

$$r(s) = e^{-\beta|s|}$$

for some $\beta > 0$, and

$$R^*(\Delta) = \frac{1}{\beta^2}\left(\beta\Delta + e^{-\beta\Delta} - 1\right).$$

This implies that the overdispersion of the counts is

$$Var(N_n) - E(N_n) = \frac{2\omega^2}{\beta^2}\left(\beta\Delta + e^{-\beta\Delta} - 1\right).$$

In order to investigate if this very simple type of model, is at least in the right ball park, we randomly chose four days in 1995. The four days are rather different in intensity levels, the lowest of the four was 21 February where the average duration was 62.6 seconds while the highest was on 27 June where the average duration was 31.7 seconds, almost twice as busy. In order not to have to deal with individual day effects we for each of these four days calculated the empirical values of $Var(N_n) - E(\lambda_n)$ for $1 \leq \Delta \leq 300$ and estimated the two parameters β and ω^2. The results can be found in Figure 8. For all the days the shapes of the curves seem to capture the relationship quite well. There is some variation in both β and ω^2 showing different levels of over-dispersion and memory. The estimated values of β correspond to half-lives of between 80 and 127 seconds. This is not quite enough to capture all of the memory we have seen in the intensity process in Figures 5 and 6.

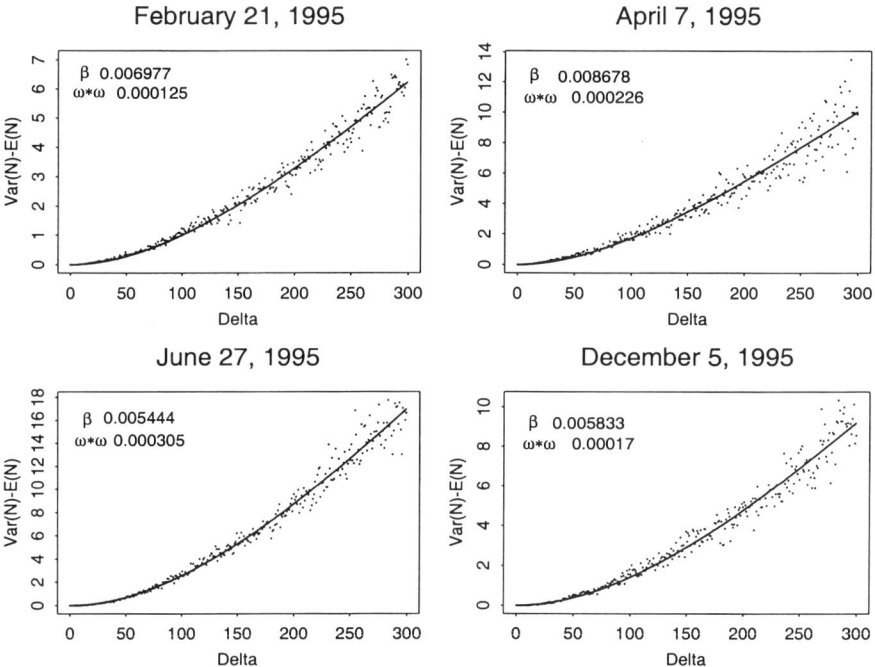

Figure 8: *The observed relationship between* $\mathrm{Var}(N_n) - \mathrm{E}(N_n)$ *and* Δ *(represented as a* $\cdot$*) and the estimated curve resulting from the assumption that the intensity process can be modelled as an Ornstein–Uhlenbeck process (represented as a solid line).*

One way to go forward from here could be to assume that the intensity process follows a sum of Ornstein–Uhlenbeck processes, studied in e.g. (6) as a way of modelling the volatility process. For such processes the autocorrelation function is given by

$$r(s) = \sum_{i=1}^{k} w_i e^{-\beta_i |s|}$$

for some $\beta_i > 0$ and $\sum_{i=1}^{k} w_i = 1$. Furthermore,

$$R^* (\Delta) = \sum_{i=1}^{k} w_i R_i^* (\Delta)$$

where

$$R_i^* (\Delta) = \frac{1}{\beta_i^2} \left(\beta_i \Delta + e^{-\beta_i \Delta} - 1 \right).$$

This implies that

$$\mathrm{Var}\,(N_n) - \mathrm{E}(N_n) = \sum_{i=1}^{k} \frac{\varpi_i}{\beta_i^2} \left(\beta_i \Delta + e^{-\beta_i \Delta} - 1 \right),$$

where $\varpi_i = 2\omega^2 w_i$. The different decay rates could then represent the persistence of different types of shocks to the intensity level.

7.2 Price changes

In Section 6 we studied a simple model which did allow for serial dependence between price changes in that they where assumed to follow a first order moving average process. We furthermore assumed that the arrivals were stochastically independent of the price changes. In this setting we have that the variance of p_n can be estimated by

$$\widehat{\mathrm{Var}(p_n)} = \hat{\sigma}_Z^2 \left[\hat{\lambda}_1 \Delta \left\{ 1 + 2\hat{\rho}_Z (1) \right\} - 2\hat{\rho}_Z (1) \left(1 - e^{-\hat{\lambda}_1 \Delta} \right) \right], \qquad (7.1)$$

where $\hat{\sigma}_Z^2$ is the estimate of the variance of the price changes, $\hat{\rho}_Z (1)$ is the estimate of the first order correlation and $\hat{\lambda}_1$ is the estimate of the mean intensity when $\Delta = 1$.

In Figure 9 we have plotted $\mathrm{Var}\,(p_n)$ for $1 \leq \Delta \leq 300$ and the estimated curve resulting from the assumption that the variance can be described by (7.1), again for four randomly chosen days. The estimate from (7.1) does not seem too far off, but it is obvious that we have to build a more sophisticated model for the $\{Z_t\}$ process (this is the topic of (38)). We have also plotted the approximations given by (6.5) and (6.4). From the graph it is seen that (6.5) works reasonably well for $\Delta < 20$ and that (6.4) seriously underestimates the variance when $-\hat{\rho}_Z (1)$ is large.

8   Conclusion

In this article we have set out a framework for the study of the trade-by-trade price movements of speculative assets traded on major stock markets. The emphasis has been on the discrete nature of prices, irregularly spaced trading and returns (price movements) over intervals of length Δ. We have used compound processes as the basis of our modelling, with trading occurring at times determined by a Cox process. We studied the use of signal extraction methods for estimating the time varying intensity of this process. In Section 7.1 of the article we studied the implications of this style of model for the

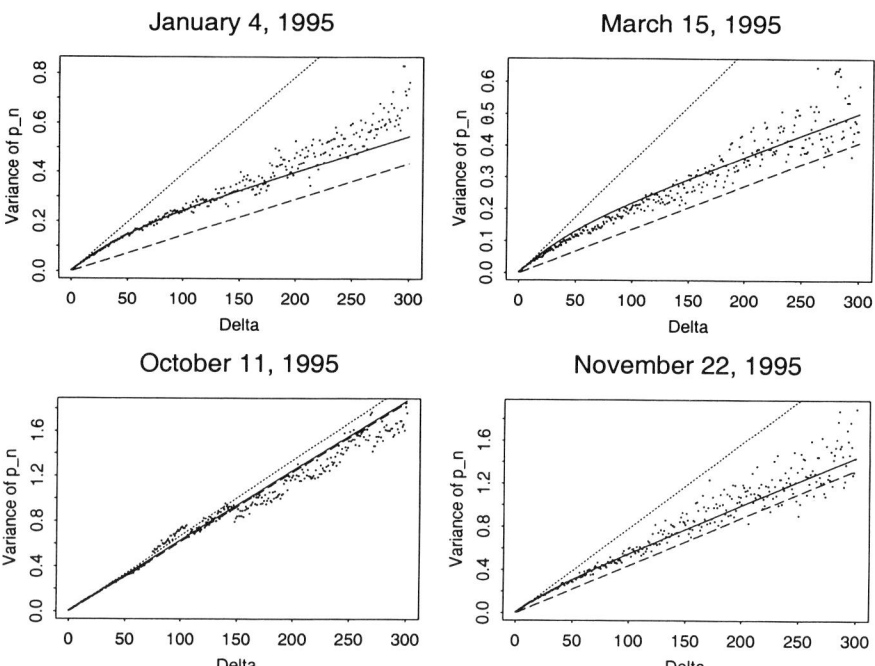

Figure 9: *The observed relationship between* Var(p_n) *and* Δ *(represented as a* ·*) and the estimated curve resulting from the assumption that the variance can be described by* (7.1) *(represented as a solid line). The approximations given by* (6.5) *and* (6.4) *are plotted as a dotted and a dashed line.*

dynamics of price changes in the market. The empirical evidence suggests that we are on the right track but that the models have to be somewhat more sophisticated. Further development of the structure for the arrival rate process was hinted at in Section 3.2 and is developed in (39). Also the model for the price changes only included the most dominant feature, the first order moving average term, this is extended in (38), where also other explanatory variables, such as duration and volume, are included.

9   Acknowledgments

Tina Rydberg thanks the Danish Natural Science Research Council for their financial support through a post-doctoral fellowship. Both authors are grateful for support from The Centre for Analytical Finance, Aarhus, Denmark. Neil Shephard's research is supported by the ESRC through the grant 'Econometrics of trade-by-trade price dynamics,' which is coded R000238391 by the ESRC. The initial draft of this article was written in December 1997 and presented at a seminar at Oxford University that month and at the Workshop on Mathematical Finance, University of Bremen, Germany, February 1998. Subsequently it was presented at numerous other international meetings. We thank Richard Spady for various helpful comments on this line of research, Richard L. Smith and Bent Nielsen for detailed comments on an earlier draft, while we have a big debt to Ole Barndorff-Nielsen for his work on SV models (with Neil Shephard) which allowed us to better understand the behaviour of $\{N_n\}$. We alone are responsible for any errors in the article.

References

[1] Ait-Sahalia, Y. (1996). Nonparametric pricing of interest rate derivative securities. *Econometrica 64*, 527–560.

[2] Ammeter, H. (1948). A generalization of the collective theory of risk in regard to fluctuating basic probabilities. *Skand. Aktuar Tidskr. 31*, 171–198.

[3] Andersen, T. G. and T. Bollerslev (1997). Intraday periodicity and volatility persistence in financial markets. *J. Empirical Finance 4*, 115–58.

[4] Andersen, T. G. and T. Bollerslev (1998). Deutschemark–dollar volatility: Intraday activity patterns, macroeconomic announcements, and longer run dependencies. *J. Finance 53*, 219–265.

[5] Barndorff-Nielsen, O. E. and N. Shephard (1998). Aggregation and model construction for volatility models. Unpublished discussion paper: Nuffield College, Oxford.

[6] Barndorff-Nielsen, O. E. and N. Shephard (1999). Non-Gaussian OU based models and some of their uses in financial economics. Unpublished discussion paper: Nuffield College, Oxford.

[7] Black, F. and M. Scholes (1973). The pricing of options and corporate liabilities. *J. Political Economy 81*, 637–654.

[8] Bollerslev, T. (1986). Generalised autoregressive conditional heteroskedasticity. *J. Econometrics 51*, 307–327.

[9] Bollerslev, T., R. F. Engle, and D. B. Nelson (1994). ARCH models. In R. F. Engle and D. McFadden (Eds.), *The Handbook of Econometrics, Volume 4*, pp. 2959–3038. Amsterdam: North-Holland.

[10] Campbell, J. Y., A. W. Lo, and A. C. MacKinlay (1997). *The Econometrics of Financial Markets*. Princeton, NJ: Princeton University Press.

[11] Clark, P. K. (1973). A subordinated stochastic process model with fixed variance for speculative prices. *Econometrica 41*, 135–156.

[12] Cox, D. R. (1972). The statistical analysis of dependencies in point processes. In P. A. P. Lewis (Ed.), *Symposium on Point Processes*, pp. 55–66. New York: Wiley and Sons.

[13] Cox, D. R. and H. D. Miller (1965). *The Theory of Stochastic Processes.* London: Chapman & Hall.

[14] Cox, D. R. and D. Oakes (1984). *Analysis of Survival Data.* London: Chapman & Hall.

[15] Dothan, M. U. (1990). *Prices in Financial Markets.* New York: Oxford University Press.

[16] Duffie, D. (1992). *Dynamic Asset Pricing Theory.* Princeton, NJ: Princeton University Press.

[17] Engle, R. F. (1982). Autoregressive conditional heteroskedasticity with estimates of the variance of the United Kingdom inflation. *Econometrica 50*, 987–1007.

[18] Engle, R. F. and J. R. Russell (1998). Forecasting transaction rates: the autoregressive conditional duration model. *Econometrica 66*, 1127–1162.

[19] Frey, R. and W. Runggaldier (1998). Nonlinear filtering techniques for estimation and risk management in partially observed stochastic volatility models. Unpublished paper: Mathematics Department, ETH.

[20] Ghysels, E., A. C. Harvey, and E. Renault (1996). Stochastic volatility. In C. R. Rao and G. S. Maddala (Eds.), *Statistical Methods in Finance*, pp. 119–191. Amsterdam: North-Holland.

[21] Ghysels, E., J. Jasiak, and C. Gourieroux (1998). Stochastic volatility duration models. Unpublished paper: Pennsylvania State University. Presented at Second international conference on high frequency data in finance, Zurich, Switzerland, April.

[22] Grandell, J. (1991). *Aspects of Risk Theory.* Berlin, Heidelberg, New York: Springer-Verlag.

[23] Grandell, J. (1997). *Mixed Poisson Processes*, Volume 77 of *Monographs on Statistics and Applied Probability.* London: Chapman and Hall.

[24] Green, P. and B. W. Silverman (1994). *Nonparameteric Regression and Generalized Linear Models: A Roughness Penalty Approach.* London: Chapman & Hall.

[25] Guillaume, D. M., M. M. Dacorogna, R. R. Dave, U. A. Muller, R. B. Olsen, and O. V. Pictet (1997). From the bird's eye view to the microscope: a survey of new stylized facts of the intra-daily foreign exchange markets. *Finance and Stochastics 2*, 95–130.

[26] Hasbrouck, J. (1996). Modeling market microstructure time series. In C. R. Rao and G. S. Maddala (eds.), *Statistical Methods in Finance*, pp. 647–692. Amsterdam: North-Holland.

[27] Hasbrouck, J. (1999a). The dynamics of discrete bid and ask quotes. *J. Finance 54*. Forthcoming.

[28] Hasbrouck, J. (1999b). Trading fast and slow: security market events in real time. Unpublished paper: Stern Business School, New York University.

[29] Hull, J. and A. White (1988). An analysis of the bias in option prices caused by stochastic volatility. *Advances in Futures and Options Pricing Research 3*, 29–61.

[30] Lancaster, T. (1990). *The Econometric Analysis of Transition Data.* Cambridge: Cambridge University Press.

[31] Meddahi, N. and E. Renault (1996). Aggregation and marginalization of GARCH and stochastic volatility models. Unpublished paper: CREST-INSEE.

[32] Meddahi, N., E. Renault, and B. Werker (1998). Modelling high frequency data in continuous time. Unpublished paper: CIRANO, CRDE, Montreal University.

[33] Merton, R. C. (1976). Option pricing when underlying stock returns are discontinuous. *J. Financial Economics 3*, 125–144.

[34] O'Hara, M. (1995). *Market Microstructure Theory.* Oxford: Blackwell Publishers.

[35] Pitt, M. K. and N. Shephard (1999). Filtering via simulation: auxiliary particle filter. *J. American Statistical Association 94*, 590–599.

[36] Rogers, L. C. G. and O. Zane (1998). Designing and estimating models of high frequency data. Unpublished paper: Department of Mathematics, University of Bath. Presented at Workshop on Mathematical Finance, University of Bremen, Germany, February.

[37] Russell, J. R. and R. F. Engle (1998). Econometric analysis of discrete-valued, irregularly-spaced financial transactions data using a new autoregressive conditional multinomial models. Unpublished paper: Graduate School of Business, University of Chicago. Presented at Second international conference on high frequency data in finance, Zurich, Switzerland, April.

[38] Rydberg, T. H. and N. Shephard (1998). Dynamics of trade-by-trade price movements: decomposition and models. Working paper, Nuffield College, Oxford. Presented at Workshop on Econometrics and Finance, Isaac Newton Institute, Cambridge University, October 1998.

[39] Rydberg, T. H. and N. Shephard (1999). BIN models for trade-by-trade data. Modelling the number of trades in a fixed interval of time. Working paper, Nuffield College, Oxford.

[40] Shephard, N. (1996). Statistical aspects of ARCH and stochastic volatility. In D. R. Cox, D. V. Hinkley, and O. E. Barndorff-Nielson (eds.), *Time Series Models in Econometrics, Finance and Other Fields*, pp. 1–67. London: Chapman & Hall.

[41] Stein, E. M. and J. Stein (1991). Stock price distributions with stochastic volatility: an analytic approach. *Rev. Financial Studies 4*, 727–752.

[42] Synder, D. L. and M. I. Miller (1991). *Random Point Processes in Time and Space* (2 ed.). New York: Springer-Verlag.

[43] Wold, H. (1948). On stationary point processes and Markov chains. *Skand. Aktuar Tidskr. 31*, 229–240.

The Sample Autocorrelations of Financial Time Series Models*

Richard A. Davis and Thomas Mikosch

1 Introduction

Over the past few years heavy-tailed phenomena have attracted the interest of various researchers in time series analysis, extreme value theory, econometrics, telecommunications, and various other fields. The need to consider time series with heavy-tailed distributions arises from the observation that traditional models of applied probability theory fail to describe jumps, bursts, rapid changes and other erratic behavior of various real-life time series.

Heavy-tailed distributions have been considered in the financial time series literature for some time. This includes the GARCH processes whose marginal distributions can have surprisingly heavy (Pareto-like) tails. There is plenty of empirical evidence (see for example Embrechts *et al.* [22] and the references cited therein) that financial log-return series of stock indices, share prices, exchange rates, etc., can be reasonably modeled by processes with infinite 3rd, 4th or 5th moments. In order to detect nonlinearities, the econometrics literature often recommends us to consider not only the time series itself but also powers of the absolute values. This leads to some serious problems: if we accept that the underlying time series has infinite 2nd, 3rd, 4th,... moments we have to think about the meaning of the classical tools of time series analysis. Indeed, the sample autocovariance function (sample ACVF), sample autocorrelation function (sample ACF) and the periodogram are meaningful estimators of their deterministic counterparts ACVF, ACF and spectral density only if the 2nd moment structure of the underlying time series is well defined. If we detect that a log-return series has infinite 4th moment it is questionable to use the sample ACF of the squared time series in order to make statements about the dependence structure of the underlying stationary model. For example consider plots of the sample ACF of the squares of the Standard & Poor's index for the periods 1961–1976, and 1977–1993 displayed in Figure 1.1. For the first half of the data, the ACF of the squares appears to decay slowly, while for the second half the ACF is not significant past lag 9. The discrepancies in the appearances in the two graphs suggest that either the process is non-stationary or it exhibits heavy tails.

The same drawback for the sample ACF is also present for the periodogram. The latter estimates the spectral density, a quantity that does not exist for

*This research supported in part by NSF DMS Grant No. DMS-9972015

the squared process if the 4th moments are infinite. Thus one should exercise caution in the interpretation of the periodogram of the squares for heavy-tailed data.

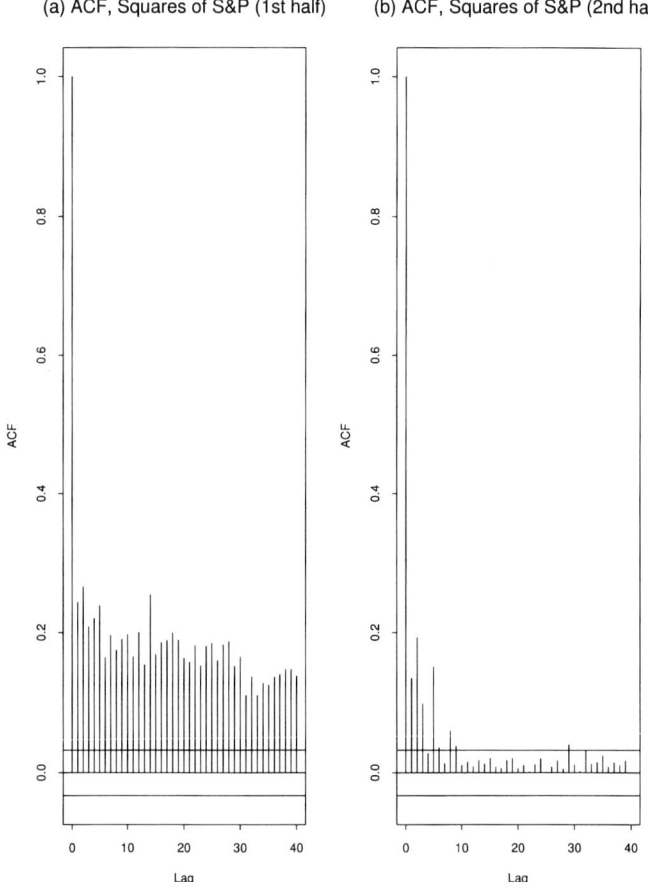

Figure 1.1 *Sample ACF of the squares of the S&P index for the periods (a) 1961–1976 and (b) 1977-1993.*

Since it has been realized that heavy tails are present in many real-life situations the research on heavy-tailed phenomena has intensified over the years. Various recent publications and monographs, such as Samorodnitsky and Taqqu [38] on infinite variance stable processes; Embrechts *et al.* [22] on extremes in finance and insurance; and Adler *et al.* [1] on heavy tails, demonstrate the emerging interest and importance of the topic.

It is the aim of this article to reconsider some of the theory for the sample ACVF and sample ACF of some classes of heavy-tailed processes. These include linear processes with regularly varying tails, solutions to stochastic recurrence equations (SREs) and stochastic volatility models. The latter two classes are commonly used for the econometric modeling of financial time series in order to describe the following empirically observed facts: nonlinearity, dependence and heavy tails. We also included the class of linear processes because of its enormous practical importance for applications but also because heavy tails and linear processes do actually interact in an 'optimal' way. This means that the sample ACF still estimates some notion of a *population* ACF, even if the variance of the underlying time series is infinite, and the rate of convergence is faster than the classical $\sqrt{n}$ asymptotics. The situation can change abruptly for nonlinear processes. In this case, the sample ACF can have a non-degenerate limit distribution – a fact which makes the interpretation of the sample ACF impossible – or the rate of convergence to the ACF can be extremely slow even when it exists. Such cases include GARCH processes and, more generally, solutions to SREs. However, not all nonlinear models exhibit unpleasant behavior of their sample ACFs. A particularly 'good' example in this context is the class of stochastic volatility models whose behavior of the sample ACF is close to the linear process case.

Fundamental to the study of all these heavy-tailed processes is the fact that their finite-dimensional distributions are multivariate regularly varying. Therefore we start in Section 2 with a short introduction to this generalization of power law tails to the multivariate setting. We also define stable random distributions which constitute a well-studied class of infinite variance distributions with multivariate regularly varying tails. In Section 3 we consider the sample ACF of linear processes, followed by the sample ACF of solutions to SREs in Section 4 and stochastic volatility models in Section 5. The interplay between the tails and the dependence is crucial for the understanding of the asymptotic behavior of the sample ACF. Therefore we first introduce in every section the corresponding model and discuss some of its basic properties. Then we explain where the heavy tails in the process come from and, finally, we give the theory for the sample ACF of these processes. One may distinguish between two types of models. The first type consists of models with a heavy-tailed input (noise) resulting in a heavy-tailed output. This includes the linear and stochastic volatility models. The second type consists of models where light- or heavy-tailed input results in heavy-tailed output. Solutions to SREs belong to the latter type. They are mathematically more interesting in the sense that the occurrence of the heavy tails has to be explained by a deeper understanding of the nonlinear filtering mechanism.

2 Preliminaries

2.1 Multivariate regular variation

Recall that a non-negative function f on $(0, \infty)$ is said to be regularly varying at infinity if there exists an $\alpha \in \mathbb{R}$ such that

$$f(x) = x^{\alpha} \, L(x),$$

and L is slowly varying, i.e.

$$\lim_{x \to \infty} \frac{L(cx)}{L(x)} = 1, \quad \forall\, c > 0.$$

We refer to Bingham *et al.* [5] for an encyclopedic treatment of regular variation.

For many applications in probability theory we need to define regular variation of random variables and random vectors.

Definition 2.1 We say that the random vector $\mathbf{X} = (X_1, \dots, X_d)$ with values in $\mathbb{R}^d$ and its distribution are *regularly varying in* $\mathbb{R}^d$ if there exist $\alpha \geq 0$ and a probability measure P_Θ on the Borel σ-field of the unit sphere $\mathbb{S}^{d-1}$ of $\mathbb{R}^d$ such that the following limit exists for all $x > 0$:

$$\frac{P(|\mathbf{X}| > t\,x,\, \mathbf{X}/|\mathbf{X}| \in \cdot)}{P(|\mathbf{X}| > t)} \xrightarrow{v} x^{-\alpha} P_\Theta(\cdot), \quad t \to \infty, \tag{1}$$

where $\xrightarrow{v}$ denotes vague convergence on the Borel σ-field of $\mathbb{S}^{d-1}$. The distribution P_Θ is called the *spectral measure* of $\mathbf{X}$, and α is the *index* of $\mathbf{X}$.

We refer to Kallenberg [31] for a detailed treatment of vague convergence of measures. We also mention that (1) can be expressed in various equivalent ways. For example, (1) holds if and only if there exist a sequence of positive constants a_n and a measure μ such that

$$n\, P(a_n\, \mathbf{X} \in \cdot) \xrightarrow{v} \mu(\cdot)$$

on the Borel σ-field of $\mathbb{R}^d$. In this case, one can choose (a_n) and μ such that, for every Borel set B and $x > 0$,

$$\mu((x, \infty) \times B) = x^{-\alpha} P_\Theta(B).$$

For $d = 1$, we see that X is regularly varying with index α if and only if

$$P(X > x) \sim p\, x^{-\alpha} L(x) \quad \text{and} \quad P(X \leq -x) \sim q\, x^{-\alpha} L(x), \quad x \to \infty, \tag{2}$$

where p, q are non-negative numbers such that $p + q = 1$ and L is slowly varying. Notice that the spectral measure is just a two-point distribution on $\{-1, 1\}$. Condition (2) is usually referred to as the *tail balancing condition*.

Note that regular variation of $\mathbf{X}$ in $\mathbb{R}^d$ implies regular variation of $|\mathbf{X}|$ and of any linear combination of the components of X. The measure P_Θ can be concentrated on a lower-dimensional subset of $\mathbb{S}^{d-1}$. For example, if the random variable X is regularly varying with index α then $\mathbf{X} = (X, 1, \dots, 1)$ with values in $\mathbb{R}^d$ is regularly varying. If $\mathbf{X}$ has independent components it is easily seen that the spectral measure P_Θ has support on the intersections with the axes. For further information on multivariate regular variation we refer to de Haan and Resnick [26] or Resnick [37]. We also refer to the Appendix of Davis *et al.* [16] for some useful results about equivalent definitions of regular variation in $\mathbb{R}^d$ and about functions of regularly varying vectors.

In what follows, we will frequently make use of a result by Breiman [12] about the regular variation of products of independent non-negative random variables ξ and η. Assume ξ is regularly varying with index $\alpha > 0$ and $E\eta^{\alpha+\epsilon} < \infty$ for some $\epsilon > 0$. Then $\xi\eta$ is regularly varying with index α:

$$P(\xi\eta > x) \sim E\eta^\alpha \, P(\xi > x), \qquad x \to \infty. \tag{3}$$

A multivariate version of Breiman's result can be found in Davis *et al.* [16].

2.2 Stable distributions

For further use we introduce the notion of α-stable distribution. The following definition is taken from Samorodnitsky and Taqqu [38] which we recommend as a general reference on stable processes and their properties.

Definition 2.2 Let $0 < \alpha < 2$. Then $\mathbf{X} = (X_1, \ldots, X_d)$ is an *α-stable random vector in $\mathbb{R}^d$* if there exist a finite measure Γ on the unit sphere $\mathbb{S}^{d-1}$ of $\mathbb{R}^d$ and a vector $\mathbf{x}_0$ in $\mathbb{R}^d$ such that:

1. If $\alpha \neq 1$, then $\mathbf{X}$ has characteristic function

$$E \exp\{i\,(\mathbf{y}, \mathbf{X})\} = \exp\left\{-\int_{\mathbb{S}^{d-1}} |(\mathbf{y}, \mathbf{x})|^\alpha (1 - i\,\mathrm{sign}((\mathbf{y}, \mathbf{x})))\Gamma(d\mathbf{x}) + i\,(\mathbf{y}, \mathbf{x}_0)\right\}.$$

2. If $\alpha = 1$, then $\mathbf{X}$ has characteristic function

$$E \exp\{i\,(\mathbf{y}, \mathbf{X})\} = \exp\left\{-\int_{\mathbb{S}^{d-1}} |(\mathbf{y}, \mathbf{x})|\left(1 + i\,\frac{2}{\pi}\mathrm{sign}((\mathbf{y}, \mathbf{x}))\,\log|(\mathbf{y}, \mathbf{x})|\right)\Gamma(d\mathbf{x})\right.$$
$$\left. +i\,(\mathbf{y}, \mathbf{x}_0)\right\}.$$

It can be shown that the pair $(\Gamma, \mathbf{x}_0)$ is unique. Moreover, the vector $\mathbf{X}$ is regularly varying with index α, and the measure Γ determines the form of the spectral measure P_Θ.

The characteristic function of an α-stable vector $\mathbf{X}$ is particularly simple if it is symmetric in the sense that $\mathbf{X}$ and $-\mathbf{X}$ have the same distribution and $\mathbf{x}_0 = 0$. In this case, we say that $\mathbf{X}$ has a *symmetric α-stable distribution* (SαS). The characteristic function of an SαS vector $\mathbf{X}$ is particularly simple:

$$E \exp\{i\,(\mathbf{y}, \mathbf{X})\} = \exp\left\{-\int_{\mathbb{S}^{d-1}} |(\mathbf{y}, \mathbf{x})|^\alpha \Gamma(d\mathbf{x})\right\}.$$

For $d = 1$, this formula is even simpler:

$$E e^{i\,y\,X} = e^{-\sigma_\alpha\,|y|^\alpha} \qquad \text{for so me } \sigma_\alpha > 0.$$

3   The linear process

3.1   Definition

Recall the definition of a *linear process*:

$$X_t = \sum_{j=-\infty}^{\infty} \psi_j Z_{t-j}, \quad t \in \mathbb{Z}, \tag{1}$$

where (Z_t) is an i.i.d. sequence of random variables, usually called the *noise* or *innovations* sequence, and (ψ_t) is a sequence of real number. For the a.s. convergence of the infinite series (1) one has to impose special conditions on the sequence (ψ_j) which depend on the distribution of Z. The formulation of these conditions will be specified below. It is worth noting that stationary ARMA and fractionally integrated ARMA processes have such a linear representation. We refer to Brockwell and Davis [13] as a general reference on the theory and statistical estimation of linear processes.

Linear processes, in particular the ARMA models, constitute perhaps the most studied class of time series models. Their theoretical properties are well understood and estimation techniques are covered by most standard texts on the subject. By choosing appropriate coefficients ψ_j, the ACF of a linear process can approximate the ACF of any stationary ergodic process, a property that helps explain the popularity and modeling success enjoyed by linear processes. Moreover, the tails of a linear process can be made as heavy as one wishes by making the tails of the innovations heavy. The latter property is an attractive one as well; the coefficients ψ_j occur in the tails only as a scaling factor. Thus the tails and the ACF behavior of a linear process can be modeled almost independently of each other: the tails are essentially determined by the tails of the innovations, whereas the ACF only depends on the choice of the coefficients. This will be made precise in what follows.

3.2   Tails

The distribution of X can have heavy tails only if the innovations Z_t have heavy tails. This follows from some general results for regularly varying and subexponential Z_t's; see Appendix A3.3 in Embrechts *et al.* [22]. For the sake of completeness we state a result from Mikosch and Samorodnitsky [33] which requires the weakest conditions on (ψ_j) known in the literature.

Proposition 3.1 (1)   *Assume that Z satisfies the tail balancing conditition (2) (with $X = Z$) for some $p \in (0,1]$ and $\alpha > 0$. If $\alpha > 1$ assume $EZ = 0$. If the coefficients ψ_j satisfy*

$$\begin{cases} \sum_{j=-\infty}^{\infty} \psi_j^2 < \infty & \textit{if } \alpha > 2 \\ \sum_{j=-\infty}^{\infty} |\psi_j|^{\alpha-\epsilon} < \infty & \textit{for some } \epsilon > 0 \quad \textit{if } \alpha \leq 2, \end{cases}$$

then the infinite series (1) converges a.s. and the following relation holds:

$$\frac{P(X > x)}{P(|Z| > x)} \sim \sum_{j=-\infty}^{\infty} |\psi_j|^{\alpha} [p \, I_{\{\psi_j > 0\}} + q \, I_{\{\psi_j < 0\}}] =: \|\psi\|_{\alpha}^{\alpha}. \tag{2}$$

(2) *Assume Z satisfies the tail balancing condition (2) for some $p \in (0,1]$ and $\alpha \in (0,2]$, that the infinite series (1) converges a.s.,*

$$\sum_{j=-\infty}^{\infty} |\psi_j|^\alpha < \infty,$$

and one of the conditions

$$L(\lambda_2) \le c\,L(\lambda_1) \text{ for } \lambda_0 < \lambda_1 < \lambda_2, \text{ some constants } c, \lambda_0 > 0,$$

and

$$L(\lambda_1 \lambda_2) \le c\,L(\lambda_1)\,L(\lambda_2) \text{ for } \lambda_1, \lambda_2 \ge \lambda_0 > 0, \text{ some constants } c, \lambda_0 > 0,$$

is satisfied. Then relation (2) holds.

This proposition implies that heavy-tailed input (regularly varying noise (Z_t)) results in heavy-tailed output. Analogously, one can show that light-tailed input forces the linear process to be light-tailed as well. For example, if the Z_t's are i.i.d. Gaussian, then the output time series (X_t) is Gaussian. This is clearly due to the linearity of the process: an infinite sum of independent random variables cannot have lighter tails than any of its summands.

It can be shown, using an extension of the proof of Proposition 3.1, that the finite-dimensional distributions of the process (X_t) are also regularly varying with index α, meaning that the vectors $(X_0, \dots, X_d)$, $d \ge 1$, are regularly varying in $\mathbb{R}^d$ with index α and spectral measure determined by the coefficients ψ_j.

3.3 Limit theory for the sample ACF

The limit theory for the sample ACVF and ACF of linear processes with infinite variance was derived in Davis and Resnick [17, 18, 19]. The limit theory for finite variance linear processes is very much the same as for Gaussian processes; see for example Brockwell and Davis [13]. For the sake of simplicity and for ease of presentation we restrict ourselves to the case of infinite variance symmmetric α-stable (SαS) noise (Z_t); see Section 2.2. In this case, one can show that Z has Pareto-like behavior in the sense that

$$P(Z > x) \sim \text{const } x^{-\alpha}, \quad x \to \infty.$$

Define the sample ACF as follows:

$$\widetilde{\rho}_{n,X}(h) := \frac{\sum_{t=1}^{n-h} X_t X_{t+h}}{\sum_{t=1}^{n} X_t^2}, \quad h = 1, 2, \dots . \tag{3}$$

If (Z_t) were an i.i.d. Gaussian $N(0, \sigma^2)$ noise sequence with the same coefficient sequence (ψ_j), (X_t) would be a Gaussian linear process with ACF

$$\rho_X(h) := \frac{\sum_{j=-\infty}^{\infty} \psi_j \psi_{j+h}}{\sum_{j=-\infty}^{\infty} \psi_j^2}, \quad h = 1, 2, \dots .$$

If (X_t) is generated from i.i.d. SαS noise it is by no means clear that $\tilde{\rho}_{n,X}(h)$ is even a consistent estimator of $\rho_X(h)$. However, from the following surprising result of Davis and Resnick [19] we find that $\tilde{\rho}_{n,X}(h)$ not only is consistent but has other good properties as an estimator of $\rho_X(h)$. (The following theorem can also be found in Brockwell and Davis [13], Theorem 13.3.1.)

Theorem 3.2 *Let (Z_t) be an i.i.d. sequence of SαS random variables and let (X_t) be the stationary linear process (1), where*

$$\sum_{j=-\infty}^{\infty} |j|\, |\psi_j|^\delta < \infty \quad \text{for some } \delta \in (0,\alpha) \cap [0,1].$$

Then for any positive integer h,

$$\left(\frac{n}{\log n}\right)^{1/\alpha} (\tilde{\rho}_{n,X}(1) - \rho_X(1), \ldots, \tilde{\rho}_{n,X}(h) - \rho_X(h)) \xrightarrow{d} (Y_1, \ldots, Y_h),$$

where

$$Y_k = \sum_{j=1}^{\infty} [\rho_X(k+j) + \rho_X(k-j) - 2\rho_X(j)\rho_X(k)]\, \frac{S_j}{S_0}, \tag{4}$$

and $S_0, S_1, \ldots$ are independent stable random variables, S_0 is positive stable with characteristic function

$$Ee^{i\lambda S_0} = \exp\left\{-\Gamma(1-\alpha/2)\, \cos(\pi\alpha/4)\, |u|^{\alpha/2}\, (1 - \mathrm{sign}(u)\, \tan(\pi\alpha/4))\right\}$$

and $S_1, S_2, \ldots$ are i.i.d. SαS with characteristic function $Ee^{iyS_1} = e^{-\sigma_\alpha |y|^\alpha}$, where

$$\sigma_\alpha = \begin{cases} \Gamma(1-\alpha)\cos(\pi\alpha/2), & \alpha \neq 1, \\ \dfrac{\pi}{2}, & \alpha = 1. \end{cases}$$

Remark 3.3 If $\alpha > 1$ the theorem remains valid for the mean corrected sample ACF, i.e. when $\tilde{\rho}_{n,X}(h)$ is replaced by

$$\hat{\rho}_{n,X}(h) := \frac{\sum_{t=1}^{n-h} \left(X_t - \overline{X}_n\right)\left(X_{t+h} - \overline{X}_n\right)}{\sum_{t=1}^{n} \left(X_t - \overline{X}_n\right)^2}, \tag{5}$$

where $\overline{X}_n = n^{-1}\sum_{t=1}^{n} X_t$ denotes the sample mean.

Remark 3.4 It follows at once that

$$\tilde{\rho}_{n,X}(h) - \rho_X(h) = O_P([n/\log n]^{-1/\alpha}).$$

This rate of convergence to zero compares favorably with the slower rate, $O_P(n^{-1/2})$, for the difference $\tilde{\rho}_{n,X}(h) - \rho_X(h)$ in the finite variance case.

Remark 3.5 If $EZ^2 < \infty$ and $EZ = 0$, a modification of Theorem 3.2 holds with the S_j's, $j \geq 1$, replaced by i.i.d. $N(0,1)$ random variables and S_0 by the constant 1. Notice that the structure of relation (4) is the reason for the so-called Bartlett formula; see Brockwell and Davis [13]. Thus (4) is an analogue to Bartlett's formula in the infinite variance case.

The proof of this result depends heavily on point process convergence results. However, in order to give some intuition for why $\tilde{\rho}_{n,X}(h)$ is a consistent estimator of $\rho_X(h)$, consider the simplest case of a linear process as given by the MA(1) process

$$X_t = \theta Z_{t-1} + Z_t, \quad t \in \mathbb{Z}.$$

The limit behavior of the sample ACF is closely connected with the large sample behavior of the corresponding sample ACVF. Define

$$\tilde{\gamma}_{n,X}(h) := \frac{1}{n} \sum_{t=1}^{n-h} X_t X_{t+h}, \quad h = 0, 1, \ldots, \tag{6}$$

and choose the sequences (a_n) and (b_n) such that

$$P(|Z| > a_n) \sim n^{-1} \quad \text{and} \quad P(|Z_1 Z_2| > b_n) \sim n^{-1}, \quad n \to \infty.$$

A simple calculation shows that $a_n \sim c_1 n^{1/\alpha}$ and $b_n \sim c_2 (n \log n)^{1/\alpha}$ for certain constants c_i, where we have made use of the fact that

$$P(Z_1 Z_2 > x) \sim c_3 x^{-\alpha} \log x. \tag{7}$$

Now, a point process convergence result shows that

$$n \left(a_n^{-2} \tilde{\gamma}_{n,Z}(0), b_n^{-1} \tilde{\gamma}_{n,Z}(1), b_n^{-1} \tilde{\gamma}_{n,Z}(2) \right) \xrightarrow{d} c_4 (S_0, S_1, S_2), \tag{8}$$

for some constant c_4, where S_0, S_1, S_2 are independent stable as described above. Now, consider the difference

$$\Delta_n := \tilde{\rho}_{n,X}(1) - \rho_X(1) = \frac{\sum_{t=1}^{n} X_t X_{t-1} - \rho(1) \sum_{t=1}^{n} X_t^2}{\sum_{t=1}^{n} X_t^2}.$$

Recalling that $\rho_X(1) = \theta/(1 + \theta^2)$ and (8), it is not difficult to see that

$$\left(\frac{n}{\log n} \right)^{1/\alpha} \Delta_n$$

$$= \left(\frac{n}{\log n} \right)^{1/\alpha} \frac{\left(\left[\theta \sum_{t=1}^{n} Z_t^2 + (1 + \theta^2) \sum_{t=1}^{n} Z_t Z_{t+1} + \theta \sum_{t=1}^{n} Z_{t-1} Z_{t+1} \right] - \theta \sum_{t=1}^{n} Z_t^2 \right)}{(1 + \theta^2) \sum_{t=1}^{n} Z_t^2}$$

$$+o_P(1)$$

$$= \left(\frac{n}{\log n}\right)^{1/\alpha} \frac{(1+\theta^2)\sum\limits_{t=1}^{n-1} Z_t Z_{t+1} + \theta \sum\limits_{t=1}^{n-2} Z_t Z_{t+2}}{(1+\theta^2)\sum\limits_{t=1}^{n} Z_t^2} + o_P(1)$$

$$\xrightarrow{d} S_0^{-1}[S_1 + \rho_X(1) S_2] = Y_1 .$$

From this limit relation one can see that the consistency of the estimator $\tilde{\rho}_{n,X}(1)$ is due to a special cancellation that eliminates all terms involving $\sum_{t=1}^{n} Z_t^2$ in the numerator which, otherwise, would determine the rate of convergence. Since the summands $Z_t Z_{t+1}$ and $Z_t Z_{t+2}$ have tails lighter than those of Z_t^2 (see (7)) the faster rate of convergence follows from (8) and the continuous mapping theorem.

Clearly, the cancellation effect described above is due to the particular structure of the linear process. For general stationary sequences (X_t) such extraordinary behavior cannot be expected. This will become clear in the following section.

Despite their flexibility for modeling tails and ACF behavior, linear processes are not considered good models for log-returns. Indeed, the sample ACF of the S&P index for the years 1961–1993 suggests that this log-return series might be well modeled by an MA(1) process. However the innovations from such a fitted model could not be i.i.d. since the sample ACFs of the absolute log-returns and their squares (see Figure 1.1) suggest dependence well beyond lag 1. This kind of sample ACF behavior shows that the class of *standard linear models* are not appropriate for describing the dependence of log-return series and therefore various nonlinear models have been proposed in the literature. In what follows, we focus on two standard models, the GARCH and the stochastic volatility processes. We investigate their tails and sample ACF behavior.

The latter two models are multiplicative noise models that have the form $X_t = \sigma_t Z_t$, where (σ_t) is referred to as the stochastic volatility process and is assumed to be independent of the noise (Z_t). The sequence (Z_t) is often assumed to be i.i.d. with $EZ = 0$ and $EZ^2 = 1$. GARCH models take σ_t to be a function of the 'past' of the process, whereas one specifies a stochastic model for (σ_t) in the case of a stochastic volatility model.

We start by investigating the GARCH model in the more general context of stochastic recurrence equations (SREs).

4   Stochastic recurrence equations

4.1   Definition

In what follows, we consider processes which are given by an SRE of the form

$$\mathbf{Y}_t = \mathbf{A}_t \mathbf{Y}_{t-1} + \mathbf{B}_t, \quad t \in \mathbb{Z}, \tag{1}$$

where $((\mathbf{A}_t, \mathbf{B}_t))$ is an i.i.d. sequence ($\mathbf{A}_t$ and $\mathbf{B}_t$ can be dependent), the $\mathbf{A}_t$'s are $d \times d$ random matrices and the random vectors $\mathbf{B}_t$ assume values in $\mathbb{R}^d$. For ease of presentation, we often use the convention that $\mathbf{A} = \mathbf{A}_1$, $\mathbf{B} = \mathbf{B}_1$, $\mathbf{Y} = \mathbf{Y}_1$, etc.

Example 4.1 (ARCH(1) process)
An important example of a process (Y_t) satisfying (1) is given by the squares (X_t^2) of an ARCH(1) process (autoregressive conditionally heteroscedastic processes of order 1). It was introduced by Engle [23] as an econometric model for log-returns of speculative prices (foreign exchange rates, stock indices, share prices, etc.). Given non-negative parameters α_0 and α_1, (X_t) is defined as

$$X_t = \sigma_t \, Z_t, \quad t \in \mathbb{Z}, \tag{2}$$

where (Z_t) is an i.i.d. sequence, and

$$\sigma_t^2 = \alpha_0 + \alpha_1 X_{t-1}^2, \quad t \in \mathbb{Z}.$$

Clearly,

$$Y_t = X_t^2, \quad A_t = \alpha_1 Z_t^2, \quad B_t = \alpha_0 Z_t^2,$$

satisfy the SRE (1).

An ARCH process (X_t) of order p (ARCH(p)) is defined in an analogous way: it satisfies the equation (2) with σ_t^2 given by

$$\sigma_t^2 = \alpha_0 + \sum_{i=1}^{p} \alpha_i \, X_{t-i}^2, \quad t \in \mathbb{Z},$$

where $\alpha_i \geq 0$ are certain parameters and $\alpha_p \neq 0$. We would like to stress that, for $p > 1$, neither (X_t^2) nor (σ_t^2) can be given as solutions to a one-dimensional SRE but have to be embedded in a multivariate SRE; see Example 4.3 below.

Example 4.2 (GARCH(1,1) process)
Since the fit of ARCH processes to log-returns was not completely satisfactory (a good fit to real-life data requires a large number of parameters α_j), Bollerslev [6] introduced a more parsimonious family of models, the GARCH (generalized ARCH) processes. A GARCH(1,1) (GARCH of order (1,1)) process (X_t) is given by relation (2), where

$$\sigma_t^2 = \alpha_0 + \alpha_1 X_{t-1}^2 + \beta_1 \sigma_{t-1}^2, \quad t \in \mathbb{Z}. \tag{3}$$

The process (X_t^2) cannot be written in the form (1) for one-dimensional Y_t's. However, an iteration of (3) yields

$$\sigma_t^2 \;=\; \alpha_0 + \alpha_1 \sigma_{t-1}^2 Z_{t-1}^2 + \beta_1 \sigma_{t-1}^2 = \alpha_0 + \sigma_{t-1}^2 [\alpha_1 Z_{t-1}^2 + \beta_1],$$

and so the sequence (σ_t^2) satisfies (1) with

$$Y_t = \sigma_t^2, \quad A_t = \alpha_1 Z_{t-1}^2 + \beta_1, \quad B_t = \alpha_0, \quad t \in \mathbb{Z}.$$

The GARCH(1,1) model is capable of capturing the main distinguishing features of log-returns of financial assets and, as a result, has become one of the mainstays of econometric models. In addition to the model's flexibility in describing certain types of dependence structure, it is also able to model tail heaviness, a property often present in observed data. A critical discussion of the GARCH model, and the GARCH(1,1) in particular, is given in Mikosch and Stărică [34, 35].

Example 4.3 (GARCH(p, q) process)
A GARCH(p, q) process (GARCH of order (p, q)) is defined in a similar way. It is given by (2) with

$$\sigma_t^2 = \alpha_0 + \sum_{i=1}^{p} \alpha_i X_{t-i}^2 + \sum_{j=1}^{q} \beta_j \sigma_{t-j}^2, \quad t \in \mathbb{Z}, \tag{4}$$

where the integers $p, q \geq 0$ determine the order of the process. Write

$$\mathbf{Y}_t = (X_t^2, \ldots, X_{t-p+1}^2, \sigma_t^2, \ldots, \sigma_{t-q+1}^2)'. \tag{5}$$

This process satisfies (1) with matrix-valued $\mathbf{A}_t$'s and vector-valued $\mathbf{B}_t$'s:

$$\mathbf{A}_t = \begin{pmatrix} \alpha_1 Z_t^2 & \cdots & \alpha_{p-1} Z_t^2 & \alpha_p Z_t^2 & \beta_1 Z_t^2 & \cdots & \beta_{q-1} Z_t^2 & \beta_q Z_t^2 \\ 1 & \cdots & 0 & 0 & 0 & \cdots & 0 & 0 \\ \vdots & \ddots & \vdots & \vdots & \vdots & \vdots & \vdots & \vdots \\ 0 & \cdots & 1 & 0 & 0 & \cdots & 0 & 0 \\ \alpha_1 & \cdots & \alpha_{p-1} & \alpha_p & \beta_1 & \cdots & \beta_{q-1} & \beta_q \\ 0 & \cdots & 0 & 0 & 1 & \cdots & 0 & 0 \\ \vdots & \vdots & \vdots & \vdots & \vdots & \ddots & \vdots & \vdots \\ 0 & \cdots & 0 & 0 & 0 & \cdots & 1 & 0 \end{pmatrix}, \tag{6}$$

$$\mathbf{B}_t = (\alpha_0 Z_t^2, 0, \ldots, 0, \alpha_0, 0 \ldots, 0)'. \tag{7}$$

Example 4.4 (The simple bilinear process)
The simple bilinear process

$$X_t = a X_{t-1} + b X_{t-1} Z_{t-1} + Z_t, \quad t \in \mathbb{Z},$$

for positive a, b and an i.i.d. sequence (Z_t) can be embedded in the framework of an SRE of type (1). Indeed, notice that $X_t = Y_{t-1} + Z_t$, where (Y_t) satisfies (1) with

$$A_t = a + b Z_t \quad \text{and} \quad B_t = A_t Z_t.$$

This kind of process has been treated in Basrak *et al.* [3].

One of the crucial problems is to find conditions for the existence of a strictly stationary solution to (1). These conditions have been studied for a long time, even under less restrictive assumptions than $((\mathbf{A}_t, \mathbf{B}_t))$ being i.i.d.; see for example Brandt [9], Kesten [32], Vervaat [39], Bougerol and Picard [7]. The following result gives some conditions which are close to necessity; see Babillot *et al.* [2].

Recall the notion of operator norm of a matrix $\mathbf{A}$ with respect to a given norm $|\cdot|$:

$$\|\mathbf{A}\| = \sup_{|\mathbf{x}|=1} |\mathbf{A}\mathbf{x}|.$$

For an i.i.d. sequence $(\mathbf{A}_n)$ of i.i.d. $d \times d$ matrices,

$$\gamma = \inf \left\{ \frac{1}{n} E \log \|\mathbf{A}_1 \cdots \mathbf{A}_n\| : \quad n \in \mathbb{N} \right\} \tag{8}$$

is called the *top Lyapunov exponent associated with* $(\mathbf{A}_n)$. If $E \log^+ \|\mathbf{A}_1\| < \infty$, it can be shown (see Furstenberg and Kesten [25]) that

$$\gamma = \lim_{n \to \infty} \frac{1}{n} \log \|\mathbf{A}_1 \cdots \mathbf{A}_n\| \quad \text{a.s.} \tag{9}$$

With a few exceptions (including the ARCH(1,1) and GARCH(1,1) cases) one cannot calculate γ explicitly.

Theorem 4.5 *Assume* $E \log^+ \|\mathbf{A}_1\| < \infty$, $E \log^+ |\mathbf{B}_1| < \infty$ *and* $\gamma < 0$. *Then the series*

$$\mathbf{Y}_n = \mathbf{B}_n + \sum_{k=1}^{\infty} \mathbf{A}_n \cdots \mathbf{A}_{n-k+1} \mathbf{B}_{n-k} \tag{10}$$

converges a.s., and the so defined process $(\mathbf{Y}_n)$ *is the unique causal strictly stationary solution of* (1).

Notice that $\gamma < 0$ holds if $E \log \|\mathbf{A}_1\| < 0$. The condition on γ in Theorem 4.5 is particularly simple in the case $d = 1$ since then

$$\frac{1}{n} E \log |A_1 \cdots A_n| = E \log |A_1| = \gamma.$$

Corollary 4.6 *Assume* $d = 1$, $-\infty \leq E \log |A_1| < 0$ *and* $E \log^+ |B_1| < \infty$. *Then the unique stationary solution of* (1) *is given by* (10).

Example 4.7 (Conditions for stationarity)
(1) The process (σ_t^2) of an ARCH(1) process has a stationary version if $\alpha_0 > 0$ and $E \log(\alpha_1 Z^2) < 0$. If Z is $N(0, 1)$, one can choose a positive $\alpha_1 < 2e^{\gamma_0} \approx 3.568...$, where γ_0 is Euler's constant. See Goldie [28]; cf. Section 8.4 in Embrechts *et al.* [22]. Notice that the stationarity of (σ_t^2) also implies the stationarity of the ARCH(1) process (X_t).

(2) The process (σ_t^2) of a GARCH(1,1) process has a stationary version if $\alpha_0 > 0$ and $E \log(\alpha_1 Z^2 + \beta_1) < 0$. Also in this case, stationarity of (σ_t^2) implies stationarity of the GARCH(1,1) process (X_t).

We mention at this point that it is very difficult to make any statements about the stationarity of solutions to general SREs and GARCH(p, q) processes in particular. For general GARCH(p, q) processes, precise necessary and sufficient conditions for $\gamma < 0$ in terms of explicit and calculable conditions on the parameters α_j, β_k and the distribution of Z are not known; see Bougerol and Picard [8] for the most general sufficient conditions which amount to certain restrictions on the distribution of Z, the following assumptions on the parameters,

$$\alpha_0 > 0 \quad \text{and} \quad \sum_{j=1}^{p} \alpha_j + \sum_{k=1}^{q} \beta_k \leq 1, \tag{11}$$

and some further technical conditions. We also mention that the X_t's have a second finite moment if $EZ = 0$, $EZ^2 = 1$ and one has strict inequality in (11). See Davis *et al.* [16] for further discussion and details. In the latter reference it is mentioned that the case of multivariate GARCH processes could be treated in an analogous way, but the theoretical difficulties are then even more significant.

4.2 Tails

Recall the definition of multivariate regular variation from Section 2.1. It is quite surprising that the stationary solutions to SREs have finite-dimensional distributions with multivariate regularly varying tails under very general conditions on $((\mathbf{A}_t, \mathbf{B}_t))$. This is due to a deep result on the renewal theory of products of random matrices given by Kesten [32] in the case $d \geq 1$. The one-dimensional case was considered by Goldie [28]. We state a modification of Kesten's fundamental result (Theorems 3 and 4 in [32]; the formulation of Theorem 4.8 below is taken from Davis *et al.* [16]). In these results, $\|\cdot\|$ denotes the operator norm defined in terms of the Euclidean norm $|\cdot|$.

Theorem 4.8 *Let* $(\mathbf{A}_n)$ *be an i.i.d. sequence of* $d \times d$ *matrices with non-negative entries satisfying:*

- *For some* $\epsilon > 0$, $E\|\mathbf{A}\|^\epsilon < 1$.

- $\mathbf{A}$ *has no zero rows a.s.*

- *The group generated by*

$$\{\log \rho(\mathbf{a}_n \cdots \mathbf{a}_1) : \mathbf{a}_n \cdots \mathbf{a}_1 > 0 \text{ for some } n \text{ and } \mathbf{a}_i \in \mathrm{supp}(P_\mathbf{A})\} \tag{12}$$

is dense in $\mathbb{R}$, *where* $\rho(\mathbf{C})$ *is the spectral radius of the matrix* $\mathbf{C}$, $\mathbf{C} > 0$ *means that all entries of this matrix are positive,* $P_\mathbf{A}$ *is the distribution of* $\mathbf{A}$, *and* $\mathrm{supp}(P_\mathbf{A})$ *its support.*

- *There exists a $\kappa_0 > 0$ such that*

$$E \left(\min_{i=1,\ldots,d} \sum_{j=1}^{d} A_{ij} \right)^{\kappa_0} \geq d^{\kappa_0/2} \tag{13}$$

and

$$E \left(\|\mathbf{A}\|^{\kappa_0} \log^+ \|\mathbf{A}\| \right) < \infty. \tag{14}$$

Then there exists a unique solution $\kappa_1 \in (0, \kappa_0]$ to the equation

$$0 = \lim_{n \to \infty} \frac{1}{n} \log E \|\mathbf{A}_n \cdots \mathbf{A}_1\|^{\kappa_1}. \tag{15}$$

If $(\mathbf{Y}_n)$ is the stationary solution to the SRE in (1) with coefficient matrices $(\mathbf{A}_n)$ satisfying the above conditions and $\mathbf{B}$ has non-negative entries with $E|\mathbf{B}|^{\kappa_1} < \infty$, then $\mathbf{Y}$ is regularly varying with index κ_1. Moreover, the finite-dimensional distributions of the stationary solution $(\mathbf{Y}_t)$ of (1) are regularly varying with index κ_1.

A combination of the general results for SREs (Theorems 4.5 and 4.8) specified to GARCH(p, q) processes yields the following result which is given in Davis et al. [16].

Theorem 4.9 *Consider the SRE (1) with $\mathbf{Y}_t$ given by (5), $\mathbf{A}_t$ by (6) and $\mathbf{B}_t$ by (7).*

(A) (Existence of stationary solution)
Assume that the following condition holds:

$$E \log^+ |Z| < \infty \text{ and the Lyapunov exponent } \gamma < 0. \tag{16}$$

Then there exists a unique causal stationary solution of the SRE (1).

(B) (Regular variation of the finite-dimensional distributions)
Let $|\cdot|$ denote the Euclidean norm and $\|\cdot\|$ the corresponding operator norm. In addition to the Lyapunov exponent γ (see (8)) being less than 0, assume the following conditions:

1. *Z has a positive density on $\mathbb{R}$ such that either $E|Z|^h < \infty$ for all $h > 0$ or $E|Z|^{h_0} = \infty$ for some $h_0 > 0$ and $E|Z|^h < \infty$ for $0 \leq h < h_0$.*

2. *Not all of the parameters α_j and β_k vanish.*

Then there exists a $\kappa_1 > 0$ such that $\mathbf{Y}$ is regularly varying with index κ_1.

A consequence of the theorem is the following:

Corollary 4.10 *Let (X_t) be a stationary* GARCH(p, q) *process. Assume the conditions of part B of Theorem 4.9 hold. Then there exists a $\kappa > 0$ such that the finite-dimensional distributions of the process $((\sigma_t, X_t))$ are regularly varying with index κ.*

Example 4.11 (ARCH(1) and GARCH(1,1))
For these two models we can give an explicit equation for the value of κ. Indeed, (15) for $d = 1$ degenerates to $E|A|^{\kappa_1} = 1$. Recall from Example 4.1 that $A_t = \alpha_1 Z_t^2$. Hence the tail index κ of X is given by the solution to the equation $E(\alpha_1 Z^2)^{\kappa/2} = 1$. Similarly, in the GARCH(1,1) case of Example 4.2 we have $A_t = \alpha_1 Z_{t-1}^2 + \beta_1$ which gives the tail index κ for σ by solving $E(\alpha_1 Z^2 + \beta_1)^{\kappa/2} = 1$. Then, by Breiman's results (3) it follows that

$$P(|X| > x) = P(|Z|\sigma > x) \sim \text{const } P(\sigma > x) \sim \text{const } x^{-\kappa}.$$

Unfortunately, these are the *only* two cases where one can give an explicit formula for κ in terms of the parameters of the GARCH process and the distribution of the noise.

The above results show that there is quite an intriguing relation between the parameters of a GARCH(p, q) process, the distribution of the noise (Z_t) and the tails of the process. In particular, it is rather surprising that the finite-dimensional distributions are regularly varying. Indeed, although the input noise (Z_t) may have light tails (exponential, normal) the resulting output (X_t) has Pareto-like tails. This is completely different from the linear process case where we discovered that the tails and the ACF behavior are due to totally different sources: the coefficients ψ_j and the tails of the noise. In the GARCH(p, q) case the parameters α_j, β_k and the whole distribution of Z, not only its tails, contribute to the heavy tailedness of marginal distribution of the process.

The squares of a GARCH(p, q) process can be written as the solution to an ARMA equation with a martingale difference sequence as noise *provided the 2nd moment of X_t is finite*. However, the analogy between an ARMA and a GARCH process can be quite misleading especially when discussing conditions for stationarity and the tail behavior of the marginal distribution. The source of the heavy tails of GARCH processes does not come directly from the martingale difference sequence, but rather from the nonlinear mechanism that connects the output with the input.

The interaction between the parameters of the GARCH(p, q) process and its tails is illustrated in the form of the invariant distribution of the process which contains products of the matrices $\mathbf{A}_t$ in front of the 'noise' $\mathbf{B}_t$ (see (10)). This is in contrast to a linear process (1) where the coefficents in front of the innovations Z_t are constants. Notice that it is the presence of sums of products of an increasing number of $\mathbf{A}_t$'s which causes the heavy tails of the distribution of $\mathbf{Y}_t$. For example, if one assumes that (Z_t) is i.i.d. Gaussian noise in the definition of a GARCH(p,q) process and considers the corresponding $\mathbf{Y}_t$'s, $\mathbf{A}_t$'s and $\mathbf{B}_t$'s (see (5)–(7)), then it is readily seen that a truncation of the infinite series (10) yields a random variable which has all finite power moments.

The interaction between the tails and dependence structure, in particular the nonlinearity of the process, is also responsible for the sample ACF behavior of solutions to SREs. In contrast to the linear process case of Section 3.3, we show in the next section that the cancellation effect which was explained in

Section 3.3 does not occur for this class of processes. This fact makes the limit theory of the sample ACF for such processes more difficult to study.

4.3 Limit theory for the sample ACF

The limit theory for the sample ACF, ACVF and cross-correlations of solutions to SRE's heavily depends on point process techniques. We refrain here from discussing those methods and refer to Davis *et al.* [16] for details. As mentioned earlier, because of the nonlinearity of the processes, we cannot expect that a theory analogous to linear processes holds, in particular we may expect complications for the sample ACF behavior if the tail index of the marginal distribution is small. This is the content of the following results.

We start with the sample autocovariances of the first component process (Y_t) say of $(\mathbf{Y}_t)$; the case of sample cross-correlations and the joint limits for the sample autocorrelations of different component processes can be derived as well.

Recall the definition of the sample ACVF $\tilde{\gamma}_{n,Y}$ from (6) and the corresponding sample ACF from (3). We also write

$$\gamma_Y(h) = EY_0 Y_h \quad \text{and} \quad \rho_Y(h) = \gamma_Y(h)/\gamma_Y(0), \quad h \geq 0,$$

for the autocovariances and autocorrelations, respectively, of the sequence (Y_t), when these quantities exist. Also recall the notion of an infinite variance stable random vector from Section 2.2.

Theorem 4.12 *Assume that $(\mathbf{Y}_t)$ is a solution to (1) satisfying the conditions of Theorem 4.8.*

(1) If $\kappa_1 \in (0,2)$, then

$$\left(n^{1-2/\kappa_1}\gamma_{n,Y}(h)\right)_{h=0,\dots,m} \quad \xrightarrow{d} \quad (V_h)_{h=0,\dots,m},$$

$$(\rho_{n,Y}(h))_{h=1,\dots,m} \quad \xrightarrow{d} \quad (V_h/V_0)_{h=1,\dots,m},$$

where the vector $(V_0,\dots,V_m)$ is jointly $(\kappa_1/2)$-stable in $\mathbb{R}^{m+1}$.

(2) If $\kappa_1 \in (2,4)$ and for $h = 0, \ldots, m$,

$$\lim_{\epsilon \to 0} \limsup_{n \to \infty} \ \mathrm{var}\left(n^{-2/\kappa_1} \sum_{t=1}^{n-h} Y_t Y_{t+h} I_{\{|Y_t Y_{t+h}| \le a_n^2 \epsilon\}}\right) = 0, \qquad (17)$$

then

$$\left(n^{1-2/\kappa_1}(\gamma_{n,Y}(h) - \gamma_Y(h))\right)_{h=0,\ldots,m} \xrightarrow{d} (V_h)_{h=0,\ldots,m}, \qquad (18)$$

$$\left(n^{1-2/\kappa_1}(\rho_{n,X}(h) - \rho_X(h))\right)_{h=1,\ldots,m} \xrightarrow{d}$$
$$\gamma_X^{-1}(0)\,(V_h - \rho_X(h)\,V_0)_{h=1,\ldots,m}, \qquad (19)$$

where $(V_0, \ldots, V_m)$ is jointly $(\kappa_1/2)$-stable in $\mathbb{R}^{m+1}$.

(3) If $\kappa_1 > 4$ then (18) and (19) hold with normalization $n^{1/2}$, where $(V_1, \ldots, V_m)$ is multivariate normal with mean zero and covariance matrix $[\sum_{k=-\infty}^{\infty} \mathrm{cov}(Y_0 Y_i, Y_k Y_{k+j})]_{i,j=1,\ldots,m}$ and $V_0 = E(Y_0^2)$.

The limit random vectors in parts (1) and (2) of the theorem can be expressed in terms of the limiting points of appropriate point processes. For more details, see Davis and Mikosch [14] where the proofs of (1) and (2) are provided and also Davis *et al.* [16]. Part (3) follows from a standard central limit theorem for strongly mixing sequences; see for example Doukhan [20].

The distributional limits of the sample ACF and ACVF of GARCH(p,q) processes (X_t) do not follow directly from Theorem 4.12 since only the the squares of the process satisfy the SRE (1). However, an application of the point process convergence in Davis and Mikosch [14] guarantees that similar results can be proved for the processes (X_t), $(|X_t|)$ and (X_t^2) or any power $(|X_t|^p)$ for some $p > 0$. The limit results of Theorem 4.12 remain qualitatively the same for $Y_t = X_t, |X_t|, X_t^2, \ldots$, but the parameters of the limiting stable laws have to be changed. See Davis *et al.* [16] for details.

Theorems 3.2 and 4.12 demonstrate quite clearly the differences between the limiting behavior of the ACF for linear and nonlinear processes. In the linear case, the rate of convergence as determined by the normalizing constants is faster the heavier the tails. In the nonlinear case, the rate of convergence of the sample ACFs to their deterministic counterparts is slower the heavier the tails, and if the underlying time series has infinite variance, the sample autotcorrelations have non-degenerate limit laws.

Since it is generally believed that log-returns have heavy tails in the sense that they are Pareto-like with tail parameter between 2 and 5 (see for example Müller *et al.* [36] or Embrechts *et al.* [22], in particular Chapters 6 and 7), Theorem 4.12 indicates that the sample ACF of such data has to be treated with some care because it could mean nothing or that the classical $\pm 1.96/\sqrt{n}$ confidence bands are totally misleading. Clearly, for GARCH processes the form of the limit distribution and the growth of the scaling constants of the sample ACF depend critically on the values of the model's parameters. We will see in the next section that the sample ACF of stochastic volatility models behaves quite differently. Its limiting behavior is more in line with that for a linear process.

5   Stochastic volatility models

5.1   Definition

As evident from the preceding discussion, the theoretical development of the basic probabilistic properties of GARCH processes is thorny: conditions for stationarity are difficult to formulate and verify, the tail behavior is complicated and little is known about the dependence structure. On the other hand, estimation for GARCH processes is relatively easy by using conditional maximum likelihood based on the i.i.d. assumption of the noise; see for example Gouriéroux [29] and the references therein. The latter property is certainly one of the attractions for this kind of model and has contributed to its popularity.

Over the last few years, another kind of econometric time series has attracted some attention: the *stochastic volatility processes*. Like GARCH models, these processes are multiplicative noise models, i.e.

$$X_t = \sigma_t\, Z_t, \quad t \in \mathbb{Z}, \tag{1}$$

where (Z_t) is an i.i.d. sequence of random variables which is completely independent of another strictly stationary sequence (σ_t) of non-negative random variables. The independence of the two sequences (Z_t) and (σ_t) allows one to easily derive the basic probabilistic properties of stochastic volatility processes. For example, the dependence structure of the process is determined via the dependence in the volatility sequence (σ_t). For our purposes, we shall assume that

$$\sigma_t = e^{Y_t}, \quad t \in \mathbb{Z}, \tag{2}$$

where (Y_t) is a linear process

$$Y_t = \sum_{j=0}^{\infty} \psi_j \varepsilon_{t-j}, \quad t \in \mathbb{Z}, \tag{3}$$

with coefficients ψ_j satisfying

$$\sum_{j=0}^{\infty} \psi_j^2 < \infty \tag{4}$$

and an i.i.d. noise sequence (ε_t). For ease of presentation we assume that ε is $N(0,1)$ which, together with (4), ensures that the defining sum for Y_t in (3) is convergent a.s. The condition of Gaussianity of the ε_t's can be relaxed at the cost of more technical conditions, see Davis and Mikosch [15] for details.

Notice that the assumption (4) is the weakest possible; it allows one to use any non-deterministic Gaussian stationary time series as a model for (Y_t), in particular one can choose (Y_t) as a stationary ARMA or a FARIMA process for modeling any kind of long or short range dependence in (Y_t); see Brockwell and Davis [13] for an extensive discussion of ARMA and FARIMA processes and Samorodnitsky and Taqqu [38] for a discussion and mathematical aspects of long range dependence. Hence one can achieve any kind of ACF behavior in (σ_t) as well as in (X_t) (due to the independence of (Y_t) and (ε_t)). This latter property gives the stochastic volatility models a certain advantage over the GARCH models. The latter are strongly mixing with geometric rate under very general assumptions on the parameters and the noise sequence; see Davis *et al.* [16] for details. As a consequence of the mixing property, if the ACF of these processes is well defined, it decays to zero at an exponential rate, hence long range dependence effects (in the sense that the ACF is not absolutely summable) cannot be achieved for a GARCH process or any of its powers. Since it is believed in parts of the econometrics community that log-return series might exhibit long range dependence, the stochastic volatility models are quite flexible for modeling this behavior; see for example Breidt *et al.* [11].

In what follows, we show that the tails and the sample ACF of these models also have more attractive properties than the GARCH models even in the infinite 2nd and 4th moment cases. Ultimately, it is hoped that the dichotomy in behavior of the ACF for GARCH and stochastic volatility models will be of practical significance in choosing between the two models. This will be the subject of future research. In contrast to the well behaved properties of the ACF, estimation of the parameters in stochastic volatility models tends to be more complicated than that for GARCH processes. Often one needs to resort to simulation based methods to calculate efficient estimates. Quasi-likelihood estimation approaches are discussed in Harvey *et al* [27], Breidt and Carriquiry [10], and Breidt *et al* [11] while simulation based methods can be found in Jacquier *et al.* [30] and Durbin and Koopman [21].

5.2   Tails

By virtue of Breiman's result (3), we know that

$$P(X > x) \sim E\sigma^\alpha \, P(Z > x) \text{ and } P(X > x) \sim E\sigma^\alpha \, P(Z \le -x), \ dx \to \infty, \quad (5)$$

provided $E\sigma^{\alpha+\epsilon} < \infty$ for some $\epsilon > 0$ and Z is regularly varying with index $\alpha > 0$ and tail balancing condition

$$P(Z > x) = p \, x^{-\alpha} \, L(x) \quad \text{and} \quad P(Z > x) = q \, x^{-\alpha} \, L(x), \quad (6)$$

where L is slowly varying and $p + q = 1$ for some $p \in [0, 1]$. In what follows, we assume that (6) holds, and we also require

$$E|Z|^\alpha = \infty. \tag{7}$$

Then $Z_1 Z_2$ is also regularly varying with index α satisfying (see equations (3.2) and (3.3) in Davis and Resnick [19])

$$\frac{P(Z_1 Z_2 > x)}{P(|Z_1 Z_2| > x)} \to \tilde{p} := p^2 + (1 - p)^2 \qquad \text{as } x \to \infty. \tag{8}$$

Another application of (3) implies that $X_1 X_h$ is regularly varying with index α:

$$\begin{cases} P(X_1 X_h > x) & = P(Z_1 Z_2 \sigma_1 \sigma_h > x) \quad \sim E[\sigma_1 \sigma_h]^\alpha\, P(Z_1 Z_2 > x), \\ P(X_1 X_h \le -x) & = P(Z_1 Z_2 \sigma_1 \sigma_h \le -x) \quad \sim E[\sigma_1 \sigma_h]^\alpha\, P(Z_1 Z_2 \le -x), \end{cases} \tag{9}$$

provided $E[\sigma_1 \sigma_h]^{\alpha+\epsilon} < \infty$ for some $\epsilon > 0$. Since we assumed the exponential structure (2) for the σ_t's and that the Y_t's are Gaussian, the σ_t's are log-normal and therefore the latter moment condition holds for every $\epsilon > 0$.

An application of a multivariate version of Breiman's result (see the Appendix in Davis *et al.* [16]) ensures that the finite-dimensional distributions of (X_t) are regularly varying with the same index α. We refrain from giving details.

5.3 Limit theory for the sample ACF

In order to describe the limiting behavior of the sample ACF of a stochastic volatility process in the heavy-tailed case, two sequences of constants (a_n) and (b_n) which figure in the normalizing constants must be defined. Specifically, let (a_n) and (b_n) be the respective $(1 - n^{-1})$-quantiles of $|Z_1|$ and $|Z_1 Z_2|$ defined by

$$\begin{aligned} a_n &= \inf\{x : P(|Z_1| > x) \le n^{-1}\} \quad \text{and} \\ b_n &= \inf\{x : P(|Z_1 Z_2| > x) \le n^{-1}\}. \end{aligned} \tag{10}$$

Using point process techniques and arguments similar to the ones given in [19], the weak limit behavior for the sample ACF can be derived for stochastic volatility processes. These results are summarized in the following theorem.

Theorem 5.1 *Assume (X_t) is the stochastic volatility process satisfying (1)–(3) where Z satisfies conditions (6) and (7). Let $\tilde{\gamma}_{n,X}(h)$ and $\tilde{\rho}_{n,X}(h)$ denote the sample ACVF and ACF of the process as defined in (6) and (3) and assume that either*

(i) $\alpha \in (0, 1)$,

(ii) $\alpha = 1$ and Z_1 has a symmetric distribution,

or

(iii) $\alpha \in (1, 2)$ and Z_1 has mean 0.

Then

$$n \left(a_n^{-2} \widetilde{\gamma}_{n,X}(0), b_n^{-1} \widetilde{\gamma}_{n,X}(1), \ldots, b_n^{-1} \widetilde{\gamma}_{n,X}(r) \right) \quad \overset{d}{\to} \quad (V_h)_{h=0,\ldots,r},$$

where $(V_0, V_1, \ldots, V_r)$ are independent random variables, V_0 is a non-negative stable random variable with exponent $\alpha/2$ and $V_1, \ldots, V_r$ are identically distributed as stable with exponent α. Moreover, for all three cases we have

$$\left(a_n^2 b_n^{-1} \widetilde{\rho}_{n,X}(h) \right)_{h=1,\ldots,r} \quad \overset{d}{\to} \quad (V_h/V_0)_{h=1,\ldots,r} \ .$$

Remark 5.2 By choosing the volatility process (σ_t) to be identically 1, we can recover the limiting results obtained in Davis and Resnick [19] for the autocovariances and autocorrelations of the (Z_t) process. If $(S_0, S_1, \ldots, S_r)$ denotes the limit random vector of the sample autocovariances based on (Z_t), then there is an interesting relationship between S_k and V_k, namely,

$$(V_0, V_1, \ldots, V_r) \overset{d}{=} \left(\|\sigma_1\|_\alpha^2 S_0, \|\sigma_1 \sigma_2\|_\alpha S_1, \ldots, \|\sigma_1 \sigma_{1+r}\|_\alpha S_r \right),$$

where $\| \cdot \|_\alpha$ denotes the L_α-norm. It follows that

$$\left(a_n^2 b_n^{-1} \widetilde{\rho}_{n,X}(h) \right)_{h=1,\ldots,r} \quad \overset{d}{\to} \quad \left(\frac{\|\sigma_1 \sigma_{h+1}\|_\alpha}{\|\sigma_1\|_\alpha^2} \frac{S_h}{S_0} \right)_{h=1,\ldots,r}.$$

Remark 5.3 The conclusion of (iii) of the theorem remains valid if $\widetilde{\rho}_{n,X}(h)$ is replaced by the mean-corrected version of the ACF given by (5).

5.3.1   Other powers

It is also possible to investigate the sample ACVF and ACF of the processes $(|X_t|^\delta)$ for any power $\delta > 0$. We restrict ourselves to the case $\delta = 1$ in order to illustrate the method.

Notice that $|X_t| = |Z_t| \sigma_t$, $t = 1, 2, \ldots$, has a structure similar to the original process (X_t). Hence Theorem 5.1 applies directly to the ACF of $|X_t|$ when $\alpha < 1$ and to the ACF of the stochastic volatility model with noise $|Z_t| - E|Z_t|$ when $\alpha \in (1, 2)$.

In order to remove the centering of the noise in the $\alpha \in (1, 2)$ case, we use the following decomposition for $h \geq 1$ with $\gamma_{|X|} = E|X_0 X_k|$, $\widetilde{Z}_t = |Z_t| - E|Z|$ and $\widetilde{X}_t = \widetilde{Z}_t \sigma_t$:

$$n(\widetilde{\gamma}_{n,|X|}(h) - \widetilde{\gamma}_{|X|}(h))$$

$$= \sum_{t=1}^{n-h} \widetilde{Z}_t \widetilde{Z}_{t+h} \sigma_t \sigma_{t+h} + E|Z| \sum_{t=1}^{n-h} \widetilde{Z}_t \sigma_t \sigma_{t+h}$$

$$+E|Z| \sum_{t=1}^{n-h} \widetilde{Z}_{t+h} \sigma_t \sigma_{t+h} - (E|Z|)^2 \sum_{t=1}^{n-h} (\sigma_t \sigma_{t+h} - E\sigma_0 \sigma_h)$$

$$= I_1 + I_2 + I_3 + I_4 .$$

Since
$$n^{-1} I_1 = \widetilde{\gamma}_{n,\widetilde{X}}(h),$$

and $E\widetilde{Z} = 0$,

Theorem 5.1 (iii) is directly applicable to $(\widetilde{X}_t)$. Also notice that $na_n^{-2} \widetilde{\gamma}_{n,|X|}(0)$ converges weakly to an $(\alpha/2)$-stable distribution, for the same reasons as given for (X_t). It remains to show that

$$b_n^{-1} I_j \xrightarrow{P} 0, \quad j = 2, 3, 4. \tag{11}$$

Point process arguments can be used to show that $a_n^{-1} I_j$, $j = 2, 3$, converge to an α-stable distribution, and since $a_n/b_n \to 0$, (11) holds for $j = 2, 3$. It is straightforward to show that $\text{var}(b_n^{-1} I_4) \to 0$ for cases when the linear process in (3) has absolutely summable coefficients or when the coefficients are given by a fractionally integrated model. Thus $b_n^{-1} I_4 \xrightarrow{P} 0$ and the limit law for $\widetilde{\gamma}_{n,|X|}(h)$ is as specified in Theorem 5.1.

5.3.2 A brief simulation comparison

To illustrate the differences in the asymptotic theory for the ACF of GARCH and stochastic volatility models, a small simulation experiment was conducted. One thousand replicates of lengths 10,000 and 100,000 were generated from a GARCH(1,1) time series model with parameter values in the stochastic volatility recursion (3) given by

$$\alpha_0 = 8.6 \times 10^{-6}, \quad \alpha_1 = .110 \quad \text{and} \quad \beta_1 = .698 .$$

The noise was generated from a student's t-distribution with 4 degrees of freedom – normalized to have variance 1. With this choice of parameter values and noise distribution the marginal distribution has Pareto tails with approximate exponent 3. The sampling behavior of both the ACF of the data and its squares are depicted using box plots in Figures 5.4(a),(c) and 5.5(a),(c) for samples of size $10,000$ and $100,000$. As seen in these figures, the sample ACF of the data appears to be converging to 0 as the sample size increases (note the differences in the magnitude of the vertical scaling on the graphs). On the other hand, the sampling distributions for the ACF of the squares (Figure 5.5(a),(c)) appear to be the same for the two samples sizes $n = 10,000$ and $100,000$ reflecting the limit theory as specified by Theorem 4.12 (1). These box plots can be interpreted as estimates of the limiting distributions of the ACF.

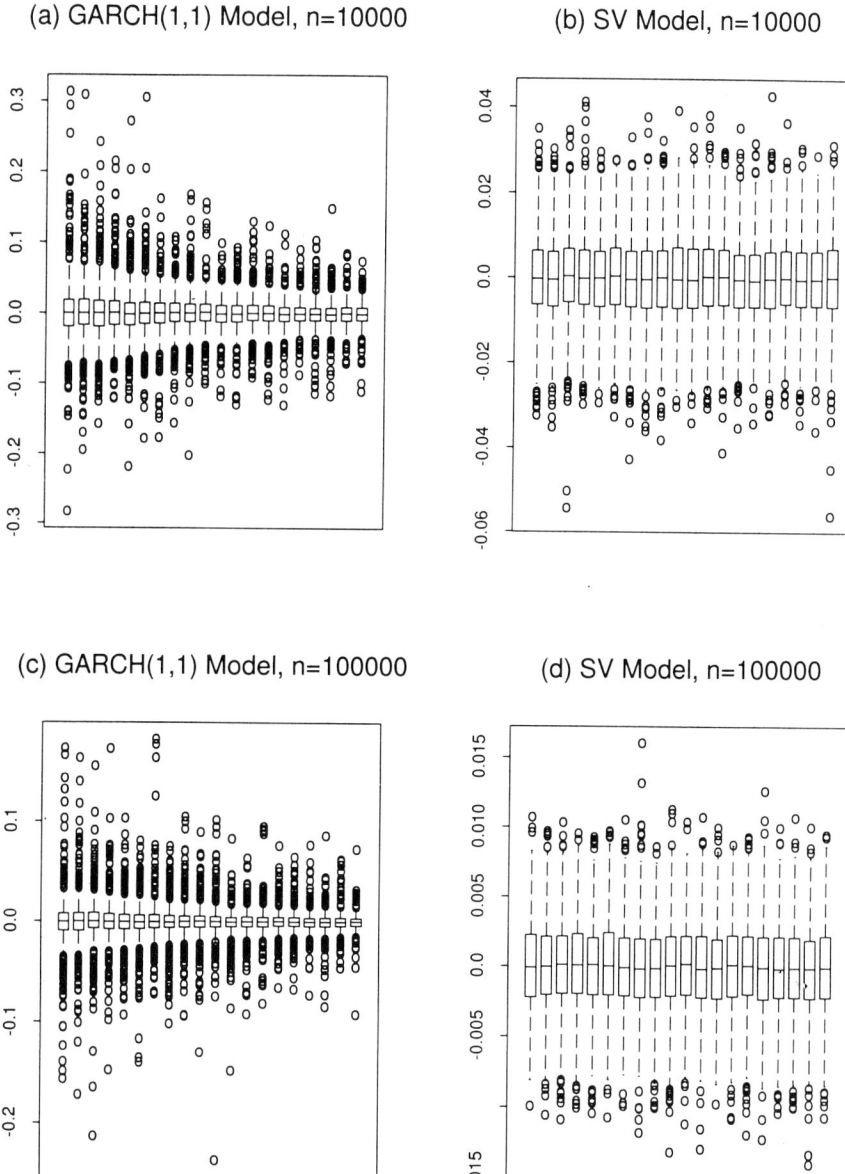

Figure 5.4 *Boxplots based on* 1000 *replications of the sample ACF (at lags* 1 *to* 20) *for data generated from a* GARCH(1,1) *model and from a stochastic volatility model.* (a) *GARCH model with sample size* $n = 10,000$; (b) *stochastic volatility model with* $n = 10,000$; (c) *GARCH model with* $n = 100,000$; *and* (d) *stochastic volatility model with* $n = 100,000$.

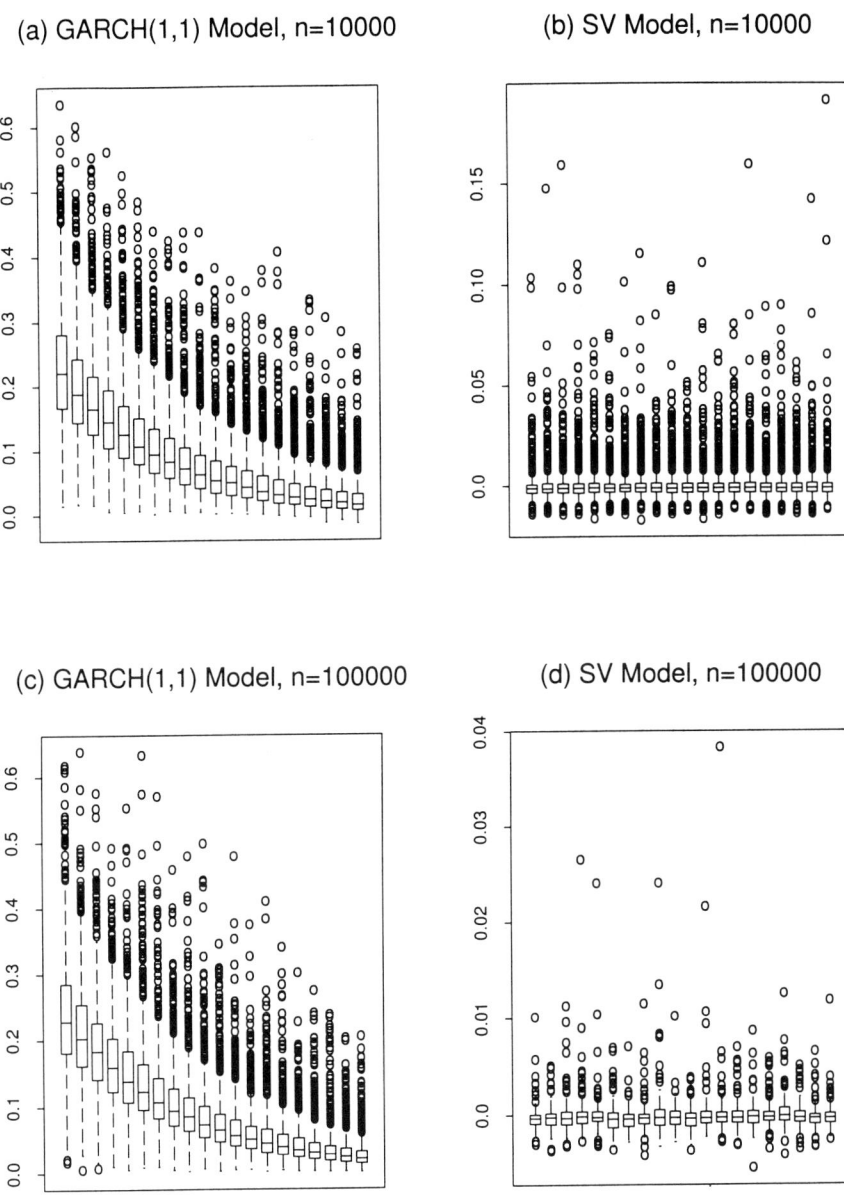

Figure 5.5 *Boxplots based on* 1000 *replications of the sample ACF (at lags* 1 *to* 20*) of the squares from a* GARCH(1,1) *model and from a stochastic volatil-ity model. (a) GARCH model with sample size* $n = 10,000$*; (b) stochastic volatility model with* $n = 10,000$*; (c) GARCH model with* $n = 100,000$*; and (d) stochastic volatility model with* $n = 100,000$*.*

Sample paths of sizes $10,000$ and $100,000$ were also generated from a stochastic volatility model, where the stochastic volatility process (σ_t) satisfies the model

$$\log \sigma_t = .85 \log \sigma_{t-1} + \epsilon_t,$$

(ϵ_t) is a sequence of i.i.d. $N(0,1)$ random variables, and the noise (Z_t) was taken to be i.i.d. with a t-distribution on three degrees of freedom. (The noise was normalized to have variance 1.) We chose this noise distribution in order to match the tail behavior of the GARCH(1,1) process. The sampling behavior of both the ACF of the data and its squares are shown in Figures 5.4(b),(d) and 5.5(b),(d). These plots demonstrate the *weak consistency* of the estimates to zero as n increase. Notice that the autocorrelations of the squares are more concentrated around 0 as predicted by the theory. Finally, these graphs illustrate the generally good performance of the sample ACF of stochastic volatility models in terms of convergence in probability to their *population* counterparts – especially when compared to the behavior found in the GARCH model.

References

[1] ADLER, R., FELDMAN, R. AND TAQQU, M.S. (EDS.) (1998) *A Practical Guide to Heavy Tails: Statistical Techniques for Analysing Heavy-Tailed Distributions.* Birkhäuser, Boston.

[2] BABILLOT, M., BOUGEROL, P. AND ELIE, L. (1997) The random difference equation $X_n = A_n X_{n-1} + B_n$ in the critical case. *Ann. Probab.* **25** 478–493.

[3] BASRAK, B., DAVIS, R.A. AND MIKOSCH, T. (1999) The sample ACF of a simple bilinear process. *Stoch. Proc. Appl.* **83** (1999), 1–14.

[4] BILLINGSLEY, P. (1968) *Convergence of Probability Measures.* Wiley, New York.

[5] BINGHAM, N.H., GOLDIE, C.M. AND TEUGELS, J.L. (1987) *Regular Variation.* Cambridge University Press, Cambridge.

[6] BOLLERSLEV, T. (1986) Generalized autoregressive conditional heteroskedasticity. *J. Econometrics* **31**, 307–327.

[7] BOUGEROL, P. AND PICARD, N. (1992) Strict stationarity of generalized autoregressive processes. *Ann. Probab.* **20**, 1714–1730.

[8] BOUGEROL, P. AND PICARD, N. (1992) Stationarity of GARCH processes and of some nonnegative time series. *J. Econom.* **52**, 115–127.

[9] BRANDT, A. (1986) The stochastic equation $Y_{n+1} = A_n Y_n + B_n$ with stationary coefficients. *Adv. Appl. Probab.* **18**, 211–220.

[10] BREIDT, F.J. AND CARRIQUIRY, A.L. (1996) Improved quasi-maximum likelihood estimation for stochastic volatility models. In: *Modelling and Prediction. Proceedings Hsinchu, 1994*, pp. 228–247. Springer, New York.

[11] BREIDT, F.J., CRATO, N. AND LIMA, P. DE (1998) The detection and estimation of long memory in stochastic volatility. *J. Econom.* **83**, 325–348.

[12] BREIMAN, L. (1965) On some limit theorems similar to the arc-sin law. *Theory Probab. Appl.* **10**, 323–331.

[13] BROCKWELL, P.J. AND DAVIS, R.A. (1991) *Time Series: Theory and Methods,* 2nd edition. Springer, New York.

[14] DAVIS, R.A. AND MIKOSCH, T. (1998) The sample autocorrelations of heavy-tailed processes with applications to ARCH. *Ann. Statist.* **26**, 2049–2080.

[15] DAVIS, R.A. AND MIKOSCH, T. (2000) The sample ACF of stochastic volatility models. In preparation.

[16] DAVIS, R.A., MIKOSCH, T. AND BASRAK, B. (1999) Limit theory for the sample autocorrelations of solutions to stochastic recurrence equations with applications to GARCH processes. Preprint. Available at www.math.rug.nl/~mikosch.

[17] DAVIS, R.A. AND RESNICK, S.I. (1985) Limit theory for moving averages of random variables with regularly varying tail probabilities. *Ann. Probab.* **13**, 179–195.

[18] DAVIS, R.A. AND RESNICK, S.I. (1985) More limit theory for the sample correlation function of moving averages. *Stoch. Proc. Appl.* **20**, 257–279.

[19] DAVIS, R.A. AND RESNICK, S.I. (1986) Limit theory for the sample covariance and correlation functions of moving averages. *Ann. Statist.* **14**, 533–558.

[20] DOUKHAN, P. (1994) *Mixing. Properties and Examples.* Lecture Notes in Statistics **85**. Springer Verlag, New York.

[21] DURBIN, J. AND KOOPMAN, S.J. (1997) Monte Carlo maximum likelihood estimation for non-Gaussian state space models. *Biometrika* **84**, 669–684.

[22] EMBRECHTS, P., KLÜPPELBERG, C. AND MIKOSCH, T. (1997) *Modelling Extremal Events for Insurance and Finance.* Springer, Berlin.

[23] ENGLE, R.F. (1982) Autoregressive conditional heteroscedastic models with estimates of the variance of United Kingdom inflation. *Econometrica* **50**, 987–1007.

[24] ENGLE, R.F. AND BOLLERSLEV, T. (1986) Modelling the persistence of conditional variances. With comments and a reply by the authors. *Econometric Rev.* **5**, 1–87.

[25] FURSTENBERG, H. AND KESTEN, H. (1960) Products of random matrices. *Ann. Math. Statist.* **31**, 457–469.

[26] HAAN, L. DE AND RESNICK, S.I. (1977) Limit theory for multivariate sample extremes. *Z. Wahrscheinlichkeitstheorie verw. Gebiete* **40**, 317–337.

[27] HARVEY, A.C., RUIZ, E., SHEPHARD, N. (1994) Multivariate stochastic variance models. *Rev. Econom. Stud.* **61**, 247–264.

[28] GOLDIE, C.M. (1991) Implicit renewal theory and tails of solutions of random equations. *Ann. Appl. Probab.* **1**, 126–166.

[29] GOURIÉROUX, C. (1997) *ARCH Models and Financial Applications.* Springer Series in Statistics. Springer, New York.

[30] JACQUIER, E., POLSON, N.G., AND ROSSI, P.E. (1994) Bayesian analysis of stochastic volatility models (with discussion). *J. Bus. Econom. Statist.* **12**, 371–417.

[31] KALLENBERG, O. (1983) *Random Measures,* 3rd edition. Akademie-Verlag, Berlin.

[32] KESTEN, H. (1973) Random difference equations and renewal theory for products of random matrices. *Acta Math.* **131**, 207–248.

[33] MIKOSCH, T. AND SAMORODNITSKY, G. (2000) The supremum of a negative drift random walk with dependent heavy-tailed steps. *Ann. Appl. Probab.*, to appear. Available as walk.ps at www.orie.cornell.edu/~/gennady/techreports.

[34] MIKOSCH, T. AND STĂRICĂ, C. (1998) Limit theory for the sample autocorrelations and extremes of a GARCH(1,1) process. *Ann. Statist.*, to appear. Available at www.math.rug.nl/~mikosch.

[35] MIKOSCH, T. AND STĂRICĂ, C. (1999) Change of structure in financial time series, long-range dependence and the GARCH model. Technical Report. University of Groningen. Available at www.math.rug.nl/~mikosch.

[36] MÜLLER, U.A., DACOROGNA, M.M. AND PICTET, O.V. (1996) Heavy tails in high-frequency financial data. Preprint. Olsen & Associates, Zürich.

[37] RESNICK, S.I. (1987) *Extreme Values, Regular Variation, and Point Processes.* Springer, New York.

[38] SAMORODNITSKY, G. AND TAQQU, M.S. (1994) *Stable Non-Gaussian Random Processes. Stochastic Models with Infinite Variance.* Chapman and Hall, London.

[39] VERVAAT, W. (1979) On a stochastic differential equation and a representation of non-negative infinitely divisible random variables. *Adv. Appl. Prob.* **11**, 750–783.

The Many Roads to Time-Frequency

Patrick Flandrin

1 Introduction

Until relatively recently, classical signal processing was faced with a paradoxical situation. On the one hand, most signals and processes—as observed in Nature or in man-made systems—were known to exhibit some form of time-dependence of their structural properties (spectral content, statistical dependencies, transfer function, ...). On the other hand however, standard methodologies were generally based on assumptions of steady-state behaviours or 'stationarity'. Since nonstationarities are by no means exceptional, and since they carry moreover the more informative features of a signal, it became essential to develop general approaches which, e.g., would go beyond Fourier-type analyses. In this respect, 'time-frequency' has progressively emerged as a natural (and better and better accepted)' paradigm, one of its salient features being the non-uniqueness of its tools, as a reflection of the many forms of possible nonstationarities and as a result of the intrinsic limitations which exist between pairs of canonically conjugate (i.e., Fourier transform of each other) variables.

As proved by their (nonlinear) historical development, it turns out that basic distributions of time-frequency analysis can be introduced from many different perspectives rooted not only in signal theory, but also in quantum mechanics, in statistics or in pseudo-differential operator theory. Each of those points of view gives some specific insight into the very same mathematical objects, as well as complementary interpretations in terms of building blocks, devices, covariances, correlations, probabilities, measures, mechanical or optical analogies, symmetries, ... It is the purpose of this article to give a comprehensive survey of most of these many roads to time-frequency distributions, emphasizing the usefulness of—and the central role played by—a few key distributions.

One can note that many quadratic distributions mentioned here (spectrogram, scalogram, Wigner–Ville, Bertrand, ...) play a central role in time-frequency analysis and would be worthy of specific discussions. The purpose of this article is however not to provide the reader with an exhaustive overview of such methods, nor to compare them (this can be found, e.g., in [8, 13, 17, 22, 28] or [3]), but rather to put emphasis on the many motivations which may lead to their introduction.

2 Atoms

In order to perform a mixed time-frequency analysis, a first and intuitive approach—which goes back to Gabor [20]—is to *linearly* decompose a signal over a set of elementary 'building blocks' which would be reasonably 'localized' in both time and frequency. More precisely, the value of a signal $x(t)$ at a given time t_0 can be equivalently expressed as

$$x(t_0) = \int_{-\infty}^{+\infty} x(t)\, \delta_t(t_0)\, dt$$

(where $\delta_t(.)$ stands for the Dirac distribution centred at t), or as

$$x(t_0) = \int_{-\infty}^{+\infty} X(f)\, e_f(t_0)\, df,$$

with the symbolic notation $e_f(t) := \exp(i2\pi ft)$. Whereas the first decomposition puts emphasis on a temporal description, the second one—in which $X(f)$ identifies the Fourier transform of the signal $x(t)$—relies on a dual description in terms of waves. Both viewpoints are equally important from the point of view of physical interpretation, but they are mutually exclusive from the mathematical point of view of the Fourier transform. The purpose of time-frequency analysis is therefore to conceal these two aspects in a mixed perspective, so that both previous decompositions $x(t)$ and $X(f)$ would be replaced by a third one, $\lambda_x(t, f)$, which would be such that

$$x(t_0) = \iint_{-\infty}^{+\infty} \lambda_x(t, f)\, g_{tf}(t_0)\, dt\, df. \tag{2.1}$$

The functions $g_{tf}(.)$ involved in (2.1) allow for a transition between the previous situations, considered as limiting cases: perfect localization in time and no localization in frequency when $g_{tf}(.) \to \delta_t(.)$; perfect localization in frequency and no localization in time when $g_{tf}(.) \to e_f(.)$. In fact, they play a role of time-frequency *atoms* in the sense that they are asked to be both *building blocks* of any signal, and as *localized* as possible, in accordance with the ultimate constraints imposed by Heisenberg–Gabor type inequalities [20]. For the decomposition (2.1) to make sense, it is naturally required that it be *invertible*, so that we have

$$\lambda_x(t, f) = \int_{-\infty}^{+\infty} x(t_0)\, \tilde{g}_{tf}^*(t_0)\, dt_0, \tag{2.2}$$

where the star denotes complex conjugation, and $\tilde{g}_{tf}(.)$ is some 'analysis' function deduced from the 'synthesis' function $g_{tf}(.)$, making of $\lambda_x(t, f)$ a (linear) time-frequency *representation* of $x(t)$.

There is of course a high degree of arbitrariness in the choice of such a representation, since it basically depends on analysis waveforms which can be controlled (almost) at will by the user. Generally, the family of atoms $g_{tf}(.)$ is determined by the action of some suitable group of transformations upon a basic and unique element $g(.)$. Choosing, e.g., $g_{tf}(s) := g(s-t) \exp(i2\pi fs)$ leads therefore to the family of *short-time Fourier transforms* with window $g(.)$, whereas $g_{tf}(s) := (f/f_0)^{1/2} g((f/f_0)(s-t))$ (with $f > 0$, $f_0 > 0$ and $g(.)$ zero-mean) gives rise to the family of *wavelet transforms* [14, 26].

3 Energy

In many applications, the physical quantities which are relevant (or which are even the only ones that can be observed) are of *energy* type, motivating the search for decompositions aimed at the energy of a signal, rather than at the signal itself.

3.1 Distributions

By definition, and from the isometry of the Fourier transform, the energy of a signal $x(t) \in L^2(\mathbb{R})$ can be expressed as

$$E_x = \int_{-\infty}^{+\infty} |x(t)|^2 \, dt = \int_{-\infty}^{+\infty} |X(f)|^2 \, df. \tag{3.1}$$

Therefore, and analogously with the approach followed in the linear case, decomposing the energy of $x(t)$ jointly in time and frequency amounts to finding an *energy distribution* $\rho_x(t, f)$ such that

$$E_x = \int\int_{-\infty}^{+\infty} \rho_x(t, f) \, dt \, df. \tag{3.2}$$

The question is how to define such a quantity, which leads to many different possibilities.

3.2 Measurement devices

A first way of deriving an admissible expression for $\rho_x(t, f)$ is to take an 'operational' point of view, which consists in considering any energy distribution as 'measurable' by some measurement device, possibly idealized. Here are some examples:

Spectrogram, sonagram, scalogram The simplest solution in this direction amounts to using the continuous basis identities (2.1)–(2.2) and to taking for definition

$$\rho_x(t, f) = |\lambda_x(t, f)|^2, \tag{3.3}$$

the elementary atom $g(.)$ being normalized so that (3.2) holds. In the case of windowed Fourier analyses, the resulting quantity can be given two complementary interpretations, both intimately connected with some measurement device that can be actually constructed. Considering in fact frequency as a function of time, one can read (3.3) as a *local spectrum energy density* (this is referred to as a *spectrogram*). On the contrary, if we consider time as a function of frequency, one is faced with the time evolution of the energy output of a filterbank (this is referred to as a *sonagram*). This second interpretation naturally allows for the many variations in which the filterbank is no longer uniform but, e.g., constant-Q (this results in such a case in a wavelet-type analysis, or *scalogram*).[1]

Page One of the weaknesses of the classical Fourier analysis is the lack of any time dependence. This can be overcome by making the computation of the signal spectrum *causal*, and by considering the time variations of the quantity computed this way. This is what was proposed by C.H. Page [29], leading to the definition:

$$P_x(t, f) = \frac{\partial}{\partial t} \left| \int_{-\infty}^{t} x(s) \, e^{-i2\pi f s} \, ds \right|^2,$$

a quantity that is physically attainable, and which satisfies (3.2).

Rihaczek A completely different point of view is to consider the local energy of a signal in a time-frequency domain centred at a given point (t, f) as the energy of interaction between the restriction of this signal over some time interval of width δT centred at t and a filtered version of the same signal in a frequency bandwidth δB centred at f. In the limit of δT and δB going both to zero, this *Gedankenexperiment* led A.W. Rihaczek to define as a *complex energy density* the quantity [32]:

$$R_x(t, f) := \lim_{\delta T \, \delta B \to 0} \frac{1}{\delta T \, \delta B} \int_{t-\delta T/2}^{t+\delta T/2} x(s) \left[\int_{f-\delta B/2}^{f+\delta B/2} X(\xi) \, e^{i2\pi \xi s} \, d\xi \right]^* ds.$$

One can easily check that it corresponds to

$$R_x(t, f) = x(t) \, X^*(f) \, e^{-i2\pi f t}$$

and that it is such that the condition (3.2) is satisfied.

These few examples are far from exhausting all the solutions that may be— or have been—proposed. Their multiplicity and diversity (in their form as well as in the way they have been obtained) raise however the natural question of the existence of links which could exist between them, i.e., the question of *classes of solutions*.

---

[1]It can even be observed that, historically, such structures were the first ones to be introduced [25].

3.3 Classes

Attempting to classify admissible solutions, to parameterize them or to group them in homogeneous families can be achieved in at least two ways. The first one is based on *observation*, and it consists in finding in existing objects some similarity features that would reveal some form of parenthood: this is basically zoology (or botany). The second way is based on *deduction* and, in opposition to the first one, amounts to constructing families of solutions on the basis of a set of postulates or constraints.

Unifications Following the first direction, a careful study of the previously given definitions leads to the conclusion that they are all quadratic forms of the signal, all of them admitting furthermore as common parameterization:

$$C_x(t, f) = \int\!\!\int\!\!\int_{-\infty}^{+\infty} \varphi(\xi, \tau)\, x\left(s + \frac{\tau}{2}\right) x^*\left(s - \frac{\tau}{2}\right) e^{i2\pi[\xi(s-t)-f\tau]}\, ds\, d\xi\, d\tau,$$

(3.4)

thanks to the introduction of some well-chosen kernel function $\varphi(\xi, \tau)$ (for instance, Page's definition is recovered by choosing $\varphi(\xi, \tau) = \exp(i\pi\xi|\tau|)$ whereas the Rihaczek distribution corresponds to $\varphi(\xi, \tau) = \exp(i\pi\xi\tau)$). This way of classifying distributions was first proposed by L. Cohen in the mid-sixties (in a context of quantum mechanics that we will discuss further in the following), giving birth to what is now referred to as *Cohen's class* [11, 13, 17].

Such a unification was an important breakthrough, since it gives easy access to the properties of any distribution within the class through a corresponding structural property of its kernel function. It permits us also to readily obtain as many new distributions as wanted by a suitable *a priori* specification of the kernel. In particular, it turns out that the simplest choice, i.e., $\varphi(\xi, \tau) = 1$, leads to the definition

$$W_x(t, f) = \int_{-\infty}^{+\infty} x\left(t + \frac{\tau}{2}\right) x^*\left(t - \frac{\tau}{2}\right) e^{-i2\pi f\tau}\, d\tau,$$

(3.5)

which we recognize as the proposal made in 1932 by E.P. Wigner [36] (in quantum mechanics) and in 1948 by J. Ville [35] (in signal theory).

Covariances The second possibility for constructing classes of solutions consists in imposing *a priori* some very general structure, and in deducing more restrictive parameterizations by imposing further contraints, considered as 'natural'. Although this is not mandatory, strictly speaking, one usually starts from a quadratic form of the signal

$$\rho_x(t, f) = \int\!\!\int_{-\infty}^{+\infty} K(s, s'; t, f)\, x(s)\, x^*(s')\, ds\, ds',$$

whose kernel *a priori* depends on four independent variables. The further constraint

$$\iint_{-\infty}^{+\infty} K(s, s'; t, f) \, dt \, df = \delta(s - s')$$

guarantees that the considered quadratic form defines a proper energy distribution, in the sense of (3.2). It then suffices to impose additional *covariance* constraints for reducing the space of admissible solutions. In short, this amounts to requiring that the equation

$$\rho_{\mathbf{H}x}(t, f) = (\tilde{\mathbf{H}}\rho)_x(t, f)$$

be satisfied, with $\mathbf{H} : L^2(\mathbb{R}) \rightarrow L^2(\mathbb{R})$ some transformation operator (and $\tilde{\mathbf{H}} : L^2(\mathbb{R}^2) \rightarrow L^2(\mathbb{R}^2)$ the corresponding operator on the time-frequency plane), i.e., to imposing that the desired distribution 'follows' the signal in the transformations that it undergoes.

The simplest example corresponds to *shifts* in time and frequency, with the consequence that the general kernel necessarily takes on the particular form

$$K(s, s'; t, f) = K_0(s - t, s' - t) \, e^{-i2\pi f(s - s')},$$

where $K_0(s, s')$ is some arbitrary function that depends on only *two* independent variables (this situation has of course to be compared to that of the covariance by shifts in time only, which is known to transform any linear operator into a linear *filter*). The key point is that the class obtained this way exactly coincides with Cohen's class (3.4), which had been previously introduced on the basis of heuristic arguments only, provided that we define

$$\varphi(\xi, \tau) := \int_{-\infty}^{+\infty} K_0 \left(t + \frac{\tau}{2}, t - \frac{\tau}{2} \right) e^{-i2\pi\xi t} \, dt.$$

Cohen's class acquires therefore a peculiar status which goes beyond a mere phenomenological description: it is in fact the (only) class of all quadratic time-frequency distributions which are covariant under shifts [17].

Generalizing the procedure, a *deductive* introduction of classes of distributions can be achieved from covariance principles based on constraints different from shifts. For instance, preserving the covariance to shifts in time while adding that of covariance to *dilations* leads—in the space of analytic signals, i.e., signals whose spectrum vanishes on the real half-line of negative frequencies—to the so-called *affine* class, an admissible formulation of which is given by [33]

$$\Omega_x(t, f) = \frac{f}{f_0} \iint_{-\infty}^{+\infty} \pi(\xi, \zeta) \, X \left(\frac{\zeta - \xi/2}{f_0/f} \right) X^* \left(\frac{\zeta + \xi/2}{f_0/f} \right) e^{-i2\pi(f/f_0)\xi t} \, d\xi \, d\zeta,$$

with $\pi(\xi, \zeta)$ some arbitrary (bifrequency) kernel. One central element of this class is the *unitary Bertrand distribution* [7], characterized by the specific choice

$$\pi(\xi, \zeta) = \frac{(\xi/2f_0)}{\sinh(\xi/2f_0)} \, \delta \left(\zeta - (\xi/2) \coth(\xi/2f_0) \right),$$

and whose definition is usually written as [7]

$$B_x(t, f) = f \int_{-\infty}^{+\infty} \sqrt{\lambda(u)\lambda(-u)} X\left(f\lambda(u)\right) X^*\left(f\lambda(-u)\right) e^{-i2\pi utf} \, du, \quad (3.6)$$

with $\lambda(u) := (u/2)e^{-(u/2)}/\sinh(u/2)$. This distribution is in many respects the natural counterpart of the Wigner–Ville distribution for (wideband) analytic signals, and it can be shown to reduce to the latter in the narrowband limit.

Other choices can be considered, at will. For instance, covariance constraints with respect to *frequency-dependent* shifts (nonlinear group delays) have been introduced, leading to the so-called *hyperbolic* class or *power* class [8, 30, 31].

4 Correlations

Whereas the Fourier transform introduces a duality between the time and frequency variables, it is also well-known that, from the point of view of *energy* or *power*, it puts forward a companion duality between the concepts of *distribution* and *correlation*. This allows for a different introduction of the aforementioned time-frequency distributions, thus offering a novel interpretation.

Wiener–Khintchine A starting point for this approach is given by 'Wiener–Khintchine type' relationships, according to which a (energy or power) spectrum density $\Gamma_x(f)$ is the Fourier transform of a (deterministic or stochastic) correlation function $\gamma_x(\tau)$:

$$\Gamma_x(f) = \int_{-\infty}^{+\infty} \gamma_x(\tau) \, e^{-i2\pi f\tau} \, d\tau.$$

In both cases, the concept of correlation is related to an idea of interaction between a signal and its shifted versions in time. As far as estimation is concerned (in the stochastic case), a general and classical method amounts to computing a *weighted* Fourier transform—based on some window $w(.)$—of an estimate of the *stochastic* correlation function, as it may be given by the corresponding *deterministic* correlation function $\hat{\gamma}_x(\tau)$ of the observed realization:

$$\hat{\Gamma}_x(f) = \int_{-\infty}^{+\infty} w(\tau)\, \hat{\gamma}_x(\tau)\, e^{-i2\pi f\tau} \, d\tau. \quad (4.1)$$

Ambiguity As it has been described, the above procedure (sometimes referred to as a 'correlogram' estimation) is of specific interest in the case of *stationary* processes, for which spectral properties are supposed not to change with time. If we are however concerned with *nonstationary* situations, and if we admit that a time-frequency distribution can be considered as a time-dependent spectrum density, it becomes natural—in order to end up with a meaningful definition—to make use of the stationary approach while adding to it an explicit time dependence. This leads naturally to generalizing the notion of correlation to both time *and* frequency, i.e., to replacing ordinary (1D) correlation functions by (2D) *ambiguity functions* [17], such functions having indeed been introduced (in the radar literature) as a measure of interaction between a signal and its time-frequency shifted versions (delay and Doppler). It is therefore remarkable that, if one chooses to define an ambiguity function under the *symmetrized* form:

$$A_x(\xi, \tau) := \int_{-\infty}^{+\infty} x\left(t + \frac{\tau}{2}\right) x^*\left(t - \frac{\tau}{2}\right) e^{i2\pi\xi t} \, dt,$$

the time-frequency extension of the procedure (4.1) reduces to

$$\rho_x(t, f) = \int\int_{-\infty}^{+\infty} \varphi(\xi, \tau) \, A_x(\xi, \tau) \, e^{i2\pi(\xi t + \tau f)} \, d\xi \, d\tau,$$

or, in other words, to the exact Cohen class (3.4).

Amongst many other interesting features, this point of view offers a *rationale* for the choice of the kernel function $\varphi(\xi, \tau)$, which is *a priori* arbitrary. In particular, the correlative structure of $A_x(\xi, \tau)$ guarantees that it be maximum (in modulus) at the origin of the plane. As a by-product, this permits us to localize, in the ambiguity plane, the regions which are associated with interferences in the time-frequency plane, and thus to reduce the importance of interference terms by a well-adapted choice of the kernel function $\varphi(\xi, \tau)$ [5, 15, 17, 23].

5   Probability

Another possibility of introduction and interpretation of time-frequency distributions is to make use of an analogy with *probability density functions*.

Marginal distributions   Coming back to Parseval's relation (3.1), one can consider the two functions under the integral sign as (1D) energy densities (in time and frequency):

$$\rho_x(t) := |x(t)|^2; \quad \rho_x(f) := |X(f)|^2.$$

For signals of unit energy, they can be viewed as well as (1D) probability density functions, associated with the time and frequency variables. In this interpretation, a time-frequency distribution becomes a (2D) joint density of time and frequency, and it may seem natural to require that its marginal densities exactly equate to the individual densities:

$$\int_{-\infty}^{+\infty} \rho_x(t, f)\, dt = \rho_x(f); \qquad \int_{-\infty}^{+\infty} \rho_x(t, f)\, df = \rho_x(t). \tag{5.1}$$

Such constraints can be translated into admissibility conditions within a given class of distributions: within Cohen's class, for instance, the conditions for a correct marginalization read $\varphi(\xi, 0) = \varphi(0, \tau) = 1$.

Conditional distributions Going further with the analogy, one can imagine defining (via Bayes' formula) *conditional densities* according to

$$\rho_x(t, f) = \rho_x(t|f)\, \rho_x(f) = \rho_x(f|t)\, \rho_x(t).$$

The objective is, in this case, to interpret the local behaviour (in time or in frequency) of a time-frequency distribution in terms of *conditional means* and, e.g., to guarantee that such quantities give a direct access to local physical quantities such as the *group delay* $t_x(f)$,

$$\int_{-\infty}^{+\infty} t\, \rho_x(t|f)\, dt = t_x(f) := -\frac{1}{2\pi}\frac{d}{df}\arg X(f),$$

and the *instantaneous frequency* $f_x(t)$,

$$\int_{-\infty}^{+\infty} f\, \rho_x(f|t)\, df = f_x(t) := \frac{1}{2\pi}\frac{d}{dt}\arg x(t).$$

Again, such constraints can be translated into admissibility conditions within Cohen's class, namely [17]

$$\frac{\partial \varphi}{\partial \xi}(0, \tau) = \frac{\partial \varphi}{\partial \tau}(\xi, 0) = 0,$$

a set of conditions which is satisfied in particular by the Wigner-Ville distribution.

Time-frequency modelling Interpreting a time-frequency energy distribution by analogy with a probability distribution function paves the way for a signal modelling which operates directly on the plane. This has been proposed in [10], on the basis of a *mixture model* in which different components of a signal are each characterized by a 2D Gaussian distribution.

Reassignment Exploiting the same analogy also justifies the introduction of nonlinear techniques, such as *reassignment* [2], aimed at improving the localization of standard time-frequency distributions. Rather than taking (3.3) as a definition, a spectrogram with window $h(.)$ can be viewed as a member of Cohen's class, and it is easy to show that—thanks to this interpretation—it admits the equivalent formulation:

$$\left| \int_{-\infty}^{+\infty} x(s)\, h^*(s-t)\, e^{-i2\pi fs}\, ds \right|^2 = \int\!\!\int_{-\infty}^{+\infty} W_x(s,\xi)\, W_h(s-t,\xi-f)\, ds\, d\xi.$$

$$(5.2)$$

The meaning of this reformulation is that a spectrogram is nothing but a smoothed Wigner–Ville distribution, with the twofold consequence that (*i*) negative values and oscillatory contributions are smoothed out and (*ii*) localized components are smeared. Whereas interference reduction is beneficial, the spreading of localized components can be considered as an undesirable by-product of the smoothing process. A more careful analysis of (5.2) indicates however that the value of a spectrogram at a given time-frequency point (t, f) is a number (the total local energy) that summarizes the information displayed by a whole distribution (the underlying Wigner–Ville distribution W_x) in some time-frequency neighbourhood (loosely defined by the essential support of W_h). In the general case where the distribution is not uniform, assigning the obtained number to the *geometrical* centre (t, f) of the neighbourhood makes no real sense. A much more meaningful location—that can also be understood from a mechanical analogy—is in fact the *centroid* of the distribution within the neighbourhood: the idea of reassignment is precisely to *move* the spectrogram value from the point (t, f) where it has been computed to its corresponding local centroïd $(\hat{t}(t, f), \hat{f}(t, f))$. Localized distributions with reduced interference can be obtained this way. The concept of reassignment was first put forward in [24], but for the spectrogram only. It has been generalized later to much broader classes of distributions (including Cohen's class and the affine class), and efficient ways of computing local centroids have been developed [2, 9].

Limitations The probabilistic analogy does however present some limitations. For instance, it would seem natural to measure a *dispersion* around local mean values with the help of *conditional variances*. Unfortunately, one can show that, in general, such quantities may take on negative values, thus impairing their physical interpretation. This drawback is mainly due to the existence of negative values in most distributions. The same limitation is observed when *entropies* are considered as tentative measures of complexity on the time-frequency plane: the occurrence of negative values precludes the use of the standard Shannon entropy—which naturally involves the logarithm of the distribution—and adequate substitutes have to be found, e.g., by means of Rényi entropies [4].

One must nevertheless admit that this situation has to be accepted as a necessary fact, the requirement of positivity being exclusive of many other desirable properties. In particular, a theorem due to Wigner establishes that there is no quadratic energy distribution which has correct marginals while being non-negative [17].

6 Operators and measurements

The aforementioned limitations are basically related to local (and, *a fortiori*, pointwise) interpretations of time-frequency distributions. It must be noted that this does not preclude the use of a distribution whose interpretation may be problematic by itself, as long as it is only considered—from an operational perspective—as a suitable means of getting access to quantities which are physically sensible. In other words, independently of the values that a distribution may take (be they positive or not), it is much more the signification of a local measurement (and its possibility) which has to be questioned. In this respect, the formalism of quantum mechanics may serve as a guide [13, 17, 21].

Observables   In quantum mechanics, an observable measurement (the only thing which has a physical significance) is described as the expectation value of some operator (assumed to represent a given physical quantity) with respect to all possible outcomes of a system. In this context, introducing a joint representation is aimed at describing the very same result as an ensemble average of an ordinary function (attached to the considered physical quantity), with respect to a 'quasi-probability density function' of states. Replacing the variables of 'position' and 'momentum' by those of 'time' and 'frequency,' respectively, this amounts to *defining* a joint distribution $\rho_x(t, f)$ *via* the equivalence

$$\langle \mathbf{G} \rangle_x := \int_{-\infty}^{+\infty} (\mathbf{G}x)(t)\, x^*(t)\, dt = \int\int_{-\infty}^{+\infty} G(t, f)\, \rho_x(t, f)\, dt\, df,$$

in which the function $G(t, f)$ is *de facto* connected with the operator $\mathbf{G}$.

The non-uniqueness of $\rho_x(t, f)$ can therefore be re-expressed in terms of the non-uniqueness of any 'correspondence rule' associating an operator to a function, when the latter is based on variables whose elementary operators do not commute. This is of course the case for time and frequency, for which one has

$$(\mathbf{T}x)(t) = t\, x(t); \quad (\mathbf{F}x)(t) = \frac{1}{i2\pi} \frac{dx}{dt}(t),$$

leading therefore to the non-commuting relation

$$[\mathbf{T}, \mathbf{F}] := \mathbf{TF} - \mathbf{FT} = \frac{i}{2\pi} \mathbf{I},$$

where $\mathbf{I}$ stands for the identity operator.

Correspondence rules   In the Cohen class formalism, the arbitrariness in the writing of **G** is directly linked to the kernel function $\varphi(\xi, \tau)$. Precisely, one can show [11, 17] that we have

$$\mathbf{G}_\varphi = \iint_{-\infty}^{+\infty} \varphi(\xi, \tau)\, g(\xi, \tau)\, e^{i2\pi(\xi\mathbf{T}+\tau\mathbf{F})}\, d\xi\, d\tau,$$

with

$$g(\xi, \tau) := \iint_{-\infty}^{+\infty} G(t, f)\, e^{-i2\pi(\xi t + \tau f)}\, dt\, df.$$

In the Wigner–Ville case ($\varphi(\xi, \tau) = 1$), one recovers the well-known rule proposed by H. Weyl (the so-called 'Weyl quantization' [19]), but other choices are possible. One can, for instance, mention that a correspondence rule, proposed in 1925 by M. Born and P. Jordan in one of their seminal papers on quantum mechanics, defines in fact—although implicitly—the parameterization $\varphi(\xi, \tau) = \sin \pi\xi\tau / \pi\xi\tau$, and therefore a distribution (referred to, nowadays, as the 'Born–Jordan distribution') whose geometrical properties are very close to those of the so-called 'Choï–Williams distribution,' proposed in 1989 on the basis of interference arguments of a completely different nature [15, 17].

Kernels and symbols   Operators defined according to a given correspondence rule can be characterized by their *kernel* $\gamma_\varphi(t, s)$, defined through

$$(\mathbf{G}_\varphi x)(t) = \int_{-\infty}^{+\infty} \gamma_\varphi(t, s)\, x(s)\, ds.$$

In the framework of Cohen's class, it turns out [12, 17] that all such kernels may be expressed as

$$\gamma_\varphi(t, s) = \int_{-\infty}^{+\infty} F\left(\frac{t+s}{2} - \theta, t - s\right) \gamma(\theta, t - s)\, d\theta,$$

with

$$\gamma(t, \tau) := \int_{-\infty}^{+\infty} g(\xi, \tau)\, e^{i2\pi\xi t}\, d\xi = \int_{-\infty}^{+\infty} G(t, f)\, e^{i2\pi f\tau}\, df$$

and

$$F(t, \tau) := \int_{-\infty}^{+\infty} \varphi(\xi, \tau)\, e^{i2\pi\xi t}\, d\xi.$$

This offers a new and unified perspective in terms of mean values of the function $\gamma(., t - s)$ on the interval $[\min(t, s), \max(t, s)]$. In the specific cases of the Born-Jordan distribution and of (the real part of) the Rihaczek distribution, one obtains, respectively,

$$\varphi(\xi, \tau) = \frac{\sin \pi\xi\tau}{\pi\xi\tau} \quad \Rightarrow \quad \gamma_\varphi(t, s) = \frac{1}{|t - s|} \int_{\min(t,s)}^{\max(t,s)} \gamma(\theta, t - s)\, d\theta,$$

$$\varphi(\xi, \tau) = \cos \pi\xi\tau \quad \Rightarrow \quad \gamma_\varphi(t, s) = \frac{\gamma(t, t - s) + \gamma(s, t - s)}{2},$$

whereas, in the Wigner–Ville case, one has

$$\varphi(\xi, \tau) = 1 \quad \Rightarrow \quad \gamma_\varphi(t, s) = \gamma\left(\frac{t+s}{2}, t-s\right).$$

In this latter case, one gets furthermore

$$G(t, f) = \int_{-\infty}^{+\infty} \gamma_\varphi\left(t + \frac{\tau}{2}, t - \frac{\tau}{2}\right) e^{-i2\pi f\tau}\, d\tau$$

and, in the language of pseudo-differential calculus [19], the function $G(t, f)$ is the *Weyl symbol* of the operator $\mathbf{G}$ (passing from an operator to its symbol is the converse of the correspondence rule that associates an operator to a function).

Let us finally remark that, in the case of a *covariance* operator, the corresponding Weyl symbol exactly equates to what is referred to as the *Wigner–Ville spectrum* of a nonstationary process [17, 27].

7 Geometries

Amongst other ways of introducing time-frequency energy distributions, one can finally mention those based on *geometrical* principles.

Tomography A first approach amounts to generalizing the constraint of marginal distributions (5.1) to integrations in the time-frequency plane along directions which are not necessarily parallel to either time or frequency. If we consider for instance a linear 'chirp' $c_{f_0,\beta}(t)$, such a signal is naturally associated with a straight line on the plane (its 'instantaneous frequency' $f = f_0 + \beta t$), and it becomes therefore natural to require that the equality

$$\int_{-\infty}^{+\infty} \rho_x(t, f_0 + \beta t)\, dt = \left|\int_{-\infty}^{+\infty} x(t)\, c_{f_0,\beta}^*(t)\, dt\right|^2 \tag{7.1}$$

be satisfied for any analysed signal $x(t)$.[2]

Formulated this way, the problem reduces therefore to a Radon inversion, and it has been shown [6] that its solution is restricted to the Wigner–Ville distribution (3.5) when (7.1) is required to be satisfied for all possible chirp rates β. It becomes therefore possible to generalize this elegant tomographic construction just by changing the underlying geometry, implicitly based on straight lines in the plane. This can be achieved, e.g., if straight lines are replaced by *hyperbolae*, thus leading unequivocally to the unitary Bertrand distribution (3.6).

---

[2]The right-hand side of (7.1) measures in fact the energy of interaction between the analysed signal and the chirp, and it can be considered as well as related to a *fractional Fourier transform* [1].

Symmetries   Another geometrical approach can be phrased in terms of *symmetries*. The reason is that, if we introduce the (Glauber) *displacement operator*

$$\mathbf{D}_{t,f} = e^{i2\pi(f\mathbf{T}-t\mathbf{F})},$$

one can show by a direct calculation that [34]

$$W_x(t,f) = 2\langle \mathbf{D}_{t,f}\,\mathbf{\Pi}\,\mathbf{D}_{-t,-f}\rangle x,$$

where $\mathbf{\Pi}$ is the *parity* operator, defined by

$$(\mathbf{\Pi}x)(t) = x(-t); \quad (\mathbf{\Pi}X)(f) = X(-f).$$

It follows that the value of a Wigner–Ville distribution, at a given point of the time-frequency plane, can be interpreted as the expectation value of a parity operator centred around the considered point. From this perspective too, the idea of symmetry can be preserved, while suitably modified in its definition. In this respect, one can show [16, 17] that replacing the usual central symmetry (related to the notion of *arithmetic* means) by, e.g., an *inversion* (based on the notion of *geometric* means) leads to a specific form of *affine* distribution.

¿From a reversed point of view, the idea of symmetry is intimately connected with that of *localization* on specific curves of the time-frequency plane. In fact, another way of saying that the value of a distribution at a given point results from contributions that are symmetric with respect to this point is that any two contributions interact to create a third contribution centred at a suitably defined 'mid-point' between the interacting components. It follows that localization appears as a natural by-product of such an 'interference geometry' since it can only occur on curves which coincide with the locus of all of their mid-points. In the Wigner–Ville case, the underlying geometry is the usual one, governed by ordinary symmetry, and localization is achieved on straight lines of the plane (tones, impulses and linear chirps), but the very same principle can be extended—in a constructive way—to more general nonlinear situations [16, 18].

8   Conclusion

Basic tools of (non-parametric) time-frequency analysis have been introduced in a number of different ways. Although no definition can be advocated as the unique and ultimate one for all purposes, it has been shown that their potential infinity can be substantially reduced on the basis of interpretation arguments, and that some order naturally emerges when multiplying 'independent' perspectives on the problem. Intrinsic limitations to Fourier transform pairs force us to admit that there are multiple ways of wedding time and frequency, and that none of them is perfect in all respects. This diversity is also believed to be a richness, and the fact that apparently unrelated motivations lead *in fine* to similar subsets of admissible candidates should help the user in choosing solutions tailored to specific needs.

References

[1] Almeida, L. (1994). 'The fractional Fourier transform and time-frequency representations,' *IEEE Trans. on Signal Proc.* **SP-42**, 3084–3091.

[2] Auger, F., Flandrin, P. (1995). 'Improving the readability of time-frequency and time-scale representations by using the reassignment method,' *IEEE Trans. on Signal Proc.* **SP-43**, 1068–1089.

[3] Auger, F., Flandrin, P., Gonçalvès, P., Lemoine, O. (1997). *A Time-Frequency Toolbox, for Use with Matlab.*
http://www.physique.ens-lyon.fr/ts/tftb.html.

[4] Baraniuk, R.G., Flandrin, P., Janssen, A.J.E.M., Michel, O. (1998). 'Measuring information content and complexity in time-frequency using the Rényi entropies,' Isaac Newton Institute Technical Report No. NI 97034–NSP, University of Cambridge (UK).

[5] Baraniuk, R.G., Jones, D.L. (1993). 'A signal-dependent time-frequency representation: optimal kernel design,' *IEEE Trans. on Signal Proc.* **SP-41**, 1589–1602.

[6] Bertrand, J., Bertrand, P. (1987). 'A tomographic approach to Wigner's function,' *Found. Phys.* **17**, 397–405.

[7] Bertrand, J., Bertrand, P. (1992). 'A class of affine Wigner distributions with extended covariance requirements,' *J. Math. Phys.* **33**, 2515–2527.

[8] Boudreaux-Bartels, G.F. (1996). 'Mixed time-frequency signal transformations,' in *The Transforms and Applications Handbook* (A.D. Poularikas, *ed.*), CRC Press, Boca Raton (FL), 829–885.

[9] Chassande-Mottin, E. (1998). *Méthodes de réallocation dans le plan temps-fréquence pour l'analyse et le traitement de signaux non stationnaires*, Thèse de Doctorat, Univ. de Cergy-Pontoise.
http://www.physique.ens-lyon.fr/ts/these/ecm98.html.

[10] Coates, M.J. (1999). *Time-Frequency Modelling*, PhD Dissertation, Univ. of Cambridge (UK).

[11] Cohen, L. (1966). 'Generalized phase-space distribution functions,' *J. Math. Phys.* **7**, 781–786.

[12] Cohen, L. (1970). 'Hamiltonian operators via Feynman path integrals,' *J. Math. Phys.* **11**, 3296–3297.

[13] Cohen, L. (1995). *Time-Frequency Analysis*, Prentice-Hall, Englewood Cliffs (NJ).

[14] Daubechies, I. (1992). *Ten Lectures on Wavelets*, SIAM, Philadelphia (PA).

[15] Flandrin, P. (1984). 'Some features of time-frequency representations of multicomponent signals", IEEE Int. Conf. on Acoust., Speech and Signal Proc. ICASSP'84, San Diego (CA), 41.B.4.1–41.B.4.4., IEEE Press Piscataway (NJ).

[16] Flandrin, P., Gonçalvès, P. (1996). 'Geometry of affine time-frequency distributions,' *Appl. Comp. Harm. Anal.* **3**, 10–39.

[17] Flandrin, P. (1999). *Time-Frequency/Time-Scale Analysis*, Academic Press, San Diego (CA).

[18] Flandrin, P. (1999). 'La notion de localisation dans le plan temps-fréquence,' *Traitement du Signal* **15**, 483–492.

[19] Folland, G.B. (1989). *Harmonic Analysis in Phase Space*, Ann. of Math. Studies **122**, Princeton Univ. Press, Princeton (NJ).

[20] Gabor, D. (1946). 'Theory of communication,' *J. IEE* **93**, 429–457.

[21] Hillery, M., O'Connell, R.F., Scully, M.O., Wigner, E.P. (1984). 'Distribution functions in physics: fundamentals,' *Physi. Reports* **106**, 121–167.

[22] Hlawatsch, F., Boudreaux-Bartels, G.F. (1992). 'Linear and quadratic time-frequency signal representations,' *IEEE Signal Proc. Mag.* **9**, 21–27.

[23] Hlawatsch, F., Flandrin, P. (1998). 'The interference structure of the Wigner distribution and related time-frequency signal representations,' in [28, pp. 59–133].

[24] Kodera, K., de Villedary, C., Gendrin, R. (1976). 'A new method for the numerical analysis of non-stationary signals", *Phys. Earth and Plan. Int.* **12**, 142–150.

[25] Kœnig, R., Dunn, H.K., Lacy, L.Y. (1946). 'The sound spectrograph,' *J. Acoust. Soc. Amer.* **18**, 19–49.

[26] Mallat, S. (1997). *A Wavelet Tour of Signal Processing*, Academic Press, San Diego (CA).

[27] Martin, W., Flandrin, P. (1985). 'Wigner-Ville spectral analysis of non-stationary processes,' *IEEE Trans. on Acoust., Speech and Signal Proc.* **ASSP-33**, 1461–1470.

[28] Mecklenbräuker, W.F.G., Hlawatsch, F., *eds.* (1998). *The Wigner Distribution – Theory and Applications in Signal Processing*, Elsevier, Amsterdam.

[29] Page, C.H. (1952). 'Instantaneous power spectra,' *J. Appl. Phys.* **23**, 103–106.

[30] Papandreou, A., Hlawatsch, F., Boudreaux-Bartels, G.F. (1993). 'The hyperbolic class of quadratic time-frequency representations – Part I: constant-Q warping, the hyperbolic paradigm, properties and members,' *IEEE Trans. on Signal Proc.* **SP-41**, 3425–3444.

[31] Papandreou-Suppappola, A., Hlawatsch, F., Boudreaux-Bartels, G.F. (1998). 'Quadratic time-frequency representations with scale covariance and generalized time-shift covariance: a unified framework for the affine, hyperbolic, and power classes,' *Digital Signal Proc.* **8**, 3–48.

[32] Rihaczek, A.W. (1968). 'Signal energy distribution in time and frequency,' *IEEE Trans. on Info. Theory* **14**, 369–374.

[33] Rioul, O., Flandrin, P. (1992). 'Time-scale energy distributions: a general class extending wavelet transforms", *IEEE Trans. on Acoust., Speech and Signal Proc.* **ASSP-40**, 1746–1757.

[34] Royer, A. (1977). 'Wigner function as expectation value of a parity operator,' *Phys. Rev. A* **15**, 449–450.

[35] Ville, J. (1948). 'Théorie et applications de la notion de signal analytique,' *Câbles et Transm.* **2ème A.**, 61–74.

[36] Wigner, E.P. (1932). 'On the quantum correction for thermodynamic equilibrium,' *Phys. Rev.* **40**, 749–759.

Multiple Window Time-Varying Spectrum Estimation

Metin Bayram and Richard Baraniuk

1 Summary

We overview a new non-parametric method for estimating the time-varying spectrum of a non-stationary random process. Our method extends Thomson's powerful multiple window spectrum estimation scheme to the time-frequency and time-scale planes. Unlike previous extensions of Thomson's method, we identify and utilize optimally concentrated Hermite window and Morse wavelet functions and develop a statistical test for extracting chirping line components. Examples on synthetic and real-world data illustrate the superior performance of the technique.

2 Introduction

Many methods exist for estimating the power spectra of stationary signals [1]. However, these methods are insufficient for the non-stationary signals that occur in important applications such as radar, sonar, acoustics, biology, and geophysics. These applications demand *time-frequency representations* that indicate how the power spectrum changes over time. To date, research in time-frequency analysis has mainly focused on deterministic signals. Only more recently has attention turned to non-stationary random processes [2, 3, 4, 5, 6, 7, 8, 9, 10, 11, 12, 13].

Unlike the power spectrum for stationary random processes, there is no unique definition for the time-varying spectrum of a non-stationary random process $\mathbf{x}$. Perhaps the best compromise is the *Wigner–Ville spectrum* (WVS) $\mathsf{W}_\mathbf{x}$ [9]. Given the instantaneous auto-correlation function

$$r_\mathbf{x}(t,\tau) := \mathsf{E}[\mathbf{x}^*(t-\tau/2)\,\mathbf{x}(t+\tau/2)], \qquad (2.1)$$

the WVS is defined as its Fourier transform

$$\mathsf{W}_\mathbf{x}(t,f) := \int r_\mathbf{x}(t,\tau)\,e^{-j2\pi f\tau}\,d\tau. \qquad (2.2)$$

Alternatively, the WVS can be defined as the expected value of the empirical *Wigner distributions* (WDs) $\mathbf{W}_\mathbf{x}$ [14, 15] of the realizations of the process:

$$\mathsf{W}_\mathbf{x}(t,f) = \mathsf{E}[\mathbf{W}_\mathbf{x}(t,f)] = \mathsf{E}\left[\int \mathbf{x}^*(t-\tau/2)\,\mathbf{x}(t+\tau/2)e^{-j2\pi f\tau}\,d\tau\right]. \qquad (2.3)$$

In this framework, the problem of time-varying spectrum estimation can be stated as estimating the WVS W_x given only one realization of the non-stationary process **x**.

A number of different WVS estimates have been proposed. The simplest is the empirical WD $\mathbf{W_x}$ itself. However, while it is unbiased, it has very large (infinite in theory) variance and cross-components – artifacts produced due to its quadratic nature [9]. Figure 1 plots a test signal composed of a deterministic FM chirp with sinusoidal instantaneous frequency [14, 15] submerged in a realization of an additive bandpass Gaussian process of linearly rising center frequency, its ideal time-varying spectrum, and four different WVS estimates. The large variance and the cross-components cloud the WD's interpretation, as can be seen in Figure 1(c).

To reduce the variance of the WVS estimate at the expense of some bias, the WD can be smoothed through convolution with a 2D kernel function [9]. This bias-variance tradeoff is well illustrated by the *spectrogram* estimator of Figure 1(d). (The spectrogram smoothing kernel is the Wigner distribution of the analysis window [14, 15].)

Unfortunately, the large amount of smoothing required to obtain a low variance WVS estimate damages the resolution of deterministic chirping signals of the form $e^{j2\pi\gamma(t)}$, whose ideal time-frequency representations have the form $\delta(f - \gamma'(t))$. In this article, we focus on estimating the WVS of mixed stochastic/deterministic signals of the form

$$\mathbf{x}(t) = \mathbf{y}(t) + \sum_i \mu_i(t)\, e^{j2\pi\gamma_i(t)}, \tag{2.4}$$

with **y** a zero-mean, non-stationary, Gaussian random process and $\mu_i(t)\, e^{j2\pi\gamma_i(t)}$ a deterministic 'chirp' signal with instantaneous amplitude $\mu_i(t)$ and instantaneous phase $\gamma_i(t)$. For inspiration, we turn to the seminal stationary spectrum estimation work of Thomson [16].

Realizing that random and deterministic spectral components must be dealt with separately, Thomson introduced a powerful multiple window (MW) spectrum estimator for stationary signals in [16] to obtain a low variance spectrum without degrading the resolution of line components. His method uses a statistical significance test to detect and extract stationary deterministic line components (sinusoids) from the data, computes an MW spectrum estimate of the sinusoid-free data using a set of optimal windows, and reshapes the spectrum to account for the excised sinusoids.

Because of its excellent performance, several groups have applied this technique, ad hoc, to non-stationary signals in a sliding-window fashion [2, 3, 4, 17]. There are two potential problems to such an approach: (1) the windows used by Thomson are not optimal in a joint time-frequency setting and (2) the chirping rates of the line components must be very low (for them to be well approximated as piecewise sinusoidal). Figure 1(e) shows the sliding-window

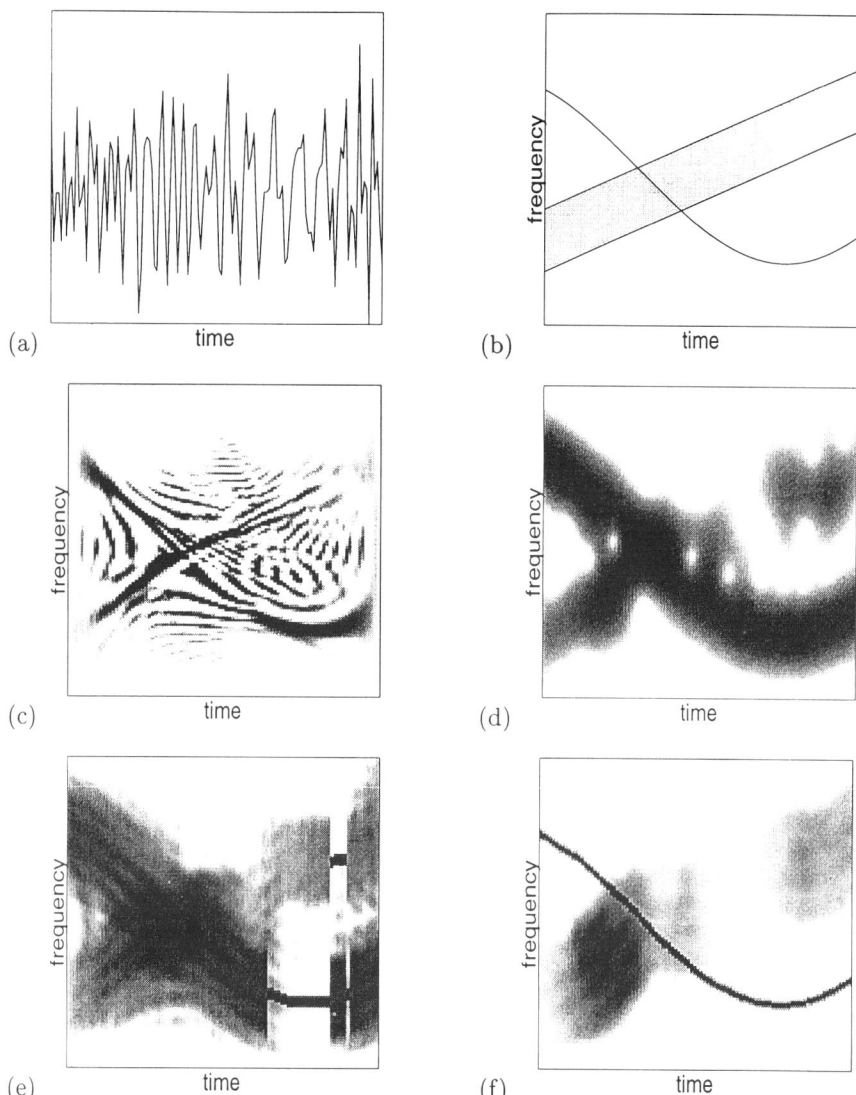

Figure 1: (a) Test signal **x** composed of a chirp with sinusoidal instantaneous frequency plus an additive bandpass Gaussian noise of linearly rising center frequency. (b) Ideal representation. (c) Empirical Wigner distribution $\mathbf{W_x}$. (d) Spectrogram using a Gaussian window. (e) Sliding-window Thomson's method as in [2, 3, 4]. (f) Multiple window (MW) time-frequency distribution.

Thomson's method applied to the test signal. We see clearly that the method comes short of detecting and extracting the non-stationary line component (except where its chirp rate becomes small).

In this article we overview a refinement of the previous extensions of Thomson's method into an improved, unified time-varying MW spectrum estimate for non-stationary signals of the form (2.4) [5, 6, 7, 8]. We will identify the optimal windows and develop a statistical test to detect and extract the time-varying line components. Our method preserves the resolution of line components, has low variance, and offers fine control over the bias-variance tradeoff. Figure 1(f) shows the method applied to the test signal.

This article is organized as follows. In Section 3 we give a brief review of Thomson's MW method for stationary signals and explain the essence of his statistical test for sinusoids. Section 4 discusses MW time-frequency analysis and identifies the optimal (Hermite) windows to use in the MW method. Section 5 extends the significance test to include rapidly varying line components of the form $e^{j2\pi\gamma(t)}$. Section 6 extends the ideas in Section 4 to the time-scale plane, again identifying the optimal (Morse) windows. In Section 7 we demonstrate the performance of the estimators and, in Section 8, we discuss a number of related techniques and recent extensions of our approach. Section 9 concludes the article with a discussion.

3 Thomson's Multiple Window Method

Here we overview the key elements of Thomson's method for stationary power spectrum estimation, reformulated in continuous time. The stationary version of the signal model (2.4) reads

$$\mathbf{x}(t) = \mathbf{y}(t) + \sum_i \mu(f_i)\, e^{j2\pi f_i t + \rho_i} \tag{3.1}$$

with $\mathbf{y}$ a zero-mean, stationary, Gaussian random process having a continuous power spectrum and $\mu(f_i)\, e^{j2\pi f_i t + \rho_i}$ a deterministic sinusoid having a line power spectrum.

The classical spectrum estimator for stationary signals, the *periodogram*, is defined as simply the squared magnitude of the Fourier transform of a windowed version of the data:

$$\widehat{\mathsf{P}}_{\mathbf{x}}(f) := \left| \int \mathbf{x}(t)\, w(t)\, e^{-j2\pi f t}\, dt \right|^2 \tag{3.2}$$

with $w(t)$ the window function. While the periodogram suffers from large variance, this variance can be reduced by cutting the data into blocks, computing a periodogram of each block, and then averaging the periodograms [18]. However, this procedure also smears and biases the resulting spectrum estimate. The bias/variance tradeoff is clear: reducing the variance necessitates averaging over a larger number of shorter blocks, which increases the bias.

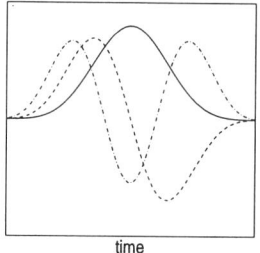

time

Figure 2: The first three prolate spheroidal wave functions in the time domain.

3.1 Summary of Thomson's Method

Inspired by the notion of averaging but displeased with the resulting bias, Thomson suggested computing several periodograms of the *entire signal* using a set of different windows and then averaging the resulting periodograms [16]. For a low variance, low bias estimate, he demanded that the windows be orthogonal (to minimize variance) and optimally concentrated in frequency (to minimize bias). The optimal windows satisfying these requirements for signals of finite extent are the *prolate spheroidal wave functions* [16, 19] (see Figure 2). These orthogonal functions are the eigenfunctions of a localization operator that band limits and then time limits functions. As windows, they are perfectly suited to stationary spectrum estimation, because they are simultaneously compactly supported in time and optimally concentrated in frequency. In addition to averaging over multiple windows, Thomson also introduced a separate pre-estimate for deterministic sinusoidal components.

Thomson's MW method consists of three main steps [16]:

1. Detect and extract all significant sinusoids (stationary deterministic line components) in the data $\mathbf{x}$ using a statistical significance test (see Section 3.2) to obtain an estimate $\widehat{\mathbf{y}}$ of the part having a continuous spectrum:

$$\widehat{\mathbf{y}} = \mathbf{x} - \{sinusoids\}. \tag{3.3}$$

2. Average K "orthogonal" periodogram estimates of $\widehat{\mathbf{y}}$ using prolate spheroidal data windows $\{v_k\}$ [16, 19][1]

$$\widehat{\mathsf{P}}_T(f) := \frac{1}{K} \sum_{k=0}^{K-1} \left| \int \widehat{\mathbf{y}}(t)\, v_k(t)\, e^{-j2\pi ft}\, dt \right|^2. \tag{3.4}$$

The concentration of the prolate windows in frequency results in a low bias estimate of the spectrum.

3. Reshape the spectrum estimate $\widehat{\mathsf{P}}_T$ to account for the sinusoids excised in Step 1.

---

[1] Thomson actually weights the periodogram computed with window v_k with the reciprocal of the corresponding prolate spheroidal eigenvalue λ_k [16]. However, since these are typically very close to 1, we will neglect them until Section 8.

3.2   Thomson's F-test for Sinusoids

Before we can extract the significant sinusoids from the data $\mathbf{x}$ as in (3.3), we must detect their presence and estimate their parameters. Assume the signal model (3.1) and define the kth *eigenspectrum* χ_k as the Fourier transform of the windowed original data:

$$\chi_k(f) := \int \mathbf{x}(t)\, v_k(t)\, e^{-j2\pi ft}\, dt. \tag{3.5}$$

The expected value of χ_k at frequency f_i is given by

$$\mathsf{E}[\chi_k(f_i)] = \mu(f_i)\, V_k(0), \tag{3.6}$$

with V_k the Fourier transform of v_k. Using a simple linear regression, the complex amplitude $\mu(f_i)$ of each possible sinusoid can thus be estimated as

$$\widehat{\mu}(f_i) = \frac{\displaystyle\sum_{k=0}^{K-1} V_k(0)\, \chi_k(f_i)}{\displaystyle\sum_{k=0}^{K-1} V_k^2(0)}. \tag{3.7}$$

The eigenspectra yield a simple statistical test for whether sinusoids are actually present in the data. Assuming that a sinusoid is present at frequency f_i with complex amplitude $\widehat{\mu}(f_i)$, we subtract it from the data to obtain an estimate of the 'background' continuous spectrum around f_i. Comparing the power in the background spectrum with the power in the assumed sinusoid yields an F variance-ratio test with 2 and $2K - 2$ degrees of freedom for the significance of the estimated line component [16]. Defining

$$F(f_i) := \frac{(K-1)\, |\widehat{\mu}(f_i)|^2 \displaystyle\sum_{k=0}^{K-1} V_k(0)^2}{\displaystyle\sum_{k=0}^{K-1} |\chi_k(f_i) - \widehat{\mu}(f_i)\, V_k(0)|^2}, \tag{3.8}$$

if $F(f_i)$ exceeds a significance threshold then we say that a sinusoid exists at frequency f_i.

The probability of missing a sinusoid increases with the threshold. On the other hand, the false alarm probability increases with decreasing threshold. False alarms give rise to *spurious peaks* in the spectrum estimate. For very closely spaced sinusoids, the above F-test fails. For such situations Thomson suggests a double F-test that searches for the existence of two sinusoids at a time instead of one [16].

Summary: Averaging orthogonal periodogram estimates reduces the variance of the MW power spectrum estimate by approximately K times compared to the variance of a single periodogram (in which $K = 1$) [16]. Furthermore, the concentrated prolate windows and sinusoid extraction ensure high resolution. These properties make Thomson's MW method the tool of choice for estimating the power spectra of stationary random processes.

4   Multiple Window Time-Frequency Analysis

The excellent performance of Thomson's MW method for stationary signals has led several groups to apply the method to time-varying spectrum estimation by simply sliding the estimatoralong the signal [2, 3, 4, 17]. While reasonably effective for certain classes of piecewise stationary signals, this approach suffers from two primary drawbacks. First, prolate spheroidal window functions have no inherent optimality properties in the joint time-frequency domain. Second, Thomson's F-test sinusoid extraction procedure fails on chirping line components of rapidly changing instantaneous frequency (as we saw in Figure 1(e)). In this section, we will begin a more thorough extension of Thomson's MW method to the time-frequency plane by first identifying an optimal set of windows [5, 6, 7, 8].

4.1   Hermite windows

One of Thomson's key insights is to smooth the spectrum estimate using orthogonal windows that are concentrated in the smallest possible region in the domain of interest (the frequency domain in his case). The foundation of the stationary MW method rests on the fact that the prolates are the most frequency-concentrated of all sets of orthogonal, time-limited windows. For time-frequency signal analysis, it is clear that we should average over multiple orthogonal windows that are optimally concentrated in an appropriate time-frequency domain. The optimality of the prolate functions is not as natural in time-frequency, since these functions treat the time-frequency plane as two separate spaces rather than as one geometric whole [20, 21, 22, 23].

More natural for time-frequency are the *Hermite functions*, defined by

$$h_k(t) := \pi^{-1/4}(2^k k!)^{-1/2} \left(t - \frac{d}{dt} \right)^k e^{-t^2/2}, \qquad k = 0, 1, 2, \dots . \qquad (4.1)$$

The zeroth-order Hermite function is the Gaussian. The Hermite functions are optimally concentrated in the circular time-frequency region

$$\left\{ (t, f): \quad t^2 + f^2 \le R^2 \right\} \qquad (4.2)$$

of area $A = \pi R^2$ (see Figure 3(a)) and thus treat all time-varying spectral features in the same fashion [20, 21, 22, 23]. Hermite functions optimally concentrated in elliptical time-frequency regions are easily obtained by compressing or dilating the h_k.

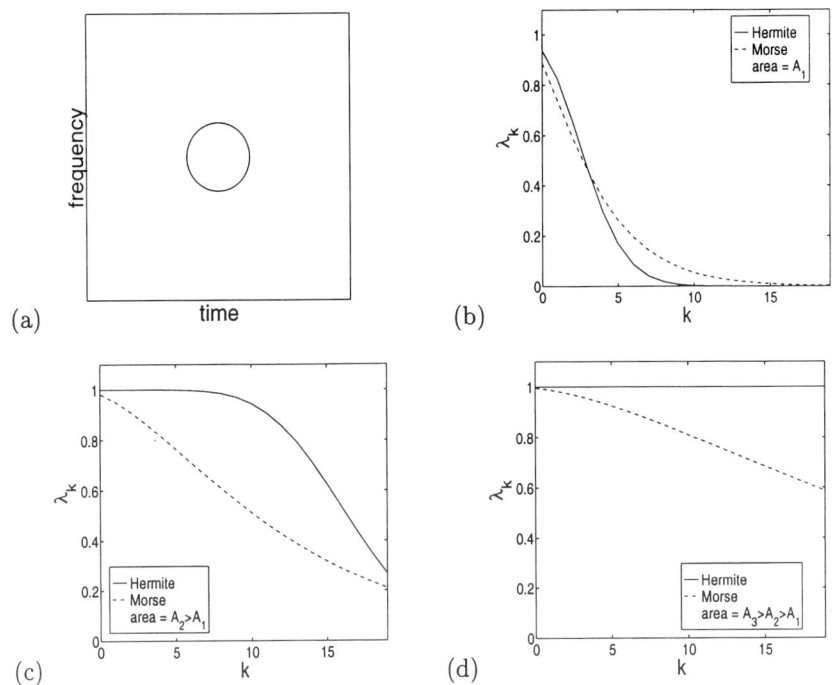

Figure 3: (a) Circular concentration region (4.2) for the Hermite functions. (b)–(d) Solid: eigenvalues of the Hermite functions over this circular region for three different areas $A_1 < A_2 < A_3$. Dashed: eigenvalues of the Morse wavelets (to come in Section 6.1) over the tear-drop region (6.2) shown in Figure 9 for the same three areas. Clearly more Hermite functions than Morse wavelets are concentrated in a region of area A.

Figure 4 plots the first three Hermite functions, their Fourier transforms, and their Wigner distributions. Note the circular symmetry of the Wigner distributions that matches the circular shape of the concentration region.

The Hermite functions are eigenfunctions of the Fourier transform and also of a time-frequency localization operator over the circular time-frequency region (4.2) [20]. The eigenvalues in this latter case are a function of the area $A = \pi R^2$ of the region [20]:

$$\lambda_k(R) := 1 - e^{\frac{-R^2}{2}} \sum_{i=0}^{k} \frac{1}{i!} 2^{-i} R^{2i}. \tag{4.3}$$

In Figure 3(b)–(d) we plot the behavior of these eigenvalues with k and area A. The closer λ_k is to one, the better concentrated h_k is in the circular region (4.2). Thus, as the area increases, more Hermite functions are concentrated within the circular region.

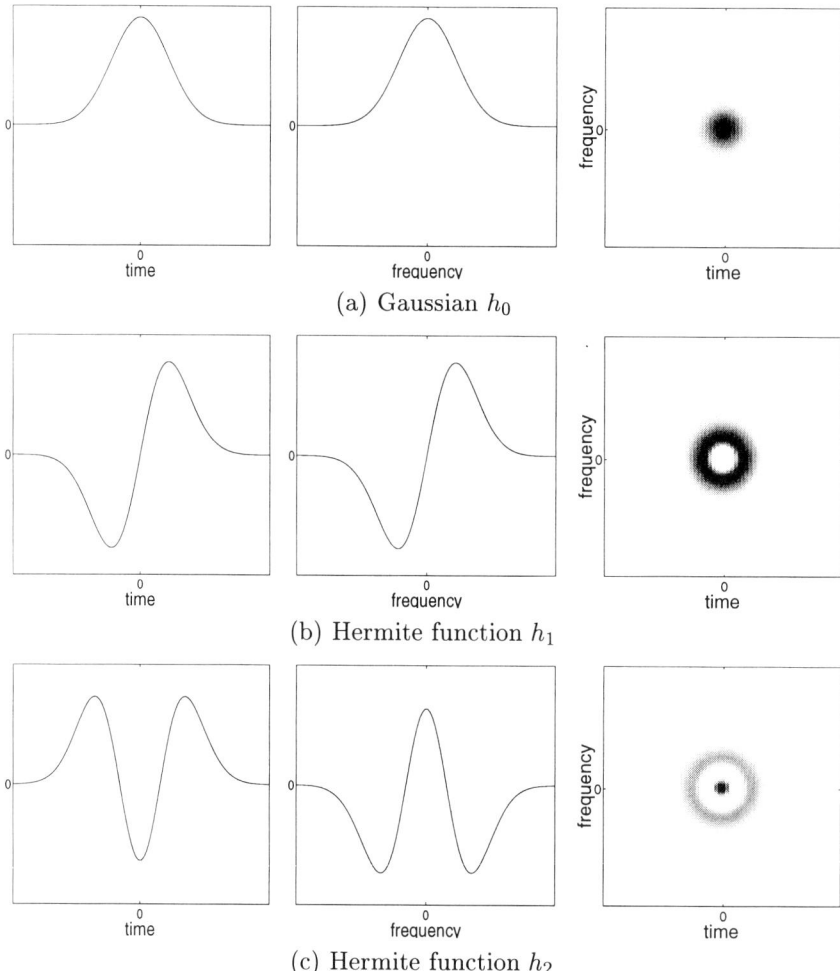

(a) Gaussian h_0

(b) Hermite function h_1

(c) Hermite function h_2

Figure 4: Hermite functions in the (left) time domain, (middle) frequency domain, and (right) time-frequency domain via Wigner distribution. (a) h_0 (Gaussian), (b) h_1 (since the Fourier transform of h_1 is purely imaginary, we plot the imaginary part), (c) h_2.

4.2 Multiple Window WVS Estimate

Under the stationary signal model (3.1), Thomson's MW spectrum average (3.4) estimates the energy content of the stationary signal $\mathbf{y}$ at frequency f by projecting onto the prolate-windowed sinusoids $v_k(t)\,e^{j2\pi ft}$. By analogy, under the non-stationary signal model (2.4), we estimate the energy content of the non-stationary signal $\mathbf{y}$ at time t and frequency f by projecting onto the sliding Hermite-windowed sinusoids $h_k(\tau - t)\,e^{j2\pi f\tau}$. The estimate can be written as the average of K Hermite-windowed *eigenspectrograms* of the data:

$$\widehat{W}_{\mathbf{y}}(t, f) = \frac{1}{K} \sum_{k=0}^{K-1} \left| \int \widehat{\mathbf{y}}(\tau) \, h_k(\tau - t) \, e^{-j2\pi f \tau} \, d\tau \right|^2. \qquad (4.4)$$

We choose K such that for a given radius R in (4.2) the first K eigenvalues in (4.3) are very close to one. The bias-variance tradeoff is clear: smaller R mean smaller K and thus lower bias in the estimate at the expense of less averaging and hence higher variance.

4.3   Cohen's Class Interpretation

The MW WVS estimate (4.4) belongs to Cohen's class of time-frequency distributions. All distributions $\mathbf{C}$ in Cohen's class can be written as [14, 15]

$$\mathbf{C}_{\mathbf{y}}(t, f) = \mathbf{W}_{\mathbf{y}}(t, f) ** \phi(t, f) \qquad (4.5)$$

with ϕ a kernel function and $**$ 2D convolution. The kernel generating the spectrogram is precisely the Wigner distribution of the window function.

The Wigner distribution of the kth-order Hermite function is the kth-order Laguerre function: [24, 25, 26]

$$\mathbf{W}_{h_k}(t, f) = L_k(t^2 + f^2) := e^{-\frac{\pi}{2}(t^2 + f^2)} \sum_{m=0}^{k} \frac{k!}{(k-m)! \, m!} \frac{[-\pi(t^2 + f^2)]^m}{m!}. \qquad (4.6)$$

Therefore, we have a closed form expression for the kernel ϕ corresponding to the MW WVS estimate (4.4) as a weighted sum of K Laguerre functions. In this interpretation, the MW WVS estimate (4.4) reads

$$\widehat{W}_{\mathbf{y}}(t, f) = \mathbf{W}_{\widehat{\mathbf{y}}}(t, f) ** \frac{1}{K} \sum_{k=0}^{K-1} L_k(t^2 + f^2). \qquad (4.7)$$

In Figure 5, we plot the kernel for an MW spectrum estimate using four Hermite windows.

The fact that thus far the MW WVS estimate is just a distribution from Cohen's class seems to imply that we could and should just smooth the empirical Wigner distribution with a 'top hat' function (which is unity inside a circle and zero outside) rather than go through the rigmarole of (4.4). As we will see in the next section, however, the eigenspectrograms play a key rôle in detecting/extracting the chirping line components from the signal (recall the model (2.4)).

Note that, unlike most Cohen class time-frequency distributions [14, 15], the MW WVS estimate is manifestly positive for all signals. And the connection with positive distributions does not stop here. Computation of the jackknife estimate of the variance of the MW WVS estimate [27] leads naturally to the concept of combining eigenspectrograms using a geometric rather

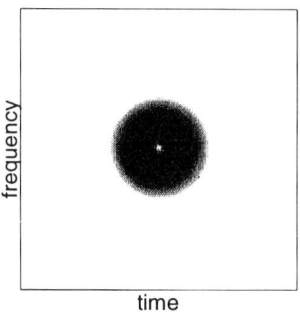 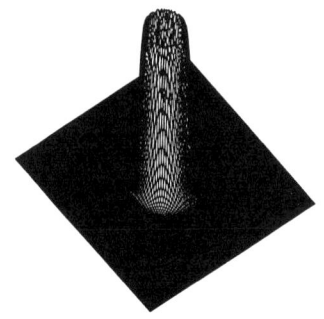

Figure 5: Cohen's class kernel function ϕ corresponding to the MW WVS estimate using the first four Hermite windows (sum of the first four Laguerre functions). Note how closely it approximates a 'top hat.'

than arithmetic mean. This is closely related to Loughlin, Pitton, and Hannaford's generation of positive time-frequency distributions using products of spectrograms [28].

5   Extracting line components

As in Thomson's method for stationary signals, the averaging inherent in (4.4) will degrade the resolution of chirping line components. Following Thomson's program, we will first detect and extract all line components in the data before performing (4.4) and then reshape the estimate accordingly.

A straightforward application of Thomson's sinusoid extraction algorithm to a signal from the model (2.4) as in [3, 4] relies on an assumption that the chirp functions $e^{j2\pi\gamma_i(t)}$ can be closely approximated locally as sinusoids. Unfortunately, this is not the case for rapidly chirping components, as we saw in Figure 1(e). In order to detect and extract highly non-stationary chirps, we now present a simple statistical significance test for *linear chirps* of the form $e^{j2\pi(f_0 t + ct^2)}$ [7]. Linear chirps can closely approximate locally all but the most rapidly changing chirp functions. Our approach can thus be interpreted as a first extension of Thomson's technique to time-frequency line components.

5.1   Algorithm to detect and extract chirp components

We make two basic assumptions about the chirp components. We assume (1) that not more than one chirp is present within the elliptical analysis region of the Hermite windows h_k and (2) that the highest chirp rate c attained by any chirp $e^{j2\pi(f_0 t + ct^2)}$ is $\frac{1}{4T}$, with T the effective time support of the Hermite windows. Within the windows' support, we will approximate the line components as piecewise linear chirps:

$$\sum_i \mu_i(t) \, e^{j2\pi\gamma_i(t)} \approx \sum_i \mu_i(t) \, e^{j2\pi(f_i(t)t + c_i(t)t^2)} \tag{5.1}$$

with time-varying offset frequency $f_i(t)$ and chirp rate $c_i(t)$. The chirp detection and extraction algorithm runs as follows:

1. Project the data $\mathbf{x}$ onto linear chirps of the form $e^{j2\pi(ft+ct^2)}$ for a fine grid of offset frequencies f and chirp rates c. This is equivalent to Thomson's F-test (3.8) applied at each frequency and each chirp rate. A 2D test statistic $F(f,c)$ results.

2. Repeat the above step at each time point to obtain the 3D test statistic $F(t,f,c)$.

If the chirp rates of the line components in (2.4) are too high, we must use shorter Hermite windows to deal with (5.1). However, short windows will not detect line components with low chirp rates, because there may not be enough oscillations within the windows for the F-test to be reliable. Therefore, for signals containing line components of both high and low chirp rates, it may be necessary to run the above algorithm for different sized windows and combine the results into one test statistic.

5.2 Suppressing spurious peaks

Due to the repeated application of the test (3.8), the number of spurious peaks in F increases far beyond that seen with stationary signals. (For stationary signals, Thomson applies the test at only one chirp rate $c = 0$, whereas we apply it for each time and chirp rate.) Roughly, if the F-test is performed at M chirp rates for each frequency f, then M times more spurious peaks will appear compared to when the test is performed at only one chirp rate ($c = 0$). These peaks must be suppressed to create a readable time-frequency image.

To suppress spurious peaks that peek above the significance threshold, we employ the following nonlinear cleaning algorithm:

1. Slice $F(t,f,c)$ along the chirp-rate dimension at several $\{c_j\}$.

2. For each c_j, apply a nonlinear order-statistic filter to $F(t,f,c_j)$ to remove peaks that have not coalesced into a region larger than the Heisenberg uncertainty principle mandates. (Intuition: spurious peaks are isolated in $F(t,f,c_j)$, while true peaks lie along curves in $F(t,f,c_j)$.) The nonlinear filter essentially counts the number of peaks that lie on a line within a region and compares the count to a threshold.

3. Combine the results from each c_j to obtain the cleaned F-test statistic.

While the linear chirp detection/cleaning/extraction algorithm is computationally expensive, it is readily parallelizable.

In Figure 6, we demonstrate the performance of this algorithm on the signal $\mathbf{x}(t) = \exp\left(j2\pi(\frac{a}{b}\sin(bt) + f_0 t)\right) + n(t)$, with a, b, and f_0 constants and $n(t)$ a

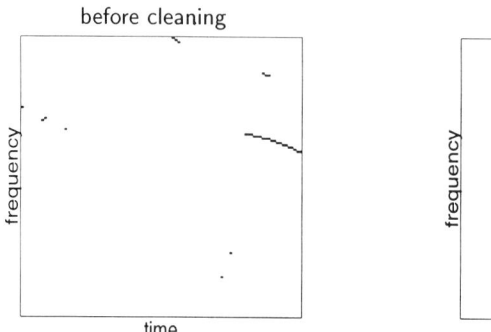

Figure 6: Test statistic $F(t, f, c_0)$ for a fixed c_0 before and after nonlinear cleaning.

stationary Gaussian white noise. The signal-to-noise ratio was set to 0.4 dB. The figure plots $F(t, f, c_0)$ for a fixed chirp rate c_0 that corresponds to spurious peaks for early t and intersects the true line component for late t. The nonlinear cleaning algorithm exploits the fact that the peaks corresponding to true line components form curves, whereas the spurious peaks lie isolated.

5.3 Impulses and closely spaced chirps

In our signal model (2.4) and the above line detection/extraction algorithm, we have not explicitly addressed impulses, which are vertical time-frequency line components of the form $\delta(t-t_0)$. Since the Fourier transform of an impulse is a complex sinusoid, we can detect and extract impulses from the data by applying the above algorithm in the Fourier domain with $c = 0$.

Multiple, crossing chirps will confuse any algorithm that seeks to extract them one at a time. The above approach can be extended à la Thomson to seek and destroy more than one chirp at each time-frequency point. However, as in the stationary case, the detection power of such a test will suffer.

6 Multiple Window Time-Scale Analysis

For random processes containing scaling phenomena (high frequency components of short duration and low frequency components of long duration), standard time-frequency techniques are not appropriate. These types of processes are better matched by the *time-scale representations* from the *affine class* [15, 29]. The smoothing kernels in the affine class change with frequency to accommodate component scaling. The smoothing regions in different parts in the time-frequency plane for Cohen's class and the affine class are shown in Figure 7. To reach higher frequencies, Cohen's class kernels translate, whereas affine class kernels scale.

To estimate the time-varying frequency spectrum of scaling processes, we will modify our Thomson-inspired estimation procedure to incorporate the

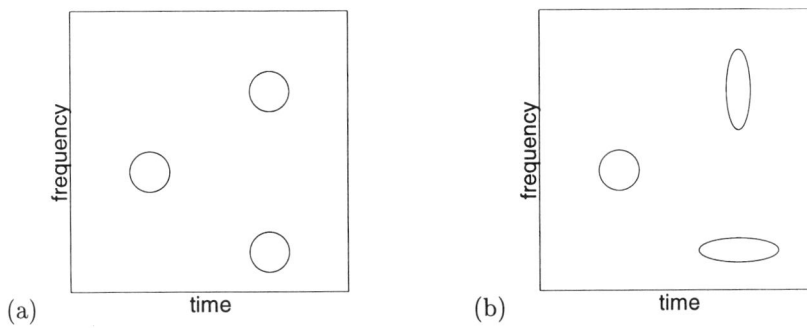

(a)   time   (b)   time

Figure 7: (a) Cohen's class kernels smooth by translating in time-frequency. (b) Affine class kernels smooth by scaling and translating in time-frequency.

wavelet transform (replaces the spectrogram) and the Morse wavelets (replace the Hermite windows) that are optimally concentrated in tear-drop shaped regions in time-frequency matched to the scaling behavior in Figure 7(b) [5, 6, 7, 8].

6.1   Morse wavelets

The *Morse wavelets* [30, 31, 32, 33] play a rôle in time-scale analogous to that of the Hermite windows in time-frequency. The kth order Morse wavelet[2] $\psi_k(t)$ is defined in the frequency domain as

$$\Psi_k(f) := f^{\beta/2} e^{-f^\gamma/2} \frac{d^\beta}{df^\beta} \left[e^{f^\gamma} \frac{d^{\beta+k}}{df^{\beta+k}} \left(f^{\beta+k} e^{-f^\gamma} \right) \right], \quad k = 0, 1, 2, \ldots, \quad (6.1)$$

with $\beta > 0$ the degree of flatness at $f = 0$ and $\gamma > 0$. The zeroth-order Morse wavelet is commonly known as the Klauder wavelet [34], although it goes by other names as well [35, p. 25], [36, 37]. Figure 8 shows the first three Morse wavelets in time, their Fourier transforms, and their Wigner distributions.

The Morse wavelets are the eigenfunctions of a localization operator over a tear-drop shaped region whose exact formula for any β and γ can be found in [32]. For the special case $\beta = \gamma = 1$, the Morse functions are mutually orthogonal and maximally concentrated in the time-frequency region [30, 32]

$$\left\{ (t, f) : \quad t^2 + \frac{9}{4f^2} + 1 \leq \frac{3C}{|f|} \right\} \tag{6.2}$$

of area $A = 3\pi(C - 2)$ [32]. Figure 9 depicts this region. Just as a circular disk contains all points equidistant from the center point in the Euclidean distance, this region contains all points equidistant from the center point in the (scale-invariant) *Lobachevsky distance* [31].

---

[2]While Morse defined only a special case of these wavelets for $\gamma = 1$ [33], we will refer to the entire class derived in [31, 32] as the Morse wavelets.

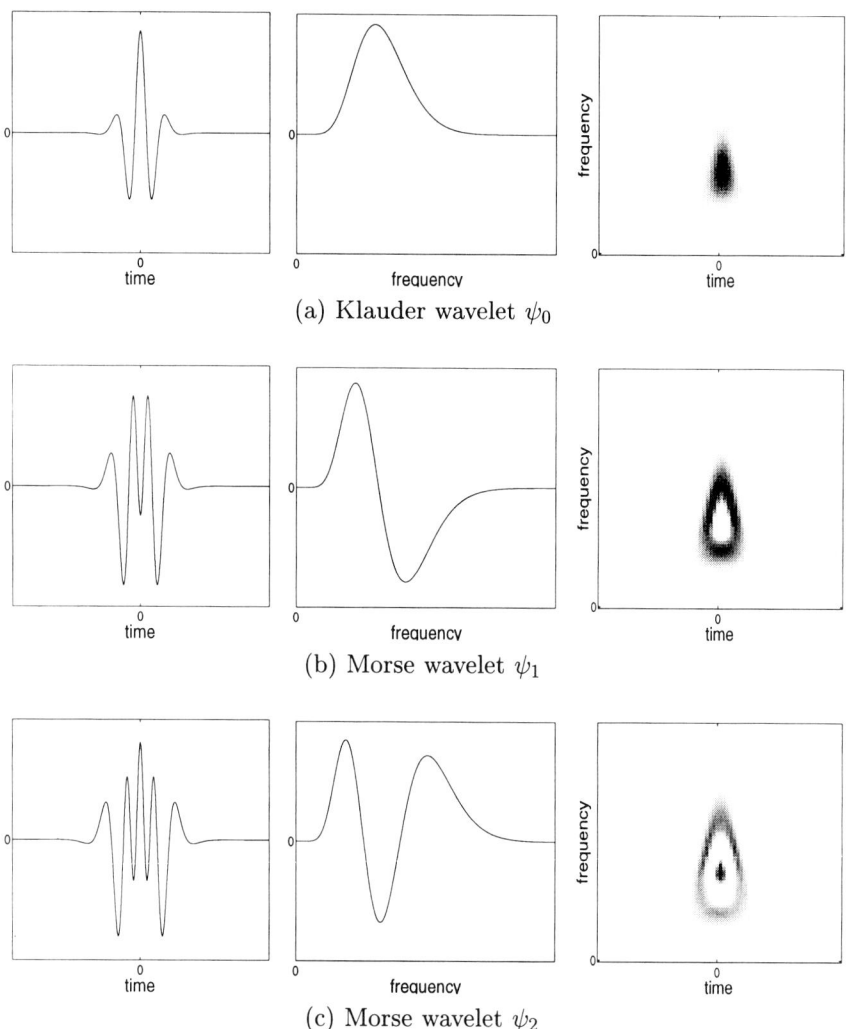

(a) Klauder wavelet ψ_0

(b) Morse wavelet ψ_1

(c) Morse wavelet ψ_2

Figure 8: Morse wavelets in the (left) time domain, (middle) frequency domain, and (right) time-frequency domain via Wigner distribution. (a) ψ_0 (Klauder wavelet), (b) ψ_1 (since the Fourier transform of ψ_1 is purely imaginary, we plot the imaginary part), (c) ψ_2.

For $\beta = \gamma = 1$, the eigenvalues of the bandpass localization operator corresponding to the Morse wavelets are given by

$$\lambda_k(C) := \frac{2(k+1)}{(C+1)(k+1)} \left(\frac{C-1}{C+1}\right)^{k+1}. \tag{6.3}$$

(No closed form expression exists for the eigenvalues for any other choices of β and γ.) As in the Hermite case, these eigenvalues indicate the degree

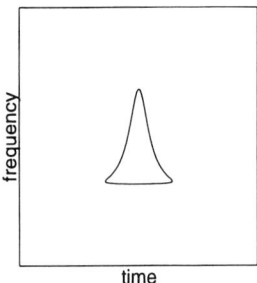

Figure 9: The tear-drop shaped concentration region (6.2) for the Morse wavelets for $\beta = \gamma = 1$.

of concentration of the corresponding Morse wavelet in the tear-drop region (6.2).

A simple comparison of the concentration properties of the Morse wavelets and the Hermite functions is easily made. In Figure 3(b)–(d), we plot the eigenvalues of the Morse wavelets and Hermite functions for three different areas of their concentration regions. Clearly, for a given area, the Hermite functions have more eigenvalues close to unity compared to the Morse wavelets. Therefore, the concentration properties of the Hermite functions on the circular region (4.2) are much better than the concentration properties of the Morse wavelets on the tear-drop region (6.2). Roughly speaking, this means that the bias introduced by averaging over K Morse wavelets should be larger than that due to averaging over K Hermite functions.

6.2 Multiple window WVS estimate

We form our time-scale MW WVS estimate as the weighted average of the squares of K *eigenscalograms* (squares of wavelet transforms) using Morse wavelets:

$$\widehat{W}_{\mathbf{y}}(t, f) = \frac{1}{K} \sum_{k=0}^{K-1} \left| \left(\frac{f}{f_0} \right)^{1/2} \int \widehat{\mathbf{y}}(\tau) \, \psi_k \left(\frac{f}{f_0}(\tau - t) \right) d\tau \right|^2. \tag{6.4}$$

Here f_0 is a reference frequency (the peak frequency of Ψ_0, for example). Again we see a clear bias-variance tradeoff: larger K implies more averaging (smaller variance) but a larger concentration region (larger bias).

A chirp line detection algorithm can be performed similarly to the time-frequency case.

6.3 Affine class interpretation

The time-scale MW WVS estimate (6.4) belongs to the *affine class* of time-scale covariant distributions [15, 29]. Each distribution Ω in this class can be

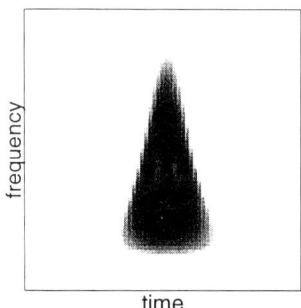 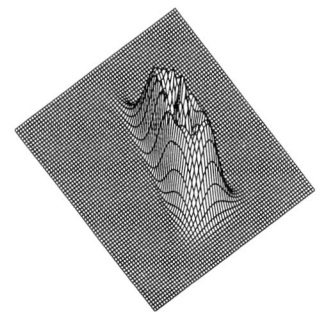

Figure 10: Affine class kernel function corresponding to the time-scale MW WVS estimate using four Morse wavelets (sum of their Wigner distributions).

interpreted as an affine-smoothed version of the Wigner distribution

$$\Omega_{\mathbf{y}}(t, f) = \iint \mathbf{W_y}(\tau, \nu) \, \Pi\left(f(\tau - t), \frac{\nu}{f}\right) d\tau \, d\nu \qquad (6.5)$$

with kernel Π centered at time zero and frequency $f_0 = 1$. The affine class can also be defined in terms of the unitary Bertrand distribution [38, 39]. Interestingly, the Klauder wavelet ψ_0 has a positive Bertrand distribution, just as the Gaussian h_0 has a positive Wigner distribution [40].

As in the time-frequency, Cohen's class case, the MW time-scale spectrum estimator can be interpreted as a member of the affine class with a kernel that is a weighted sum of Wigner (or Bertrand) distributions of Morse wavelets (see Figure 10). Unfortunately, closed form formulas for the Wigner and Bertrand distributions of any of the Morse wavelets have not been found. Nevertheless, like the time-frequency MW WVS estimate, the time-scale estimate is manifestly positive.

7 Examples

For a first example, refer to Figure 1, where we illustrate the performance of the time-frequency MW WVS estimate using a test signal composed of a chirp with sinusoidal instantaneous frequency in an additive bandpass Gaussian noise of linearly rising center frequency. The time-domain signal and its ideal representation in time-frequency are shown in Figure 1(a) and (b). It is not possible to identify the components of the test signal from the empirical Wigner distribution due to its high variance. The spectrogram smooths the Wigner distribution to reduce the variance, but smears the line component excessively. A sliding version of Thomson's method as proposed in [2, 3, 4] does not perform well for this non-stationary data, since a local sine approximation to the chirping line component is inadequate. In contrast, the MW estimate of Figure 1(f) has both high resolution and low variance. The empirical variance of the MW WVS estimate is approximately $\frac{1}{4}$ that of the spectrogram, which agrees with the fact that four windows were employed in its computation.

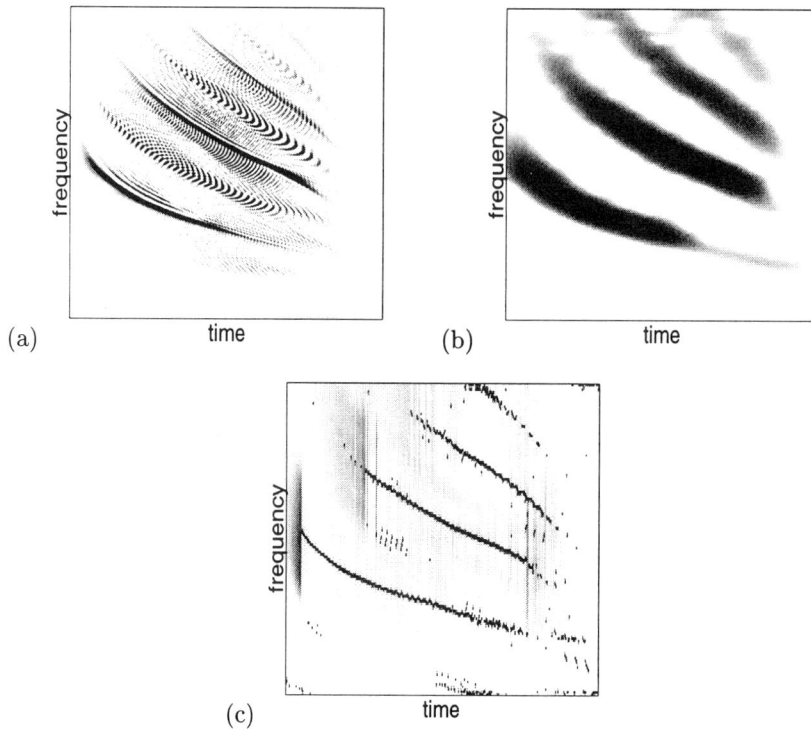

(a) time   (b) time

(c) time

Figure 11: Three WVS estimates of the echo-location pulse emitted by the Large Brown Bat: (a) empirical Wigner distribution, (b) spectrogram with Gaussian window, and (c) time-frequency MW estimate.

In Figure 11, we demonstrate the ability of the linear chirp detection/extraction algorithm to detect four hyperbolic chirps simultaneously. The data is a digitized 2.5 msec echo-location pulse emitted by the Large Brown Bat, *Eptesicus fuscus*. There are 400 samples, and the frequency range spanned is approximately $[0, 70]$ kHz. Comparing the time-frequency MW method against the Wigner distribution and spectrogram, we see that the detection algorithm successfully pulls out even the weakest high frequency line component. The method even reveals aliasing in this component due to under-sampling in data acquisition (note the 'wraparound' in frequency). Inspection of the early part of each chirp in the MW estimate reveals a threshold effect before which the line extraction algorithm locks on to each component.

In Figure 12, we illustrate the performance of the time-scale MW method using a 256-point test signal containing two Hölder singularities [41, 42] in additive white Gaussian noise $n(t)$:

$$\mathbf{x}(t) = |t - 64|^{-0.1} + |t - 180|^{-0.1} + n(t). \qquad (7.1)$$

Unlike the scalogram in Figure 12(c), the MW estimate of Figure 12(d) clearly

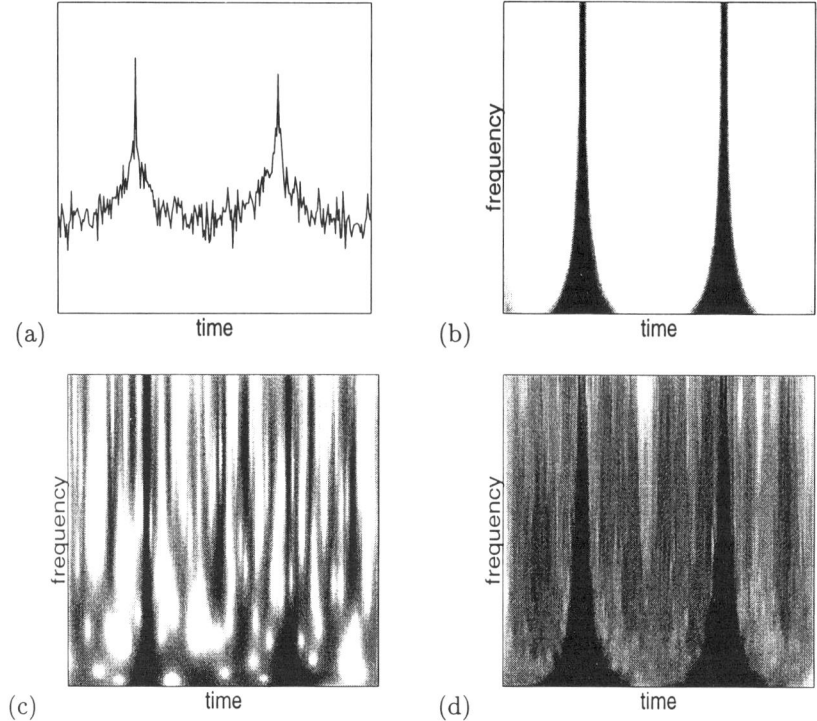

Figure 12: (a) Test signal composed of two singularities in additive white Gaussian noise. (b) Scalogram of noise-free signal with Klauder wavelet. (c) Scalogram of noisy signal with Klauder wavelet. (d) MW time-scale estimate of noisy signal.

captures the cone-like time-frequency structure of the singularities even in the presence of significant noise.

8   Related Work

Our primary contributions to MW time-varying spectral analysis have been identifying the Hermite windows for the time-frequency estimate, introducing the time-scale estimate and its Morse wavelets, and extending the line component F test to the linear chirps [5, 6, 7, 8]. Since our original papers were published, a number of interesting extensions and improvements have been made to the technique. We will review these here, as well as point to a large body of related work.

Different sets of orthogonal window functions: Xu, Haykin, and Racine compare and contrast prolate and Hermite windows in [43]. Pitton has developed new sets of concentrated time-frequency windows functions that balance the advantages of prolate and Hermite windows [44, 45, 46]. Lilly and Park have also considered multi-wavelet time-scale spectrum estimation using a specially designed set of wavelets [47].

Adaptive weighting algorithms: We have oversimplified our explanation of Thomson's method. Thomson does not weight each eigenspectrum by K^{-1} as we have; rather, he adaptively changes the weights to optimize the bias/variance tradeoff of the estimator. The improvement can be dramatic [16]. In the time-frequency setting, Çakrak and Loughlin adaptively weight Hermite eigenspectra using a least squares procedure [48, 49], while Pitton exploits knowledge of the windows' leakage characteristics [45]. Other stationary adaptation schemes such as that of Hansson [50] and Walden *et al.* [51] could also prove useful in the time-frequency setting.

Extended chirp extraction algorithms: Many alternatives exist to our simple linear chirp extraction algorithm. Çakrak and Loughlin employ a multiple window estimate to estimate the instantaneous frequencies of polynomial-phase line components [52]. Pitton has furthermore extended the F test in [46]. Further afield, we could extract chirps using the polynomial phase transform [53], the reassignment method [54, 55, 56, 57], the ridge and snakes method [58], or the squeezing method [59].

Alternative frameworks: Multiple window estimates succeed when the time-frequency spectrum can be approximated as 'locally stationary' within time-frequency regions larger than the concentration region of the orthogonal windows. As such, they can be viewed as special cases of the more general estimation frameworks of Sayeed and Jones [10] and Kozek *et al.* [11, 12, 60, 61]. Rather than assuming a parametric Cohen's class kernel that is the sum of several Wigner distributions of Hermite functions (recall (4.5)–(4.7)), Sayeed and Jones [10] design an optimal kernel ϕ that minimizes the mean-square error between the true WVS and the estimate. Other estimation procedures for locally stationary time-frequency spectra that fit within these general frameworks include those of Mallat *et al.* [13] and von Sachs *et al.* [62, 63].

9 Conclusions

In this article, we have overviewed two multiple window time-frequency and time-scale spectrum estimators that extend Thomson's seminal work [16] on multiple window spectrum estimation. The hallmarks of our approach are:

1. Averaging over sets of orthogonal, optimally concentrated windows, the Hermite functions for time-frequency analysis and the Morse wavelets for time-scale analysis. A low bias/low variance estimate results.

2. Detecting and extracting non-stationary line components by approximating them as piece-wise linear chirps. This pre-processing preserves the resolution of the line components.

As we saw in Section 8, much progress has been made recently in extending this framework. But many interesting open issues remain in the theory, implementation, and application of these time-varying spectrum estimators.

Acknowledgements

Thanks to Patrick Flandrin, Paulo Gonçalvès, Akbar Sayeed, and David Thomson for many stimulating discussions and to Al Feng for the bat data and permission to use it in this article. Kudos to Douglas Jones for discussions involving the application of the JAM in this context. This work was supported by the Isaac Newton Institute of Cambridge University (Rosenbaum Fellowship), NSF grants MIP–9457438 and CCR–9973188, DARPA/AFOSR grant F49620–97–1–0513, ONR grants N00014–99–1–0813 and N00014–99–1–0813, the State of Texas Advanced Technology Program, and the Texas Instruments Leadership University Program.

References

[1] S. M. Kay, *Modern Spectral Estimation: Theory and Application.* Englewood Cliffs, NJ: Prentice-Hall, 1988.

[2] G. Frazer and B. Boashash, "Multiple window spectrogram and time-frequency distributions," in *Proc. IEEE Int. Conf. Acoust., Speech, Signal Processing — ICASSP '94*, vol. IV, pp. 293–296, 1994.

[3] K. A. Farry, *Issues in Myoelectric Teleoperation of Complex Artificial Hands.* Ph.D. dissertation, Dep. Elec. Comput. Eng., Rice University, 1994.

[4] K. A. Farry, I. D. Walker, and R. G. Baraniuk, "Myoelectric teleoperation of a complex robotic hand," *IEEE Trans. Robotics, Automation*, vol. 12, pp. 775–788, Aug. 1996.

[5] M. Bayram and R. G. Baraniuk, "Multiple window time-frequency analysis," in *Proc. IEEE Int. Symp. Time-Frequency and Time-Scale Analysis*, (Paris, France), pp. 511–514, June 1996.

[6] M. Bayram and R. G. Baraniuk, "Multiple window time-varying spectrum estimation," in *Conf. Info. Sci. and Sys. (CISS)*, vol. 30, Mar. 1996.

[7] M. Bayram and R. G. Baraniuk, "Multiple window time-frequency and time-scale analysis," in *Proc. SPIE Int. Soc. Opt. Eng.*, 1996.

[8] M. Bayram, "Multiple window time-frequency analysis," Master's thesis, Dep. Elec. Comput. Eng., Rice University, May 1996.

[9] W. Martin and P. Flandrin, "Wigner-Ville spectral analysis of nonstationary random processes," *IEEE Trans. Acoust., Speech, Signal Processing*, vol. 33, pp. 1461–1470, Dec. 1985.

[10] A. M. Sayeed and D. L. Jones, "Optimal kernels for nonstationary spectral estimation," *IEEE Trans. Signal Processing*, vol. 43, pp. 478–491, Feb. 1995.

[11] W. Kozek, "Optimally Karhunen-Loève-like STFT expansion of nonstatonary processes," in *Proc. IEEE Int. Conf. Acoust., Speech, Signal Processing — ICASSP '93*, vol. IV, (Minneapolis), pp. 428–431, 1993.

[12] W. Kozek and K. Riedel, "Quadratic time-varying spectral estimation for underspread processes," in *Proc. IEEE Int. Symp. Time-Frequency and Time-Scale Analysis*, pp. 460–463, 1994.

[13] S. Mallat, G. C. Papanicolaou, and Z. Zhang, "Adaptive covariance estimation of locally stationary processes," *Ann. Statist.*, vol. 26, pp. 1–47, 1998.

[14] L. Cohen, *Time-Frequency Analysis*. Englewood Cliffs, NJ: Prentice-Hall, 1995.

[15] P. Flandrin, *Time-Frequency and Time-Scale Analysis*. Academic Press, 1999.

[16] D. J. Thomson, "Spectrum estimation and harmonic analysis," *Proc. IEEE*, vol. 70, pp. 1055–1096, Sept. 1982.

[17] D. J. Thomson. Personal Communication.

[18] S. M. Kay and G. F. Boudreaux-Bartels, "On the optimality of the Wigner distribution for detection," in *Proc. IEEE Int. Conf. Acoust., Speech, Signal Processing — ICASSP '85*, pp. 1017–1020, 1985.

[19] D. Slepian and H. O. Pollack, "Prolate spheroidal wave functions, Fourier analysis and uncertainty," *Bell Syst. Tech. J.*, vol. 40, pp. 43–64, Jan. 1961.

[20] I. Daubechies, "Time-frequency localization operators: A geometric phase space approach," *IEEE Trans. Inform. Theory*, vol. 34, pp. 605–612, July 1988.

[21] P. Flandrin, "Maximum signal energy concetration in a time-frequency domain," in *Proc. IEEE Int. Conf. Acoust., Speech, Signal Processing — ICASSP '88*, pp. 2176–2179, 1988.

[22] T. W. Parks and R. G. Shenoy, "Time-frequency concentrated basis functions," in *Proc. IEEE Int. Conf. Acoust., Speech, Signal Processing — ICASSP '90*, vol. 5, pp. 2459–2462, 1990.

[23] J. Ramanathan and P. Topiwala, *in Wavelets and Their Applications*, pp. 313–325. Kluwer Academic Pulishers, 1994.

[24] G. B. Folland, *Harmonic Analysis in Phase Space*. Princeton, NJ: Princeton University Press, 1989.

[25] H. J. Groenewold, "On the principles of elementary quantum mechanics," *Physica*, vol. 21, pp. 405–460, 1946.

[26] N. G. de Bruijn, "Uncertainty principles in Fourier analysis," in *Inequalities*, Academic Press, 1967.

[27] D. J. Thomson, "Jackknifing multiple-window spectra," in *Proc. IEEE Int. Conf. Acoust., Speech, Signal Processing — ICASSP '94*, vol. 1, pp. 73–76, 1994.

[28] P. Loughlin, J. Pitton, and B. Hannaford, "Approximating time-frequency density functions via optimal combinations of spectrograms," *IEEE Signal Processing Letters*, vol. 1, pp. 199–202, Dec. 1994.

[29] O. Rioul and P. Flandrin, "Time-scale energy distributions: A general class extending wavelet transforms," *IEEE Trans. Signal Processing*, vol. 40, pp. 1746–1757, July 1992.

[30] I. Daubechies, *Ten Lectures on Wavelets*. New York: SIAM, 1992.

[31] I. Daubechies, J. R. Klauder, and T. Paul, "Wiener measures for path integrals with affine kinematic variables," *J. Math. Phys.*, vol. 28, pp. 85–102, Jan. 1987.

[32] I. Daubechies and T. Paul, "Time-frequency localization operators — a geometric phase space approach: II. The use of dilations," *Inverse Problems*, no. 4, pp. 661–680, 1988.

[33] P. Morse, "Diatomic molecules according to the wave mechanics II. Vibrational levels," *Physical Review*, vol. 34, pp. 57–64, July 1929.

[34] J. R. Klauder, "Path integrals for affine variables," in *Functional Integration. Theory and Applications* (J. P. Antoine and E. Tirapagui, eds.), Plenum, New York, 1980.

[35] Y. Meyer, *Wavelets: Algorithms and Applications*. SIAM, 1993.

[36] H. O. Rasmussen, "The wavelet Gibbs phenomenon," in *Wavelets, Fractals, and Fourier Transforms* (M. Farge, J. C. R. Hunt, and J. C. Vassilicos, eds.), pp. 123–142, Oxford, UK: Clarendon Press, 1993.

[37] M. Holschneider, "On the wavelet transformation of fractal objects," *J. Stat. Phys.*, vol. 50, no. 5/6, pp. 963–993, 1988.

[38] J. Bertrand and P. Bertrand, "A class of affine Wigner functions with extended covariance properties," *J. Math. Phys.*, vol. 33, pp. 2515–2527, July 1992.

[39] R. G. Baraniuk, "Beyond time-frequency analysis: Energy densities in one and many dimensions," *IEEE Trans. Signal Processing*, vol. 46, pp. 2305–2314, Sept. 1998.

[40] P. Flandrin, "On separability, positivity and minimum uncertainty in time-frequency energy distributions," *J. Math. Phys.*, 1998.

[41] P. Gonçalvès and P. Flandrin, "Sur la localisation et la géométrie des distributions affines," in *14ème Colloque GRETSI*, (Juan Les Pins, France), pp. 355–367, 1993.

[42] P. Gonçalvès and R. G. Baraniuk, "Pseudo affine Wigner distributions: Definition and kernel formulation," *IEEE Trans. Signal Processing*, vol. 46, pp. 1505–1516, June 1998.

[43] Y. Xu, S. Haykin, and R. J. Racine, "Multiple window time-frequency analysis of EEG using Slepian sequences and Hermite functions," Nov. 1997. Preprint.

[44] J. W. Pitton, "Nonstationary spectrum estimation and time-frequency concentration," in *Proc. IEEE Int. Conf. Acoust., Speech, Signal Processing — ICASSP '98*, 1998.

[45] J. W. Pitton, "Time-frequency spectrum estimation: An adaptive multitaper method," in *Proc. IEEE Int. Symp. Time-Frequency and Time-Scale Analysis*, (Pittsburgh), pp. 665–668, 1998.

[46] J. W. Pitton, "Adaptive multitaper time-frequency spectrum estimation," in *Proc. SPIE Int. Soc. Opt. Eng.*, (Denver), 1999.

[47] J. M. Lilly and J. Park, "Multiwavelet spectral and polarization analyses of seismic records," *Geophys. J. Int.*, no. 122, pp. 1001–1021, 1995.

[48] F. Çakrak and P. Loughlin, "Multiple window nonlinear time-frequency analysis," in *Proc. IEEE Int. Conf. Acoust., Speech, Signal Processing — ICASSP '98*, pp. 2409–2412, 1998.

[49] F. Çakrak and P. Loughlin, "Multiple window time-varying spectral analysis," *IEEE Trans. Signal Processing*, 1999. Submitted.

[50] M. Hansson, "Optimized weighted average of peak matched multiple window spectrum estimates," *IEEE Trans. Signal Processing*, vol. 47, pp. 1141–1146, Apr. 1999.

[51] A. T. Walden, D. B. Percival, and E. J. McCoy, "Spectrum estimation by wavelet thresholding of multitaper estimators," *IEEE Trans. Signal Processing*, vol. 46, pp. 3153–3165, Dec. 1998.

[52] F. Çakrak and P. Loughlin, "Instantaneous frequency estimation of polynomial phase signals," in *Proc. IEEE Int. Symp. Time-Frequency and Time-Scale Analysis*, pp. 549–552, 1998.

[53] S. Peleg and B. Friedlander, "The discrete polynomial-phase transform," *IEEE Trans. Signal Processing*, vol. 43, pp. 1901–1914, Aug. 1995.

[54] K. Kodera, C. D. Villedary, and R. Gendrin, "A new method for the numerical analysis of non-stationary signals," *Phys. Earth and Plan. Int.*, vol. 12, pp. 142–150, 1976.

[55] K. Kodera, R. Gendrin, and C. D. Villedary, "Analysis of time-varying signals with small BT values," *IEEE Trans. Acoust., Speech, Signal Processing*, vol. ASSP-26, no. 1, pp. 64–76, 1978.

[56] F. Auger and P. Flandrin, "Improving the readability of time-frequency and time-scale representations by the reassignment method," *IEEE Trans. Signal Processing*, vol. 43, pp. 1068–1089, May 1995.

[57] F. Auger, E. Chassande-Mottin, I. Daubechies, and P. Flandrin, "Differential reassignment," *IEEE Signal Processing Letters*, vol. 4, pp. 293–294, Oct. 1997.

[58] R. A. Carmona, W. L. Hwang, and B. Torrèsani, "Characterization of signals by the ridges of their wavelet transforms." Mar. 1995.

[59] I. Daubechies and S. Maes, "A nonlinear squeezing of the continuous wavelet transform based on auditory nerve models," in *Wavelets in Medicine and Biology* (A. Aldroubi and M. Unser, eds.), pp. 527–546, CRC Press, 1996.

[60] W. Kozek, F. Hlawatsch, H. Kirchauer, and U. Trautwein, "Correlative time-frequency analysis and classification of nonstationary random processes," in *Proc. IEEE Int. Symp. Time-Frequency and Time-Scale Analysis*, pp. 417–420, 1994.

[61] G. Matz, F. Hlawatsch, and W. Kozek, "Generalized evolutionary spectrum analysis and the Weyl spectrum of nonstationary random processes," *IEEE Trans. Signal Processing*, vol. 45, pp. 1520–1533, June 1997.

[62] R. von Sachs and K. Schneider, "Wavelet smoothing of evolutionary spectra by nonlinear thresholding," *Appl. Comp. Harm. Anal.*, vol. 3, pp. 268–282, 1996.

[63] M. H. Neumann and R. von Sachs, "Wavelet thresholding in anisotropic function classes and application to adaptive estimation of evolutionary spectra," *Ann. Statist.*, vol. 25, pp. 38–76, 1997.

Multitaper Analysis of Nonstationary and Nonlinear Time Series Data

David J. Thomson

1 Introduction

One of the basic problems in the analysis of scientific data is to estimate the structure of an observed time series. To do this, it is commonly *assumed* that data is stationary and often Gaussian, so that the problem reduces to estimating the spectral density function, or power spectrum, from a finite data sample. In practice, however, most of the signals encountered are nonstationary or, less commonly, not adequately described by second moments. Because of the complexities entailed by admitting this, one commonly finds such processes treated as being stationary or, at best, pseudostationary, *i.e.*, the spectral density changes slowly with time. If the change is slow enough, the spectrogram is an adequate description. In recent years, wavelets have been advocated as a description of nonstationary processes, but as their statistical properties are inadequate, wavelets are not useful for the problems considered here. Moreover, a major shortcoming of both spectrograms and wavelets as tools for describing nonstationarity is that they ignore the presence of common signal elements at different frequencies. For this, the Loève, or dual frequency, spectrum is an indispensable tool.

This article outlines some multitaper[1] approaches to the analysis of nonlinear and nonstationary time series and discusses analysis of environmental data, with a viewpoint between that of statistics and that of signal processing. The data sets used here are chosen primarily to illustrate analysis methods, but have scientific interest in their own right. I have previously (Thomson, 1995) discussed the Central England temperature series and here I show methods to describe some of its nonstationary aspects. These appear to be largely of solar origin. Specifically, the *statistical structure* of the process changes systematically. Analysis of

---

[1]The original name for these methods was *multiple-window*, but because of the confusion with computer operating systems, *multitaper* is less ambiguous.

these changes shows that they are dominated by periods associated with the sun. This nonstationary structure is distinct from the environmental changes seen in the 20th century from increasing atmospheric CO_2 levels. I have discussed this part of the problem in a series of papers applying signal processing methods to study the relationship between atmospheric CO_2 and global warming: Kuo *et al.* (1990), Thomson (1995), Karl *et al.* (1996), Thomson (1997). The warming is mostly due to CO_2 and, what is more disturbing, the seasonal cycle is also being disrupted by human use of fossil fuels.

Space physics data in the form of magnetic fields and electron fluxes provides interesting examples of nonstationary data. My co-authors and I, Thomson *et al.* (1995), have proposed the dissenting view that much of the structure of the solar wind is a result of normal modes of the sun, as opposed to turbulence, chaos, or the like. The test described for nonlinearity in the presence of nonstationarity was motivated by this work.

Finally, to illustrate applicability of multitaper methods to fast non-stationarity, I use seismic data from a Central Asian earthquake.

I begin with some background on spectrum estimation of nonstation-ary processes, and the multitaper method. This is followed by several sections on the variance of multitaper estimates in different situations. The maxim is that when windowed Fourier transforms of reasonably sized data sets depart too far from the distributions expected under a stationary Gaussian hypothesis, one should consider alternative hypotheses. Thus one must understand the variability expected from stationary processes.

The remaining sections then progress from slow (and systematic) non-stationarity, to the Loève spectrum, to bispectra and a related test for nonlinearity in nonstationary data, and conclusions.

2   Background on Spectrum Estimation

Given N observations, $x(t)$ for $t = 0, 1, \cdots, N-1$, equally spaced[2] in time at $\Delta t = 1$, how does one estimate the spectrum? As the inventor of multitaper estimates, Thomson (1982), I have a biased[3] opinion on the

---

[2]Although I am assuming a sampling rate of unity, it is assumed that the real sampling interval, Δt, has been chosen appropriately for the problem under consideration.

[3]Several papers illuminate this bias. Lindberg (1986), Lindberg and Thomson (1990), Bronez (1992), Seymour and Haykin (1993), Riedel *et al.* (1994a), Walden

answer to this question, but, taking a direct quote from Tauxe and Wu (1990):

> "Spectral analysis has recently undergone a revolution with the development at Bell Labs of sophisticated techniques in which the data are multiplied in turn by a set of tapers which are designed to maximize resolution and minimize bias [Thomson, 1982]. In addition to minimizing the bias while maintaining a given resolution, the multi-taper approach allows an estimate of the statistical significance of certain features (such as spectral lines) in the power spectrum by comparing the character of the DFT's of different data windows. These techniques are now in routine use ... "

To understand the origins of this process, remember that the first commonly used estimate of the spectrum, the *periodogram*, suggested by Stokes in 1879 and first applied by Schuster in 1898,

$$
P(f) = \frac{1}{N} \left| \sum_{t=0}^{N-1} x(t) e^{-i2\pi ft} \right|^2 ,
\tag{2.1}
$$

is just the square of the discrete Fourier transform (DFT) of the observations, scaled by $1/N$. The frequency f is normally taken in the Nyquist band $-\frac{1}{2} \le f < \frac{1}{2}$. As an estimate of the spectrum, the periodogram is both biased and inconsistent. In practice this means it almost always gets an unstable wrong answer. There are statements in the statistical spectrum estimation literature saying that, as a function of sample size, the periodogram is 'asymptotically unbiased'; these should be ignored. Experience teaches that the only reason people go to the trouble and expense of collecting more data is because they are going to ask more difficult questions, so one is always working with small-sample problems, not with asymptotics. In real data this bias can overwhelm the signals of interest. In Thomson (1977), I showed data where the periodogram was in error by more than a factor of 10^{10} over most of the frequency range. The periodogram is *inconsistent* because its variance, $\approx \mathbf{E}^2\{P(f)\}$, where $\mathbf{E}$ denotes expected value, does not decrease with sample size. In the cited example, the variance was too large by a factor of more than 10^{20} over most of the frequency range. Although

---

(1995), Hansson and Salomonsson (1997), McCoy *et al.* (1998), also find multitaper methods outperform classical alternatives.

these problems with the periodogram[4] were known before World War I,[5] many researchers still persist in using periodograms. A glance at the current literature on spectrum estimation theory and practice confirms that evolution is a slow process.

In the signal processing community, it is common to follow the recipe in Tukey (1967) for computing an estimate of the spectrum: choose a suitable *data window* $D(t)$ and compute

$$S_D(f) = \left| \sum_{t=0}^{N-1} x(t)D(t)e^{-i2\pi ft} \right|^2 \qquad (2.2)$$

and smooth, often by convolving $S_D(f)$ with a second window. Good data windows ((Harris, 1978)) give a much less biased estimate of the spectrum than the periodogram which, comparing (2.1) and (2.2), is a direct estimate with $D(t) \equiv 1/\sqrt{N}$. Because $S_D(f)$ is the sum of two squares (the real and imaginary parts of the DFT at frequency f), it has a chi-squared distribution with two degrees-of-freedom (df). Thus $S_D(f)$ is still inconsistent and the smoothing part of the recipe is necessary to obtain a useful estimate. Smoothing aside, however, this estimate poses a more fundamental problem: what motivated the introduction of the data window? Specifically, given that one has decided to use an estimate of the form (2.2), it is easy to show that one window is better or worse than another, but this is a different problem than showing why one should use (2.2) in the first place.

Multiple-window, or *multiple-taper*, or multitaper spectrum estimates were introduced in Thomson (1982) in an attempt to correct many of the shortcomings with 'standard' spectrum estimation procedures. Here one chooses an analysis bandwidth W for the estimate, $0 < W \leq \frac{1}{4}$ with $NW \approx 4$ to 6 a typical choice. The dimensionality of a signal with bandwidth W and a time duration of N samples is $K \approx 2NW$ (Slepian and

---

[4]Be cautioned that estimates that are based on the periodogram or raw DFTs have similar bias problems. Sample autocorrelations are just the Fourier transform of the periodogram so Blackman–Tukey, autoregressive, maximum-entropy, and other spectrum estimates that depend *directly* on sample autocorrelations should not be used, see the section 'Estimates to be avoided' in Thomson (1990a). 'Singular-spectrum analysis' ((Vautard *et al.*, 1992)) begins with the sample autocorrelations and computes the eigenvectors of the Toeplitz matrix formed from them and so inherits periodogram properties. Similarly, estimates of the analytic signal derived from Hilbert transforms using unwindowed DFTs have periodogram bias.

[5]Problems with the excessive variability of the periodogram were noted by Rayleigh (1912). Parzen (1957) comments 'Various authors have pointed out that the sample spectral density function, or periodogram, is not a consistent estimate of the spectral density function.'

Pollak 1961; Landau and Pollak 1961). Because our goal is to estimate the energy in a frequency band $(f - W, f + W)$ as accurately as possible, choose the K sequences of duration N whose energy concentration in this band is the best possible. These sequences are the *discrete prolate spheroidal sequences*, or *Slepian sequences* ((Slepian, 1978)). Now compute the expansion or *eigencoefficients*

$$y_k(f) = \sum_{t=0}^{N-1} x(t) v_t^{(k)}(N, W) e^{-i2\pi ft} \qquad (2.3)$$

where $v_t^{(k)}(N, W)$ is the kth Slepian sequence with parameters N and W for $k = 0, 1, \ldots, K - 1$. ¿From these the crudest multitaper spectrum estimate is the average,

$$\bar{S}(f) = \frac{1}{K} \sum_{k=0}^{K-1} |y_k(f)|^2 , \qquad (2.4)$$

simply an average of K estimates of the form (2.2) made using the same data, but with different tapers. Because the different tapers are orthogonal, the different terms in (2.4) are approximately uncorrelated. Each contributes two degrees-of-freedom, so (2.4) has a χ^2_{4NW} distribution. For fixed W the variance decreases with N, so the estimate is consistent. In practice, an adaptively weighted average of the $|y_k(f)|^2$'s is preferred to (2.4), see Thomson (1982, 1990b, 1994b), and Section 3.3.

As will be summarized in Section 3.1, the multitaper theory also explains the origins of the data window, or taper (Thomson, 1982). Beginning with an integral equation connecting the observations and the spectral representation of the process, one finds an approximate solution in terms of the eigenfunctions of the kernel. These frequency-domain eigenfunctions become the time-domain data windows or tapers. ¿From this perspective, Tukey's estimate (2.2) is approximately the first term of the series solution (2.4).

As part of this theoretical development, an effort was made to separate the deterministic (periodic) components of the process from the nondeterministic background, which the older estimates had lumped together. With multitapers, detection of sinusoids is commonly done with an F-test, a ratio of the energy explained by a periodic component at frequency f, to the remainder of the energy in the frequency band $(f - W, f + W)$; see Thomson (1982, 1990a,b), Jonsson and Steinhardt (1993), Thomson *et al.* (1995) for this topic.

In the original multitaper estimate, an approximate *linear* inversion of the integral equation was used, and the spectrum obtained by local averages of its magnitude; the newer *quadratic-inverse* theory (Thomson, 1990b) gives minimum-variance unbiased expansions of the spectrum, and represents a step in the process of eliminating the dependence on the choice of the bandwidth W. Multitaper correlation estimates, the Fourier transform of (2.4), have been studied in Van Veen and Scharf (1990) and McWhorter and Scharf (1998) and again outperform simple autocorrelation estimates.

Generally, multitaper methods have become the estimate of choice for serious spectrum estimation problems, and are becoming routine in geophysics (Tauxe, 1993). With an excellent text on the subject (Percival and Walden, 1993) and availability in *MATLAB* and the *SSA* toolkit (Dettinger *et al.*, 1995), they appear to be becoming so in other fields, (Lindberg and Thomson 1995; Kirsteins *et al.* 1998; Mitra and Peseran 1999; Xu *et al.* 1999). Stoica and Sundlin (1999) show that multitaper estimates are approximately maximum-likelihood nonparametric estimates of the spectrum. Multitaper theory has been extended to arbitrarily sampled spatial data (Bronez 1988; Liu and Van Veen 1992), but for regularly sampled time series with missing samples, interpolation may be a better alternative (Thomson and Schild, 1993, 1997).

3 Nonstationary Processes

We are given a single, finite realization of N contiguous samples of a discrete-time process $x(t)$ for $t = 0, \ldots, N - 1$. We assume that the process is harmonizable ((Loève, 1963)) so that it has the Cramér, or spectral, representation

$$x(t) = \int_{-\frac{1}{2}}^{\frac{1}{2}} e^{i2\pi\nu t} dX(\nu) \tag{3.1}$$

where $dX(\nu)$ is the generalized Fourier transform of the process. Denoting complex conjugate with a superscript $*$, this process has the covariance function

$$\Gamma_{\mathrm{L}}(t_1, t_2) = \mathbf{E}\{x(t_1)x^*(t_2)\} \tag{3.2}$$

$$= \iint e^{i2\pi(t_1 f_1 - t_2 f_2)} \gamma_{\mathrm{L}}(f_1, f_2) df_1 df_2 \tag{3.3}$$

where the generalized spectral density is given by

$$\gamma_{\mathrm{L}}(f_1, f_2) df_1 df_2 = \mathbf{E}\{dX(f_1)dX^*(f_2)\}. \tag{3.4}$$

This equation describes the essential feature of nonstationary processes, namely, that there is correlation between different frequencies. If the process is stationary, the process must depend only on $t_1 - t_2$ and not explicitly on t_1 or t_2 (Doob, 1952), so the Loève spectrum $\gamma_L(f_1, f_2)$, collapses to

$$\mathbf{E}\{dX(f_1)dX^*(f_2)\} = S(f_1)\delta(f_1 - f_2)df_1df_2. \tag{3.5}$$

The corresponding nonstationary process has a constant (white) spectrum and a time-varying covariance $\delta(t_1 - t_2)D(t_1, f)$ where $D(t, f)$ is the expected power at time t, in this case independent of the frequency f. Thus neither the spectrum nor covariance should be expected to be 'smooth' and continuity properties depend on direction in the f_1, f_2 or t_1, t_2 planes. These problems are more easily dealt with by rotating both the time and frequency coordinates of the generalized correlations (3.3) and spectral densities (3.4) respectively by 45°. In the time domain we define the new coordinates to be a 'center', t_0, and a delay, or lag, τ by $t_1 + t_2 = 2t_0$, $t_1 - t_2 = \tau$. We denote the covariance function in the rotated coordinates by $\Gamma(\tau, t_0) = \Gamma_L(t_0 + \frac{\tau}{2}, t_0 - \frac{\tau}{2})$. Similarly, in the rotated frequency domain, define a 'stationary frequency', f, and a 'nonstationary frequency', g, with $f_1 + f_2 = 2f$ and $f_1 - f_2 = g$. Denote the rotated spectrum by $\gamma(g, f) = \gamma_L(\frac{f+g}{2}, \frac{f-g}{2})$.

With these definitions (3.3) becomes

$$\Gamma(t_0, \tau) = \iint e^{i2\pi(\tau f + t_0 g)}\gamma(g, f)df\,dg. \tag{3.6}$$

The 'ordinary' frequency, f, is associated with the time lag, τ, and the 'nonstationary' frequency, g, with the average time, t_0.

Now consider continuity of γ as a function of f and g. On the line $g = 0$, γ is just the ordinary spectrum with the usual continuity (or lack thereof) conditions normally applying to stationary spectra. As a function of g, however, one expects to find a δ-function discontinuity at $g = 0$ because all data contains some stationary additive noise. Consequently, smoothing estimates of Γ in the g, f plane should *not* be isotropic, but have higher resolution along the nonstationary frequency coordinate than along the ordinary frequency axis, f. One solution is to smooth entirely along the f axis and leave g unsmoothed.[6] This was done in the estimate proposed in Thomson (1982) as well as in Gerr and Allen (1994).

---

[6]As with most rules in time series, remember that there is almost certainly a physically interesting process where the appropriate smoothing should be on g. So far, however, all the examples I have seen go in the first category.

A slightly less arbitrary way of handling the g coordinate is to Fourier transform (3.6) with respect to the 'nonstationary' frequency, g, and define

$$D(t_0, f) = \int e^{i2\pi t_0 g} \gamma(g, f) dg \qquad (3.7)$$

as the theoretical 'dynamic spectrum' of the process. The motivation is to transform the very rapid variation expected around $g = 0$ into a slowly varying function of t_0 while leaving the usual dependence on f. Recalling the definition of Γ_{L}, this becomes

$$D(t_0, f) = \int e^{-i2\pi\tau f} \mathbf{E}\left\{ x\left(t_0 + \frac{\tau}{2}\right) x^*\left(t_0 - \frac{\tau}{2}\right) \right\} d\tau \qquad (3.8)$$

so $D(t_0, f)$ is simply the expected value of the Wigner distribution ((Cohen, 1995)). This relation has been noted previously, see, *e.g.*, Martin (1982). Note carefully that, unlike the standard Wigner distribution, D is defined to be an expected value and is non-negative definite. The validity of replacing the expectation operator in (3.8) with an average requires knowledge, or testable assumptions, about the mixing process involved. One method is to test the 'downsamples', to be described in Section 3.3, for lack of serial correlation.

3.1   Multitaper estimates

The next two sections outline basic multitaper theory for a nominally stationary process. For the most part, this material applies to estimates made from a single data section. Section 4 examines the variance of such estimates, while Sections 5 to 8 cover progressively more complicated forms of nonstationarity. This progression is, roughly, from a spectrum, to a spectrogram (implicitly assuming 'slow' nonstationarity, Priestley (1981)), to Loève's spectrum for a fully nonstationary process.

Because the processes under study are nonstationary, I commonly subdivide long data sets into sections, segments, or blocks (the terms are used interchangeably) of length N samples and step the block positions along in time. The time-steps, or offsets, between blocks are usually equal, but when there are serious data gaps or outliers, one makes exceptions and abandons equal spacing for robustness and efficient use of data. The base time of a block is denoted by b.

The nonstationary quadratic-inverse estimates described in Section 5 is an intermediate form that allows 'time-derivatives' of the spectrum to be estimated using data from a single block.

One should note that there are three variable parameters in this approach, given here in order of importance:

1. The bandwidth W. The bandwidth must be chosen on the basis of the fundamentals (usually physics) of the problem. Be cautioned, however, that assumptions about physical processes can be seriously awry,[7] so that exploring data over a range of bandwidths is recommended.

2. The block size, N, or, equivalently, the time-bandwidth product. Typically $NW \approx 4$ to 6 is a good starting point. If NW is too small the estimator will have poorer sidelobe performance than that obtainable with a larger value of NW and, in addition, the estimates will be statistically unstable.

3. The offset between blocks, Δb. This parameter is usually chosen for reasons of robustness or on the basis of experience with non-stationarity. The high-resolution spectrogram described in Section 6 makes it a secondary parameter that can be chosen primarily on the basis of robustness considerations. A good initial choice of Δb is $N/2$.

Multitaper estimates of the spectrum ((Thomson, 1982)), are a class of estimates based on approximately solving the integral equation that expresses the projection of $dX(f)$ onto the Fourier transform of the data, $y(f)$. Taking the discrete Fourier transform of the observed data

$$y(f) = \sum_{t=0}^{N-1} x(t)e^{-i2\pi ft} \tag{3.9}$$

and using the spectral representation (3.1) for $x(t)$ one has the fundamental equation of spectrum estimation

$$y(f) = \int_{-\frac{1}{2}}^{\frac{1}{2}} \mathbf{K}_N(f - \xi)dX(\xi) \tag{3.10}$$

---

[7]In my personal experience, assumptions about data purported to come from turbulent, chaotic, or fractal processes should be regarded with exceptional skepticism. Similarly, assumptions that have been 'verified' with Monte Carlo simulations should usually be ignored. Simulations are 'doomed to succeed', and, apart from checking one's programs and calculus, the time and effort required to do them is almost always better spent on exploratory data analysis.

where the Dirichlet kernel is given by

$$\mathbf{K}_N(f) = \frac{\sin N\pi f}{\sin \pi f} e^{-i2\pi f \frac{N-1}{2}}. \tag{3.11}$$

There are several points that must be remembered about this equation. Because one may take the inverse Fourier transform of $y(f)$ and recover $x(t)$ for $0 \le t \le N-1$, $y(f)$ is a trivially sufficient statistic and completely equivalent to the original data. The *finite* Fourier transform $y(f)$ is **not** equivalent to the spectral generator $dX(\xi)$. Remember that $dX(\xi)$ is assumed to generate the entire data sequence for *all* t, not just the portion observed. Despite definitions given in many elementary texts $\frac{1}{N}|y(f)|^2$ is *not* the spectrum, even in the limit of large N. It is the periodogram (2.1) and is biased and inconsistent. While (3.10) is formally a convolution of dX with a Dirichlet kernel, it is more constructive to think of it as a Fredholm integral equation of the first kind. As such, it does not have a unique solution. It does, however, have useful approximate solutions. We mentioned above that 'multitaper estimates' does not refer to a particular estimate, but rather to a class of estimates with the class being defined by the method used to form the necessarily approximate solution of the integral equation. Thought of in this way, spectrum estimation is a form of inverse theory. As a specific example, Riedel and Sidorenko (1995, 1996) developed a multitaper 'sine' estimate, which uses a different error norm for solving the integral equation. While the 60dB range of their 'sine' tapers lack the crushing sidelobe performance of the Slepian tapers, they are adequate for many applications and easier to compute. Other choices for tapers are discussed in Mullis and Scharf (1991) and Bartholomew and Tague (1995).

3.2   Slepian Functions and Sequences

The key to obtaining 'reasonable' solutions for (3.10) is a set of special functions known as *Discrete Prolate Spheroidal Wave Functions*, or, in honor of their main proponent in signal processing, David Slepian *Slepian Functions and Sequences*, (Slepian 1953; Slepian and Pollak 1961; Slepian 1964, 1978, 1983). Slepian sequences are defined as the real, unit-energy sequences on $[0, N-1]$ having the greatest energy in a bandwidth W. I use Slepian's (1978) notation $v_n^{(k)}(N, W)$ for sample n of the kth sequences or, for short, $v_n^{(k)}$. These sequences are solutions of the symmetric

Toeplitz matrix eigenvalue equation

$$\lambda_k v_n^{(k)} = \sum_{m=0}^{N-1} \frac{\sin 2\pi W(n-m)}{\pi(n-m)} v_m^{(k)} \tag{3.12}$$

for $0 \le n \le N - 1$ and are defined by this equation for n outside this interval. One should note that solving this equation directly is *not* the way to compute Slepian sequences. Use the tridiagonal form, Equation (14) of Slepian (1978) as in Appendix B of Thomson (1990b), or see Gruenbacher and Hummels (1994). The Slepian functions, the Fourier transforms of the corresponding sequences, are

$$V_k(f) = \sum_{n=0}^{N-1} v_n^{(k)} e^{-i2\pi nf} \tag{3.13}$$

and satisfy the homogeneous integral equation

$$\lambda_k V_k(f) = \int_{-W}^{W} \mathbf{K}_N(f - \nu) V_k(\nu) d\nu. \tag{3.14}$$

The V's differ from the U's defined in Slepian (1978) in the definition of the Fourier transform, see Appendix A of Thomson (1990b).

For reasons given in Landau and Pollak (1961), the eigenvalues of this equation determine the dimensionality of the space $(-W, W) \times [0, N-1]$. The N eigenvalues are bounded between 0 and 1 with

$$K \approx \lfloor 2NW \rfloor \tag{3.15}$$

'large' eigenvalues, *i.e.*, nearly 1, so the dimensionality of the subspace is approximately $2NW$. Note the eigenvalues give the fraction of energy in the bandwidth $(-W, W)$ of the corresponding sequence so $1 - \lambda_k$ is the 'leakage' of the kth window. *The leakage properties of these windows are the best possible.* It does not seem to be generally recognized that if one approximates

$$e^{-i2\pi ft} \approx \sum_{k=0}^{K-1} v_t^{(k)} V_k(f) \tag{3.16}$$

on $L_2(-W, W) \times [0, N - 1]$, then, because of the energy concentration properties of the Slepian sequences, the approximation is better than with any other set of orthogonal windows. This becomes important when considering the efficiency of overlapped, spectogram-like estimates. The sequences are orthonormal and the functions orthonormal on $[-\frac{1}{2}, \frac{1}{2})$ and *also* orthogonal on $(-W, W)$,

$$\int_{-W}^{W} V_j(f) V_k^*(f) df = \lambda_j \delta_{jk}. \tag{3.17}$$

The difference between the latter two orthogonality relations shows that the Slepian functions are also orthogonal on $[-\frac{1}{2}, -W) \cup (W, \frac{1}{2})$, the outer interval, and using this double orthogonality allows one to estimate bias from the outer interval. It is useful to denote the orthonormal eigenfunctions on the inner interval by

$$\mathcal{V}_k(\xi) = \frac{1}{\sqrt{\lambda_k}} V_k(\xi). \tag{3.18}$$

The inverse Fourier transform of $\mathcal{V}_k(\xi)$ is used extensively and from (A10) of Thomson (1990b), one has

$$\sqrt{\lambda_k}\, v_t^{(k)} = \int_{-W}^{W} \mathcal{V}_k(\xi) e^{i2\pi t\xi} d\xi, \quad \forall t. \tag{3.19}$$

3.3   Spectrum Estimation as an Inverse Problem

Given the fundamental equation (3.10), the basic multitaper approach is to make a frequency-localized least-squares eigenfunction approximation to the solution of the integral equation on the interval $(f - W, f + W)$. One then varies the center frequency f to obtain solutions across the whole frequency range. This is done by assuming that $d\tilde{X}$, the *observable*[8] portion of dX, has the expansion

$$d\tilde{X}(f \ominus \xi) \approx \sum_{k=0}^{K-1} x_k(f) \cdot \mathcal{V}_k^\star(\xi) d\xi \tag{3.20}$$

where the signs $\ominus$ and $\oplus$ are used as a reminder of the restriction $|\xi| < W$.[9] The coefficients $x_k(f)$ in equation (3.20) are *defined* by

$$x_k(f) = \int_{-W}^{W} \mathcal{V}_k(\xi) dX(f - \xi). \tag{3.21}$$

---

[8]'Observable' is being used in the optimistic sense of what would be recoverable with an N point Slepian function if dX were available. Writing dX in this way is a slight abuse of notation. Strictly, one should replace $d\xi$ with $d\mathcal{W}(\xi)$, a white, unit power, orthogonal increment process.

[9]I usually write equations such as (3.22) with $\ominus$ when they can be used as convolutions. There are occasions when a positive frequency offset is clearer. The same equations apply with $\oplus$ if $\mathcal{V}_k^\star(\xi)$ is replaced with $\mathcal{V}_k(\xi)$.

Although these coefficients are strictly unobservable, they are what we attempt to estimate. The preceding two equations form a projection and, for $0 \leq k \leq K - 1$, the effect of the projections of dX and $d\tilde{X}$ are identical. Denoting the estimates of $x_k(f)$ by $\hat{x}_k(f)$ one forms an estimate of dX in the same form as (3.20), that is,

$$d\widehat{X}(f \ominus \xi) \overset{\bullet}{=} \sum_{k=0}^{K-1} \hat{x}_k(f) \cdot V_k^{\star}(\xi)d\xi. \tag{3.22}$$

To estimate the $\hat{x}_k(f)$'s one begins with a standard, formal eigenfunction solution of the integral equation (3.10) and computes the raw eigencoefficients (2.3)

$$y_k(f) = \frac{1}{\lambda_k} \int_{-W}^{W} V_k(\xi)y(f - \xi)d\xi. \tag{3.23}$$

Replacing y with the fundamental integral equation (3.10) and using the integral equation for the Slepian functions (3.14) gives

$$y_k(f) = \int_{-\frac{1}{2}}^{\frac{1}{2}} V_k(\xi)dX(f - \xi). \tag{3.24}$$

Again, starting with (3.23), replacing y with the DFT (3.9), and the inverse Fourier transform of V with (3.19), one has

$$y_k(f) = \sum_{n=0}^{N-1} e^{-i2\pi fn} v_n^{(k)}(N, W) x(n). \tag{3.25}$$

The expansion coefficients are obtained by windowing the data with a Slepian sequence and Fourier transforming. In form, (3.25) is identical to the conventional direct spectrum estimate, (2.2), *except* that the latter uses only the first, $k = 0$, term in this expansion. Obviously, a fast Fourier transform (FFT) may be used for efficient computation of the expansion coefficients. *These expansion coefficients represent the information in the signal projected onto the local frequency domain.* We use the $K = 2NW$ coefficients corresponding to functions with eigenvalues $\lambda_k \approx 1$ for subsequent statistical inference.

Because the Slepian sequences are time-limited, they cannot be strictly band-limited, and the kth sequence has a fraction $1 - \lambda_k(N, W)$ outside the interval $(-W, W)$. Comparison of the ideal eigencoefficients (3.21) with the raw eigencoefficients (3.24) shows that, apart from the trivial factor $\sqrt{\lambda_k}$, the major difference between $x_k(f)$ and $y_k(f)$ is a bias term

$$b_k(f) = \fint V_k(\xi)dX(f - \xi) \tag{3.26}$$

where the 'cut' integral is defined by

$$\fint = \int_{-\frac{1}{2}}^{\frac{1}{2}} - \int_{-W}^{W} \quad . \tag{3.27}$$

Uncorrected, this out-of-band, or broad-band, bias can be severe for the higher-order, or transition eigencoefficients, that is, those of order $k \approx K$, particularly when the spectral range, the ratio of the maximum to minimum of the spectrum, is large. There are at least three ways of dealing with this exterior bias:

1. For 'reasonable' time-bandwidth products the lowest-order sequences are effectively band-limited, so truncating the series at $K \approx \lfloor 2NW \rfloor - 2$ to $\lfloor 2NW \rfloor - 4$ or so limits bias at the expense of variance. The bias may be bounded assuming nothing but properties of the Slepian functions and the Cauchy inequality. Squaring (3.26) one has

$$|b_k(f)|^2 \leq \fint |V_k(\xi)|^2 d\xi \fint |dX(f - \xi)|^2. \tag{3.28}$$

The first integral is $1 - \lambda_k$. The second is clearly bounded by the integral over the full range so

$$\mathbf{E}\{|b_k(f)|^2\} \leq (1 - \lambda_k)\sigma^2. \tag{3.29}$$

Also, as the variance of a spectrum estimate is proportional to the square of the biased estimate, the overall mean-squared-error is often less with truncation than if the higher terms are included.

2. Although the derivation of the adaptive weighting procedure given in Thomson (1982) used the orthogonal increment properties of stationary processes, the procedure is, nonetheless, effective for controlling bias in general. This is the preferred method for an initial estimate and we define

$$\hat{x}_k^{(1)}(f) = d_k(f)y_k(f) \tag{3.30}$$

where the weights, $d_k(f)$, are defined in Section V of Thomson (1982) and also in Eqn. (5.2) of Thomson (1990b).

3. The best method for reducing bias found to date is by coherent sidelobe subtraction as outlined in Thomson (1994b). Form an estimate of $dX(f \ominus \xi)$ from the adaptively weighted estimates of the eigencoefficients,

$$d\widehat{X}^{(1)}(f \ominus \xi) \approx \sum_{k=0}^{K-1} \hat{x}_k^{(1)}(f)\, \mathcal{V}_k^{\star}(\xi)d\xi, \qquad (3.31)$$

valid for $|\xi| < W$. For any frequency, f_0, there is a range of frequencies $(f_0 - W, f_0 + W)$ giving an estimate of $d\widehat{X}^{(p)}(f_0)$, specifically,

$$d\widehat{X}^{(p)}(f_0 : \xi) \approx \sum_{k=0}^{K-1} \hat{x}_k^{(p)}(f_0 \ominus \xi)\, \mathcal{V}_k^{\star}(\xi)d\xi, \qquad (3.32)$$

nominally independent of the free parameter ξ. Here $\hat{x}_k^{(p)}(f)$ is the estimate of $x_k(f)$ at the pth iteration. We use a weighted sum of the free-parameter expansions to form an estimate of dX,

$$d\widehat{X}^{(p)}(f) = \frac{1}{2W} \int_{-W}^{W} Q(\xi) \sum_{k=0}^{K-1} \hat{x}_k^{(p)}(f - \xi)\, \mathcal{V}_k^{\star}(\xi)d\xi \qquad (3.33)$$

where the weighting function Q may reflect nothing more than that the convergence of the orthogonal expansions is generally poorer near the ends of the domain than in the center, or that, in regions where the spectrum is changing rapidly, some expansions are less reliable than others. My preferred estimate is to weight $d\widehat{X}$ by the current estimate of Fisher information, specifically,

$$d\widehat{X}^{(p)}(f) = \frac{\frac{1}{2W} \int_{-W}^{W} \frac{Q_o(\xi)}{\widehat{S}^{(p)}(f-\xi)} \sum_{k=0}^{K-1} \hat{x}_k^{(p)}\, \mathcal{V}_k^{\star}(\xi)d\xi}{\frac{1}{2W} \int_{-W}^{W} \frac{Q_o(\xi)}{\widehat{S}^{(p)}(f-\xi)}d\xi} \qquad (3.34)$$

where $Q_o(\xi)$ is typically $1 - (\xi/W)^2$ and $\widehat{S}^{(p)}(f)$ is the current estimate of the spectrum. Next, estimate the exterior bias using the convolution over the exterior domain,

$$\hat{b}_k^{(p+1)}(f) = \fint V_k(\xi)d\widehat{X}^{(p)}(f - \xi), \qquad (3.35)$$

and subtract it from the raw eigencoefficients (2.3) to form an improved estimate

$$\hat{x}_k^{(p+1)}(f) = y_k(f) - \hat{b}_k^{(p+1)}(f). \qquad (3.36)$$

Rather crude bounds on the performance of this procedure show that it has bias bounded by $c(1 - \lambda_K)^3 \sigma^2$. In practice, the performance seems to be considerably better than this.

Finally, denote the estimated eigencoefficients by $\hat{x}_k(f)$ and collect them in the vector $\widehat{\mathbf{X}}(f)$, using superscript T to denote the transpose

$$\widehat{\mathbf{X}}(f) = [\hat{x}_0(f), \hat{x}_1(f), \cdots, \hat{x}_{K-1}(f)]^T. \tag{3.37}$$

Using † for transpose conjugate, the corresponding spectrum estimate is

$$\widehat{S}(f) = \frac{1}{K}\widehat{\mathbf{X}}^\dagger(f)\widehat{\mathbf{X}}(f) = \frac{1}{K}\sum_{k=0}^{K-1}|\hat{x}_k(f)|^2. \tag{3.38}$$

To see the dependence on the bandwidth W of the estimate, recall that there are $K \approx \lfloor 2NW \rfloor$ windows with eigenvalues near 1. If the spectrum is flat *within the local domain*, the coefficients are uncorrelated because the windows are orthogonal and each contributes 2 df, so estimates of the form (3.38) have $2K$ df. If W is too small, one has poor statistical stability, but if W is too large, the estimate has poor frequency resolution. Typically W is chosen between $1.5/N$ and $20/N$ with a time-bandwidth product of 4 or 5 being a common starting point. Thus $W = 4/N$ or $5/N$ with corresponding $K = 6$ or 8 gives estimates with 12 or 16 df.

We must emphasize, however, that these only apply to the simplest forms of estimates and both quadratic-inverse estimates, see Thomson (1990b, 1993, 1994b), and free-parameter estimates of the type described above give high-resolution estimates that are, within reason, largely independent of the choice of W. These estimates also give implicit extrapolations of the time series.

To conclude this section, note that (3.22) can be Fourier transformed over $(-W, W)$ using (3.19) to give a complex demodulate of the series centered at frequency f,

$$\hat{x}(t; f) = \int_{-W}^{W} e^{i2\pi\nu t} d\widehat{X}(f + \nu) \tag{3.39}$$

$$= \sum_{k=0}^{K-1} \hat{x}_k(f)\sqrt{\lambda_k}\, v_t^{(k)}. \tag{3.40}$$

This is simply the Slepian sequence expansion of the baseband version of the signal from the band $(f - W, f + W)$. The narrowband tracking filters, Lanzerotti *et al.* (1995), and projection filters, Thomson (1994c,

1995), are based on similar Slepian sequence expansions with coherent sideband cancellation. Because $\hat{x}(t; f)$ is strictly band-limited, it is sufficient to sample it at steps $\Delta = \frac{1}{2W}$. This downsampling leaves K complex numbers that we denote by

$$x^{\circ}(j; f) = \hat{x}(t_j; f) \tag{3.41}$$

for $j = 1, 2, \ldots, K$ with $t_j = (j - \frac{1}{2})\Delta$ so there are, again, $2NW$ samples. The strict bandlimiting used in (3.39) implies that the Slepian sequences are entire functions of t, so $\hat{x}(t; f)$ is defined for all t, not just $0 \le t \le N - 1$. This is used when doing the convolutions in coherent sidelobe cancellation. Extrapolation properties of Slepian sequences have been extensively studied, see, *e.g.*, Papoulis (1975), Cadzow (1979), Huang *et al.* (1984) among many others, and Landau (1985) for a nice analytic summary. In addition to this extrapolation, Fourier transforms of the *weighted* eigencoefficients (3.30) also have duration outside $[0, N - 1]$.

With multivariate data, I denote the raw data from the pth series by $\mathbf{x}_p(t)$. Starting, or base, times for different blocks are denoted by b, so the raw eigencoefficients from series p starting at base time b are

$$y_{p,k}(b, f) = \sum_{n=0}^{N-1} \mathbf{x}_p(b + n)\, v_k^{(n)} e^{-i2\pi f n} \tag{3.42}$$

and the refined eigencoefficient estimates are $\hat{x}_{p,k}(b, f)$.[10]

4   Variance of Multitaper Estimates

A recurring question with spectrum estimates is "How variable can a spectrum estimate reasonably be?". This question requires careful consideration when deciding how to analyze a particular data set. Can a given spectrum estimate reasonably to be an extreme example from a stationary process, or is the data really nonstationary? A similar problem arises in the analysis of space physics and climate data. Should one attribute an apparently highly variable spectrum to a perverse distribution, or is a process with many discrete modes a "simpler" explanation? Unfortunately, the answer to this question depends on many conditions such as the frequency range over which the variability is estimated, and the process being analyzed. Also, although spectrum estimates can be

---

[10]The ordering of arguments as (b, f) is to agree with the common 'time-frequency' convention, and similarly with those of $\gamma(g, f)$ and $D(t_0, f)$.

uncomfortably imprecise, they do not have to be as bad as people accustomed to periodogram estimates expect, and it is important to assess the variability accurately because departures from nominal can provide important clues about the nature of the process. The following 10 sections address these problems.

There are many papers dedicated to the sampling properties of spectrum estimates from the seminal papers of Bartlett (1946) and Blackman and Tukey (1959), and classics, Whittle (1953), Jones (1962), to specialized work, Groth (1975), to more recent work such as Zhurbenko (1978, 1979). Variances of multitaper spectrum estimates are considered in (Thomson, 1982, 1990b, Walden *et al.*, 1994) and elsewhere.

An important set of papers, *e.g.*, Rosenblatt (1961), Rozanov (1961), Mallows (1967), Hannan and Thomson (1967) shows that the statistics of narrow-band processes must be nearly Gaussian. Because the eigencoefficients depend either implicitly, (2.3), or explicitly, (3.42) on the starting time of the data segment, all the spectrum estimates considered here should be thought of as being derived from narrow-band processes[11].

One aspect of this problem is that confidence limits set by assuming a stationary Gaussian process with a "simple" spectrum are often ridiculously optimistic. The reasons for this often seem to be that data is nonstationary, or non-Gaussian, or both, usually in subtle ways. Clearly, if data is obviously non-stationary when plotted, or visually skewed, one should not be surprised that stationary Gaussian statistics do not work well.[12] The cases of interest here are those when the data looks "normal." How does one tell when time-series data is abnormal? Similarly, if one is estimating line parameters such as frequency, phase, or amplitude *from data*, how does one set confidence limits when parameters such as the signal-to-noise power ratio are unknown? The most effective method I know for both problems is to use a multitaper method, then resample *on the tapers*. Specifically, if one resamples by using a jackknife, simply delete each window in turn from the estimation procedure, Thomson and Chave (1991), Thomson (1994a) or, alternatively, apply random weights to the different tapers and bootstrap in the usual way. For a comparison

---

[11]Think of the eigencoefficients $y_{p,k}(b, f)$ in (3.42) as a function of the base time, b, alone while keeping p, k, and f fixed.

[12]The serial correlations present in time series data make conventional tests of "goodness-of-fit" unreliable. One approach is to subsample the series at a rate where the autocorrelation is approximately zero and do a quantile-quantile (Q-Q) plot on the subsampled data. (For background on Q-Q plots see Kleiner and Graedel (1980), Goodall (1983), Chambers *et al.* (1983).) A second approach is to use all the observations in the Q-Q plot and adjust the sample size for entropy. If there is any doubt, do a robust estimate and check for differences between it and the standard estimate.

of resampling methods applied to multitaper estimates, see Fodor and Stark (1998).

Conventionally, most researchers on spectrum estimation have divided the problem into two parts; the "line" spectrum, and the remainder, or continuum. Having emphasized this distinction myself, Thomson (1982, 1990a,b), I now view this as an oversimplification because, even in some of the problems that are primarily concerned with periodic processes, one must pay extraordinarily careful attention to the continuous parts of the spectrum. There are several reasons for this: first, the existence of discrete lines is usually an approximation to begin with, with narrow resonances perhaps a better one and, in this case, the shape of the lines is of interest; second, even when the fundamental process generates essentially mathematical sinusoids, one may only be able to observe them through some intervening process; third, the line components may be amplitude or frequency modulated; and, fourth, the background spectrum may simply be complicated.

More subtle problems arise when the process contains *many* lines. Unless enough data is available to resolve the individual lines, estimates of the spectrum of such processes can be highly variable as the various periodic components within the estimators bandwidth drift in and out of phase, Thomson *et al.* (1996). The unresolved lines can be mistaken for a complicated background so that the probability of false detections is increased.

4.1   Standard Distribution of Spectrum Estimates

For most purposes, logarithms of spectral densities are regarded as "more useful" than the densities themselves. This may be because predictability is measured by

$$\sigma_P^2 = \exp\{ \int_{-\frac{1}{2}}^{\frac{1}{2}} \ln S(f) df \} , \qquad (4.1)$$

that physical processes often have very red spectra so just being able to see the high-frequency end of a plotted spectrum requires a logarithmic scale, or simply that, since human vision and hearing respond logarithmically, our evolutionary preference is for logarithmic scales. An important *statistical* reason for using log-spectra is that the sampling variance does not depend on frequency.

Turning to details, recall that if a spectrum is estimated by averaging m independent estimates of the form $|\hat{x}_p(b, f)|^2$ for a particular frequency

f under utopian conditions[13] the estimate, $\widehat{S}$, will have a chi-square probability density function, p, with $2m$ df,

$$p(\widehat{S}) = \left[\frac{m}{S}\right]^m \frac{\widehat{S}^{m-1}}{\Gamma(m)} e^{-m\frac{\widehat{S}}{S}} \tag{4.2}$$

where S is the true spectrum. Consequently, the μth moment is

$$\mathbf{E}\{(\widehat{S})^\mu\} = \frac{\Gamma(m+\mu)}{m^\mu \Gamma(m)} S^\mu$$

and the mean and variance of $\widehat{S}$ are S and S^2/m respectively. Terms in $\widehat{S}^{-1}$ and $\widehat{S}^{-2}$ occur commonly, for example, as weights (*i.e.* 3.34), in estimates of $\frac{d\ln S(f)}{df}$, in transfer function estimates, or as a starting point for inverse autocorrelations, (Cleveland 1972; Chatfield 1979; Bhansali 1983; Rao and Gabr 1989). These are biased with expected values of $m/(m-1) \cdot S^{-1}$ and $m^2/((m-1)(m-2)) \cdot S^{-2}$, respectively, so $1/\widehat{S}$ does not have a mean with $m=1$, nor variance for $m=2$. Requiring $\widehat{S}^{-2}$ to have a variance requires $m>4$, so one should be very cautious how spectrum estimates with less than 10 df are used and, for most purposes, a minimum of 16 to 20 df are preferable.

Similarly, it has been known since at least Bartlett and Kendall (1946) that the expected value of the logarithm of a χ^2-distributed estimate is biased. That is,

$$\mathbf{E}\{\ln \widehat{S}\} = \ln S + B_\chi(m) \tag{4.3}$$

where the bias $B_\chi(m)$ is given by

$$B_\chi(m) = \psi(m) - \ln m \,, \tag{4.4}$$

ψ being the digamma function. Similarly, the variance of the logarithm of a χ^2 estimate is

$$\mathbf{Var}\{\ln \widehat{S}\} = \psi'(m) \tag{4.5}$$

where ψ' is the trigamma function. For a simple direct spectrum estimate of the form (2.2), the distribution is χ_2^2, $m=1$ and the variance, $\psi'(1) = \frac{\pi^2}{6} \approx 1.645$, is unacceptably large. Going to $m=5$, the minimum recommended above, gives a variance of $\psi'(5) \approx 0.221$, an improvement

---

[13]Utopian conditions are stationary, ergodic, Gaussian data with the spectrum constant over the analysis bandwidth.

by a factor of 7.4 over the simple estimate. Even this large a change in the variance understates the change in distribution. If one takes as an example a set of $J = 100$ spectrum estimates, the expected minimum in a χ_2^2 distribution is $S/J = 0.01S$ while the expected minimum in a χ_{10}^2 distribution is $0.239S$. A plot of the χ_2^2 estimate will consequently require an expansion of the scale by at least a factor of 20 over the χ_{10}^2 estimate with the result that significant details are commonly lost in the "grass".

The correlation between $\widehat{S}$ and $\ln \widehat{S}$ is also of interest. The integral is easily evaluated, (Erdélyi *et al.*, 1954, Vol. 1, formula 11, pg. 148) and gives the remarkably simple result

$$\mathbf{Cov}\{\widehat{S}, \ln \widehat{S}\} = \frac{S}{m}$$

so the correlation is $1/\sqrt{m\psi'(m)}$. The asymptotic formula for ψ' gives

$$\mathbf{corr}\{\widehat{S}, \ln \widehat{S}\} = \frac{1}{\sqrt{1 + \frac{1}{2m}}},$$

accurately to better than 1% by $m = 5$. As this correlation exceeds 95%, it is reasonably safe to treat the statistics of $\widehat{S}$ and $\ln \widehat{S}$ similarly for reasonable df, *i.e.*, df & 10.

As an example, Figure 1 shows excess variance in the spectrum of the Central England temperature series (CET), Manley (1974), Parker *et al.* (1992). In this example, of which more will be seen later, monthly data from Jan. 1659 to May 1997 is used. Data blocks, or segments, 44-years long were used to obtain reasonable frequency resolution and, specifically, to include 2 of the 22-year solar magnetic cycles. A 12-month periodic midmean[14] was removed from the data, and the residuals prewhitened with an AR-1 filter in all the examples shown here. The variance of the January residuals (4.02) is almost four times as large as that of the August residuals (1.14), so there is clearly cyclostationary structure present. Because the object of this paper is to study nonstationarity, I have not attempted[15] to make an *ad-hoc* correction for this effect.

---

[14]This is not strictly correct because the annual cycle does not quite follow the calendar year, (Thomson, 1995). The three-day calendar offset in 1772 has not been corrected.

[15]Another bias. One does not "correct" for observed effects until they are understood.

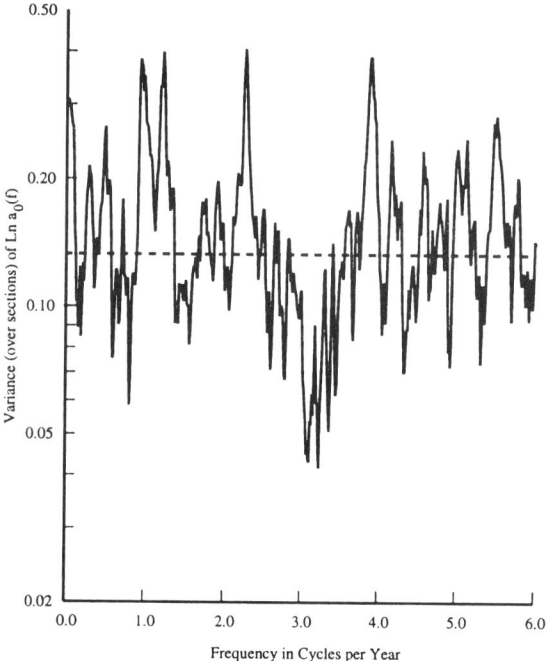

Figure 1: The variance of $\ln \hat{S}$ for the Central England monthly temperature series. Estimates were computed on 146 44-year segments, each offset by 2 years. A periodic midmean was removed from the original data, and the residuals prewhitened by a first-order autoregressive (AR-1) model. On each data block 8 windows with $NW = 5$ were used. The average variance across frequencies is 0.154, compared with 0.133 expected, (4.5), for a $\ln \chi^2_{16}$.

4.2   Bartlett's M-Statistic

Because spectrum estimates have χ^2 statistics, many of the tests used for ordinary variance estimates can be applied, with due regard for the differences, to spectra. In my experience, the most useful of these is Bartlett's test for homogeneity of variances. This test, (Bartlett 1937; Pearson and Hartley 1970) is just the log-ratio of the arithmetic and geometric means of a set of variances. If $\widehat{S}_j$ are a set of J independent spectrum estimates, each having ν df, the M statistic is

$$M = J\nu \left[\ln \left\{ \frac{1}{J} \sum_{j=1}^{J} \widehat{S}_j \right\} - \frac{1}{J} \sum_{j=1}^{J} \ln \widehat{S}_j \right] . \qquad (4.6)$$

This test has several advantages: first, it is the likelihood-ratio test for homogeneity; second, its distribution is known under both the null and alternative hypotheses; third, it is simple to compute; and fourth, M/C with $C = 1 + \frac{J+1}{3J\nu}$ is asymptotically distributed as χ^2_{J-1} for reasonable ν and J. Also, while M is known to be sensitive to departures from normality, the narrow-band arguments mentioned earlier imply that this is less of a concern in spectrum estimates than elsewhere. There are two distinct applications to spectrum estimates:

- The first is to test estimates at a given frequency, made from data on different data segments or sets, for homogeneity. Here the assumption is that nonstationarity is slow, and one is asking if variations seen in a spectrogram are "real" or just typical of the random fluctuations seen in estimates from a stationary process. Be cautioned, however, that most interpretations of a spectrogram are visual and that the human eye is usually much better at detecting patterns than simple statistical tests. A weak "chirp", for example, will frequently be missed by such a test, but easily seen. Also, "stationarity" is being used in a frequency-dependent context or meaning. I have seen many examples where an otherwise stationary process incorporates a few narrowband nonstationary components.

- The second is to test estimates within a block of frequencies in a single spectrum for homogeneity. In other words, are the observed fluctuations typical of a white noise process, or is something more complicated present? Clearly, such a test normally requires careful prewhitening or similar "detrending" of the spectrum as the coloured background spectra commonly encountered will reject most spectra from physical processes.

As an example, Figure 2 shows an estimate of the spectrum of the radial (away from the sun) component of the interplanetary magnetic field (Balogh *et al.*, 1992) measured near the ecliptic by the *Ulysses* spacecraft. I used data[16] between day 298 of 1990, just after launch, until day 33 of 1992, just prior to the spacecraft's encounter with Jupiter. The spectrum of such data is frequently assumed to be a generally featureless "power law" with a slope (on a log-log plot) of -1, (Matthaeus and Goldstein, 1986) or, frequently, of -5/3 (Goldstein *et al.*, 1995), but this view has been challenged, (Thomson *et al.*, 1995). The data is generally of high quality, but has a few gaps which were interpolated using the

---

[16]This data is publically available at the European Space Agency's web site, http://helio.estec.esa.nl/ulysses.

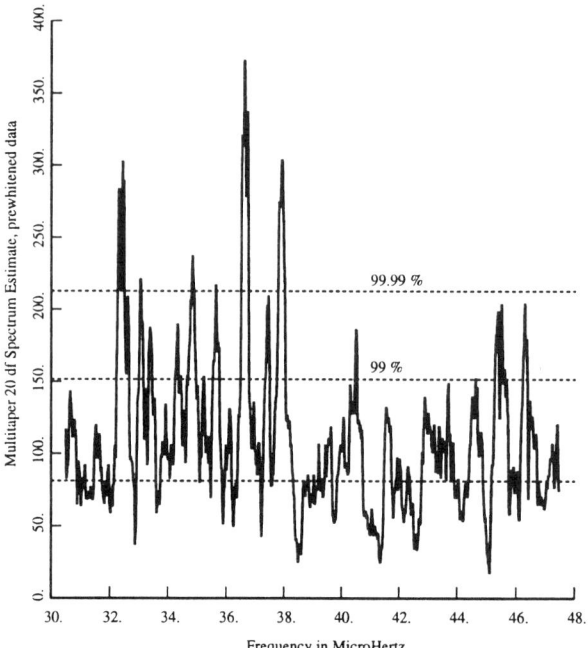

Figure 2: Portion of a spectrum of the radial component of the interplanetary magnetic field measured by the *Ulysses* spacecraft between day 298 of 1990 and day 33 of 1992. The data was prewhitened and scaled by radius. The bottom dashed line is the mean estimated from the 5% point of the data, the upper two are at the 99 and 99.99% points above it.

iterative scheme outlined in Thomson *et al.* (1995), and prewhitened, (Thomson, 1977, Kleiner *et al.*, 1979), to give a generally flat spectrum. The "power law" arguments imply that a simple filter is adequate. Because the interplanetary magnetic field decays roughly as $1/R^2$, where R is the heliographic radius, the prewhitened data were scaled by R to compensate for the change from 1 to 5 AU in radius covered during this time. Otherwise the results are biased by a trivially removable factor of 25 in power.[17] Subsamples of this data spaced 96 hours, the first zero-crossing of the data's autocorrelation, are roughly linear when plotted on a Gaussian Q-Q plot, so the eigencoefficients should not depart measurably from Gaussian. A multitaper spectrum with $NW = 6.0$ and

---

[17]Note another bias: If a simple transformation or scaling will give more stationary data, it is stupid not to use it. This, however, is conditioned on knowing the reason for the scaling.

$K = 10$ was made from 11,157 hourly samples, enough that the windowed transforms should be closely Gaussian. The dotted line in Figure 2 was obtained by fitting the lower 5% points of the spectrum estimates, and scaling by $\frac{\nu}{Q_{\chi^2}(0.05,\nu)} = 1.843$ where $Q_{\chi^2}(0.05, \nu)$ is the χ^2 quantile at the 5% point, and here $\nu = 2K = 20$. This line appears to be a reasonable fit to both ends of the spectrum (and elsewhere) so the prewhitening operates as intended. The variance of $\ln \widehat{S}$ is, however, about 2.08 times that predicted by (4.5). The series of 12 large[18] peaks between 32 and 38 μHz, reach a level of 4.6 times the base level, a highly improbable event in a χ^2_{20} distribution. (The 99.99% point is 2.62 times the mean.) Bartlett's test was applied to the spectrum estimate shown in Figure 2. The sample size of 11,157 hours gives a Rayleigh resolution of 24.9 nHz, and the time-bandwidth of 6 gives a bandwidth $B = 149$ nHz. Being slightly conservative, the 17 μHz frequency range shown gives $J = 112$ effective estimates.[19] The test rejects homogeneity at a level of about 10^{-10}; that is, this spectrum is *not* typical of what one would expect from a purely nondeterministic process with a smooth spectrum. Similarly, attempts to match the distribution of this spectrum estimate with a central χ^2_{20} were unsuccessful, but a mixture of central and noncentral distributions worked well.

4.3 Global Variance Properties of Spectrum Estimates

In this section we examine the "global" variability of a multitaper spectrum estimate, that is the variability over wide frequency ranges as opposed to the detailed behaviour within the local band $(-W, W)$. Everyone is familiar with the extreme variability of a periodogram or single-window direct estimate, and thus one's expectations of a spectrum are not high. We have, however, already noted in section 4.1, that the variance of a multitaper spectrum estimate is *much* lower than that of a single direct estimate, and this decreased variance can be used to check whether a measured spectrum is varying within "reasonable" bounds or whether one should suspect something beyond the expected level of random fluctuations.

The basic measures used are the autocorrelations and spectrum *of the*

---

[18] Above the 99% point. Even at the resolution plotted here, some double peaks cannot be easily seen.

[19] The simple fact that there are *twelve* peaks above the 99% point in the lower half of the sample with an effective sample size of 112 implies that the data is unusual.

estimated spectrum. These, as usual, are evaluated for a complex white, stationary process.[20]

4.4   Covariance of Multitaper Estimates

A common experience in spectrum estimation is to observe apparently recurrent "patterns" in the estimate across frequencies. (Such correlations are often demonstrated graphically, with the "echelle" plots used in helioseismology, Libbrecht *et al.* (1990) being a good example.) In common stationary processes, one does not expect multitaper estimates to be correlated at frequencies separated by more than $2W$, so the presence of such correlations usually indicates more complicated behaviour. Over time I have encountered several examples of such processes:

1. Manufacturing processes where rotating machinery is used. Frequently the product, waveguides, Thomson (1977), or optical fibers, Brener *et al.* (1998), will contain many components that are almost periodic because rollers and belts slip slightly, bearings precess, the product stretches nonuniformly, etc. These components are frequently equally-spaced in frequency, but offset from the frequency origin so that the carrier frequencies are not necessarily integer multiples of each other.

2. Processes associated with the sun, including space physics data, Thomson *et al.* (1995) and climate data, Kuo *et al.* (1990), Thomson (1995), Thomson *et al.* (1998). Solar activity appears to be dominated by modal structure on time scales ranging from thousands of years, Thomson (1990a), to the well-known 5-minute p-modes.

3. Seismic data: Park *et al.* (1987b), Mellors *et al.* (1996, 1998), Thomson and Vernon (1998).

4. Economic time series, *e.g.*, Ramsey and Thomson (1999).

5. Astronomical data, Thomson and Schild (1993, 1997). Here the presence of a frequency offset between the two components of a gravitionally-lensed QSO raises interesting questions.

6. Plasma data from a tokamak, Riedel *et al.* (1994a,b).

---

[20]The analysis can be carried through for a general spectrum, but the details require considerable space and are not particularly enlightening.

To establish a baseline for such correlations, begin with the simple multitaper estimate (2.4)

$$\widehat{S}(f) = \frac{1}{K} \sum_{k=0}^{K-1} \sum_{n,m=0}^{N-1} x(n)x^*(m)\, v_n^{(k)}\, v_m^{(k)} e^{-i2\pi(n-m)f}$$

and define the covariance

$$
\begin{aligned}
\Upsilon(\Delta) &= \mathbf{E}\{\widehat{S}(f)\widehat{S}^*(f-\Delta)\} &\qquad (4.7)\\
&= \frac{1}{K^2}\mathbf{E}\{x(n_1)x^*(m_1)x^*(n_2)x(m_2)\}\, v_{n_1}^{(j)}\, v_{m_1}^{(j)}\, v_{n_2}^{(k)}\, v_{m_2}^{(k)}\\
&\quad \cdot\ e^{-i2\pi(n_1-m_1)f} e^{+i2\pi(n_2-m_2)(f-\Delta)}
\end{aligned}
$$

Expanding the fourth moments for complex Gaussian data, Miller (1974), the variance term is $\mathbf{E}\{x(n_1)x^*(n_2)\}\mathbf{E}\{x^*(m_1)x(m_2)\}$ which, under the assumed white noise, gives $n_1 = n_2 = n$ and $m_1 = m_2 = m$. Using the definition (7.2) of Thomson (1982)

$$\Lambda_{jk}(\Delta) = \sum_{n=0}^{N-1} v_n^{(j)}\, v_n^{(k)} e^{i2\pi n\Delta}$$

and using tr for trace, the covariance becomes

$$\Upsilon(\Delta) = \frac{1}{K^2} tr\{\Lambda(\Delta)\Lambda^{\dagger}(\Delta)\}. \qquad (4.8)$$

Figure 3 shows $\Upsilon(\Delta)$ for $NW = 4.5$ and $K = 8$ as a typical case. The covariance drops by 4 orders of magnitude once a bandwidth of $2W$ is exceeded. Thus, when strong correlations are *observed* in a multitaper spectrum, one should suspect the data and not the spectrum estimate.

4.5 Spectra of Multitaper Spectrum Estimates

As shown in the previous section, multitaper estimates should not be correlated at frequencies spaced further apart than $2W$. Therefore, apparent correlations that are found at frequencies spaced further apart than $2W$ should be examined carefully. One method of checking for such systematic correlations is to simply take the spectrum, invariably "detrended" by prewhitening or some similar operation, consider it as a "time series", and compute its spectrum.[21] In Thomson (1977) the form of the *antespectrum* or "spectrum of the spectrum" was derived for a simple direct

---

[21]The argument exactly parallels the one for looking at spectra instead of autocorrelations in ordinary time series. It is usually easier to see effects in a spectrum.

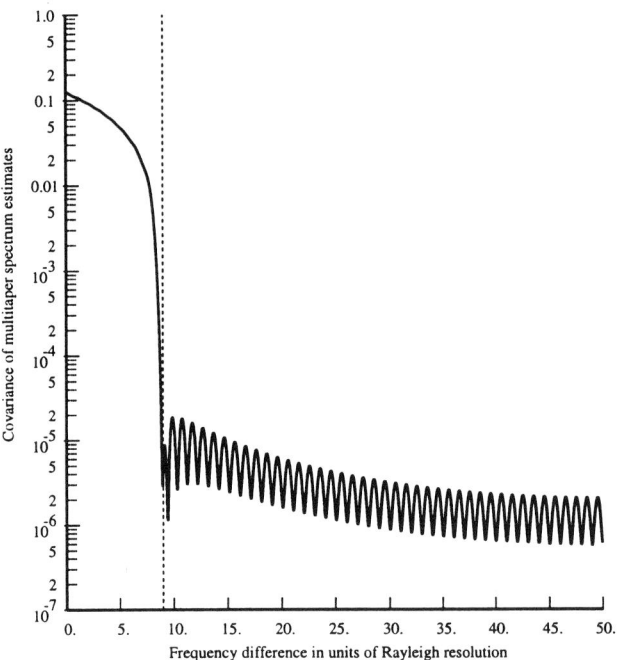

Figure 3: Theoretical covariance of multitaper spectrum estimates for $NW = 4.5$ and $K = 8$. The frequency axis is in units of Rayleigh resolutions with the vertical line at $2NW$.

estimate. However, as it predated the discovery of multitaper methods, the concept was of limited use but, as with most aspects of spectrum estimation, this has changed with multitaper estimates. What should one expect such a "spectrum of the spectrum" to look like?

Before considering this question in detail, it is necessary to distinguish clearly between three "Fourier transforms of the spectrum":

The autocorrelation, the Fourier transform of the *spectrum,*

$$R(\tau) = \int_{-\frac{1}{2}}^{\frac{1}{2}} S(f)e^{i2\pi f\tau} df$$

giving, by the Wiener-Khintchine theorem, Cox and Miller (1965) $\mathbf{E}\{x(t)x^*(t-\tau)\}$.

The cepstrum, the Fourier transform of the *log-spectrum,*

$$c_q = \int_{-\frac{1}{2}}^{\frac{1}{2}} \ln S(f)e^{i2\pi fq} df$$

discussed in Section 5.2 below.

The antespectrum, the spectrum of the *fluctuations* of a *spectrum estimate,* which is the subject of this section.

Conceptually, the antespectrum can be thought of as the spectrum of the fluctuations in a spectrum estimate. Beginning with the relative error

$$\delta S(f) = \frac{\widehat{S}(f)}{S(f)} - 1$$

define the antespectrum $\Xi(q)$ as

$$\Xi(q) = \mathbf{E}\left\{ \left| \int_{-\frac{1}{2}}^{\frac{1}{2}} \delta S(f) e^{i2\pi f q} df \right|^2 \right\}. \tag{4.9}$$

For white, complex data this becomes

$$\Xi(q) = \frac{1}{K^2} \sum_{n=0}^{N-1-|q|} \left| \sum_{j=0}^{K-1} v_n^{(j)} v_{n-|q|}^{(j)} \right|^2, \tag{4.10}$$

the Fourier transform of (4.8). $\Xi(q)$ is zero for $|q| > N - 1$, that is, the antespectrum is "bandlimited".

To re-iterate, both the autocorrelation and the cepstrum are functions of the (true) spectrum, and one tries to estimate them. The antespectrum is the spectrum of the *errors* in the estimated spectrum.

Figure 4 shows both the theoretical antespectrum, $\Xi(q)$, (4.10), and an estimate, $\widehat{\Xi}(q)$, for $\log_{10}$ of the spectrum estimate shown in Figure 2 above. The antespectrum was, perhaps obviously, estimated using another multitaper spectrum estimate, complete with a harmonic F-test. There is general agreement, but there are some distinct differences, for example: First, the empirical antespectrum is significantly larger than expected in a band 2.2 to 2.3 c/μHz. This corresponds to periods *in frequency* of 454 to 435 nHz, close to the solar sidereal rotation frequency of 456 nHz, Seidelmann (1992). This is approximately the splitting expected (between the $m = 1$ and $m = -1$ singlets) for $l = 1$ solar gravity modes, Pallé (1991). Second, there are some strongly periodic terms. Of these, the strongest is at 9.230 c/μHz, corresponding to a periodic structure in the spectrum at about 108 nHz. The source of this is unknown, but may be related to mode splitting. Third, the discrepancy between observed and expected at high frequencies is likely a result of

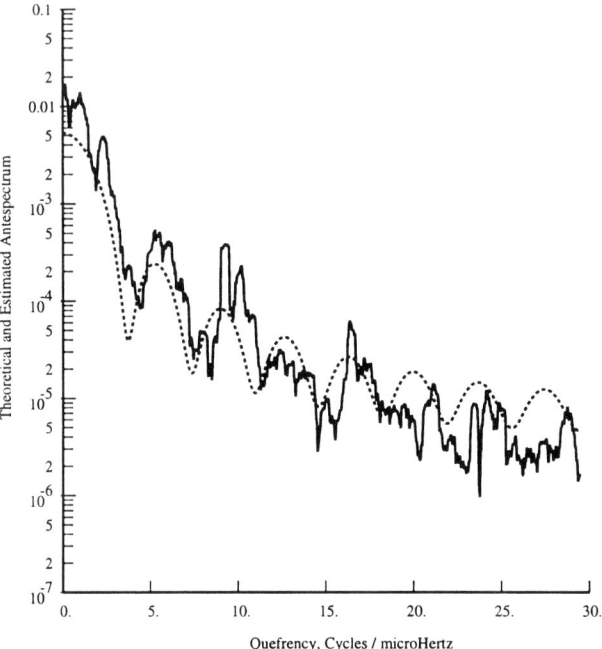

Figure 4: Theoretical (dotted) and estimated (solid) antespectra for the portion of spectrum shown in Fig. 2. The empirical curve is estimated from $\log_{10}(S)$ here with $NW = 4.5$ and $K = 8$. The F-test for the prominent line at 9.230 c/μHz is 47.5, has a nominal significance of 99.9999% at 2 and 14 df.

the adaptive weighting reducing the effect of the higher-order windows. Examining Figure 2, one can see that if the bandwidth were as large as 0.5 μHz, the peaks would not be resolved, and the estimated spectrum would have a mixture distribution. Keeping the same time-bandwidth product, $B = 0.5\,\mu$Hz requires at least 3300 hours of data.

In applying these methods, bear in mind that the spectrum is being computed on δS, not S. The Fourier transform of the latter is just the sample autocovariance, hence unlikely to be terribly informative. Moreover, the integration over the full frequency range in (4.9) should not be taken too seriously as it was done simply for analytic convenience and, as was implicit in the example, different parts of the frequency range may have different properties. Spectra derived from measurements of physical processes cover many orders-of-magnitude in both frequency and amplitude and, when used as data, should be expected to be nonstationary.

4.6 Local Variance of Spectrum Estimates

With a simple multitaper estimate where the "locally white" hypothesis is true, the eigencoefficients are uncorrelated, so $\widehat{S}$ defined in (3.38) is just the sum of K uncorrelated direct spectrum estimates. Each is χ_2^2, so the sum is χ_{2K}^2 and the variance is

$$\mathbf{Var}\{\widehat{S}(f)\} = \frac{S^2(f)}{K}$$

However, when the spectrum is varying within $(-W, W)$ the situation is more complicated. To describe the relation between eigencoefficients, define the complex vector $\mathbf{X}(f)$ of ideal eigencoefficients (3.21)

$$\mathbf{X}(f) = [x_0(f), x_1(f), \ldots, x_{K-1}(f)]^T \tag{4.11}$$

and the eigencoefficient covariance, or *ecco*, matrix

$$\mathbf{C} = \mathbf{E}\{\mathbf{X}(f)\mathbf{X}^\dagger(f)\} \tag{4.12}$$

with elements

$$C_{jk}(f) = \mathbf{E}\{x_j(f)x_k^*(f)\} . \tag{4.13}$$

The ecco matrix describes the frequency-localized structure of the process. In stationary processes, it contains spectral shape; in nonstationary processes, the temporal evolution for that frequency band.

If the ideal, low-resolution multitaper spectrum estimate, parallelling (3.38), is defined by

$$\check{S}(f) = \frac{1}{K} \sum_{j=0}^{K-1} |x_j(f)|^2 , \tag{4.14}$$

one can expand the variance

$$V = \mathbf{Var}\{\check{S}(f)\} = \mathbf{E}\{(\check{S} - S)^2\}$$

by substituting (4.14) and expanding the square. Using the fourth-moment formula for complex Gaussian processes, the variance term is

$$V = \frac{1}{K^2} \sum_{j,k=0}^{K-1} \mathbf{E}\{x_j x_k^*\}^2 = \frac{1}{K^2} tr\{\mathbf{CC}^\dagger\} \tag{4.15}$$

or

$$V = \frac{1}{K^2} \sum_{j,k=0}^{K-1} \mathbf{C}_{jk}\mathbf{C}_{kj} \; .$$

For stationary processes, the covariance of x_j and x_k is

$$\mathbf{C}_{jk} = \int_{-W}^{W} S(f-\xi)\,\mathcal{V}_j(\xi)\,\mathcal{V}_k^\star(\xi)d\xi \; . \tag{4.16}$$

To begin the detailed theory, I consider a simple, but useful example to motivate the more general quadratic-inverse theory considered later. This example is simply a brief examination of the effects of the first derivative of the spectrum on the estimated spectrum and a stable method to estimate the derivative. Here we assume that the spectrum is simple enough that it is well approximated by the first two terms of a Taylor's series

$$S(f+\xi) \approx S(f) + S'(f)\xi \tag{4.17}$$

for $|\xi| < W$. It is useful to think of this expression as an orthogonal expansion of the spectrum with basis functions 1 and ξ on $f - W < \xi < f + W$, and the coefficient of ξ as an "average" derivative rather than a strict point estimate. Using the Taylor's series in the ecco matrix (4.16), define the basis matrix corresponding to the pth power of ξ as

$$\mathbf{F}_{jk}^{(p)} = \int_{-W}^{W} \frac{\xi^p}{p!}\,\mathcal{V}_j(\xi)\,\mathcal{V}_k^\star(\xi)d\xi \; . \tag{4.18}$$

By the orthonormality of the Slepian functions, $\mathbf{F}^{(0)} = \mathbf{I}$ while symmetry gives $\mathbf{F}_{jj}^{(1)} = 0$, so $tr\{\mathbf{F}^{(1)}\mathbf{F}^{(0)}\} = 0$, *i.e.*, the basis matrices are *trace-orthogonal*. The covariance corresponding to (4.16) is

$$\mathbf{C} = S(f)\cdot\mathbf{F}^{(0)} + S'(f)\cdot\mathbf{F}^{(1)}$$

so $\mathbf{C}$ is written as a series of basis matrices with scalar coefficients. This is similar to the the expansions used in components of variance, Malley (1986), except that here the coefficients are not required to be positive and, in cross-spectra, can be complex. Returning to the variance, (4.15) becomes

$$V = \frac{1}{K}S^2 + \frac{(S')^2}{K^2}tr\{\mathbf{F}^{(1)}\mathbf{F}^{(1)\dagger}\} \; .$$

Substituting (4.18) in the trace formula, identifying the two sums of products of Slepian sequences with a Mercer's theorem expansion, see

Smithies (1962) and (A 17) of Thomson (1990b), and approximating the resulting double integral gives

$$\mathcal{D}_1 = \frac{tr\{F^{(1)}F^{(1)\dagger}\}}{K} \approx \frac{4}{15}\frac{\sqrt{6}}{\pi}\frac{2NW}{K}\cdot W^2 \quad . \tag{4.19}$$

The standardised variance is thus

$$\frac{K}{S^2}\mathbf{Var}\{\check{S}\} = 1 + \mathcal{D}_1 \cdot \left(\frac{S'(f)}{S(f)}\right)^2 . \tag{4.20}$$

The first derivative term does not change the expected value of the estimate, but increases the variance by a term proportional to $(\frac{\partial}{\partial f}\ln S(f))^2$. If one estimates the df, ν, by the moment formula using the average $\mathcal{S}$ and variance, V, of a χ^2, $\widehat{\nu} = 2S^2/V$, one obtains

$$\widehat{\nu} = \frac{2K}{1 + \mathcal{D}_1(\frac{\partial}{\partial f}\ln S(f))^2} \tag{4.21}$$

so that the effective df of the estimate are reduced.

This has been, so far, a purely probabilistic exercise, but it may be made more useful. Replace the idealized eigencoefficients $x_k(f)$ by an estimate $\widehat{x}_k(f)$ and form the vector $\widehat{\mathbf{X}}(f)$ (3.37), and a rank-1 estimate $\widehat{\mathbf{C}} = \widehat{\mathbf{X}}(f)\widehat{\mathbf{X}}^\dagger(f)$ of $\mathbf{C}(f)$. Consider

$$\widehat{\mathbf{C}}(f) \approx \tilde{S}(f)\cdot F^{(0)} + \widehat{S}'(f)\cdot F^{(1)}$$

and minimize the Frobenius norm of the approximation error as a function of $\tilde{S}(f)$ and $\widehat{S}'(f)$, that is, take

$$\min \|\widehat{\mathbf{X}}(f)\widehat{\mathbf{X}}^\dagger(f) - \tilde{S}(f)F^{(0)} - \widehat{S}'(f)F^{(1)}\|_F^2 .$$

This estimate of the spectrum, $\tilde{S}(f)$, is

$$\tilde{S}(f) = \frac{1}{tr\{F^{(0)}F^{(0)\dagger}\}}\widehat{\mathbf{X}}^\dagger(f)F^{(0)\dagger}\widehat{\mathbf{X}}(f) .$$

Recalling $F^{(0)} = \mathbf{I}$, this is the usual multitaper estimate, (3.38), that is, $\tilde{S}(f) \equiv \widehat{S}(f)$. Similarly, the estimate of $S'(f)$ is

$$\widehat{S}'(f) = \frac{1}{tr\{F^{(1)}F^{(1)\dagger}\}}\widehat{\mathbf{X}}^\dagger(f)F^{(1)\dagger}\widehat{\mathbf{X}}(f). \tag{4.22}$$

When the spectrum is locally white, the derivative estimate, $\mathbf{E}\{\widehat{S}'(f)\} = 0$ and $\mathbf{Cov}\{\widehat{S}(f), \widehat{S}'(f)\} = 0$. On first trying this estimate I was pleasantly surprised to find that it was reasonably stable, contrary to common experience with numerical derivatives of estimated quantities.

This example shows that it is possible to extract information about the spectrum *within* the band $(f - W, f + W)$ and not just the average power level, $\widehat{S}(f)$. It also suggests the existence of an orthogonal basis on the local frequency domain with a corresponding trace-orthogonal basis matrices for the ecco matrix. These give stable, uncorrelated estimates of the coefficients that are quadratic forms in the eigencoefficients.

Direct continuation of this approach by a simple Taylor series is rather unrewarding because the second and higher derivative terms do not have the trace-orthogonality property so one rapidly becomes immersed in algebra. One can replace the Taylor's series with a polynomial expansion $\check{P}_p(\xi)$ defined by orthogonalizing the trace products. Beginning with $\check{P}_0(\xi) = 1$ and $\check{P}_1(\xi) = \xi$, define $\check{P}_2(\xi) = \xi^2 + b$ and determine b by the requirement that $tr\{\mathrm{F}^{(2)}\mathrm{F}^{(0)}\} = 0$ where $\mathrm{F}^{(p)}$ is now defined by replacing the term ξ^p in (4.18) with $\check{P}_p(\xi)$. It is possible to do such a Gram-Schmidt-like procedure, but it has disadvantages: first, for reasons given in the following section, it soon runs into numerical difficulties and, second, the polynomials $\check{P}_m(\xi)$ are not orthogonal, so direct expansions of the spectrum are more difficult. There is, moreover, no obvious reason why such an *ad-hoc* procedure should produce statistically attractive results. Attempts to rectify these shortcomings led to quadratic-inverse theory, the subject of the following section.

4.7 Quadratic Inverse Theory

Returning to the problem of describing the covariance between eigencoefficients (4.12), we replace the polynomial expansions of the preceding example with an eigenvalue expansion. A method to do this was described in Thomson (1990b) and is briefly as follows. Beginning with the projection kernel implicit in (3.20) and (3.21)

$$\mathrm{P}_K(\xi, \eta) = \sum_{k=0}^{K-1} \mathcal{V}_k(\xi)\, \mathcal{V}_k^\star(\eta) \qquad (4.23)$$

define the set of real orthogonal functions $B_m(f)$ on $(-W, W)$ as the solutions of the integral equation

$$g_m B_m(\xi) = \int_{-W}^{W} |\mathrm{P}_K(\xi, \eta)|^2\, B_m(\eta) d\eta \qquad (4.24)$$

ordered by the eigenvalues $g_0 \geq g_1 \geq \cdots \geq g_{2K-1} \geq 0$ and normalized by

$$\frac{1}{2W} \int_{-W}^{W} B_l(f)B_m(f)df = \delta_{lm} . \tag{4.25}$$

A short digression on the properties of these functions is in order. First, for the usual multitaper estimates, the eigenvalues λ_k are all close to 1 so that $P_K(f, g)$ is approximately the Dirichlet kernel. The kernel of (4.24) is thus similar to the Fejér kernel, Stoica and Moses (1997), and the eigenfunctions B_l like those of

$$\tilde{g}_l \tilde{B}_l(\xi) = \int_{-W}^{W} \left[\frac{\sin N\pi(\xi - \eta)}{\sin \pi(\xi - \eta)} \right]^2 \tilde{B}_l(\eta)d\eta . \tag{4.26}$$

Because its kernel is degenerate there are $2K \approx 4NW$ non-zero eigenvalues of (4.24). For the asymptotic form (4.26) there are still about $4NW$ non-zero eigenvalues and these tend to

$$g_l \rightarrow 2NW - \frac{l}{2}, \ l = 0, 1, \ldots, 4NW . \tag{4.27}$$

Properties of these eigenvalues are similar to those of $sinc^2$ available in the optics literature where equation (4.24) characterizes incoherent imaging. For the $sinc^2$ kernel it is known, (Gori, 1974, Gori and Palma, 1975), that the eigenvalues are non-degenerate and the eigenfunctions are complete. The eigenvalues, $\tilde{g}_l$, of (4.26) have an exponentially vanishing tail starting at $l \approx 4NW$ that explains the numerical problems mentioned in the previous section.

Returning to the properties of the spectrum on the inner band, replace (4.17) with an expansion using the $B_m(\xi)$'s in place of $\xi^p/p!$

$$S(f \ominus \xi) \approx \sum_{m=0}^{L-1} b_m(f)B_m(\xi) \tag{4.28}$$

where $L \leq 2K - 1$ and $\ominus$ is used to indicate the restriction $|\xi| < W$. The standardization (4.25) of the functions $B_m(f)$ was chosen so the resulting expansion coefficients

$$b_m(f) = \frac{1}{2W} \int_{-W}^{W} S(f - \nu)B_m(\nu)d\nu$$

are directly comparable in magnitude with S. Substituting in (4.16) gives an expression for $\mathbf{C}(f)$ as a sum of basis matrices with scalar coefficients

$$\mathbf{C}(f) \approx \sum_{m=0}^{L-1} b_m(f)\mathbf{B}^{(m)} . \tag{4.29}$$

Here the $K \times K$ Hermitian basis matrices, $\mathbf{B}^{(m)}$ are given by

$$B_{jk}^{(m)} = \int_{-W}^{W} \mathcal{V}_j(\xi)\, \mathcal{V}_k^\star(\xi) B_m(\xi) d\xi \qquad (4.30)$$

and have the desired property that

$$tr\{\mathbf{B}^{(l)}\mathbf{B}^{(m)\dagger}\} = g_m \delta_{l,m} . \qquad (4.31)$$

The $\mathbf{B}^{(m)}$'s are real symmetric for m even and imaginary skew-symmetric for m odd.

If $S(f)$ is constant over the interval $(f - W, f + W)$, the diagonal elements of $\mathbf{C}(f)$ are the different eigenspectra, $\mathbf{E}\{|x_k(f)|^2\}$, and

$$\mathbf{C}(f) = S(f) \cdot \mathbf{I}$$

where $\mathbf{I}$ is the $K \times K$ identity matrix. Similarly, if W is chosen such that the spectrum varies significantly across the band $(f - W, f + W)$, $\mathbf{C}(f)$ will not be diagonal.

An estimate of $\mathbf{C}(f)$

$$\tilde{\mathbf{C}}(f) \approx \sum_{m=0}^{L-1} \hat{b}_m(f) \mathbf{B}^{(m)} \qquad (4.32)$$

by minimizing the Frobenius norm between $\widehat{\mathbf{C}}$ and $\tilde{\mathbf{C}}$ using (4.26) gives the estimated expansion coefficient

$$\hat{b}_m(f) = \frac{1}{g_m} tr\{\widehat{\mathbf{C}}(f) \mathbf{B}^{(m)\dagger}\}$$

that, for a single series, becomes the standard quadratic form

$$\hat{b}_m(f) = \frac{1}{g_m} \widehat{\mathbf{X}}^\dagger(f) \mathbf{B}^{(m)\dagger} \widehat{\mathbf{X}}(f) \qquad (4.33)$$

where $\widehat{\mathbf{X}}(f)$ is as defined in (3.31).

Empirically, the estimated b_l's are stable and their variance and distributions are of interest. Taking the expected value of (4.33) via (4.16) and assuming that the spectrum in $(f - W, f + W)$ may be expressed by the expansion (4.28) the trace-orthogonality (4.31) gives

$$\mathbf{E}\{\hat{b}_l(f)\} = b_l(f)$$

for $g_l > 0$ so that the estimate (4.33) is unbiased. To compute the variances, assume complex Gaussian statistics, expand the fourth moment (see page 82 of Miller, 1974), and discard the product of mean-value terms to obtain the expressions given in Thomson (1990b). Paralleling the development of the previous section beginning with (4.15), one has

$$V = \frac{1}{K^2} tr\{\mathbf{C}\mathbf{C}^\dagger\} = \frac{1}{K^2} \sum_{l=0} g_l b_l^2(f) . \tag{4.34}$$

If the spectrum is locally white, $\mathbf{E}\{x_h x_j^*\} = S\delta_{h,j}$, the remaining double sum is the trace defined in (4.31) and gives

$$\mathbf{Cov}\{\hat{b}_l, \hat{b}_m^*\} = \frac{S^2}{g_l}\delta_{l,m} .$$

Thus, when the spectrum is resolved, the estimated spectrum components obtained from (4.33) are uncorrelated. Because the eigenvalues, g, decrease slowly, the variances of the lower order b_l's are not significantly larger than that of $\widehat{S}$. As an example, Figure 5 shows a simulation example of a "frequency derivative", $\mathcal{C}_F b_1(f)/b_0(f)$, where the normalization constant, $\mathcal{C}_F$, used to scale $b_1(f)/b_0(f)$ to $\frac{\partial}{\partial f} \ln S(f)$ is approximately

$$\mathcal{C}_F \approx \frac{\pi}{\sqrt{2W}} .$$

Even for the perverse "gap AR-20" process used in this example the variance is relatively low, and the estimate follows the theoretical mean value.

5   Nonstationary Quadratic-Inverse Theory

The second simple case of harmonizable processes, a *non-stationary,* white, Gaussian process is defined by the covariance $\Gamma_\mathrm{L}(n, m)$, (3.2)

$$\Gamma_\mathrm{L}(n, m) = \mathbf{E}\{x(n)x^*(m)\} = D(n, f)\delta_{n,m} \tag{5.1}$$

where $D(t, f)$ is the dynamic spectrum (3.8) at time t. In this specific example, the "white" assumption implies that $D(t, f)$ is constant in f. A bandlimited version of such a process is recovered to within a bound of order $1 - \lambda_k$ by the usual eigencoefficients. In other words, the extreme band-limiting properties of the Slepian sequences make it almost irrelevant whether the white non-stationary model is valid globally, or

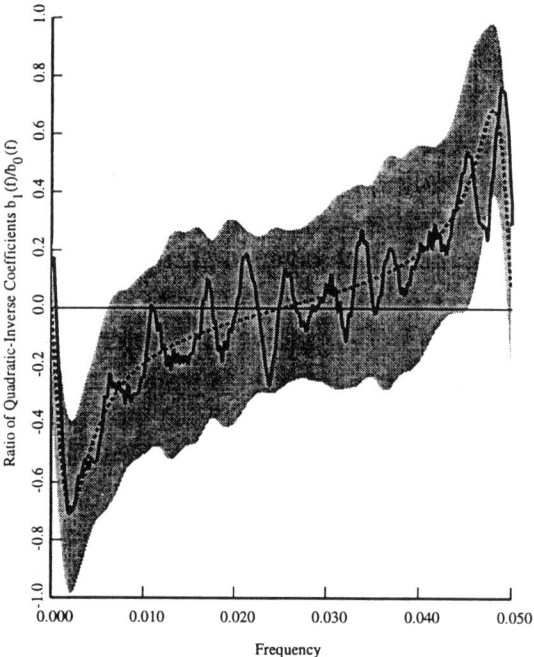

Figure 5: A simulation example of the ratio of the quadratic-inverse coefficients $b_1(f)/b_0(f)$, roughly the frequency derivative of the spectrum computed for a "gap AR-20" process, $x(t) = 0.85491x(t-20)+\eta(t)$ with η white pseudo-random noise. The estimate used $NW = 6$ and $K = 10$ tapers. The solid line is the average of 10 replications, the dotted smooth line the true value, and the shaded region is ± 1 empirical standard deviation about the true mean.

only locally within the frequency band $(f - W, f + W)$. To examine nonstationarity, it is thus convenient to use the complex demodulates (3.39) and also to expand D on a real orthogonal set of basis sequences. As in stationary quadratic-inverse theory, the basis functions giving these properties are the eigenfunctions of the squared kernel, in this case (3.12). The basis functions for the non-stationary quadratic inverse theory are the real orthonormal solutions, A_l, of the algebraic eigenvalue equation

$$\alpha_l A_l(n) = \sum_{m=0}^{N-1} \left[\frac{\sin 2\pi W(n-m)}{\pi(n-m)} \right]^2 A_l(m) . \qquad (5.2)$$

The basis matrices

$$A_{jk}^{(l)} = \sqrt{\lambda_j \lambda_k} \sum_{n=0}^{N-1} v_n^{(j)} v_n^{(k)} A_l(n) \qquad (5.3)$$

are real, symmetric, and trace-orthogonal,

$$tr\{\mathbf{A}^{(l)} \mathbf{A}^{(m)}\} = \alpha_l \delta_{l,m} . \qquad (5.4)$$

Again, there are $4NW$ such functions corresponding to significantly non-zero eigenvalues and time resolution is limited to essentially $1/4W$. Thus, corresponding to (4.24), we have

$$D(t, f) \approx \sum_{l=0} a_l(f) A_l(t) \qquad (5.5)$$

with the resulting expansion coefficients

$$a_l(f) = \sum_{t=0}^{N-1} D(t, f) A_l(t) . \qquad (5.6)$$

As before, one finds that the projection induces a corresponding set of covariance matrices on the eigencoefficients,

$$\mathbf{C}(f) = \sum_{l=0} a_l(f) \mathbf{A}^{(l)} . \qquad (5.7)$$

As before, $\mathbf{A}^{(l)}$ is a $K \times K$ matrix while $A_l(f)$ is the corresponding basis sequence. Taking a sample version of (5.7) given by

$$\widehat{\mathbf{C}}(f) = \sum_{l=0} \hat{a}_l(f) \mathbf{A}^{(l)} \qquad (5.8)$$

(5.4) gives the estimate

$$\hat{a}_l(f) = \frac{1}{\alpha_l} tr\{\widehat{\mathbf{C}}(f) \mathbf{A}^{(l)}\} = \frac{1}{\alpha_l} \widehat{\mathbf{X}}^\dagger(f) \mathbf{A}^{(l)} \widehat{\mathbf{X}}(f) \qquad (5.9)$$

as the standard quadratic form corresponding to (4.33). Again, like the stationary case, $\hat{a}_0(f) \approx \widehat{S}(f)$, but now $C_T \hat{a}_1(f) \approx \frac{\partial}{\partial t} D(t, f)$. The scale factor for the time-derivative of the spectrum is $C_T \approx \pi/(N\sqrt{2N})$.

Finally, note the difference between the stationary and non-stationary cases: the $\mathbf{A}^{(l)}$'s are symmetric, the $\mathbf{B}^{(l)}$'s Hermitian. In general the $\mathbf{A}$'s and $\mathbf{B}$'s are not orthogonal, but because $\mathbf{A}^{(0)} \approx \mathbf{B}^{(0)}$ and $tr\{\mathbf{A}^{(1)} \mathbf{B}^{(1)}\} =$

0, one can make uncorrelated estimates of S, S', and $\dot{S}$ (the "time derivative" of the spectrum) from a single data block. From an examination of these matrices, Partha Mitra[22] has suggested that

$$\partial_\phi S(f) = \Re\{e^{i\phi} \frac{1}{K-1} \sum_{k=0}^{K-2} \hat{x}_k(f)\hat{x}_{k+1}^*(f)\}$$

is an approximate estimate of the derivative of the spectrum in direction ϕ of the time-frequency plane.

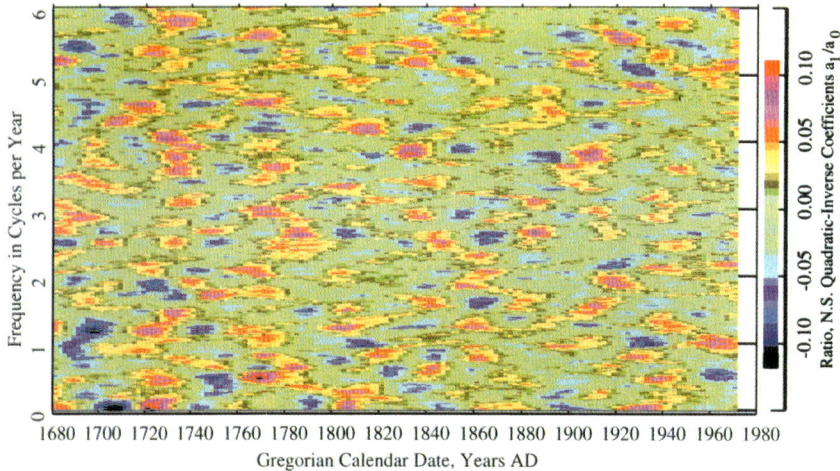

Figure 6: The time derivative of the log spectrum of the Central England temperature series, estimated by $\mathcal{C}_T\hat{a}_1(t,f)/\hat{a}_0(t,f)$. The process does not appear to be random. See Figure 10 following.

Continuing with the Central England example, Figure 6 shows the nonstationary quadratic-inverse derivative $\mathcal{C}_T\hat{a}_1(t,f)/\hat{a}_0(t,f)$, and it appears that the process has reasonably systematic nonstationarity.

5.1 A Stationarity Test

If a process is stationary, $D(t,f)$ should not depend on time. Consequently the expansion (5.5) should be equivalent to expanding a constant. Define the expansion coefficients of unity

$$e_l = \sum_{t=0}^{N-1} A_l(t) \tag{5.10}$$

---

[22]Personal communication.

and the mean power level

$$\overline{S}(f) = \frac{1}{N} \sum_{t=0}^{N-1} D(t, f) \, . \tag{5.11}$$

The mean power level corresponds to $\overline{S}(f)$. For a stationary process

$$\mathbf{E}\{\hat{a}_l(f)\} = e_l \overline{S}(f) \tag{5.12}$$

and

$$\mathbf{Cov}\{\hat{a}_l(f), \hat{a}_m(f)\} = \frac{1}{\alpha_l} \overline{S}^2(f) \delta_{l,m} \tag{5.13}$$

where the α_l's are the eigenvalues from (5.2). Thus the terms

$$\frac{[\hat{a}_l(f) - e_l \overline{S}(f)]^2}{\overline{S}^2(f)/\alpha_l} \tag{5.14}$$

should have unit variance and be approximately distributed as χ_1^2. Because the coefficients $\hat{a}_l(f)$ are uncorrelated and $\overline{S}$ is estimated, in the stationary case the sum

$$\Xi_T(f) = \sum_{l=0}^{L-1} \alpha_l \left[\frac{\hat{a}_l(f)}{\overline{S}(f)} - e_l \right]^2 \tag{5.15}$$

will have an approximate χ_{L-1}^2 distribution and can be used as a test for non-stationarity. This test is an improved version of that introduced in Thomson (1990b) for unresolved structure in the stationary version of quadratic-inverse theory. An example is shown in Thomson (1997).

5.2 Cepstrum Estimates

In this section, I discuss some multitaper estimates of the cepstrum and related quantities. These include some ideas on fitting cepstrum-like models, a new quadratic-inverse estimate of the cepstrum, and a nonstationary estimate of the time derivative of a cepstrum. The latter is of particular interest in the analysis of nonstationary processes because it is invariant to linear filtering operations. In contrast with section 4.5, where the spectrum of the *fluctuations* of an estimate about its mean value were considered, the emphasis here is on the overall structure of the process. The cepstrum was used by Wiener to factor the spectrum and get the optimum predictor for a discrete time series, Wiener (1949), was named by Tukey, Bogert *et al.* (1967), and is used routinely for speech and other signal processing chores, see Childers *et al.* (1977). The cepstrum is the Fourier transform of $\ln S(f)$. Here, one approximates the log-spectrum of a real process by the cosine series

$$\ln S(f) \approx c_0 + 2 \sum_{q=1}^{Q} c_q \cos 2\pi q f \qquad (5.16)$$

and determines the coefficients by minimizing the integrated mean-square error

$$e_c^2 = \int_{-\frac{1}{2}}^{\frac{1}{2}} \left[\ln S(f) - c_0 - 2 \sum_{q=1}^{Q} c_q \cos 2\pi q f \right]^2 df . \qquad (5.17)$$

This minimization is explicit in Wiener's method of spectral factorization Wiener (1949), Bhansali (1974), but in numerous applications, *e.g.*, Markel and Gray (1976), one simply computes the cepstrum coefficients

$$c_q = \int_{-\frac{1}{2}}^{\frac{1}{2}} \cos 2\pi f q \ln\{S(f)\} df \qquad (5.18)$$

as a formal algebraic step, see, *e.g.*, Schroeder (1981), in the process of deriving autoregressive or moving average representations. The prediction variance (4.1) is derived from the constant term, c_0, of a cepstrum expansion. The time-like or delay variable q is known as *quefrency.* In this process one is representing the spectrum as an exponential model, Bloomfield (1973),

$$S(f) \approx \exp\{c_0 + 2 \sum_{q=1}^{Q} c_q \cos 2\pi f q\} . \qquad (5.19)$$

As with other orthogonal expansions, one must choose either the truncation point, Q, or specific terms of the series. One uses the Bessel inequality, Smithies (1962), as a guide. Expanding the square in (5.17), one must choose enough coefficients to match the "energy" given by the first term

$$\mathbf{Q}_0 = \int_{-\frac{1}{2}}^{\frac{1}{2}} \ln^2\{S(f)\} df \qquad (5.20)$$

and. as the error, e_c^2, in (5.17) is non-negative, (5.18) gives

$$c_0^2 + 2\sum_{q=1}^{Q} c_q^2 \leq \mathbf{Q}_0 \ . \tag{5.21}$$

As mentioned above, the variance of $\ln \hat{S}$ is independent of frequency and given by $\psi'(K)$, thus one should choose Q so

$$c_0^2 + 2\sum_{q=1}^{Q} c_q^2 \approx \mathbf{Q}_0 - \psi'(K) \ .$$

A serious disadvantage of this recipe is that most of $\mathbf{Q}_0$ typically reflects the large dynamic range of spectra of physical processes and conveys little about the details of the spectrum. In addition, $\mathbf{Q}_0$ depends on the data units.

As an alternative approach, integrate (5.18) by parts to obtain

$$qc_q = -\frac{1}{2\pi}\int_{-\frac{1}{2}}^{\frac{1}{2}} \frac{S'(f)}{S(f)} \sin 2\pi q f df \ . \tag{5.22}$$

This method gives qc_q directly, a quantity long known empirically to be more meaningful than c_q itself, *e.g.*, Noll (1963). Parallelling the development of (5.20), one is now comparing the fit to

$$\mathbf{Q}_1 = \int_{-\frac{1}{2}}^{\frac{1}{2}} \left[\frac{S'(f)}{S(f)}\right]^2 df \tag{5.23}$$

that emphasizes the changes in $\ln S(f)$. Moreover, unlike $\mathbf{Q}_0$, $\mathbf{Q}_1$ is independent of the data units although it does depend on the sampling rate. Here the Bessel inequality becomes

$$\mathbf{Q}_1 \geq 8\pi^2 \sum_{q=1}^{Q} q^2 c_q^2 \ .$$

Here one chooses expansions in the B_l's to give minimum variance estimates of S'. Note that one uses $\frac{\widehat{S'}(f)}{\widehat{S}(f)}$ as a derivative estimate. One *does not* estimate S and differentiate it[23]. Assuming the variance terms to be small enough for a δ-expansion to be valid, one can show that

$$\mathbf{Var}\{\frac{\hat{b}_1}{\hat{b}_0}\} \approx \frac{1}{g_1} + \frac{1}{g_0}\left[\frac{S'}{S}\right]^2 \ .$$

---

[23] As noted earlier, I use $\mathcal{C}_F \hat{b}_1(f)/\hat{b}_0(f)$ as an estimate of $\frac{\partial}{\partial f} \ln S(f)$.

It is implicit here that $\mathbf{Var}\{\hat{b}_0^{-1}\}$ exists, thus, by the arguments of section 4.1, working with estimates with at least 10 df is advisable. Integrating over frequency one has the average variance

$$\overline{\mathbf{Var}}\{\frac{\hat{b}_1}{\hat{b}_0}\} = \frac{1}{g_1} + \frac{1}{g_0} \cdot \mathbf{Q}_1 \;.$$

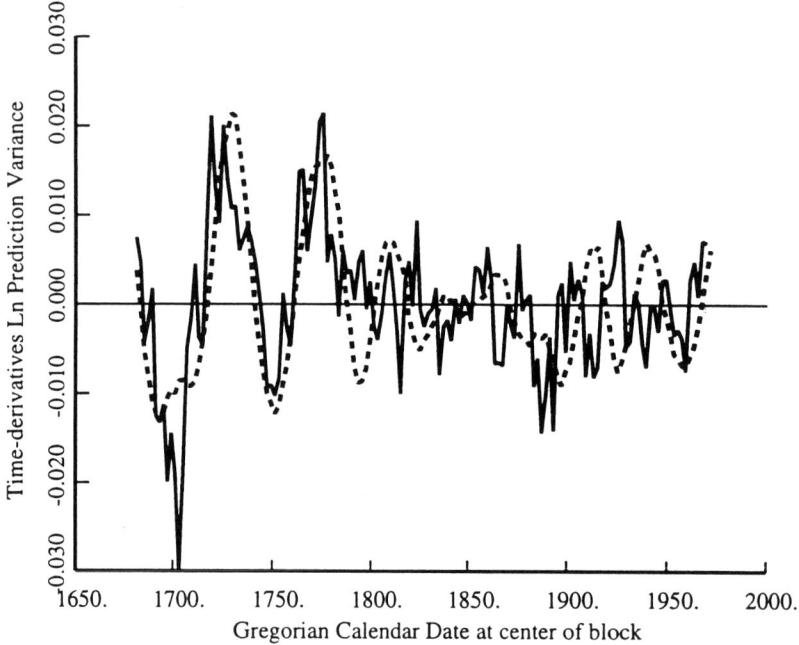

Figure 7: Two estimates of the time-derivative $\dot{c}_0(t)$ for the CET. The solid line is obtained by differencing the estimates $(\hat{c}_0(t+2) - \hat{c}_0(t))/2$ and the dashed line is the quadratic-inverse estimate (5.24). While the two estimates largely track, there are intervals, such as the first half of this century, where they do not.

In nonstationary processes, the usual cepstrum expansion must be, as it commonly is in applications, considered dynamic. Denoting the center of the estimation block by t_o, the c_q's become $c_q(t_o)$ and the time-derivatives $\dot{c}_q(t_o)$,

$$\dot{c}_q(t_o) = \int_{-\frac{1}{2}}^{\frac{1}{2}} \frac{\dot{S}(f)}{S(f)} \cos 2\pi q f \, df \qquad (5.24)$$

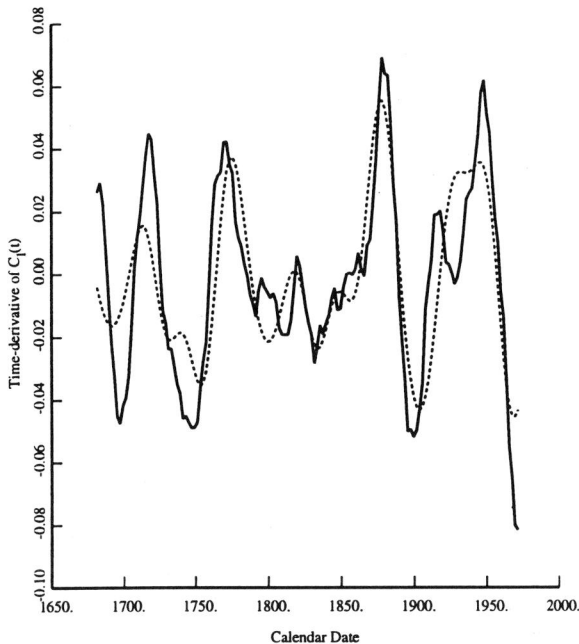

Figure 8: The $c_1(t)$ cepstrum coefficient of the CET. The dashed line is a fit by a trigonometric series with the 52, 104, and 208 year Suess periods. Because a good fit is obtained with few terms at pre-specified frequencies, I conclude that the changes are systematic.

where the time-derivative of the spectrum, $\dot{S}(f)$, is estimated by $\mathcal{C}_T \hat{a}_1(f)$. If the observed process $x(t)$ were the result of filtering some simpler process, say $z(t)$, through a linear, time-invariant filter with impulse response $h(t)$, so $x(t) = z(t) \star h(t)$ and $S_x(f) = |H(f)|^2 S_z(f)$ then the same $\dot{c}_q$ is obtained from observations of either x or z. Thus, within the limitations of observational noise, the $\dot{c}_q$'s are an invariant description of nonstationarity. Figure 7 shows two estimates of $\dot{c}_0(t)$ for the Central England temperature spectrum shown in Figure 6. The first is a simple numerical derivative of the estimate of $\hat{c}_0(t)$ shown in Figure 9 and the second the estimate (5.24) with $\frac{\partial}{\partial t} \ln S(f)$ estimated by $\mathcal{C}_T \hat{a}_1(t, f)/\hat{a}_0(t, f)$. The numerical derivative both takes more data (because it uses two data blocks) and is noisier. From Figure 7, it can be argued that the predictibility of the climate was more variable in the century following the Maunder Minimum[24] than it has been since about 1800. Arguments that this is simply

---

[24] A period between about 1645 to 1715 noted for few sunspots or aurorae.

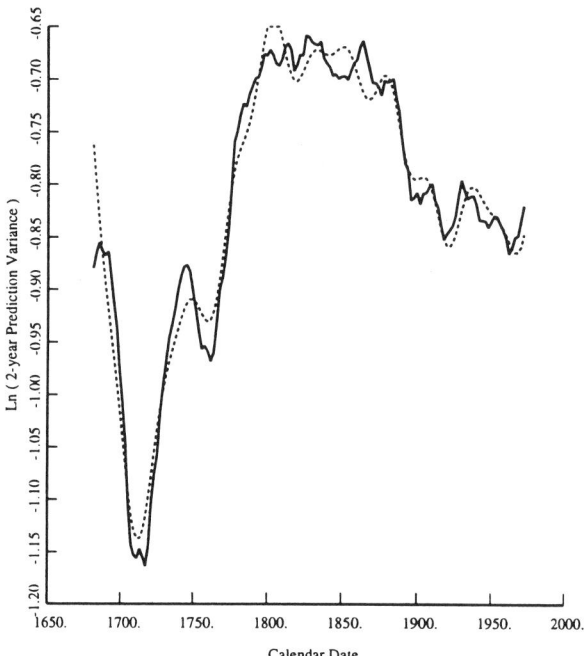

Figure 9: The log-prediction variance, $c_0(t)$ for the prewhitened Central England temperature series. The fit, shown by the dotted line, is to harmonics of the 104-year Suess line. It is of interest that the climate was more predictable in the early eighteenth century than it was in the nineteenth, or is presently.

a consequence of the older data being of poorer quality must account for the better absolute predictability of the early record seen in Figure 9 and that both $\hat{c}_0(t)$, Figure 9, and $\hat{c}_1(t)$, Figure 8, can be reasonably well-fit with a small set of periodic terms all simply related to the 104-year Suess line.[25]

From the known relations to compute predictors from the qc_q's, (Markel and Gray 1976; Schroeder 1981),

$$qa_q = qc_q - \sum_{k=1}^{q-1} kc_k a_{q-k} \qquad (5.25)$$

---

[25] "Suess wiggles" are a set of near periodic modulations of solar phenonema first identified in the ^{14}C flux by Hans Suess. The periods are simple multiples and submultiples of 103.9 years, Thomson (1990a), with 26, 34.6, 52, 69.3, 78.7, 208, and 416 years occurring commonly.

one may update the predictor using

$$\hat{c}_q(t) \approx c_q(t_o) + (t - t_o)\dot{c}_q(t_o)$$

and so obtain improved predictors.[26] Thus one *first* predicts the spectrum, then *second*, estimates a predictor from the predicted spectrum.

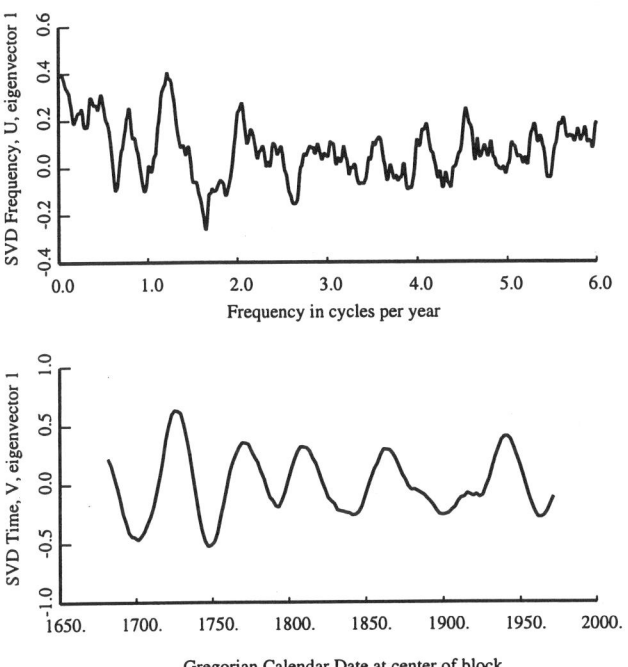

Figure 10: The first eigenvector pair from a singular value decomposition of $\mathcal{C}_T\hat{a}_1(t, f)/\hat{a}_0(t, f)$ shown in Figure 6. This eigenvector represents 14% of the variance of this signal, and, collectively, the first 12 eigenvectors account for 85%. Note, for example, the tendency of components with frequencies / 0.3c/y to behave in unison with those near 1.3 c/y.

In the CET example, like others, *e.g.*, Thomson (1990a), the nonstationarity is not "random", but reasonably systematic. The one-step prediction variance of the $c_1(t)$ coefficient shown in Figure 8 is only 3% of its variance when considered as a nondeterministic process. Because it contains several periodic terms at known (solar) frequencies, the nondeterministic prediction variance is about 0.74% of the variance. Another

---

[26]Note that the sign used for predictors in speech is opposite that used in the statistical literature and here.

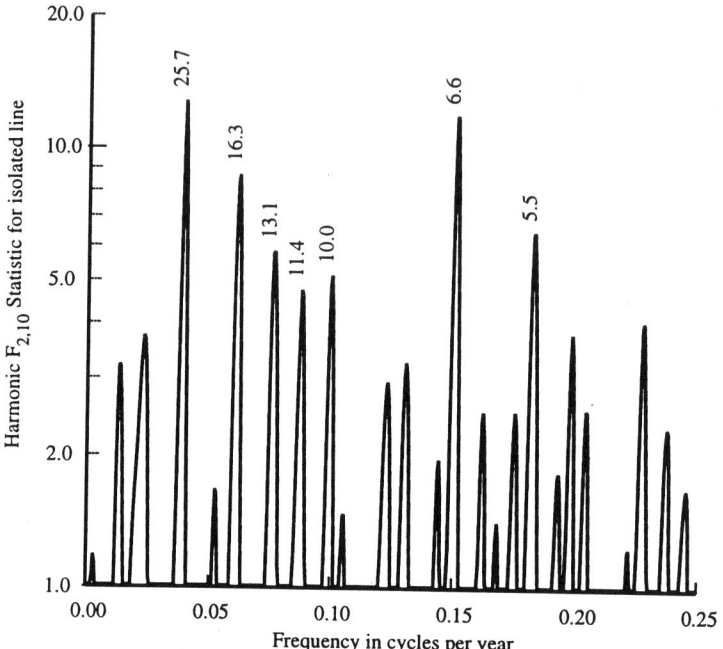

Figure 11: An F-test for periodic components in the first time eigenvector shown in Figure 10. The line frequencies appear to be harmonically related to each other and to the $\approx$ 104-year Suess line or the usual solar cycle.

useful procedure for examining such data is to compute a singular value decomposition (SVD) of the time-frequency matrix. Figure 10 shows the first left and right eigenvectors of the derivative spectrum shown in Figure 6, and Figure 11 is a harmonic F-test applied to the time eigenvector, with the periods of the larger line components marked. For a signal-to-noise power ratio ρ, the Cramér-Rao bound for the standard deviation of a period estimate is

$$sd\{\hat{\mathcal{P}}\} \approx \frac{\mathcal{P}^2}{2\pi T}\sqrt{\frac{6}{\rho}}$$

so the lines marked 25.7, 13.1, and 6.6 years are likely in a 1:2:4 frequency relationship and are probably related to the 104 year Suess line. Similar analysis applied to the cepstrum coefficients or to other dominant eigenvectors shows lines related either to the $\approx$ 104-year Suess line, the 11 and 22 year solar cycle and, occasionally, to the 18.6 year lunar cycle.[27] This

[27]An additional implication of the Cramér-Rao bound is that, when a fast Fourier transform (FFT) is being used to compute the eigencoefficients used in the harmonic

implies that much, perhaps most, of the differences between spectra computed from different data segments are not random, but closer to being deterministic. I must also emphasize that these periodic components are *not* simple additive terms, but periodic changes in the structure of the process. If one does a similar analysis on the ordinary sunspot numbers, the eigenvectors describe a process that switches between two dominant modes. The switching rate is 104 years.

6 High-resolution Spectrograms

Multitaper spectrograms have been in use since shortly after the invention of multitaper estimates, see Thomson (1990a), Lanzerotti *et al.* (1991), Kappus and Vernon (1991), Riedel *et al.* (1994b), Pitton (1998), Haykin and Thomson (1998). These were, for the most part, a standard spectrogram with the conventional spectrum estimate replaced by a multitaper one. Using the notation (3.42) with series p and q, one computes either an auto-spectrum, with $p = q$, or a cross spectrum

$$\bar{S}_{p,q}(b, f) = \frac{1}{K} \sum_{k=0}^{K-1} \hat{x}_{p,k}(b, f)\hat{x}_{q,k}^*(b, f) \tag{6.1}$$

at time $b + (N - 1)/2$ and frequency f or, alternatively, functions of the $\hat{x}_{p,k}(b, f)$'s such as the quadratic inverse time and frequency 'derivatives'. The estimate (6.1) clearly has a time resolution near N samples. This can be improved by about a factor of 2 by estimating the time derivative $\frac{\partial}{\partial t} \ln S(f)$ as in Section 5.2 and doing Hermite interpolation, Section 1.4 of Rivlin (1990), on $\ln S(f)$.

A different approach gives a positive estimate with good time resolution. Squaring the complex demodulate (3.39) and normalising gives an estimate of the power as a function of time,

$$\mathcal{F}(b \oplus t, f) = \frac{1}{K} |\hat{x}(t; b, f)|^2, \tag{6.2}$$

F-test, it must be extensively "zero padded". Specifically, if a M-point FFT is used with N data samples, $M \geq N$, for a signal-to-noise power ratio ρ, one must have

$$M > 2\pi \sqrt{\frac{\rho}{6}} \cdot N \tag{5.26}$$

for the frequency mesh spacing, $1/M$, to be as small as the statistical accuracy of the frequency estimate. Giving up a factor of two on accuracy is unattractive, so I use a FFT for initial line frequency estimates, then refine them with a search procedure.

where $\oplus$ denotes the restriction $0 \le t \le N - 1$. Integrating this distribution over a time block gives the conventional multitaper spectrogram
(6.1) as the frequency-marginal

$$\sum_{t=0}^{N-1} \mathcal{F}(b \oplus t, f) = \frac{1}{K} \sum_{k=0}^{K-1} |\hat{x}_k(b, f)|^2$$

and so is a more accurate distribution of power than other time-frequency
distributions. Similarly, integrating $\mathcal{F}(b \oplus t, f)$ over frequency, and approximating the $\hat{x}_k$'s with y_k's, (3.25), gives

$$\int \mathcal{F}(b \oplus t, f) df = \frac{1}{K} \left| \sum_{n=0}^{N-1} \sum_{k=0}^{K-1} v_t^{(k)} v_n^{(k)} \right|^2 |x(b+n)|^2$$

so the time-marginal is almost a convolution of $[\sin(2\pi W t)/(\pi t)]^2$ and
$|x(t)|^2$. The time resolution is $1/2W$. This estimate does not match the
common marginal conditions, Cohen (1995), but may be more realistic
given the large differences observed between waveforms of speech, seismic
(Vernon, 1989), geomagnetic, space physics, and other data collected
on sensors spaced short distances apart. Cross-terms are negligible for
components separated in frequency by more than $2W$, so the ambiguity
area is about 1.

The distribution of (6.2) is χ_2^2, but some stabilisation and reduction
of Gibb's ripples may be obtained by averaging the basic estimates (6.2)
over base positions b and internal block offsets t keeping $b + t$ constant.
Smoothing on both b and t can be chosen to give a smoothed estimate
$\check{\mathcal{F}}$ with stability between 2 and $2K$ df at the cost of the usual reduction
in time resolution.

The examples shown in this section use data from the *KNET*, Mellors *et al.* (1997), seismic network in Kyrgyzstan, specifically the vertical
components from stations AAK and EKS2. The specific event is a magnitude 5.1 earthquake located at 37.473° N, 59.857° E at a depth of 24km
that occurred at 21:09:16 on day 332 of 1992. The analog data had been
low-pass filtered to a 50Hz bandwidth and sampled at the corresponding
Nyquist rate of 100 Hz. The sampled data was further low-pass filtered to
a bandwidth of 2.5Hz (using a zero-phase symmetric 161-point transverse
digital filter) and resampled, or decimated, by keeping every twentieth
sample, to give an effective sampling rate of 5Hz. Such data is obviously
nonstationary, but it is not obviously non-Gaussian. In the Rayleigh
wave part of the signal, part of which is plotted in Figure 12, the sample
skewness is -0.02, and the standardized fourth moment is 2.86, close

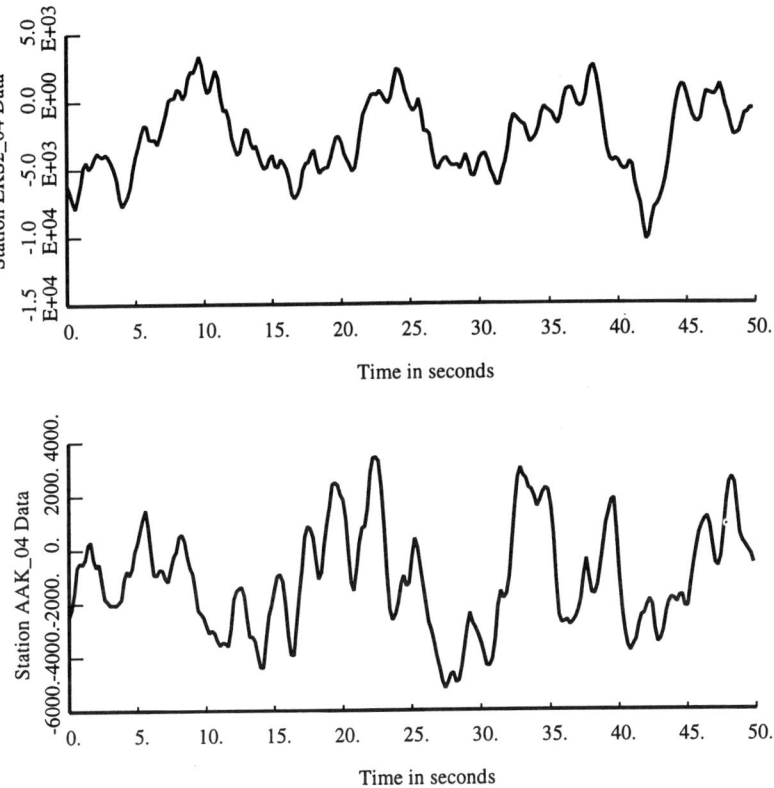

Figure 12: Data from stations EKS2 and AAK for the Ashgabat earthquake, on day 332 of 1992. The plotted data starts 315s after the P-wave arrival, with the AAK data advanced by 6.6s so the P-wave arrival times are aligned.

to the values expected for a Gaussian process. Consequently, we regard the data as Gaussian, making the problem here one of nonstationarity and not nonlinearity or nonnormality, so second moments are all that is required. In confirmation, estimates of the bispectrum, not shown, do not show significant three-frequency interactions.

Such a spectrogram is shown in Figure 13. In this example, the block length was 300 samples (60s), blocks were offset by 60 samples (12s), and the $\check{\mathcal{F}}(t, f)$ was evaluated every 10 samples. A time-bandwidth product of 4.5 was used with the first 7 Slepian sequences. The dynamic range of this spectrum is about 90 dB. The dispersive Rayleigh waves are between about 315 and 430s at low frequencies and, although a 'chirp' can be seen, it is not well resolved.

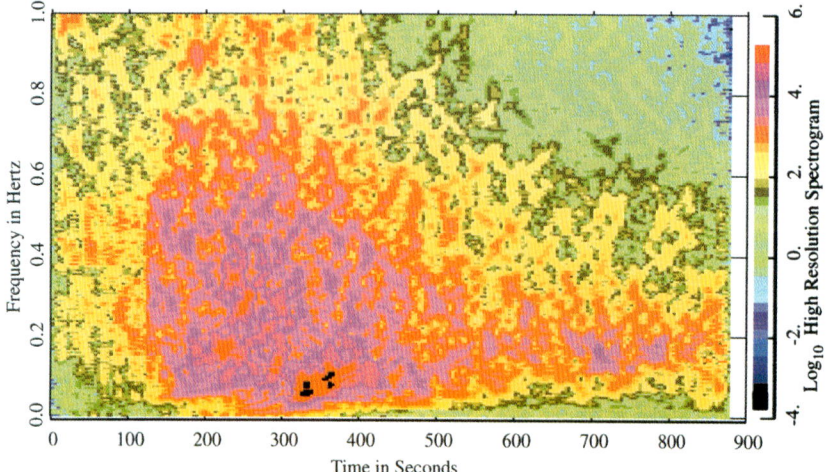

Figure 13: High-resolution multitaper spectrogram for 900 seconds of EKS2 data. The record used here starts slightly before the P-wave, and the section of data shown in Figure 12 extends from 320 to 370s in this plot. The arrivals of the P-wave near 5s and the S-wave at 130s are clear.

As mentioned in connection with (6.1), the spectrogram can be generalized to multivariate data and, in a quasi-stationary approximation, one could plot magnitude-squared coherence (MSC) and phase as a function of time and frequency. This approach, however, fails in the present example because Rayleigh waves are dispersive. As we shall see in the next section, the signals from the two stations contain highly coherent Rayleigh waves, but they are *not* at the same frequency. Thus we emphasize that, although spectrograms and other estimates such as those based on wavelets may satisfy the marginal conditions, they miss the essential feature of correlation between distinct frequencies.

7   Estimates of the Loève Spectrum

An approach that has given useful, and usually unexpected, information in several applications is to estimate the Loève, or dual-frequency, spectrum (3.4) directly. The estimate of $\gamma_{pq}(f_1, f_2)$ given in Thomson (1982) was

$$\widehat{\gamma}_{pq}(f_1, f_2) = \frac{1}{K} \sum_{k=0}^{K-1} \hat{x}_{p,k}(f_1)\hat{x}_{q,k}^*(f_2) \qquad (7.1)$$

and generally works well (Mellors *et al.* 1998; Schild and Thomson 1997; Haykin and Thomson 1998). This estimate is obtained from the estimate of dX given in (3.22) by a weighted average over $|\xi_1|, |\xi_2| < W$ around (f_1, f_2),

$$\iint_{-W}^{+W} d\widehat{X}_p(f_1 + \xi_1) d\widehat{X}_q^*(f_2 + \xi_2) \mathrm{H}(\xi_1, \xi_2) d\xi_1 d\xi_2$$

where H is the weight. Given the continuity arguments described in Section 3 or in Thomson (1998), and Haykin and Thomson (1998), an obvious choice is $\mathrm{H}(\xi_1, \xi_2) = \delta(\xi_1 - \xi_2)$ so that the stationary frequency is smoothed over a bandwidth $\pm W$, with no smoothing on the nonstationary direction. This gives the estimate (7.1) proposed in Thomson (1982) and is philosophically similar to Hurd and Gerr (1991), Gerr and Allen (1994).

It is useful to work with dual-frequency magnitude-squared coherences[28]

$$\widehat{C}_{pq}(f_1, f_2) = \frac{|\widehat{\gamma}_{pq}(f_1, f_2)|^2}{\widehat{S}_p(f_1)\widehat{S}_q(f_2)} \tag{7.2}$$

as their significance levels are easier to interpret. These coherence estimates have the same statistical properties as ordinary coherence estimates.

The reason for saying that this estimate responds to common signal elements at different frequencies is that, if one starts with the baseband complex demodulates (3.39), denoted $\hat{\mathbf{x}}_p(t; b, f)$ to include series and base-time identification, at two different frequencies and computes their covariance in the time domain

$$s_{pq}(f_1, f_2, b) = \frac{1}{K} \sum_{t=0}^{N-1} \hat{\mathbf{x}}_p(t; b, f_1) \hat{\mathbf{x}}_q^*(t; b, f_2) ,$$

the orthogonality of the Slepian sequences gives (7.1) with an additional λ_k included in the sum. As the λ_k's are close to 1 this is, in practice, the same. Applying this estimate to a 100-second block corresponding to the interval 320 to 420 seconds in Figure 13 (the first 50 seconds are shown in Figure 12) gives the magnitude-squared, dual-frequency coherence shown in Figure 14. For most frequencies there is very high coherence between the two series, *but not on the diagonal* $f_1 = f_2$. The average maximum

---

[28]Theory and examples of ordinary coherence estimates are in Thomson (1982), Vernon (1989), Kuo *et al.* (1990), Thomson and Chave (1991).

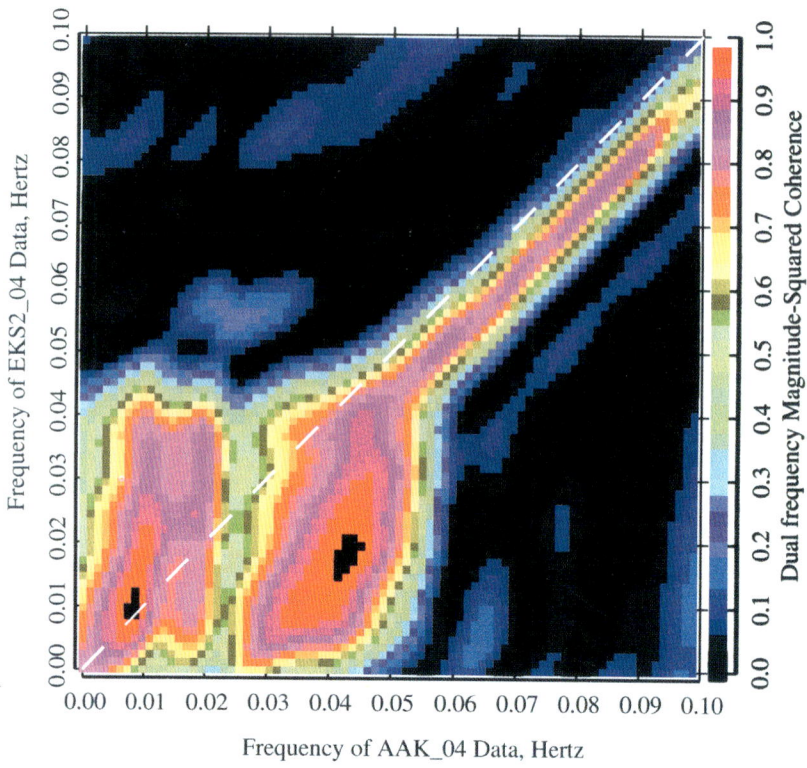

Figure 14: Dual-frequency coherence between the two data series in the Rayleigh wave region. Note that highest coherence is *not* on the diagonal.

coherence along vertical stripes in Figure 14 is 0.84 while the average along the diagonal is only 0.51.

There are numerous other applications of multitaper methods to seismology. Special windows that optimize performance with exponentially damped signals, Park *et al.* (1987b),[29] and combined time and space F-tests (Lindberg 1986; Lindberg and Park 1987) were developed for normal-mode seismology. Estimates of polarization and arrival direction, Park *et al.* (1987a), from a three-axis instrument have been made. What are effectively multitaper wavelets (Daubechies 1990; Lilly and Park 1995; Cvetković 1998) are also relevant, as are the inverse-theory reconstructions in Park (1992). Estimates of attenuation (Zhu *et al.*, 1989) and other geophysical quantities (Park *et al.* 1987c; Tauxe 1993;

---

[29]The same problem arises in magnetic resonance imaging problems and a different multitaper solution is given in (Johnson *et al.*, 1996).

Hinnov and Park 1988) have been made.

8 Multitaper Bispectrum Estimates

The following two sections address the "nonlinear" aspects of the time series problem implied by the title, this section applying to stationary processes, and the following making a start on the problem of testing for nonlinearity in the presence of nonstationarity. Higher-order spectra, or polyspectra, are used for the statistical characterization of stationary non-Gaussian stochastic processes, see *e.g.* Haubrich (1965) or Nikias and Raghuveer (1987). In many applications where the use of higher-order spectra is desirable, use of conventional methods is hampered by lack of data or, equivalently, by the evolutionary nature of the process. The multitaper complex demodulates (3.39) and their downsampled versions (3.41) can be used to compute consistent estimates of the bispectrum, (Thomson 1989; Mullins 1993; Birkelund and Hanssen 1999), and the procedure is extensible to polyspectra. A more useful extension is to *robust* estimates of bispectra.

Bispectra can be estimated from the multitaper eigencoefficients either by integrating over a three-dimensional element of the frequency domain or, equivalently, by a time-domain correlation. The former is preferred for analytic purposes, the latter computationally more efficient and, conceptually, easier to robustify. Either choice reduces to a weighted average of products of the eigencoefficients.

The bispectrum is defined for stationary processes as the third moment of dX. If f_1, f_2, and f_3 are three frequencies, then the bispectrum is defined by

$$B(f_1, f_2)df_1 df_2 = \mathbf{E}\{dX(f_1)dX(f_2)dX(f_3)\} \qquad (8.1)$$

for $f_1 + f_2 + f_3 = 0$, and zero otherwise. Corresponding to the Wiener-Khintchine theorem for second moments the bispectrum is an expansion of the third moments, Rao and Gabr (1984)

$$\mathbf{E}\{x(t)x(t+\tau_1)x(t+\tau_2)\} = \iint_{-\frac{1}{2}}^{\frac{1}{2}} B(f_1, f_2)e^{i2\pi(f_1\tau_1+f_2\tau_2)}df_1 df_2. \quad (8.2)$$

Unfortunately, while analytic properties of polyspectra are well known and, as shown by Hasselmann *et al.* (1965), are similar to second-order theory, statistical properties of estimates of polyspectra are not as well understood and do not resemble those of the second-order theory. The fundamental problem is that the estimate mimicking (2.1) is *anticonsistent*, that is its variance *increases* with sample size. For Gaussian data, the variance of the periodogram (2.1) is $1 \cdot \mathbf{E}\{P(f)\}^2$, independent of the sample size N, so the estimate is inconsistent.

The estimation analog of the periodogram, the *biperiodogram*,

$$P_2(f_1, f_2) = \frac{1}{N} y(f_1) y(f_2) y^*(f_1 + f_2) \tag{8.3}$$

is an asymptotically unbiased estimate of the bispectrum. The same cautions made about asymptotic unbiasedness of the ordinary periodogram apply here.

A minimum requirement for estimators of quantities such as the bispectrum is that they be "well behaved" when the data satisfies the null hypothesis of being stationary and Gaussian. Otherwise, needlessly complicated descriptions of the data may be invoked. For Gaussian data and distinct frequencies the variance of the biperiodogram is $N \cdot S(f_1) S(f_2) S(f_1 + f_2)$, that is, the variance is *proportional* to sample size, so the estimator is anticonsistent. This should not be surprising; one is, after all, attempting to estimate $O(N^2)$ quantities from N data samples.

The property of anticonsistency in an estimator suggests that one should find a better estimator. Because the variance of the simple biperiodogram is proportional to sample size, an obvious approach is to divide the sample into J smaller subsamples of length L, estimate a biperiodogram on each subsample, and average the results. The variance of the average is then proportional to L/J. The disadvantage of this approach is that frequency resolution is also reduced from $1/N$ to $1/L$. "Subdivide and average" estimates of the bispectrum have traditionally been calculated by three related methods: weighted Fourier transforms of sample third moments, Rao and Gabr (1984); frequency-domain averages, Lii *et al.* (1976), Alekseev (1997); and time averages of complex demodulates, Hinich and Clay (1968), Godfrey (1965), plus combinations of these. While these methods are asymptotically equivalent, their sample properties can be dramatically different and, as described in Huber *et al.* (1971), "all the pitfalls known from ordinary spectrum analysis occur here too, some of them with new twists." As usual, these problems are more serious when only a short data record is available, are exacerbated if the spectrum has a large dynamic range, has a complicated frequency dependence, or, especially, if the data is nonstationary. Moreover, estimates of higher moments are susceptible to both outliers and legitimate extreme values. These are far more common and cause more problems in the non-Gaussian data where polyspectral estimates are needed than they are with Gaussian processes, and may be partly responsible for the poor reputation of bispectra and the generally pessimistic conclusion in Brillinger (1965).

8.1 Multitaper Estimation of Bispectra

In the following we give two multiple window bispectrum estimates: the first is a time-average of products of complex demodulates and the second is an average over a cube in three-dimensional frequency space of products of the best local least-squares approximations to dX. In the former, the use of Slepian sequences allows one to generate a complex demodulate for the entire length of the series. It is consequently more efficient than a filter whose output sequence is shorter than the length of the original series by the duration of the filter's impulse response. In the latter, the three-dimensional integral is done subject to the usual bispectrum requirement that the three frequencies sum to zero, so the integration reduces to two dimensions.

8.1.1 "Time Average" Bispectrum Estimates

Consider estimating the bispectrum by averaging products of the complex demodulates (3.39) over time

$$\widehat{B}(f_1, f_2) = \frac{1}{\vartheta} \sum_{n=0}^{N-1} \hat{x}(n; f_1)\hat{x}(n; f_2)\hat{x}^*(n; f_1 + f_2) \qquad (8.4)$$

where ϑ is a normalizing constant defined below. Using representation (3.39) for the complex demodulates, this may be written in terms of the eigencoefficients as

$$\widehat{B}(f_1, f_2) = \frac{1}{\vartheta} \sum_{j,k,l=0}^{K-1} \hat{x}_j(f_1)\hat{x}_k(f_2)\hat{x}_l^*(f_1 + f_2)\mathrm{P}(j, k, l) \qquad (8.5)$$

where $\mathrm{P}(j, k, l)$ is the triple product of Slepian sequences

$$\mathrm{P}(j, k, l) = \sqrt{\lambda_j \lambda_k \lambda_l} \sum_{n=0}^{N-1} v_n^{(j)} v_n^{(k)} v_n^{(l)} . \qquad (8.6)$$

P has, via the eigenvalue equations and integral representations of the Slepian sequences, several integral representations. It is symmetric in its arguments and zero if $j + k + l$ is odd, so (8.6) is reasonably simple to compute. In practice, however, one uses the downsampled version of the $\hat{x}(n; f)$'s, the x° $(j; f)$'s of (3.41).

8.1.2 "Frequency Average" Bispectrum Estimates

Taking the inverse estimate of the orthogonal increment process (3.22) in the three bands and integrating over a volume element concentrated on $|\eta_j| < W$, another estimate of the bispectrum is

$$
\widehat{B}_f(f_1, f_2) = \iiint_{-W}^{W} d\widehat{X}(f_1 + \eta_1)d\widehat{X}(f_2 + \eta_2)d\widehat{X}(f_3 + \eta_3)C(\eta_1, \eta_2, \eta_3)
$$

(8.7)

where C is a symmetric weight function of its three arguments and $f_3 = -(f_1 + f_2)$. Substituting (3.22) in (8.7), and interchanging orders of integration and summation, this estimate may be written in the same form as (8.5), that is as a triple sum of the eigencoefficients times the integral of a weighted product of the three Slepian functions. Taking uniform weighting throughout the volume and integrating subject to the constraint shows the latter to be equivalent to (8.5).

To find the expected value of these estimates it is useful to write the eigencoefficients in terms of the integral representation (3.24) and substitute in (8.5) to obtain

$$
\widehat{B}(f_1, f_2) = \frac{1}{\vartheta} \sum_{j,k,l=0}^{K-1} P(j, k, l) \cdot \iiint_{-\frac{1}{2}}^{\frac{1}{2}} \mathcal{V}_j(\xi_1)\mathcal{V}_k(\xi_2)\mathcal{V}_l^{\star}(\xi_3)
$$
$$
\cdot d\widehat{X}(f_1 - \xi_1)d\widehat{X}(f_2 - \xi_2)d\widehat{X}^{\star}(f_1 + f_2 - \xi_3) .
$$

(8.8)

The expected value of the product of the three $d\widehat{X}$'s is, by the definition of the bispectrum (8.1),

$$
B(f_1 - \xi_1, f_2 - \xi_2)\delta(\xi_1 + \xi_2 - \xi_3)d^3\xi
$$

so

$$
\mathbf{E}\{\widehat{B}(f_1, f_2)\} = \frac{1}{\vartheta} \sum_{j,k,l=0}^{K-1} P(j, k, l) \iint_{-\frac{1}{2}}^{\frac{1}{2}} d\xi_1 d\xi_2\, \mathcal{V}_j(\xi_1)\mathcal{V}_k(\xi_2)\mathcal{V}_l^{\star}(\xi_1 + \xi_2)
$$
$$
\cdot B(f_1 - \xi_1, f_2 - \xi_2) .
$$

(8.9)

If, as assumed, the bispectrum varies slowly within the bandwidth W, the integral becomes

$$
B(f_1, f_2) \iint_{-\frac{1}{2}}^{\frac{1}{2}} \mathcal{V}_j(\xi_1)\mathcal{V}_k(\xi_2)\mathcal{V}_l^{\star}(\xi_1 + \xi_2)d\xi_1 d\xi_2 ,
\qquad (8.10)
$$

that is, the product of the bispectrum and an integral representation for $P(j, k, l)$. Defining

$$\vartheta = \sum_{j,k,l=0}^{K-1} P^2(j,k,l) \tag{8.11}$$

gives $\mathbf{E}\{\widehat{B}(f_1, f_2)\} \approx B(f_1, f_2)$, so the estimate is approximately unbiased.

The variance of (8.5) for Gaussian data and distinct frequencies, using (3.5), is approximately

$$\mathbf{E_G}\{|\widehat{B}(f_1, f_2)|^2\} = \frac{S(f_1)S(f_2)S(f_3)}{\vartheta^2} \sum_{n,m=0}^{N-1} \left[\sum_{j=0}^{K-1} \lambda_j\, v_n^{(j)}\, v_m^{(j)} \right]^3 \tag{8.12}$$

if the spectrum varies slowly within a bandwidth W of the three frequencies. Noting that the terms in brackets sums to ϑ, this becomes

$$\mathbf{E_G}\{|\widehat{B}(f_1, f_2)|^2\} \approx \frac{1}{\vartheta} S(f_1)S(f_2)S(f_1 + f_2)\,. \tag{8.13}$$

Approximating the inner sum in (8.12) by Mercer's theorem, Smithies (1962), rotating coordinates, and crudely approximating the outer sums by integrals, one obtains

$$\mathbf{E_G}\{|\widehat{B}(f_1, f_2)|^2\} \approx \frac{1}{3NW^2} S(f_1)S(f_2)S(f_1 + f_2) \tag{8.14}$$

so, for fixed W, the estimate is consistent.

Even though (8.5) is an unbiased consistent estimator of the bispectrum, the W^2 term in the denominator of (8.14) may cause the variance to be very large. Moreover, because the bispectrum of Gaussian processes is identically zero, it is unreasonable to expect a variance as small as predicted by (8.14) in cases where the bispectrum is of interest. Further, the estimates should be expected to have a "long tailed" distribution, and consequently robust estimation procedures are essential. These may be constructed from the consistent estimators given here, by, for example, using M-estimates on the results from overlapping subsections.

8.2   A Robust Estimator

We now introduce a variant of the time-domain estimate which is fast, robust, and which may be jackknifed (or bootstrapped) to obtain non-parametric error estimates. Consider the estimate (8.4) computed at the

Nyquist rate samples x° $(j; f)$ (3.41). Denote the K triple products

$$z_j = x^\circ (j; f_1) x^\circ (j; f_2) x^{\circ *} (j; f_1 + f_2)$$

for $j = 1, 2, ..., K$ with $t_j = (j - \frac{1}{2})\Delta$. In this notation, the previous estimate (8.4) is

$$\widehat{B}_{ave}(f_1, f_2) = \frac{N}{\vartheta} ave\{z_j\} \tag{8.15}$$

where ave denotes the average over the K samples. This estimate may be made robust by replacing the conventional average in (8.15) by robust estimates of location such as complex medians and midmeans, by averages of nonlinear transformations of the z_j's, and by combinations of these, $e.g.$,

$$\widehat{B}_{med}(f_1, f_2) = \frac{N}{\vartheta} median\{z_j\} \ . \tag{8.16}$$

Under the assumption that the spectrum is constant over bands $(f_j - W, f_j + W)$, the samples at the Nyquist rate are approximately uncorrelated so these estimates may be jackknifed or bootstrapped. Extension of such estimates to include averages over different data blocks in addition to the averaging within a single block is obvious.

9   Testing Nonlinearity *vs* Nonstationarity

In this section I discuss a test for nonlinear terms in nonstationary data, using electron data measured at both the *ACE* and *Ulysses* spacecraft. Theoretical discussions of the interplanetary medium usually assume that propagation is dominated by turbulence, so that discrete modal frequencies could not be preserved and the line structure shown earlier in Figure 2, reported in Thomson *et al.* (1995), and seen in Figure 1 of Ladbury (1995) should be impossible. The idea of turbulence, however, appears to be based more on theory than on observations. In late 1997 and early 1998 the ACE spacecraft, at the first Lagrange point, L_1, about 0.9 AU from the sun, and Ulysses near the ecliptic at 5.4 AU, were approximately radially aligned. Moreover, the two spacecraft have excellent and nearly identical detectors, Lanzerotti *et al.* (1992), the one on ACE being the spare from Ulysses, so that an observational test for nonlinear terms in propagation is possible.

The presence of nonlinear terms of the form $u\nabla u$ in theories of the solar wind, (Roberts *et al.*, 1996), suggest that sum and difference frequencies should be generated. This is just what a bispectrum is designed

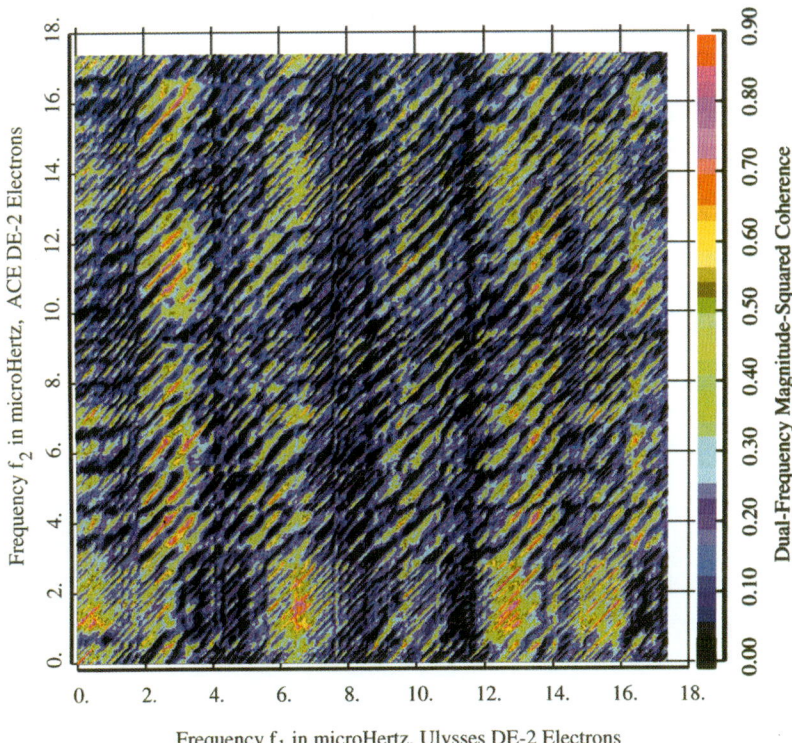

Figure 15: Dual frequency coherence between the 53-103 keV electron channels measured on the ACE and Ulysses spacecraft. Note that the MSC exceeds 0.8 for many frequency pairs. If the data were independent, or stationary, this would be (with $K = 8$) an extremely improbable outcome.

to detect, but there are problems: First, well-calibrated measures of bicoherence are unknown, so assigning significance levels is difficult. Second, as described in the preceding section, the bispectrum is described for *stationary* data, so that signals at distinct frequencies are *assumed* to be uncorrelated. It is obvious that solar wind data are nonstationary. Estimates of dual-frequency coherences invariably show high coherences between many frequency pairs. As an example, Figure 15 shows the dual-frequency magnitude-squared coherence (7.2) between the low-energy electrons measured on ACE and Ulysses, with the Ulysses data advanced by two days to allow for propagation time. Ordinary MSC, that is taken along the $f_1 = f_2$ diagonal of Figure 15, shows discrete bands with high coherence, a maximum value of 0.787 and, overall, 41% of the estimates above the 90% significance level. Figure 15, in contrast, shows many frequency pairs where the dual-frequency MSC exceeds 0.8 and, because this pattern appears to be systematic and is repeated in other energy channels and detectors, one must conclude that the processs has a rich nonstationary structure.

For this test, I take the two hypotheses:

H_0 : The process is nonstationary but linear. Because signal elements at different frequencies are correlated, this means that some part of the signal at sum and difference frequencies will be explained because the data are nonstationary.

H_1 : There is evidence for product terms in addition to those expected under H_0. In the test considered here, these will occur at $f_1 + f_2$ and $f_1 - f_2$. In engineering applications $f_3 = 2f_1 - f_2$ is common.

For the $f_1 + f_2$ terms, the test consists of the following steps:

1. Consider two distinct frequencies f_1 and f_2, and, as above, $f_3 = f_1 + f_2$. Denote the vector of downsamples at f_1 by $\mathbf{X}_1$; those at f_2 by $\mathbf{X}_2$; and those at f_3 by $\mathbf{X}_3$. Form the $K \times 3$ regressor matrix, $\mathbf{X} = (\mathbf{X}_1, \mathbf{X}_2, \mathbf{X}_3)$; and denote the vector of "output" signals at f_3 by $\mathbf{Y}$. Specifically, $\mathbf{Y} = [y^{\circ}(1; f_3), \ldots, y^{\circ}(K; f_3)]^T$ where, as usual, $y^{\circ}(j; f)$ is the downsample at $t_j = (j - \frac{1}{2})\Delta$, and so on. Because of the assumed nonstationary nature of the signals, information at the different frequencies may be correlated, so a *linear* relation could exist between $\mathbf{X}$ and $\mathbf{Y}$ which we denote by β

$$\mathbf{R}_1 = \mathbf{Y} - \mathbf{X}\beta . \tag{9.1}$$

2. Estimate β to minimize $\|\mathbf{R}_1\|$ by least squares and denote the residual sum-of-squares by

$$r_1 = \|\mathbf{R}_1\|^2 = \sum_{j=1}^{K} |R_{1,j}|^2 . \tag{9.2}$$

The linear regression may be tested by a multiple correlation coefficient

$$1 - \frac{\|\mathbf{R}_1\|^2}{\|\mathbf{Y}\|^2} \tag{9.3}$$

characterized by 6 and $2(K - 3)$ df. Note that when $f_1 = f_2$, $\mathbf{X}$ is a $K \times 1$ matrix, and $\mathbf{R}_1$ has 2 more df than when $f_1 \neq f_2$.

3. Estimate the coherence between $\mathbf{R}$ and the element-by-element, or Hadamard, product $(\mathbf{X}_1 \otimes \mathbf{X}_2)_j = x^\circ\ (j; f_1)x^\circ\ (j; f_2)$.

4. The significance of the relation between $\mathbf{R}_1$ and the product term may be done by partial-F or coherence statistics.

Note that the first term in the cross product between $\mathbf{R}_1$ and $\mathbf{X}_1 \otimes \mathbf{X}_2$ is basically the same as that of (8.4) for estimating the bispectrum. The test for the difference term is the same with $f_3 = f_1 - f_2$. To compute the partial F-test, Draper and Smith (1981), first denote the residuals

$$\mathbf{R}_2 = \mathbf{R}_1 - \mathbf{X}_1 \otimes \mathbf{X}_2 \beta_p \qquad (9.4)$$

where β_p is the coefficient of the nonlinear term. Next, denote the second residual sum-of-squares by

$$r_2 = \sum_{j=1}^{K} |R_{2,j}|^2$$

so the sum-of-squares explained by the regression on the product term is $r_1 - r_2$. This has 2 df, and the residual sum-of-squares r_2 has $2(K-4)$ df, so the partial F-test is

$$F_p = \frac{\frac{1}{2}(r_1 - r_2)}{\frac{1}{2(K-4)}r_2} . \qquad (9.5)$$

If one denotes the MSC between $\mathbf{R}$ and the product term by c, then

$$c = \frac{\left|\sum_{j=1}^{K} \mathbf{R}_{1,j}\mathbf{X}_{1,j}^*\mathbf{X}_{2,j}^*\right|^2}{\sum_{j=1}^{K} |\mathbf{R}_{1,j}|^2 \sum_{k=1}^{K} |\mathbf{X}_{1,k}\mathbf{X}_{2,k}|^2} \qquad (9.6)$$

and the partial F-test becomes

$$F_p = (K - 4)\frac{c}{1 - c} .$$

Central F-statistics with 2 and $2n$ degrees-of-freedom have particularly simple cumulative distribution functions, and the probability that such a test exceeds a level F is just

$$Q = \left(\frac{n}{n + F}\right)^n$$

or, for the partial F-test with distinct frequencies,

$$Q = (1 - c)^{K-4} .$$

In cases when two of f_1, f_2, and f_3 are the same, for example, if $f_1 = f_2$, omit the duplicate column of $\mathbf{X}$ and add two degrees-of-freedom to the residuals. It is also clearly unnecessary for both $\mathbf{X}_1$ and $\mathbf{X}_2$ to come from the same source, but if they are different, it would be necessary to include two "stationary" terms corresponding to f_3.

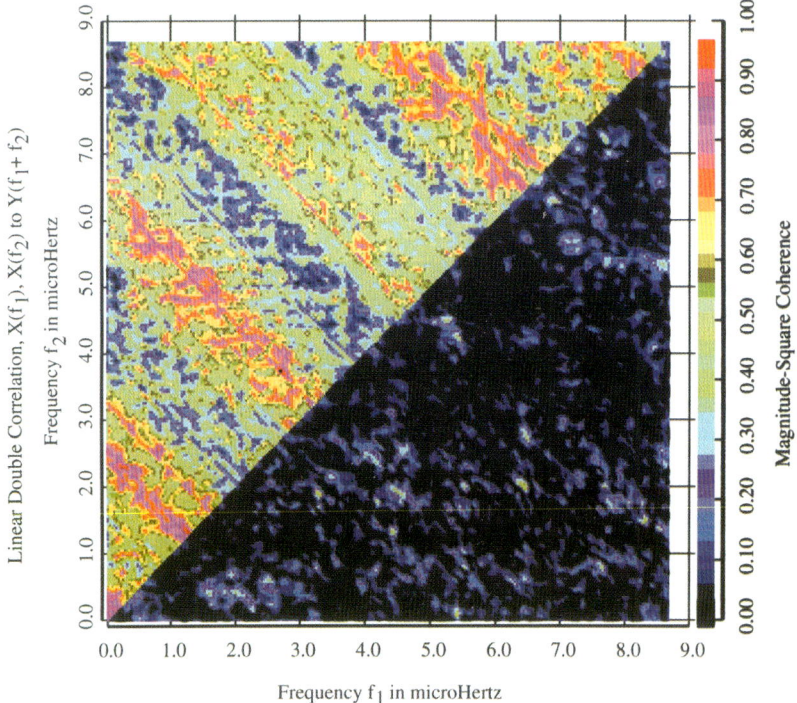

Residual Magnitude-Squared Coherence Test for Product Terms

Figure 16: Tests for linear dependence (upper-left) and nonlinearity (lower right) between ACE and Ulysses low-energy electron data

Figure 16 shows, in the upper-left, the ordinary multiple correlation coefficient (9.3) and, in the lower right, the product MSC, (9.6).[30] In this test all three inputs, $\mathbf{X}_1$, $\mathbf{X}_2$, and $\mathbf{X}_3$ were from ACE data and

---

[30]Normally, I plot results from $f_3 = f_1 - f_2$ and $f_3 = f_1 + f_2$ in the two triangles, but here the object was to contrast the large partial coherences explained by the linear terms with the relatively insignificant results from the nonlinear terms.

the output, **Y** was from Ulysses. Clearly the multiple coherence, in agreement with the dual-frequency coherence shown in Figure 15, shows highly significant test results at many frequencies with an apparently systematic organization.

In stark contrast, the one point showing high significance for the non-linear term is at $f_1 = 19.1\mu$Hz, $f_2 = 17.6\mu$Hz so $f_3 = 36.7\mu$Hz. Although none of these frequencies stand out as well-known, a peak at 36.7μHz can be seen in the radial magnetic field spectrum, Figure 2. The electron spectrum (not shown) is similar with a 4.6σ peak at 36.7μHz, a 2.8σ peak at 19.4μHz and a *local minima* at 17.6μHz. Following the arguments of section 4.6 (4.21), I suspect that the rapidly varying spectrum has reduced the effective degrees-of-freedom of the test. Thus, without further confirmation, this peak is perhaps best dismissed as a random fluctuation. If the frequencies were specified *a-priori*, the probability of obtaining a coherence of 0.65 by chance is about 0.015; however, as the results of about 6400 approximately independent tests are shown in Figure 16, the chance of observing one such test value is a near certainty. Thus one must conclude that, in this part of space and this frequency range, solar electrons propagate in an almost linear fashion. One might also conclude that poor analysis of earlier data, made without proper consideration of the rich nonstationary structure of solar phenonema, led to erroneous conclusions on the physics of the solar wind.

10   Discussion

This article has described some multitaper methods that give some insight into nonstationary and, to a lesser degree, nonlinear processes. The approach here has been to quantify the variance of a spectrum estimate for a stationary process, and describe various methods for testing for significant departures from this. Some suggestions of methods to describe causes of departures from standard assumptions have been made. Many of the processes encountered in climate and space physics problems are systematically nonstationary and, in some significant fraction of the cases I have examined, the *structure* of the process varies periodically. Moreover, most of the data from these two fields[31] appears to have very complicated spectra with *many* periodic, or modulated periodic, components.

The main points of this article have been:

---

[31]Because both climate and space physics data depend on the sun, these are unlikely to be independent examples.

- The periodogram should not be used for the analysis of scientific data under any circumstances. Although it has been known to have serious faults since before World War I, current publications show that these lessons have not been learned.

- Estimates that are mathematically equivalent to the periodogram, such as those based on sample autocorrelations, should be avoided.

- For general scientific spectrum estimation one should use multitaper estimates, and these should be made on a variety of time and bandwidth scales.

- More effort should be spent checking that the variance of the estimated spectrum is consistent with one's hypothesis than is spent making the spectrum estimate itself. Bartlett's M-test, resampling methods, the sample antespectra, and quadratic-inverse methods are all useful ways of testing variability of a spectrum estimate.

- Spectrum estimates that depart unreasonably from the expected distribution should be cause to examine assumptions. As an example, the Bartlett M-test associated with Figure 2 rejected a 'simple' spectrum at a level of about 10^{-10}. Because more than a year of data was involved, one implication is that such a result should not be observed by chance in the age of the Earth; I regard this as 'unreasonable' and hence reject the hypothesis that the interplanetary magnetic field has a simple power-law spectrum.

- The Loève, or dual-frequency, spectrum, is an excellent method for analyzing complicated nonstationary data.

- Because more data appears to be nonstationary than nonlinear, the residual bicoherence test may be more reliable than a bispectrum. Robust multitaper estimates of the bispectrum were also introduced.

Drawing valid scientific conclusions from observations of climate and space physics processes requires better analysis methods than those in common use.

Acknowledgements

This article describes some of my work presented at the 'Environmental Modelling and Signal Processing' and 'Data Analysis' sessions at the

Isaac Newton Institute's 1998 program on 'Nonlinear and Nonstationary Signal Processing'.

I would like to thank Louis J. Lanzerotti, Robert E. Gold, and members of the Ulysses and ACE teams for use of the electron data. Andre Balogh and collaborators provided, via the ESTEC data center, the Ulysses magnetometer data. Frank L. Vernon III provided the seismic data from the Kyrgyzstan network, and Phil Jones the Central England temperature data. My wife, Maja-Lisa, showed great patience in trying to counteract my tendency to leave details to the reader's imagination.

References

Alekseev, V.G. (1997). New modifications of second- and third-order periodograms. *Theory Prob. Appl*, **42**, 559–567.

Balogh, A., Forsyth, R.J., Hedgecock, P.C., Marquedant, R.J., Smith, E.J., Southworth, D.J., and Tsurutani, B.T. (1992). The magnetic field investigation on the Ulysses mission: Instrumentation and preliminary scientific results. *Astron. and Astrophysics, Suppl. Series*, **92**, 221–236.

Bartholomew, D.L. and Tague, J.A. (1995). Quadratic power spectrum estimation with orthogonal frequency division multiple windows. *IEEE Trans. on Signal Processing*, **43**, 1279–1282.

Bartlett, M.S. (1937). Properties of sufficiency and statistical tests. *Proc. Royal Soc. Lond.*, **A 160**, 268–282.

Bartlett, M.S. (1946). On the theoretical specification and sampling properties of autocorrelated time series. *J. Royal Statist. Soc. Suppl.*, **8**, 27–41.

Bartlett, M.S. and Kendall, D.G. (1946). The statistical analysis of variance-heterogeneity and the logarithmic transformation. *J. Royal Statist. Soc. Suppl.*, **8**, 128–38.

Bhansali, R.J. (1974). Asymptotic properties of the Wiener-Kolmogorov predictor. *J. Royal Statist. Soc.*, **B 36**, 61–73.

Bhansali, R.J. (1983). The inverse partial correlation function of a time series and its application. *J. Multivariate Analy.*, **13**, 310–327.

Birkelund, Y. and Hanssen, A. (1999). Multitaper estimators for bispectra. In *Proc. IEEE Signal Processing Workshop on Higher-Order Statistics*, pages 207–211, Caesarea, Israel. IEEE Catalog Number 0-7695-0140-0/99.

Blackman, R.B. and Tukey, J.W. (1959). *The Measurement of Power Spectra.* Dover, New York. Originally published in the *Bell System Tech. J.* Vol. XXXVII, 1958.

Bloomfield, P. (1973). An exponential model for the spectrum of a scalar time series. *Biometrika*, **60**, 217–226.

Bogert, R.P., Healy, M.J., and Tukey, J.W. (1967). The quefrency alanysis of time series for echoes: Cepstrum, pseudo-autocovariance, cross-cepstrum and saphe cracking. In B. Harris, editor, *Spectral Analysis of Time Series*, pages 209–243. John Wiley and Sons.

Brener, I., Mitra, P.P., Thomson, D.J., and Philen, D.L. (1998). High resolution zero-dispersion wavelength mapping in single-mode fiber. *Optica Letters*, **23**, 1520–1522.

Brillinger, D.R. (1965). An introduction to polyspectra. *Ann. Math. Statist.*, **36**, 1351–74.

Bronez, T.P. (1988). Spectral estimation of irregularly sampled multidimensional processes by generalized prolate spheroidal sequences. *IEEE Trans. on Signal Processing*, **36**, 862–873.

Bronez, T.P. (1992). On the performance advantage of multitaper spectral analysis. *IEEE Trans. on Signal Processing*, **40**, 2941–2946.

Cadzow, J.A. (1979). An extrapolation procedure for band-limited signals. *IEEE Trans. on Acoustics, Speech, and Signal Processing*, **27**, 4–12.

Chambers, J., Cleveland, W.S., Kleiner, B., and Tukey, P. (1983). *Graphical Methods for Data Analysis.* Wadsworth, Pacific Grove, CA.

Chatfield, C. (1979). Inverse autocorrelations. *J. Royal Statist. Soc.*, **A 142**, 363–377.

Childers, D.G., Skinner, D.P., and Kemerait, R.C. (1977). The cepstrum: A guide to processing. *Proc. IEEE*, **65**, 1428–1443. Correction, Vol. 66, 1288 (1978).

Cleveland, W.S. (1972). The inverse autocorrelations of a time series and their applications. *Technometrics*, **14**, 277–293.

Cohen, L. (1995). *Time-Frequency Analysis.* Prentice-Hall, Englewood Cliffs, NJ.

Cox, D.R. and Miller, H.D. (1965). *The theory of stochastic processes.* Chapman and Hall, London.

Cvetković, Z. (1998). Short-time Fourier analysis - a novel window design procedure. In *Proc. ICASSP*, volume 3, pages 1773–1776.

Daubechies, I. (1990). The wavelet transform, time-frequency localization, and signal analysis. *IEEE Trans. on Information Theory*, **36**, 961–1005.

Dettinger, M.D., Ghil, M., Strong, C.M., Weibel, W., and Yiou, P. (1995). Software expedites singular-spectrum analysis of noisy time series. *EOS, Trans. AGU*, **76**, 12,14,21.

Doob, J.L. (1952). *Stochastic Processes*. John Wiley and Sons, New York.

Draper, N.R. and Smith, H. (1981). *Applied Regression Analysis*. John Wiley and Sons, New York.

Erdélyi, A., Magnus, W., Oberhettinger, F., and Tricomi, F.G. (1954). *Tables of integral transforms*. McGraw-Hill, New York.

Fodor, I.K. and Stark, P.B. (1998). Multitaper spectrum estimates. In *Proc. SOHO 6 / GONG 98 workshop on helioseismology*. in press.

Gerr, N.L. and Allen, J.C. (1994). The generalized spectrum and spectral coherence of a harmonizable time series. *Digital Signal Processing*, **4**, 222–238.

Godfrey, M.D. (1965). An exploratory study of the bi-spectrum of economic time series. *Appl. Statist.*, **14**, 48–69.

Goldstein, B.E., Smith, E.J., Balogh, A., Horbury, T.S., Goldstein, M.L., and Roberts, D.A. (1995). Properties of magnetohydrodynamic turbulence in the solar wind as observed by Ulysses at high heliographic latitudes. *Geophys. Res. Let.*, **22**, 3393–3396.

Goodall, C. (1983). Examining residuals. In D.C. Hoaglin, F. Mosteller, and J.W. Tukey, editors, *Understanding Robust and Exploratory Data Analysis*, pages 211–246. John Wiley and Sons, New York.

Gori, F. (1974). Integral equations for incoherent imagery. *J. Opt. Soc. Amer.*, **64**, 1237–1243.

Gori, F. and Palma, C. (1975). On the eigenvalues of $sinc^2$ kernel. *J. Phys. A: Math. Gen.*, **8**, 1709–1719.

Groth, E.J. (1975). Probability distributions related to power spectra. *Astrophys J. Suppl.*, **29**, 285–302.

Gruenbacher, D.M. and Hummels, D.R. (1994). A simple algorithm for generating discrete prolate spheroidal sequences. *IEEE Trans. on Signal Processing*, **42**, 3276–3278.

Hannan, E.J. and Thomson, P.J. (1967). Spectral inference over narrow bands. *J. Appl. Prob.*, **8**, 157–169.

Hansson, M. and Salomonsson, G. (1997). A multiple window method for estimation of peaked spectra. *IEEE Trans. on Signal Processing*, **45**, 778–781.

Harris, F.J. (1978). On the use of windows for harmonic analysis with the discrete Fourier transform. *Proc. IEEE*, **66**, 51–83.

Hasselmann, K., Munk, W., and MacDonald, G. (1965). Bispectra of ocean waves. In M. Rosenblatt, editor, *Time Series Analysis*, pages 125–39. John Wiley and Sons, New York.

Haubrich, R.A. (1965). Earth noise, 5 to 500 millicycles per second, 1: Spectral stationarity, normality, and nonlinearity. *J. Geophys. Res.*, **70**, 1415–27.

Haykin, S. and Thomson, D.J. (1998). Signal detection in a nonstationary environment reformulated as an adaptive pattern classification problem. *Proc. IEEE*, **86**, 2325–2344.

Hinich, M.J. and Clay, C.S. (1968). The application of the discrete Fourier transform in the estimation of power spectra, coherence, and bispectra of geophysical data. *Rev. of Geophysics*, **6**, 347–63.

Hinnov, L.A. and Park, J. (1988). Multi-windowed spectrum estimates of the ILS polar motion. In A.K. Babcock and G.A. Wilkins, editors, *The Earth's Rotation and Reference Frames for Geodesy and Geodynamics*, pages 221–226. Kluwer Academic, Dordrecht.

Huang, T.S., Sanz, J.L.C., Fan, H., Shafii, J., and Tsai, B.-M. (1984). Numerical comparison of several algorithms for band-limited signal extrapolation. *Applied Optics*, **23**, 307–317.

Huber, P.J., Kleiner, B., Gasser, T., and Dumermuth, G. (1971). Statistical methods for investigating phase relations in stationary stochastic processes. *IEEE Trans. on Audio and Electroacoustics*, **19**, 78–86.

Hurd, H.L. and Gerr, N.L. (1991). Graphical methods for determining the presence of periodic correlation. *J. of Time Series Analysis*, **12**, 337–350.

Johnson, J., Thomson, D.J., Wu, E.X., and Williams, S.C.R. (1996). Multiple-window spectrum estimation applied to *in vivo* NMR spectroscopy. *J. Mag. Resonance*, **B 110**, 138–149.

Jones, R.H. (1962). Spectral estimates and their distributions. *Skandinavisk Aktuatietidskrift*, **45**, Part I, 39–69; Part II, 135–53.

Jonsson, J.O. and Steinhardt, A.O. (1993). The total p_{FA} of the multiwindow harmonic detector and its application to real data. *IEEE Trans. on Signal Processing*, **41**, 1702–1705.

Kappus, M.E. and Vernon, F.L. (1991). Acoustic signature of thunder from seismic records. *J. Geophys. Res.*, **96**, 10,989–11,006.

Karl, T.R., Jones, P.D., Knight, R.W., White, O.R., Mende, W., Beer, J., and Thomson, D.J. (1996). Testing for bias in the climate record. *Science*, **271**, 1879–1883.

Kirsteins, I.P., Mehta, S.K., and Fay, J. (1998). Adaptive separation of unknown narrowband and broadband time series. In *Proc. ICASSP*, volume 4, pages 2525–2529.

Kleiner, B. and Graedel, T.E. (1980). Exploratory data analysis in the geophysical sciences. *Rev. Geophys.*, **18**, 699–717.

Kleiner, B., Martin, R.D., and Thomson, D.J. (1979). Robust estimates of spectra (with discussion). *J. Royal Statist. Soc.*, **B-41**, 313–351.

Kuo, C., Lindberg, C., and Thomson, D.J. (1990). Coherence established between atmospheric carbon dioxide and global temperature. *Nature*, **343**, 709–714. (Reprinted in pp 395-400 of *Coherence and Time Delay Estimation*, G. C. Carter, Ed., IEEE Press, 1993.).

Ladbury, R. (1995). Is the answer blowing in the solar wind? *Physics Today*, **48**(9), 17–18.

Landau, H.J. (1985). An overview of time and frequency limiting. In J.F. Price, editor, *Fourier Techniques and Applications*, pages 201–220. Plenum.

Landau, H.J. and Pollak, H.O. (1961). Prolate spheroidal wave functions, Fourier analysis and uncertainty-II. *Bell System Tech. J.*, **40**, 65–84.

Lanzerotti, L.J., Gold, R.E., Thomson, D.J., Decker, R.E., Maclennan, C.G., and Krimigis, S.M. (1991). Statistical properties of shock-accelerated ions in the outer heliosphere. *Astrophys. J.*, **380**, L93–L96.

Lanzerotti, L.J., Gold, R.E., Anderson, K.A., Armstrong, T.P., Lin, R.P., Krimigis, S.M., Pick, M., Roelof, E.C., Sarris, E.T., Simnett, G.M., and Frain, W.E. (1992). Heliosphere instrument for spectra, composition, and anisotropy at low energies. *Astron. and Astrophysics Suppl.*, **92**, 349–363.

Lanzerotti, L.J., Armstrong, T.P., Gold, R.E., Maclennan, C.G., Roelof, E.C., Simnett, G.M., Thomson, D.J., Anderson, K.A., Hawkins, S.E., Krimigis, S.M., Lin, R.P., Pick, M., Sarris, E.T., and Tappin, S.J. (1995). Over the southern solar pole: low-energy interplanetary charged particles. *Science*, **268**, 1010–1013.

Libbrecht, K.G., Woodard, M.F., and Kaufman, J.M. (1990). Frequencies of solar oscillations. *Astrophys. J. Suppl.*, **74**, 1129–1149.

Lii, K.S., Rosenblatt, M., and Van Atta, C. (1976). Bispectral measurements in turbulence. *J. Fluid Mech.*, **77 part 1**, 45–62.

Lilly, J. and Park, J. (1995). Multiwavelet spectral and polarization analysis of seismic records. *Geophys. J. Intl.*, **122**, 1001–1021.

Lindberg, C.R. (1986). *Multiple taper spectral analysis of terrestrial free oscillations*. Ph.D. thesis, Univ. Calif., San Diego.

Lindberg, C.R. and Park, J. (1987). Multiple-taper spectral analysis of terrestial free oscillations: part II. *Geophys. J. Royal Astr. Soc.*, **91**, 795–836.

Lindberg, C.R. and Thomson, D.J. (1990). Comment on "a new method of spectral analysis and its application to the earth's free oscillations: the 'sompi' method", by S. Hori *et al. J. Geophys. Res.*, **95**, 12,785–12,788.

Lindberg, C.R. and Thomson, D.J. (1995). Method and apparatus for detecting control signals. U.S. Patent 5,442,696.

Liu, T.-C. and Van Veen, B.D. (1992). Multiple window based minimum variance spectrum estimation for multidimensional random fields. *IEEE Trans. on Signal Processing*, **40**, 578–589.

Loève, M. (1963). *Probability Theory*. D. Van Nostrand.

Malley, J.D. (1986). *Optimal Unbiased Estimation of Variance Components*. Springer-Verlag, Berlin.

Mallows, C.L. (1967). Linear processes are nearly Gaussian. *J. Appl. Prob.*, **4**, 313–329.

Manley, G. (1974). The mean temperature of central England, 1698-1952. *Q. J. R. Meteorol. Soc.*, **100**, 242–261.

Markel, J.D. and Gray, A.H. (1976). *Linear prediction of speech*. Springer-Verlag, Berlin.

Martin, W. (1982). Time-frequency analysis of random signals. In *Proc. ICASSP*, pages 1325–1328.

Matthaeus, W.H. and Goldstein, M.L. (1986). Low-frequency $1/f$ noise in the interplanetary magnetic field. *Physical Rev. Let.*, **57**, 495–498.

McCoy, E.J., Walden, A.T., and Percival, D.B. (1998). Multitaper spectral estimation of power law processes. *IEEE Trans. on Signal Processing*, **46**, 655–668.

McWhorter, L.T. and Scharf, L.L. (1998). Multiwindow estimators of correlation. *IEEE Trans. on Signal Processing*, **46**, 440–448.

Mellors, R., Vernon, F.L., and Thomson, D.J. (1996). Detection of dispersive signals using multi-taper dual-frequency coherence. In *Proc. of the 18th Seismic Research Symposium on Monitoring a Comprehensive Test Ban Treaty*, pages 745–753, Annapolis, Maryland.

Mellors, R., Vernon, F.L., Pavlis, G., Abers, G.A., Ghose, S., Hamburger, M.W., Mishatkin, V., and Iliasov, B. (1997). The Ms=7.3 1992 Suusamyr, Kyrgyzstan: 1: Constraints on fault geometry and source parameters based on aftershocks and body-wave modeling. *Bull. Seismol. Soc. Amer.*, **87**, 11–22.

Mellors, R., Vernon, F.L., and Thomson, D.J. (1998). Detection of dispersive signals using multitaper double frequency coherence. *Geophys. J. Intl.*, **135**, 146–154.

Miller, K.S. (1974). *Complex Stochastic Processes*. Addison Wesley, Reading, MA.

Mitra, P.P. and Peseran, B. (1999). Analysis of dynamic brain imaging data. *Biophysical J.*, **76**, 691–708.

Mullins, C.F. (1993). Multiple window cumulant estimation. In *Proc. IEEE SP workshop on Higher-Order statistics*, page M1.4, South Lake Tahoe, CA.

Mullis, C.T. and Scharf, L.L. (1991). Quadratic estimators of the power spectrum. In S. Haykin, editor, *Advances in Spectrum Analysis and Array Processing*, volume 1, pages 1–57. Prentice-Hall.

Nikias, C.L. and Raghuveer, M.R. (1987). Bispectrum estimation: A digital signal processing framework. *Proc. IEEE*, **75**, 869–91.

Noll, A.M. (1963). Cepstrum pitch determination. *J. Acoust. Soc. Amer.*, **36**, 293–309.

Pallé, P.L. (1991). The search for solar gravity modes. *Adv. Space Res.*, **11**, (4)29–(4)38.

Papoulis, A. (1975). A new algorithm in spectral analysis and band-limited extrapolation. *IEEE Trans. on Circuits and Systems*, **22**, 735–742.

Park, J. (1992). Envelope estimation for quasi-periodic geophysical signals in noise: a multitaper approach. In A.T. Walden and P. Guttorp, editors, *Statistics in the Environmental and Earth Sciences*, pages 189–219. Edward Arnold, London.

Park, J., Vernon, F.L., and Lindberg, C.R. (1987a). Frequency dependent polarization analysis of high-frequency seismograms. *J. Geophys. Res.*, **92**, 12,664–12,674.

Park, J., Lindberg, C.R., and Thomson, D.J. (1987b). Multiple-taper spectral analysis of terrestial free oscillations: part I. *Geophys. J. Royal Astr. Soc.*, **91**, 755–794.

Park, J., Lindberg, C.R., and Vernon, F.L. (1987c). Multitaper spectral analysis of high-frequency seismograms. *J. Geophys. Res.*, **92**, 12,675–12,684.

Parker, D.E., Legg, T.P., and Folland, C.K. (1992). A new daily central England temperature series, 1772-1991. *Int. J. Climatology*, **12**, 317–342.

Parzen, E. (1957). On consistent estimates of the spectrum of a stationary time series. *Ann. Math. Statist.*, **28**, 329–348.

Pearson, E.S. and Hartley, H.O. (1970). *Biometrika Tables for Statisticians*. Cambridge Univ. Press.

Percival, D.B. and Walden, A.T. (1993). *Spectral Analysis for Physical Applications; Multitaper and Conventional Univariate Techniques*. Cambridge Univ. Press.

Pitton, J.W. (1998). Nonstationary spectrum estimation and time-frequency concentration. *Proc. ICASSP*, **4**, 2425–2428.

Priestley, M.B. (1981). *Spectral Analysis and Time Series*. Academic Press.

Ramsey, J.B. and Thomson, D.J. (1999). A reanalysis of the spectral properties of some economic and financial time series. In P. Rothman, editor, *Nonlinear Time Series Analysis of Economic and Financial Data*, pages 45–85. Kluwer Academic Publishers, Dordrecht.

Rao, T.S. and Gabr, M.M. (1984). *An Introduction to Bispectral Analysis and Bilinear Time Series Models*. Springer-Verlag, New York.

Rao, T.S. and Gabr, M.M. (1989). The estimation of spectrum, inverse spectrum, and inverse autocovariances of a stationary time series. *J. Time Series Analy.*, **10**, 183–202.

Lord Rayleigh (1912). Remarks concerning Fourier's theorem as applied to physical problems. *Philosophical Magazine*, **XXIV**, 864–869. (in *Scientific Papers by Lord Rayleigh*, Dover Publications, New York, 1964).

Riedel, K.S. and Sidorenko, A. (1995). Minimum bias multiple taper spectral estimation. *IEEE Trans. on Signal Processing*, **43**, 188–195.

Riedel, K.S. and Sidorenko, A. (1996). Adaptive smoothing of the log-spectrum with multiple tapering. *IEEE Trans. on Signal Processing*, **44**, 1794–1800.

Riedel, K.S., Sidorenko, A., and Thomson, D.J. (1994a). Spectral estimation of plasma fluctuations. 1. Comparison of methods. *Phys. Plasma*, **1**, 485–500.

Riedel, K., Sidorenko, A., Bretz, N., and Thomson, D.J. (1994b). Spectral estimation of plasma fluctuations. II. Nonstationary analysis of edge localized mode spectra. *Phys. Plasma*, **1**, 501–514.

Rivlin, T.J. (1990). *Chebyshev Polynomials*. John Wiley and Sons, New York, second edition.

Roberts, D.A., Ogilvie, K.W., and Goldstein, M.L. (1996). The nature of the solar wind. *Nature*, **381**, 31–32.

Rosenblatt, M. (1961). Some comments on narrow band-pass filters. *Quart. Appl. Math.*, **18**, 387–393.

Rozanov, Y.A. (1961). On the applicability of the central limit theorem to stationary processes which have passed through a linear filter. *Theor. Prob. Appl.*, **6**, 321–322.

Schild, R.E. and Thomson, D.J. (1997). The Q0957+561 time delay, quasar structure, and microlensing. In D. Maoz, editor, *Astronomical time series*, pages 73–84. Kluwer Academic Publishers, Dordrecht.

Schroeder, M.R. (1981). Direct (nonrecursive) relations between cepstrum and predictor coefficients. *IEEE Trans. on Signal Processing*, **29**, 297–301.

Seidelmann, P.K. (1992). *Explanatory supplement to the Astronomical Almanac*. University Science Books, Mill Valley, CA.

Seymour, M.S. and Haykin, S. (1993). ISAR using Thomson's multiwindow adaptive spectrum estimation method. *IEEE Trans. Aero. and Elec. Sys.*, **29**, 1065–1070.

Slepian, D. (1953). Estimation of signal parameters in the presence of noise. *IRE Trans. on Information Theory*, **3**, 68–89.

Slepian, D. (1964). Prolate spheroidal wave functions, Fourier analysis and uncertainty -IV. *Bell System Tech. J.*, **43**, 3009–3057.

Slepian, D. (1978). Prolate spheroidal wave functions, Fourier analysis, and uncertainty V: the discrete case. *Bell System Tech. J.*, **57**, 1371–1429.

Slepian, D. (1983). Some comments on Fourier analysis, uncertainty, and modeling. *SIAM Rev.*, **25**, 379–393.

Slepian, D. and Pollak, H.O. (1961). Prolate spheroidal wave functions, Fourier analysis and uncertainty -I. *Bell System Tech. J.*, **40**, 43–64.

Smithies, F. (1962). *Integral Equations*. Cambridge Univ. Press.

Stoica, P. and Moses, R. (1997). *Introduction to Spectral Analysis*. Prentice-Hall, Upper Saddle River, NJ.

Stoica, P. and Sundlin, T. (1999). On nonparametric spectral estimation. *Circuits Systems Signal Process.*, **18**, 169–181.

Tauxe, L. (1993). Sedimentary records of relative paleointensity of the geomagnetic field: theory and practice. *Rev. Geophysics*, **31**, 319–354.

Tauxe, L. and Wu, G. (1990). Normalized remanence in sediments of the western equatorial Pacific: relative paleointensity of the geomagnetic field? *J. Geophys Res.*, **95**, 12,337–12,350.

Thomson, D.J. (1977). Spectrum estimation techniques for characterization and development of WT4 waveguide. *Bell System Tech. J.*, **56**, *Part I*, 1769–1815, *Part II*, 1983–2005.

Thomson, D.J. (1982). Spectrum estimation and harmonic analysis. *Proc. IEEE*, **70**, 1055–1096.

Thomson, D.J. (1989). Multi-window bispectrum estimates. In *Proc. IEEE Workshop on Higher-order spectral analysis*, pages 19–23, Vail, Colorado.

Thomson, D.J. (1990a). Time series analysis of holocene climate data. *Phil. Trans. R. Soc. Lond.*, **A 330**, 601–616.

Thomson, D.J. (1990b). Quadratic-inverse spectrum estimates: applications to paleoclimatology. *Phil. Trans. R. Soc. Lond.*, **A 332**, 539–597.

Thomson, D.J. (1993). Non-stationary fluctuations in "stationary" time series. *Proc. SPIE*, **2027**, 236–244.

Thomson, D.J. (1994a). Jackknifing multiple-window spectra. *Proc. ICASSP*, **VI**, 73–76.

Thomson, D.J. (1994b). An overview of multiple-window and quadratic-inverse spectrum estimation methods. *Proc. ICASSP*, **VI**, 185–94.

Thomson, D.J. (1994c). Projection filters for data analysis. In *Proc. Seventh IEEE SP Workshop on Stat. Sig. and Array Proc.*, pages 39–42, Quebec.

Thomson, D.J. (1995). The seasons, global temperature, and precession. *Science*, **268**, 59–68.

Thomson, D.J. (1997). Dependence of global temperatures on atmospheric CO_2 and solar irradiance. *Proc. Natl. Acad. Sci. USA*, **94**, 8370–8377.

Thomson, D.J. (1998). Multiple-window spectrum estimates for non-stationary data. In *Proc. Ninth IEEE SP Workshop on Statistical Signal and Array Processing*, pages 344–347, Portland, Oregon. IEEE.

Thomson, D.J. and Chave, A.D. (1991). Jackknifed error estimates for spectra, coherences, and transfer functions. In S. Haykin, editor, *Advances in Spectrum Analysis and Array Processing*, volume 1, pages 58–113. Prentice-Hall.

Thomson, D.J. and Schild, R. (1993). Processes with level-dependent delay. In *Proc. IEEE SP workshop on Higher-Order statistics*, pages 374–378, South Lake Tahoe, CA.

Thomson, D.J. and Schild, R. (1997). Time delay estimates for Q0957+561 A, B. In T.S. Rao, M.B. Priestley, and O. Lessi, editors, *Applications of time series analysis in astronomy and meteorology*, pages 187–204. Chapman and Hall, London.

Thomson, D.J. and Vernon, F.L. (1998). Signal extraction via multitaper spectra of nonstationary data. In *Proc. of the Thirty-Second Asilomar Conf. on Signals, Systems, and Computers*, pages 271–275, Madison, WI. Omnipress. IEEE Cat. No. 98CH36284.

Thomson, D.J., Maclennan, C.G., and Lanzerotti, L.J. (1995). Propagation of solar oscillations through the interplanetary medium. *Nature*, **376**, 139–144.

Thomson, D.J., Maclennan, C.G., and Lanzerotti, L.J. (1996). The nature of the solar wind. *Nature*, **381**, 32.

Thomson, D.J., Lanzerotti, L.J., and Maclennan, C.G. (1998). Low frequency ($\sim$ 2.19 day period) mode in records of interplanetary and central England temperature data. In *Proc. of the SOHO 6/GONG 98 Workshop, 'Structure and Dynamics of the Sun and Sun-like Stars'*, volume ESA SP-418 Vol. 2, pages 967–971.

Tukey, J.W. (1967). An introduction to the calculations of numerical spectrum analysis. In B. Harris, editor, *Spectral Analysis of Time Series*, pages 25–46. John Wiley and Sons.

Van Veen, B.D. and Scharf, L.L. (1990). Estimation of structured covariance matrices and multiple window spectrum analysis. *IEEE Trans. on Signal Processing*, **38**, 1467–1471.

Vautard, R., Yiou, P., and Ghil, M. (1992). Singular-spectrum analysis: A toolkit for short, noisy chaotic signals. *Physics D*, **58**, 95–126.

Vernon, F.L. (1989). *Analysis of data recorded on the ANZA seismic network*. Ph.D. thesis, Univ. Calif., San Diego.

Walden, A.T. (1995). Multitaper estimation of the innovation variance of a stationary time series. *IEEE Trans. on Signal Processing*, **43**, 181–187.

Walden, A.T., McCoy, E.J., and Percival, D.B. (1994). The variance of multitaper spectrum estimates for real Gaussian processes. *IEEE Trans. on Signal Processing*, **42**, 479–482.

Whittle, P. (1953). Estimation and information in stationary time series. *Arkiv För Matematik*, **2**, 423–434.

Wiener, N. (1949). *Extrapolation, Interpolation, and Smoothing of Stationary Time Series*. M.I.T. Press, Cambridge, MA.

Xu, Y., Haykin, S., and Racine, R.J. (1999). Multiple window time-frequency distribution and coherence of EEG using Slepian sequences and Hermite functions. *IEEE Trans. on Biomedical Engineering*, **46**, 861–866.

Zhu, T., Chun, K.-P., and West, G.F. (1989). High-frequency p-wave attenuation determination using multiple-window spectral analysis method. *Bull. Seism. Soc. Am.*, **79**, 1054–1069.

Zhurbenko, I.G. (1978). On a statistic for the spectral density of a stationary sequence. *Soviet Math. Dokl.*, **19**, 263–266.

Zhurbenko, I.G. (1979). Local properties of estimate of spectral function. *Prob. Infor. Trans.*, **14**, 218–222.

Signal and Image Denoising via Wavelet Thresholding: Orthogonal and Biorthogonal, Scalar and Multiple Wavelet Transforms

Vasily Strela and Andrew Walden

1   Introduction

The method of signal denoising via wavelet thresholding was popularised by Donoho and Johnstone (1994, 1995) and is now widely applied in science and engineering. It is based on thresholding of wavelet coefficients arising from the standard scalar orthogonal discrete wavelet transform (DWT). Recently this approach has been extended to incorporate thresholding coefficients arising from the discrete multiple wavelet transform (DMWT). Complications are introduced by the fact that non-orthogonal prefilters are required before the DMWT. Strela *et al.* (1999) applied scalar thresholding to the output from the Geronimo *et al.* (1994) DMWT with prefiltering, while Downie and Silverman (1998) gave a multivariate thresholding approach which takes account of the correlation in the DMWT coefficients induced by the prefiltering.

In this article we study wavelet thresholding in the context of scalar orthogonal, scalar biorthogonal, multiple orthogonal and multiple biorthogonal wavelet transforms. Two types of multiwavelet thresholding are considered: scalar and vector. Both of them take into account the covariance structure of the transform. The form of the universal threshold is carefully formulated and is the key to the excellent results obtained in the extensive numerical simulations of signal and image denoising reported here.

Sections 2 to 5 are concerned with the deterministic formulation of relevant components of multiwavelet analysis. In Section 2 we give a summary of multiresolution analysis and semi-orthogonal, orthogonal and biorthogonal multiwavelet functions. The rest of the article concentrates on the most used practical cases of multiplicity 1 (scalar) and 2 (two scaling functions). Section 3 introduces two classes of orthogonal multiwavelets which may be used in multifilter banks — the Geronimo-Hardin-Massopust (GHM) and Chui-Lian (CL) classes, while Section 4 discusses biorthogonal multifilter banks, denoted BiGHM and BiHermite. For multifilter banks the given scalar input signal must be associated with a sequence of length-2 vectors: oversampling and critical sampling preprocessing are carefully studied in Section 5.

Sections 6 and 7 look at two important aspects of the processing of 1D stochastic signals, the covariance structure of the output, and denoising via thresholding. If stochastic noise is the input to preprocessing followed by discrete multiwavelet transform, very variable covariance structures of the output can result; in particular the average input variance of white noise can be inflated or deflated on output as shown in Section 6 for the GHM and CL orthogonal multiwavelet transforms. For biorthogonal multiwavelet transforms the variance inflation/deflation are seen to be almost independent of sample size, but can be strongly dependent on the level to which the transform is taken. Denoising of signals via universal thresholding of the wavelet coefficients resulting from preprocessing and the discrete multiwavelet transform can be carried out by treating the output in a scalar or vector sense, as in Section 7. For the scalar case the variance inflation/deflation can be used to carefully formulate the appropriate universal threshold level, while for the vector case the covariance structure of the noise can be used likewise. Extensive simulations using four standard 1D test signals are reported. Multiwavelets outperform scalar wavelets for three out of four noisy test signals, and the Chui-Lian scaling functions and wavelets combined with repeated row preprocessing appears to be a good general method. Vector thresholding does not always outperform scalar thresholding.

Attention is turned to the denoising of 2D signals in Section 8. Firstly the algorithm for the 2D case is carefully set out. In particular the formulation of the first step of the multiwavelet transform in the horizontal and vertical directions is explicitly illustrated, and it is shown how the resulting coefficients arise via sums of Kronecker products of the matrices of filter coefficients. For thresholding in 2D images, the variance inflation/deflation factor can be simply formulated, and is used to derive the appropriate universal threshold for scalar thresholding. Extensive simulations using four standard 2D test images are discussed. Multiwavelets generally outperform scalar wavelets for image denoising for all four noisy test images, and the results are visually very impressive. Only for 'Lenna' and 'fingerprints' with signal-to-noise ratios of 2 do scalar wavelets perform best. As for 1D signal processing, Chui-Lian scaling functions and wavelets combined with repeated row preprocessing appear to be a good general method.

One of the most encouraging aspects of our methodology is that for both 1D and 2D cases, the reconstructed signals derived from such a good general method demonstrate much reduced noise levels — an average noise reduction gain of 7.5dB is achieved over test examples.

2   Multiresolution analysis

2.1   Multiscaling functions

The concept of multiresolution analysis can be extended from the scalar case to general dimension $r \in \mathbb{N}$, (e.g., Goodman and Lee, 1994). Let $\Phi(t) =$

$[\phi_1(t), \ldots, \phi_r(t)]^T \in L^2(\mathbb{R})^r$ be a vector of functions. It generates a multiresolution analysis of multiplicity r for $L^2(\mathbb{R})$ if the subspaces

$$V_j \equiv \overline{\text{span}\left\{2^{-j/2}\phi_i\left(\frac{t}{2^j} - k\right) : 1 \leq i \leq r, k \in \mathbb{Z}\right\}}$$

are nested—

$$\cdots \subset V_3 \subset V_2 \subset V_1 \subset V_0 \subset V_{-1} \subset \cdots \tag{2.1}$$

—and the following conditions hold:

$$\overline{\bigcup_{j \in \mathbb{Z}} V_j} = L^2(\mathbb{R}) \quad \text{and} \quad \bigcap_{j \in \mathbb{Z}} V_j = \{0\}. \tag{2.2}$$

Equations (2.1) and (2.2) ensure that the projections of any finite energy signal onto subspaces V_j will give successive approximations of the signal. If (2.1) and (2.2) hold, together with

- $x(\cdot)$ is a member of V_0 if and only if $x(t-k)$ is a member of V_0, for $k \in \mathbb{Z}$,

- $x(\cdot)$ is a member of V_0 if and only if $x(\frac{t}{2^j})/2^{j/2}$ is a member of V_j, for $j \in \mathbb{Z}$, and

- $\{\phi_i(t-k) : 1 \leq i \leq r, k \in \mathbb{Z}\}$ form a Riesz basis for V_0,

then the vector $\Phi(t)$ is a *multiscaling function*.

All scaling functions $\phi_1(t), \ldots, \phi_r(t)$ are in $V_0 \subset V_{-1}$, the latter having basis $\{\sqrt{2}\,\phi_i(2t-k) : 1 \leq i \leq r, k \in \mathbb{Z}\}$. Thus the multiscaling function must satisfy the two-scale dilation equation

$$\Phi(t) = \sqrt{2}\sum_k \mathbf{G}_k \Phi(2t - k), \tag{2.3}$$

where $\mathbf{G}_{k \in \mathbb{Z}} \in \ell^2(\mathbb{Z})^{r \times r}$ is an $r \times r$ matrix of filter coefficients.

The collection of scaling functions $\{\phi_{i,j,k}(t) : 1 \leq i \leq r, k \in \mathbb{Z}\}$, with $\phi_{i,j,k}(t) = 2^{-j/2}\phi_i(\frac{t}{2^j} - k)$, is a Riesz basis of V_j. We also write

$$[\phi_{1,j,k}(t), \ldots, \phi_{r,j,k}(t)]^T = \Phi_{j,k}(t).$$

For the special case of orthonormal multiscaling functions, the basis $\{\phi_i(t-k) : 1 \leq i \leq r, k \in \mathbb{Z}\}$ is not just a Riesz basis, but orthonormal,

$$\langle \Phi(t), \Phi(t-k)\rangle = \int \Phi(t)\Phi^T(t-k)dt = \mathbf{I}_r \delta_{k,0}, \ \ k \in \mathbb{Z}, \tag{2.4}$$

where $\mathbf{I}_r$ is the $r \times r$ identity matrix. By substituting (2.3) in the integral the orthonormality condition implies the following for the matrices of filter coefficients:

$$\sum_k \mathbf{G}_k \mathbf{G}_{2l+k}^T = \mathbf{I}_r \delta_{0,l}, \quad l \in \mathbb{Z}. \tag{2.5}$$

In the scalar case this tells us that the sum of squares of the low-pass filter coefficients is unity, and the filter is orthogonal to its even translates.

2.2   Multiwavelet functions

Now let W_j denote a space complementing V_j in V_{j-1}, so that

$$V_{j-1} = V_j \oplus W_j, \text{ and } V_j \cap W_j = \emptyset$$

where $\oplus$ denotes a direct sum. Hence W_j contains the 'detail' required in going from an approximation at level j to one at level $j-1$. In view of (2.1) and (2.2) we have

$$\bigoplus_{j \in \mathbb{Z}} W_j = L^2(\mathbb{R}).$$

Let $\Psi(t) = [\psi_1(t), \ldots, \psi_r(t)]^T \in L^2(\mathbb{R})^r$. For $j \in \mathbb{Z}$ we define

$$W_j \equiv \overline{\text{span} \left\{ 2^{-j/2} \psi_i \left(\frac{t}{2^j} - k \right) : 1 \le i \le r, k \in \mathbb{Z} \right\}}.$$

If $\{\psi_i(t-k) : 1 \le i \le r, k \in \mathbb{Z}\}$ is a Riesz basis of W_0, then $\Psi(t)$ is a multiwavelet function. The collection of wavelet functions $\{\psi_{i,j,k}(t) : 1 \le i \le r, k \in \mathbb{Z}\}$, with $\psi_{i,j,k}(t) = 2^{-j/2} \psi_i(\frac{t}{2^j} - k)$, is a Riesz basis of W_j and $\{\psi_{i,j,k}(t) : 1 \le i \le r; j, k \in \mathbb{Z}\}$ is a Riesz basis of $L^2(\mathbb{R})$. As for multiscaling functions it is convenient to write $[\psi_{1,j,k}(t), \ldots, \psi_{r,j,k}(t)]^T = \Psi_{j,k}(t)$.

Now all wavelet functions $\psi_1(t), \ldots, \psi_r(t)$ are in $W_0 \subset V_{-1}$, the latter having basis $\{\sqrt{2}\, \phi_i(2t-k) : 1 \le i \le r, k \in \mathbb{Z}\}$. Thus the multiscaling and multiwavelet functions must satisfy the two-scale wavelet equation

$$\Psi(t) = \sqrt{2} \sum_k \mathbf{H}_k \Phi(2t - k), \tag{2.6}$$

where $\mathbf{H}_{k \in \mathbb{Z}} \in \ell^2(\mathbb{Z})^{r \times r}$ is an $r \times r$ matrix of filter coefficients.

An orthogonal multiresolution analysis is one where the wavelet spaces W_j, as defined above, are the *orthogonal* complement of V_j in V_{j-1}:

$$V_{j-1} = V_j \oplus W_j, \text{ and } V_j \perp W_j.$$

For $j = 0$ we thus know

$$\langle \Phi(t-k), \Psi(t-n) \rangle = \int \Phi(t-k) \Psi^T(t-n) dt = \mathbf{0}_r, \quad k, n \in \mathbb{Z}.$$

Substituting (2.3) and (2.6) in the integral leads to

$$\sum_k \mathbf{G}_k \mathbf{H}_{2l+k}^T = \mathbf{0}_r, \quad l \in \mathbb{Z}. \tag{2.7}$$

In the scalar case this tells us that the low-pass filter is orthogonal to the even translates of the high-pass filter. A multiwavelet function $\Psi(t)$ which gives an orthogonal multiresolution analysis is called a *semi-orthogonal* multiwavelet (Cotronei *et al.* 1998). If in addition, functions in the same space are orthonormal, then the corresponding multiwavelets are *orthonormal*. For $j = 0$,

$$\langle \Psi(t), \Psi(t-k) \rangle = \int \Psi(t) \Psi^T(t-k)dt = \mathbf{I}_r \delta_{k,0} \quad k \in \mathbb{Z},$$

and $\{\psi_i(t-k) : 1 \le i \le r, k \in \mathbb{Z}\}$ is an orthonormal basis of W_0. $\{\psi_{i,j,k}(t) : 1 \le i \le r; j, k \in \mathbb{Z}\}$ is an orthonormal basis of $L^2(\mathbb{R})$. Substituting (2.6) into the integral, the orthonormality condition means that the matrices of filter coefficients must satisfy

$$\sum_k \mathbf{H}_k \mathbf{H}_{2l+k}^T = \mathbf{I}_r \delta_{0,l}, \quad l \in \mathbb{Z}. \tag{2.8}$$

In the scalar case this tells us that the sum of squares of the high-pass filter coefficients is unity, and the filter is orthogonal to its even translates.

2.3 Orthogonal Discrete Multiple Wavelet Transform

Now let $x(t) \in V_0$ with $V_1 \subset V_0$. Then define

$$\mathbf{v}_{0,k}^T = \langle x(t), \Phi_{0,k}(t) \rangle \quad \text{and} \quad \mathbf{v}_{1,k}^T = \langle x(t), \Phi_{1,k}(t) \rangle,$$

where the integrals defined by the inner products are component-wise. Note that $\mathbf{v}_{0,k}$ and $\mathbf{v}_{1,k}$ are $r \times 1$ column vectors. From (2.3) we have

$$\Phi_{1,k}(t) = \sum_m \mathbf{G}_m \Phi_{0,2k+m}(t),$$

so that

$$
\begin{aligned}
\mathbf{v}_{1,k}^T &= \int x(t) \sum_m \Phi_{0,2k+m}^T(t) \mathbf{G}_m^T dt \\
&= \sum_m \left[\int x(t) \Phi_{0,2k+m}^T(t)dt \right] \mathbf{G}_m^T = \sum_m \mathbf{v}_{0,2k+m}^T \mathbf{G}_m^T \\
\implies \mathbf{v}_{1,k} &= \sum_m \mathbf{G}_{m-2k} \mathbf{v}_{0,m}.
\end{aligned}
$$

In general we obtain

$$\mathbf{v}_{j,k} = \sum_m \mathbf{G}_{m-2k} \mathbf{v}_{j-1,m} = \langle x(t), \Phi_{j,k}(t) \rangle^T. \tag{2.9}$$

Now let

$$\mathbf{w}_{1,k}^T = \langle x(t), \Psi_{1,k}(t) \rangle,$$

where $\mathbf{w}_{1,k}$ is an $r \times 1$ column vector. From (2.6) we have

$$\Psi_{1,k}(t) = \sum_m \mathbf{H}_m \Phi_{0,2k+m}(t),$$

so that

$$
\mathbf{w}_{1,k}^T = \int x(t) \sum_m \Phi_{0,2k+m}^T(t) \mathbf{H}_m^T dt
$$

$$
= \sum_m \left[\int x(t) \Phi_{0,2k+m}^T(t) dt \right] \mathbf{H}_m^T = \sum_m \mathbf{v}_{0,2k+m}^T \mathbf{H}_m^T
$$

$$
\implies \mathbf{w}_{1,k} = \sum_m \mathbf{H}_{m-2k} \mathbf{v}_{0,m}.
$$

In general we obtain

$$
\mathbf{w}_{j,k} = \sum_m \mathbf{H}_{m-2k} \mathbf{v}_{j-1,m}. \tag{2.10}
$$

Equations (2.9) and (2.10) are the *analysis* equations.

For orthonormal multiscaling and multiwavelet functions, $\{2^{-1/2}\phi_i(\frac{t}{2} - k) : 1 \le i \le r, k \in \mathbb{Z}\}$ is an orthonormal basis for V_1 and $\{2^{-1/2}\psi_i(\frac{t}{2} - k) : 1 \le i \le r, k \in \mathbb{Z}\}$ is an orthonormal basis for W_1 and hence $x(t) \in V_0 = V_1 \oplus W_1$ can be written

$$
x(t) = \sum_m \mathbf{v}_{1,m}^T \Phi_{1,m}(t) + \sum_m \mathbf{w}_{1,m}^T \Psi_{1,m}(t). \tag{2.11}
$$

But $\mathbf{v}_{0,k}^T = \langle x(t), \Phi_{0,k}(t) \rangle$ so that

$$
\mathbf{v}_{0,k}^T = \left\langle \sum_m \mathbf{v}_{1,m}^T \Phi_{1,m}(t) + \sum_m \mathbf{w}_{1,m}^T \Psi_{1,m}(t), \Phi_{0,k}(t) \right\rangle
$$

$$
= \sum_m \mathbf{v}_{1,m}^T \langle \Phi_{1,m}(t), \Phi_{0,k}(t) \rangle + \sum_m \mathbf{w}_{1,m}^T \langle \Psi_{1,m}(t), \Phi_{0,k}(t) \rangle
$$

$$
= \sum_m \mathbf{v}_{1,m}^T \mathbf{G}_{k-2m} + \sum_m \mathbf{w}_{1,m}^T \mathbf{H}_{k-2m}
$$

$$
\implies \mathbf{v}_{0,k} = \sum_m \mathbf{G}_{k-2m}^T \mathbf{v}_{1,m} + \sum_m \mathbf{H}_{k-2m}^T \mathbf{w}_{1,m}.
$$

In general the reconstruction formula for orthonormal multiscaling and multiwavelet functions takes the form

$$
\mathbf{v}_{j-1,k} = \sum_m \mathbf{G}_{k-2m}^T \mathbf{v}_{j,m} + \sum_m \mathbf{H}_{k-2m}^T \mathbf{w}_{j,m}, \tag{2.12}
$$

which is the *synthesis* equation.

2.4 Biorthogonal Discrete Multiple Wavelet Transform

In the scalar case ($r = 1$) there are no compactly supported, symmetric and orthogonal wavelets with approximation order higher than 1. For multiwavelets, the several scaling functions give more freedom, but attempts to make translates $\Phi(t - k)$ orthogonal generally lead to scaling functions with infinite support. It is helpful then to replace one orthogonal basis by two biorthogonal ones: one for decomposition ('analysis'), and one for reconstruction ('synthesis'). We introduce another MRA of $L^2(\mathbb{R})$. For $j \in \mathbb{Z}$ we define

$$\widetilde{V}_j \equiv \overline{\text{span} \left\{ 2^{-j/2} \tilde{\phi}_i \left(\frac{t}{2^j} - k \right) : 1 \leq i \leq r, k \in \mathbb{Z} \right\}}$$

with nested subspaces

$$\cdots \subset \widetilde{V}_3 \subset \widetilde{V}_2 \subset \widetilde{V}_1 \subset \widetilde{V}_0 \subset \widetilde{V}_{-1} \subset \cdots,$$

and

$$\widetilde{W}_j \equiv \overline{\text{span} \left\{ 2^{-j/2} \tilde{\psi}_i \left(\frac{t}{2^j} - k \right) : 1 \leq i \leq r, k \in \mathbb{Z} \right\}}$$

such that

$$\widetilde{V}_{j-1} = \widetilde{V}_j \oplus \widetilde{W}_j, \text{ and } \widetilde{V}_j \cap \widetilde{W}_j = \emptyset.$$

We require that

$$\widetilde{W}_j \perp V_j \iff \left\langle \tilde{\Psi} \left(\frac{t}{2^j} - k \right), \Phi \left(\frac{t}{2^j} - n \right) \right\rangle = \mathbf{0}_r, \quad j, k, n \in \mathbb{Z},$$

and

$$W_j \perp \widetilde{V}_j \iff \left\langle \Psi \left(\frac{t}{2^j} - k \right), \tilde{\Phi} \left(\frac{t}{2^j} - n \right) \right\rangle = \mathbf{0}_r, \quad j, k, n \in \mathbb{Z},$$

and also that

$$\left\langle \tilde{\Phi} \left(\frac{t}{2^j} - k \right), \Phi \left(\frac{t}{2^j} - n \right) \right\rangle = \mathbf{I}_r \delta_{k,n}, \quad j, k, n \in \mathbb{Z},$$

and

$$\left\langle \tilde{\Psi} \left(\frac{t}{2^j} - k \right), \Psi \left(\frac{t}{2^m} - n \right) \right\rangle = \mathbf{I}_r \delta_{j,m} \delta_{k,n}, \quad j, k, m, n \in \mathbb{Z},$$

this last equation defining $\tilde{\Psi}(t)$ to be the dual of $\Psi(t)$. As they define a multiresolution analysis, the dual functions satisfy

$$\tilde{\Phi}(t) = \sqrt{2} \sum_k \tilde{\mathbf{G}}_k \tilde{\Phi}(2t - k) \text{ and } \tilde{\Psi}(t) = \sqrt{2} \sum_k \tilde{\mathbf{H}}_k \tilde{\Phi}(2t - k).$$

Since we no longer are considering orthonormal multiscaling and multiwavelet functions, $y(t) \in \widetilde{V}_0 = \widetilde{V}_1 \oplus \widetilde{W}_1$ can be written

$$y(t) = \sum_m \mathbf{r}_{1,m}^T \tilde{\Phi}_{1,m}(t) + \sum_m \mathbf{s}_{1,m}^T \tilde{\Psi}_{1,m}(t),$$

where the coefficient vectors may be found as follows:

$$
\begin{aligned}
\langle y(t), \Phi_{1,k}(t) \rangle &= \sum_m \mathbf{r}_{1,m}^T \langle \tilde{\Phi}_{1,m}(t), \Phi_{1,k}(t) \rangle \\
&\quad + \sum_m \mathbf{s}_{1,m}^T \langle \tilde{\Psi}_{1,m}(t), \Phi_{1,k}(t) \rangle \\
&= \sum_m \mathbf{r}_{1,m}^T \mathbf{I}_r \delta_{m,k} = \mathbf{r}_{1,k}^T.
\end{aligned}
$$

Hence, using the same approach as that leading to (2.9), we obtain

$$\mathbf{r}_{j,k} = \sum_m \mathbf{G}_{m-2k}\mathbf{r}_{j-1,m}.$$

Similarly, we get

$$\mathbf{s}_{j,k} = \sum_m \mathbf{H}_{m-2k}\mathbf{r}_{j-1,m}.$$

Finally, we note that the synthesis equation takes the form

$$
\begin{aligned}
\mathbf{r}_{0,k}^T = \langle x(t), \Phi_{0,k}(t)\rangle \quad &= \quad \sum_m \mathbf{r}_{1,m}^T \langle \tilde{\Phi}_{1,m}(t), \Phi_{0,k}(t)\rangle \\
&\quad + \sum_m \mathbf{s}_{1,m}^T \langle \tilde{\Psi}_{1,m}(t), \Phi_{0,k}(t)\rangle \\
&= \quad \sum_m \mathbf{r}_{1,m}^T \tilde{\mathbf{G}}_{k-2m} + \sum_m \mathbf{s}_{1,m}^T \tilde{\mathbf{H}}_{k-2m} \\
\Longrightarrow \mathbf{r}_{0,k} \quad &= \quad \sum_m \tilde{\mathbf{G}}_{k-2m}^T \mathbf{r}_{1,m} + \sum_m \tilde{\mathbf{H}}_{k-2m}^T \mathbf{s}_{1,m},
\end{aligned}
$$

and in general,

$$\mathbf{r}_{j-1,k} = \sum_m \tilde{\mathbf{G}}_{k-2m}^T \mathbf{r}_{j,m} + \sum_m \tilde{\mathbf{H}}_{k-2m}^T \mathbf{s}_{j,m}.$$

3   Orthonormal multifilter banks

3.1   The scalar case

In the scalar case $r = 1$ and $\mathbf{G}_k$ and $\mathbf{H}_k$ are just single elements, g_k and h_k say, and the filters satisfy

$$\sum_k g_k g_{2l+k} = \sum_k h_k h_{2l+k} = \delta_{0,l}, \quad \text{and} \quad \sum_k g_k h_{2l+k} = 0, \quad l \in \mathbb{Z}.$$

There are many examples of filters designed for orthonormal scalar wavelet transforms. In this article we make use of two types, due to Daubechies (1992), namely the four-coefficient 'extremal phase' design, denoted D(4), and the eight-coefficient 'least-asymmetric' design, denoted LA(8). Boundaries are handled in the standard way using circular periodisation.

3.2   Multiscaling and wavelet functions

In practice multiscaling and wavelet functions are concerned with multiplicity $r = 2$. An important example was constructed by Geronimo, Hardin and Massopust (1994), which we shall refer to as the GHM system. GHM multiscaling functions have four-coefficient matrices $\mathbf{G}_0, \mathbf{G}_1, \mathbf{G}_2$ and $\mathbf{G}_3$,

$$
\mathbf{G}_0 = \begin{bmatrix} \frac{3}{5\sqrt{2}} & \frac{4}{5} \\ -\frac{1}{20} & -\frac{3}{10\sqrt{2}} \end{bmatrix}, \quad
\mathbf{G}_1 = \begin{bmatrix} \frac{3}{5\sqrt{2}} & 0 \\ \frac{9}{20} & \frac{1}{\sqrt{2}} \end{bmatrix},
$$

$$
\mathbf{G}_2 = \begin{bmatrix} 0 & 0 \\ \frac{9}{20} & -\frac{3}{10\sqrt{2}} \end{bmatrix}, \quad
\mathbf{G}_3 = \begin{bmatrix} 0 & 0 \\ -\frac{1}{20} & 0 \end{bmatrix},
$$

and four-coefficient matrices $\mathbf{H}_0, \mathbf{H}_1, \mathbf{H}_2$ and $\mathbf{H}_3$,

$$\mathbf{H}_0 = \frac{1}{10} \begin{bmatrix} -\frac{1}{2} & -\frac{3}{\sqrt{2}} \\ \frac{1}{\sqrt{2}} & 3 \end{bmatrix}, \qquad \mathbf{H}_1 = \frac{1}{10} \begin{bmatrix} \frac{9}{2} & -\frac{10}{\sqrt{2}} \\ -\frac{9}{\sqrt{2}} & 0 \end{bmatrix},$$

$$\mathbf{H}_2 = \frac{1}{10} \begin{bmatrix} \frac{9}{2} & -\frac{3}{\sqrt{2}} \\ \frac{9}{\sqrt{2}} & -3 \end{bmatrix}, \qquad \mathbf{H}_3 = \frac{1}{10} \begin{bmatrix} -\frac{1}{2} & 0 \\ -\frac{1}{\sqrt{2}} & 0 \end{bmatrix}.$$

These matrices satisfy the conditions in (2.5), (2.7) and (2.8). GHM scaling functions have four remarkable properties showing that multiwavelets can combine more useful features than scalar wavelets:

- Both scaling functions have short supports [0,1] and [0,2]. In the scalar case one would expect support [0, 3], for a scaling function satisfying a two-scale equation with four coefficients.

- The system has second order of approximation (constant and linear functions can be represented exactly by a linear combination of translates $\phi_1(t-k), \phi_2(t-k), \ k \in \mathbb{Z}$).

- Translates of the scaling functions are orthogonal.

- Both scaling functions and the wavelets are symmetric.

In the scalar case, symmetry, orthogonality and approximation order higher than 1 cannot be combined. The GHM multiscaling and multiwavelet functions are also quite smooth — almost differentiable.

Another example of symmetric orthogonal multiwavelets with approximation order 2 is due to Chui and Lian (1996). Both scaling functions are supported on [0, 2], which is slightly longer than GHM. For the CL system, only three $\mathbf{G}$ matrices and three $\mathbf{H}$ matrices are required:

$$\mathbf{G}_0 = a_1 \begin{bmatrix} 1 & -1 \\ \frac{\sqrt{7}}{2} & -\frac{\sqrt{7}}{2} \end{bmatrix}, \quad \mathbf{G}_1 = a_1 \begin{bmatrix} 2 & 0 \\ 0 & 1 \end{bmatrix}, \quad \mathbf{G}_2 = a_1 \begin{bmatrix} 1 & 1 \\ -\frac{\sqrt{7}}{2} & -\frac{\sqrt{7}}{2} \end{bmatrix};$$

$$\mathbf{H}_0 = a_2 \begin{bmatrix} 2 & -2 \\ -1 & 1 \end{bmatrix}, \quad \mathbf{H}_1 = a_2 \begin{bmatrix} -4 & 0 \\ 0 & 2\sqrt{7} \end{bmatrix}, \quad \mathbf{H}_2 = a_2 \begin{bmatrix} 2 & 2 \\ 1 & 1 \end{bmatrix}.$$

Here $a_1 = 1/(2\sqrt{2})$ and $a_2 = 1/(4\sqrt{2})$. CL scaling functions and wavelets are less smooth than GHM ones.

The multiwavelet analysis equations (2.9) and (2.10) can be realized as a matrix filter bank as in Fig. 1. With $r = 2$, as assumed hereafter, the two-channel matrix filter bank operates on two input data streams, filtering them into four output data streams, each of which is downsampled by a factor of 2. Although multifiltering is a convolution with dowsampling, the filter coefficients are matrices and the input has vector form. This novelty raises a problem which is unusual for the scalar case. In order to start the cascade algorithm one must preprocess scalar data in order to get vector input. There are different ways to do this preprocessing; see Section 5.

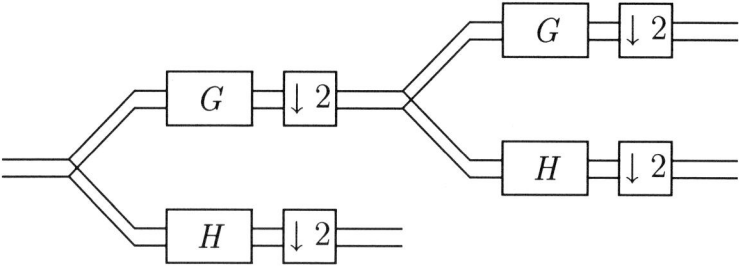

Figure 1: A multiwavelet filter bank, iterated once.

4   Biorthogonal multifilter banks

4.1   The scalar case

Here we used the biorthogonal wavelets described in Daubechies (1992) as 'variations on the spline case'. We chose the particular case with nine and seven filter coefficients for g_k and $\tilde{g}_k$; the filter coefficients are given in Daubechies (1992, Table 8.3). We shall refer to this as the Bi9-7 transform.

4.2   Multi-scaling and -wavelet functions and duals

Strela (1998) showed how to construct biorthogonal multiscaling functions. For the GHM system, one approximation order is taken from the analysis part of the multifilter bank and added to the synthesis part. GHM functions have two approximation orders, so the biorthogonal GHM (BiGHM) scaling functions $\tilde{\phi}_1(t), \tilde{\phi}_2(t)$ have approximation order 3, and the dual ones $\phi_1(t), \phi_2(t)$ have approximation order 1. Naturally therefore, $\tilde{\phi}_1(t), \tilde{\phi}_2(t)$ are smoother than $\phi_1(t), \phi_2(t)$. The scaling and wavelet coefficient matrices for the BiGHM analysis and synthesis multifilters are given by

$$\mathbf{G}_0 = a_1 \begin{bmatrix} -\frac{1}{20} & \frac{1}{20} \\ -\frac{1}{20} & \frac{1}{20} \end{bmatrix}, \quad \mathbf{G}_1 = a_1 \begin{bmatrix} \frac{1}{2} & -1 \\ \frac{1}{2} & -1 \end{bmatrix}, \quad \mathbf{G}_2 = a_1 \begin{bmatrix} \frac{11}{10} & 0 \\ 0 & \frac{11}{10} \end{bmatrix},$$

$$\mathbf{G}_3 = a_1 \begin{bmatrix} \frac{1}{2} & 1 \\ -\frac{1}{2} & -1 \end{bmatrix}, \quad \mathbf{G}_4 = a_1 \begin{bmatrix} -\frac{1}{20} & -\frac{1}{20} \\ \frac{1}{20} & \frac{1}{20} \end{bmatrix};$$

$$\mathbf{H}_0 = a_1 \begin{bmatrix} -1 & 3 \\ -\frac{2}{5} & \frac{7}{5} \end{bmatrix}, \quad \mathbf{H}_1 = a_1 \begin{bmatrix} 2 & 0 \\ 0 & 2 \end{bmatrix}, \quad \mathbf{H}_2 = a_1 \begin{bmatrix} -1 & -3 \\ \frac{2}{5} & \frac{7}{5} \end{bmatrix};$$

$$\tilde{\mathbf{G}}_0 = a_2 \begin{bmatrix} 10 & -4 \\ 15 & -7 \end{bmatrix}, \quad \tilde{\mathbf{G}}_1 = a_2 \begin{bmatrix} 20 & 0 \\ 0 & 10 \end{bmatrix}, \quad \tilde{\mathbf{G}}_2 = a_2 \begin{bmatrix} 10 & 4 \\ -15 & -7 \end{bmatrix};$$

$$\tilde{\mathbf{H}}_0 = a_1 \begin{bmatrix} \frac{1}{16} & -\frac{3}{80} \\ \frac{1}{8} & -\frac{3}{40} \end{bmatrix}, \quad \tilde{\mathbf{H}}_1 = a_1 \begin{bmatrix} -\frac{1}{4} & \frac{1}{8} \\ -\frac{1}{2} & \frac{1}{4} \end{bmatrix}, \quad \tilde{\mathbf{H}}_2 = a_1 \begin{bmatrix} \frac{3}{8} & 0 \\ 0 & \frac{9}{20} \end{bmatrix},$$

$$\tilde{\mathbf{H}}_3 = a_1 \begin{bmatrix} -\frac{1}{4} & -\frac{1}{8} \\ \frac{1}{2} & \frac{1}{4} \end{bmatrix}, \quad \tilde{\mathbf{H}}_4 = a_1 \begin{bmatrix} \frac{1}{16} & \frac{3}{80} \\ -\frac{1}{8} & -\frac{3}{40} \end{bmatrix}.$$

Here $a_1 = 1/\sqrt{2}$, and $a_2 = 1/(20\sqrt{2})$. As usual, the smoother scaling functions are used for the synthesis step.

Another biorthogonal multifilter bank (BiHermite) comes from Hermite cubic 'finite elements'. These C^1 piecewise cubics $\tilde{\phi}_1(t)$ and $\tilde{\phi}_2(t)$ are supported on $[0, 2]$. Their value and slope are specified in the integers and their second derivative jumps at unity. The Hermite multiscaling function $\tilde{\phi}(t) = [\tilde{\phi}_1(t), \tilde{\phi}_2(t)]^T$ satisfies a dilation equation with three coefficient matrices (Strela and Strang, 1995):

$$\tilde{G}_0 = \frac{\sqrt{2}}{16}\begin{bmatrix} 4 & -2 \\ 3 & -1 \end{bmatrix}, \quad \tilde{G}_1 = \frac{\sqrt{2}}{16}\begin{bmatrix} 8 & 0 \\ 0 & 4 \end{bmatrix}, \quad \tilde{G}_2 = \frac{\sqrt{2}}{16}\begin{bmatrix} 4 & 2 \\ -3 & -1 \end{bmatrix}.$$

Translates $\tilde{\phi}_1(t - k)$ and $\tilde{\phi}_2(t - k)$ form a (non-orthogonal) basis for the space V_0 of all C^1 piecewise cubics on unit intervals. Strela (1998) constructed a dual multiscaling function $\phi(t) = [\phi_1(t), \phi_2(t)]^T$ supported on $[0, 2]$ and with approximation order 4. Unfortunately it does not belong to L^2. In order to construct a more regular pair two approximation orders were given up and the obtained free parameters used towards numerical maximisation of the smoothness of $\phi_1(t), \phi_2(t)$. (This can be done by minimisation of the spectral radius of the transition operator). The scaling functions obtained have approximation order 2 and are continuous (belonging to Sobolev subspace $H^{0.828}$). The scaling coefficient matrices are as follows:

$$G_0 = \frac{1}{\sqrt{2}}\begin{bmatrix} -\frac{73}{648} & \frac{77}{324} \\ -\frac{773}{3240} & \frac{3229}{6480} \end{bmatrix}, G_1 = \frac{1}{\sqrt{2}}\begin{bmatrix} \frac{1}{2} & -\frac{89}{162} \\ \frac{187}{180} & -\frac{91}{81} \end{bmatrix}, G_2 = \frac{1}{\sqrt{2}}\begin{bmatrix} \frac{397}{324} & 0 \\ 0 & \frac{6091}{3240} \end{bmatrix},$$

$$G_3 = \frac{1}{\sqrt{2}}\begin{bmatrix} \frac{1}{2} & \frac{89}{162} \\ -\frac{187}{180} & -\frac{91}{81} \end{bmatrix}, G_4 = \frac{1}{\sqrt{2}}\begin{bmatrix} -\frac{73}{648} & -\frac{77}{324} \\ \frac{773}{3240} & \frac{3229}{6480} \end{bmatrix}.$$

The corresponding wavelet coefficient matrices for the BiHermite biorthogonal pair are

$$H_0 = a_1\begin{bmatrix} -4 & 6 \\ -2 & 2 \end{bmatrix}, \quad H_1 = a_1\begin{bmatrix} 8 & 0 \\ 0 & 8 \end{bmatrix}, \quad H_2 = a_1\begin{bmatrix} -4 & -6 \\ 2 & 2 \end{bmatrix},$$

and

$$\tilde{H}_0 = a_2\begin{bmatrix} \frac{67}{3840} & -\frac{7}{11520} \\ \frac{95}{5184} & -\frac{1}{2592} \end{bmatrix}, \tilde{H}_1 = a_2\begin{bmatrix} -\frac{1}{16} & \frac{187}{2880} \\ -\frac{89}{1296} & \frac{91}{1296} \end{bmatrix}, \tilde{H}_2 = a_2\begin{bmatrix} \frac{173}{1920} & 0 \\ 0 & \frac{13}{72} \end{bmatrix},$$

$$\tilde{H}_3 = a_2\begin{bmatrix} -\frac{1}{16} & -\frac{187}{2880} \\ \frac{89}{1296} & \frac{91}{1296} \end{bmatrix}, \quad \tilde{H}_4 = a_2\begin{bmatrix} \frac{67}{3840} & \frac{7}{11520} \\ -\frac{95}{5184} & -\frac{1}{2592} \end{bmatrix}.$$

Here $a_1 = 1/(16\sqrt{2})$ and $a_2 = 8\sqrt{2}$. Again, we use the smoother Hermite cubics for the synthesis part.

5 Preprocessing and sampling

The aim of preprocessing is to associate the given scalar input signal of length N to a sequence of length-2 vectors $\{\mathbf{v}_{0,k}\}$ for input to the analysis algorithm shown in Fig. 1. Here N is assumed to be a multiple of a power of 2, and so is of even length. The signal $\{X_k\}$ is taken to be a function $x(\cdot)$ observed at integer time points. If the preprocessing produces N length-2 vectors it is said to be an oversampling scheme, while if it produces $N/2$ length-2 vectors the result is a critical sampling. After the wavelet reconstruction (synthesis) step, e.g., inverse DMWT, a postfilter is applied. Clearly prefiltering, wavelet transform, inverse transform, and postfiltering should recover the input signal exactly if nothing else has been done.

We say that a preprocessing preserves m approximation orders if an input sequence $X_k = k^j$, $j = 0, \ldots, m - 1$, after preprocessing and filtering, gives zero high-pass output (neglecting end effects).

5.1 Repeated row preprocessing: oversampling

Here the input length-2 vectors are formed from the original time series via

$$\mathbf{v}_{0,k} = \begin{bmatrix} v_{0,k}^{(0)} \\ v_{0,k}^{(1)} \end{bmatrix} = \begin{bmatrix} X_k \\ \alpha X_k \end{bmatrix}, \quad k = 0, \ldots, N - 1.$$

Here α is a constant; it is typically chosen so that if $X_k = C$, a constant, for all k, then the output from the high-pass multifilter is zero. As shown in Heil *et al.* (1996) this can always be done if the system has approximation order higher than zero. For the GHM case, this gives $\alpha = 1/\sqrt{2}$, since

$$\begin{bmatrix} w_{1,k}^{(0)} \\ w_{1,k}^{(1)} \end{bmatrix} = [\mathbf{H}_0 + \mathbf{H}_1 + \mathbf{H}_2 + \mathbf{H}_3] \begin{bmatrix} C \\ \frac{C}{\sqrt{2}} \end{bmatrix} = \frac{1}{10} \begin{bmatrix} 8 & -\frac{16}{\sqrt{2}} \\ 0 & 0 \end{bmatrix} \begin{bmatrix} C \\ \frac{C}{\sqrt{2}} \end{bmatrix} = \begin{bmatrix} 0 \\ 0 \end{bmatrix}.$$

The output from the low-pass multifilter is simply a scaled version of the input:

$$\begin{bmatrix} v_{1,k}^{(0)} \\ v_{1,k}^{(1)} \end{bmatrix} = [\mathbf{G}_0 + \mathbf{G}_1 + \mathbf{G}_2 + \mathbf{G}_3] \begin{bmatrix} C \\ \frac{C}{\sqrt{2}} \end{bmatrix} = \frac{1}{5} \begin{bmatrix} \frac{6}{\sqrt{2}} & 4 \\ 4 & \sqrt{2} \end{bmatrix} \begin{bmatrix} C \\ \frac{C}{\sqrt{2}} \end{bmatrix} = \sqrt{2} \begin{bmatrix} C \\ \frac{C}{\sqrt{2}} \end{bmatrix}.$$

In the case of the CL, BiGHM and BiHermite systems $\alpha = 0$, reflecting the fact that one of the scaling functions is antisymmetric. Obviously repeated row preprocessing doubles the number of input data points, and preserves only one approximation order: if $X_k = k$ then after repeated row preprocessing the output of the high-pass multifilter will not be zero even if the corresponding multiwavelets have two vanishing moments.

If we represent the preprocessing step by a matrix multiplication, $\mathcal{P}\mathbf{X} = \mathbf{V}_0$, where $\mathcal{P}$ is $2N \times N$, $\mathbf{X}$ is $N \times 1$, and $\mathbf{V}_0$ is $2N \times 1$, then in detail we have

$$
\begin{bmatrix}
1 & 0 & 0 & \cdots \\
\alpha & 0 & 0 & \cdots \\
0 & 1 & 0 & \cdots \\
0 & \alpha & 0 & \cdots \\
& & & \ddots
\end{bmatrix}
\begin{bmatrix}
X_0 \\
X_1 \\
X_2 \\
X_3 \\
\vdots
\end{bmatrix}
=
\begin{bmatrix}
X_0 \\
\alpha X_0 \\
\cdot X_1 \\
\alpha X_1 \\
\vdots
\end{bmatrix}
=
\begin{bmatrix}
v_{0,0}^{(0)} \\
v_{0,0}^{(1)} \\
v_{0,1}^{(0)} \\
v_{0,1}^{(1)} \\
\vdots
\end{bmatrix}.
$$

5.2   Matrix (approximation) preprocessing: critical sampling

For the GHM multifilter bank we can conveniently use another type of preprocessing which preserves the sampling rate ('critical sampling'), and approximation order 2. This method is called 'approximation' preprocessing (Strela *et al.* 1999; Xia *et al.* 1996). Let $x(t)$ belong to V_0 generated by translates of the GHM scaling functions, so that

$$
x(t) = \sum_k \left[v_{0,k}^{(0)} \phi_1(t-k) + v_{0,k}^{(1)} \phi_2(t-k) \right].
$$

Suppose that the input sequence samples $x(\cdot)$ at the half-integers, $X_{2k} = x(k)$, $X_{2k+1} = x(k+1/2)$. Now $\phi_1(t)$ vanishes at all integer points and the only integer at which $\phi_2(t)$ is non-zero is unity. Hence,

$$
\begin{aligned}
X_{2k} &= v_{0,k-1}^{(1)} \phi_2(1), \\
X_{2k+1} &= v_{0,k}^{(0)} \phi_1(1/2) + v_{0,k-1}^{(1)} \phi_2(3/2) + v_{0,k}^{(1)} \phi_2(1/2), \\
X_{2k+2} &= v_{0,k}^{(1)} \phi_2(1).
\end{aligned}
$$

Thus,

$$
\begin{aligned}
v_{0,k-1}^{(1)} &= X_{2k}/\phi_2(1), \\
v_{0,k}^{(1)} &= X_{2k+2}/\phi_2(1), \\
v_{0,k}^{(0)} &= \frac{X_{2k+1} - v_{0,k-1}^{(1)} \phi_2(3/2) - v_{0,k}^{(1)} \phi_2(1/2)}{\phi_1(1/2)} \\
&= \frac{X_{2k+1}}{\phi_1(1/2)} - \frac{X_{2k}\phi_2(3/2)}{\phi_1(1/2)\phi_2(1)} - \frac{X_{2k+2}\phi_2(1/2)}{\phi_1(1/2)\phi_2(1)} \\
&= \frac{X_{2k+1}\phi_2(1) - \phi_2(1/2)[X_{2k} + X_{2k+2}]}{\phi_1(1/2)\phi_2(1)},
\end{aligned}
$$

$$(5.1)$$

$$(5.2)$$

where we have used the fact that $\phi_2(1/2) = \phi_2(3/2)$. Equations (5.1) and (5.2) give the required vector $\mathbf{v}_{0,k}$ for $k = 0, \ldots, N/2 - 1$.

In fact approximation prefiltering is just a special case of matrix prefiltering,

$$
\mathbf{v}_{0,k} = \sum_{m=0}^{M} \mathbf{P}_m \begin{bmatrix} X_{2(m+k)} \\ X_{2(m+k)+1} \end{bmatrix},
$$

where $\mathbf{P}_0, \mathbf{P}_1, \ldots, \mathbf{P}_M$ are 2×2 matrices, for which $M = 1$ and

$$\mathbf{P}_0 = \begin{bmatrix} \frac{-\phi_2(1/2)}{\phi_1(1/2)\phi_2(1)} & \frac{1}{\phi_1(1/2)} \\ 0 & 0 \end{bmatrix} = \begin{bmatrix} \frac{3}{8\sqrt{6}} & \frac{10}{8\sqrt{6}} \\ 0 & 0 \end{bmatrix},$$

$$\mathbf{P}_1 = \begin{bmatrix} \frac{-\phi_2(1/2)}{\phi_1(1/2)\phi_2(1)} & 0 \\ \frac{1}{\phi_2(1)} & 0 \end{bmatrix} = \begin{bmatrix} \frac{3}{8\sqrt{6}} & 0 \\ \frac{1}{\sqrt{3}} & 0 \end{bmatrix}.$$

(5.3)

For CL, BiGHM and BiHermite systems, a single matrix preprocessor, preserving two approximation orders, can be constructed. The single matrix is given by

$$\mathbf{P}_0 = \begin{bmatrix} \frac{1}{4} & \frac{1}{4} \\ \frac{1}{b} & \frac{1}{-b} \end{bmatrix}$$

(5.4)

where $b = 1 + \sqrt{7}$ for CL, $b = -6$ for BiGHM, $b = -3$ for BiHermite. One can easily check that in response to constant $X_k = 1$ and linear $X_k = k$ inputs, this preprocessor produces output

$$\mathbf{v}_{0,k} = \mathbf{P}_0 \begin{bmatrix} X_{2k} \\ X_{2k+1} \end{bmatrix},$$

such that after high-pass filtering $\sum_m \mathbf{H}_{m-2k}\mathbf{v}_{0,m}$ the result is zero (disregarding end effects). Furthermore, it can be shown that for any multiwavelet system with more than one vanishing moment, one scaling function symmetric, the other antisymmetric and both of them supported on $[0,2]$, such a matrix prefilter exists. Parameter b in $\mathbf{P}_0$ should be equal to twice the value of the derivative of the antisymmetric scaling function at $t = 1$. Zero-length ($M = 0$) prefilters are very desirable because they do not introduce correlation between pairs $\mathbf{v}_{0,k}$; such prefiltering will also be called 'approximation' here.

For general length prefilters, if we represent the matrix preprocessing by a single matrix multiplication, $\mathcal{P}\mathbf{X} = \mathbf{V}_0$, where now $\mathcal{P}$ is $N \times N$, $\mathbf{X}$ is $N \times 1$, and $\mathbf{V}_0$ is $N \times 1$, then in detail we have

$$\begin{bmatrix} \mathbf{P}_0 & \mathbf{P}_1 & \cdots & \mathbf{P}_M & \mathbf{0}_2 & \cdots & \cdots & \cdots \\ \mathbf{0}_2 & \mathbf{P}_0 & \mathbf{P}_1 & \cdots & \mathbf{P}_M & \mathbf{0}_2 & \cdots & \cdots \\ & & & & & & \ddots \end{bmatrix} \begin{bmatrix} X_0 \\ X_1 \\ X_2 \\ X_3 \\ \vdots \end{bmatrix} = \begin{bmatrix} v_{0,0}^{(0)} \\ v_{0,0}^{(1)} \\ v_{0,1}^{(0)} \\ v_{0,1}^{(1)} \\ \vdots \end{bmatrix},$$

where of course the $\mathbf{P}_m$ matrices are 2×2 and $\mathbf{0}_2$ is a 2×2 matrix of zeros.

5.3   Boundaries

Boundaries are handled by symmetric data extension for both critically sampled and oversampled schemes. As pointed out in Strela *et al.* (1999), symmetric extension – which is particularly well matched to approximation pre-processing – can be implemented only with symmetric and/or antisymmetric filter banks, exactly the types under study here.

5.4   Energy preservation

Let us represent whichever wavelet transformation is chosen (scalar, multi-, orthogonal or biorthogonal), by the matrix transformation $\mathcal{M}$. If repeated row preprocessing is used, $\mathcal{M}$ is $2N \times 2N$ while if critical sampling is applied, $\mathcal{M}$ is $N \times N$.

- If $\mathcal{M}$ is an orthonormal transform, then the energy in the sequence $\mathcal{P}\mathbf{X}$ (i.e., the sum of squares of the elements), is unchanged after application of the transform, $||\mathcal{M}\mathcal{P}\mathbf{X}||^2 = ||\mathcal{P}\mathbf{X}||^2$. In this case the only opportunity for a change in scale is the preprocessing, $||\mathbf{X}||^2 \to ||\mathcal{P}\mathbf{X}||^2$.

- If the transform is biorthogonal (scalar or multiwavelet), then both the preprocessing and matrix transformation will change the input energy. See Section 6 for details.

These facts have important ramifications when studying the variance of stochastic sequences subjected to preprocessing and the DMWT, as will now be investigated.

6   Stochastic sequences and transform covariance

Suppose, as is often the case, the observed signal comprises true (deterministic) signal $\{D_k\}$ plus (stochastic) zero-mean noise, $\{\epsilon_k\}$, i.e., $X_k = D_k + \epsilon_k$, and then the effect of preprocessing is given by $\mathcal{P}\mathbf{X} = \mathcal{P}\mathbf{D} + \mathcal{P}\epsilon$. The preprocessing is followed by the DMWT in one of the manifestations discussed in Sections 3 and 4. Hence,

$$\mathcal{M}\mathcal{P}\mathbf{X} = \mathcal{M}\mathcal{P}\mathbf{D} + \mathcal{M}\mathcal{P}\epsilon. \tag{6.1}$$

The stochastic part, $\mathcal{M}\mathcal{P}\epsilon$, denoted $\mathbf{Z}$ for convenience, has variance-covariance matrix given by $\text{cov}\{\mathbf{Z}\} = \mathcal{M}\mathcal{P}\boldsymbol{\Sigma}\mathcal{P}^T\mathcal{M}^T$, where $\boldsymbol{\Sigma}$ is the variance-covariance matrix of the noise $\{\epsilon_k\}$. Henceforth we let the input be white noise with mean zero and variance σ_ϵ^2, so that $\boldsymbol{\Sigma}$ is the identity matrix multiplied by σ_ϵ^2, and

$$\text{cov}\{\mathbf{Z}\} = \mathcal{M}[\sigma_\epsilon^2 \mathcal{P}\mathcal{P}^T]\mathcal{M}^T = \sigma_\epsilon^2 \mathcal{M}\mathcal{P}\mathcal{P}^T\mathcal{M}^T. \tag{6.2}$$

Note that the matrix $\sigma_\epsilon^2 \mathcal{P}\mathcal{P}^T$ is due to the input white noise and preprocessing; if this matrix were the identity, then an orthonormal DMWT would imply that $\text{cov}\{\mathbf{Z}\}$ would be the identity, since for an orthonormal transform $\mathcal{M}\mathcal{M}^T$ would be equal to the identity matrix.

Apart from the scale factor σ_ϵ^2, the variance of the white noise, the covariance matrix $\text{cov}\{\mathbf{Z}\}$ depends only on the form of preprocessing used, through

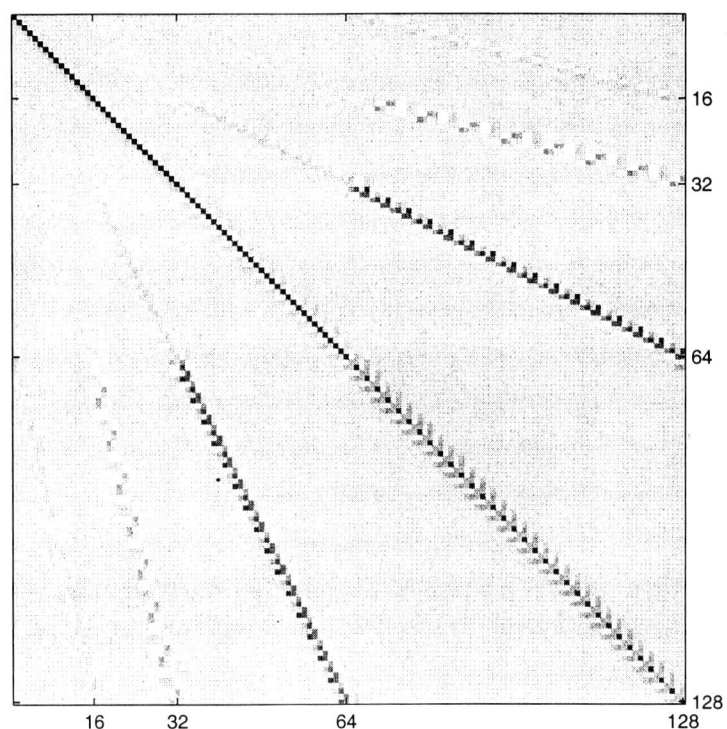

Figure 2: Grey-scale covariance matrix cov$\{\mathbf{Z}\}$ of size 128×128 for the GHM system using repeated row preprocessing with $\sigma_\epsilon^2 = 1$. Positive values are darker than the background, negative values are lighter.

$\mathcal{P}$, and the choice of wavelet transform, through $\mathcal{M}$. We have computed this covariance matrix for a number of useful cases, with σ_ϵ^2 set to unity. The covariance matrices are all 128×128. Rows numbered 65-128 hence correspond to level $j = 1$, those numbered 33-64 are for level $j = 2$, and 17-32 for level $j = 3$. For repeated row preprocessing N is 64, while for critical sampling $N = 128$.

6.1 GHM multiwavelet system

Let us look first at the GHM system using repeated row preprocessing. To understand the structure of the $\sigma_\epsilon^2 \mathcal{P}\mathcal{P}^T$ matrix, consider first the case $N = 2$. Then the covariance matrix of $[v_{0,0}^{(0)}, v_{0,0}^{(1)}, v_{0,1}^{(0)}, v_{0,1}^{(1)}]^T$ is given by $\sigma_\epsilon^2 \mathcal{P}\mathcal{P}^T$, i.e.,

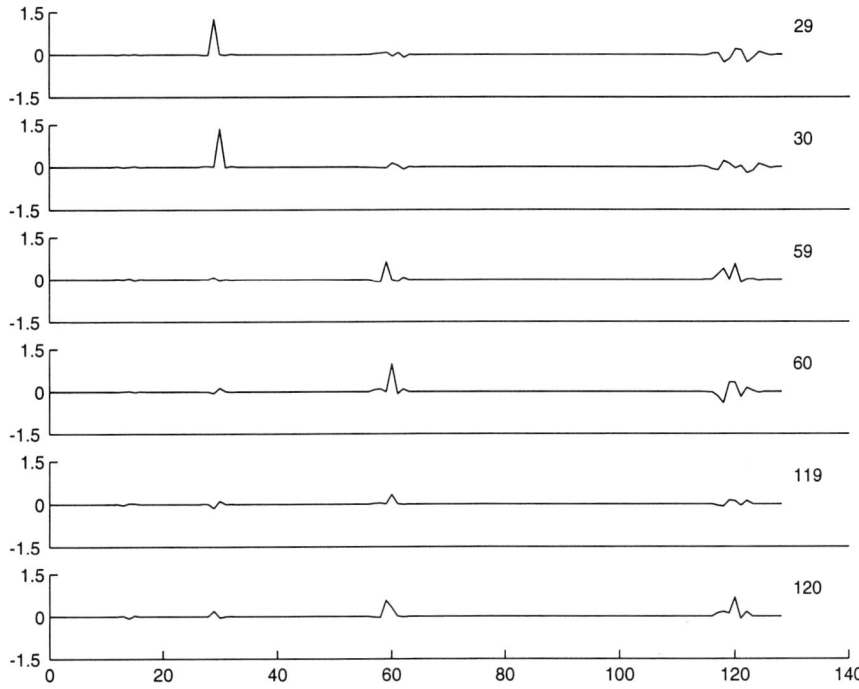

Figure 3: Various rows through the matrix.

by

$$
\sigma_\epsilon^2
\begin{bmatrix}
1 & 0 \\
\alpha & 0 \\
0 & 1 \\
0 & \alpha
\end{bmatrix}
\begin{bmatrix}
1 & \alpha & 0 & 0 \\
0 & 0 & 1 & \alpha
\end{bmatrix}
= \sigma_\epsilon^2
\begin{bmatrix}
1 & \alpha & 0 & 0 \\
\alpha & \alpha^2 & 0 & 0 \\
0 & 0 & 1 & \alpha \\
0 & 0 & \alpha & \alpha^2
\end{bmatrix}.
$$

With $\alpha = 1/\sqrt{2}$, we see that the variances on the main diagonal are alternately σ_ϵ^2 and $\sigma_\epsilon^2/2$, i.e., an 'average' variance of $0.75\sigma_\epsilon^2$, while the off-diagonal terms are each $\sigma_\epsilon^2/\sqrt{2}$. Such a covariance matrix is far from the identity matrix multiplied by σ_ϵ^2, so that even though the GHM DMWT is orthogonal, the covariance matrix of the result, cov$\{\mathbf{Z}\}$ will be far from a scaled identity matrix. Fig. 2 shows the covariance cov$\{\mathbf{Z}\}$ for the GHM system using repeated row preprocessing. Looking first at the variances along the diagonal in Fig. 2, we note these increase with level j. Fig. 3 shows various rows through the matrix; we see that at level 1, row 120, that there is significant covariance just off the main diagonal, but on the adjacent line 119 this is markedly different due to the alternating variances on the inputs to the GHM DMWT. Moving through the levels, we see that the covariance just off the main diagonal essentially disappears; see for example the main diagonal spikes on lines 29 and 30 ($j = 3$.) There is notable covariance between all scales.

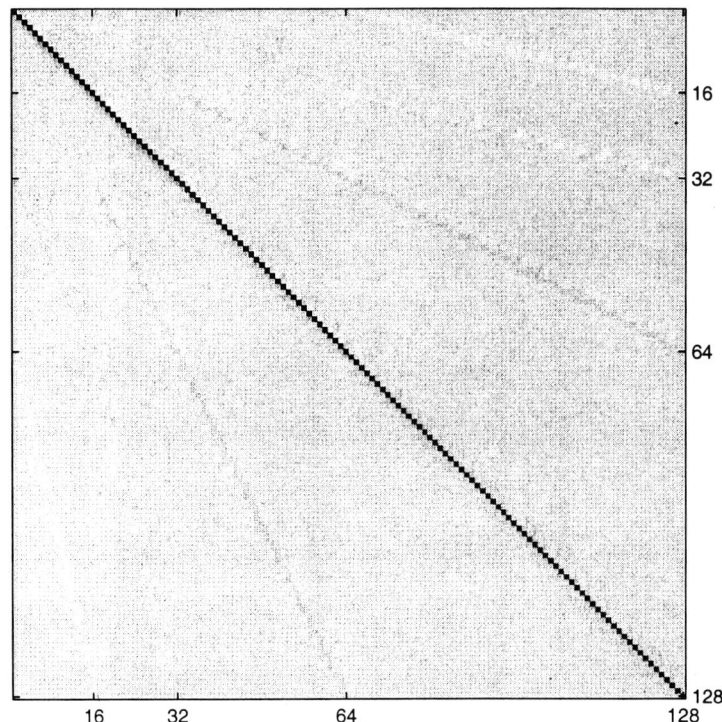

Figure 4: Grey-scale covariance matrix cov$\{\mathbf{Z}\}$ of size 128×128 for the GHM system using approximation preprocessing with $\sigma_\epsilon^2 = 1$.

If we let c denote a variance deflation, or inflation, factor, by which σ_ϵ^2 is changed to an average variance $c\sigma_\epsilon^2$, then we see that for GHM repeated row preprocessing $c = 0.75$. Since the GHM DMWT is orthonormal, it will not change the scale of the stochastic input. If we add the variances of the coefficients at all levels in Fig. 2, and divide by the total number of coefficients, we obtain an 'average' variance deflation factor of $c = 0.75$, as expected.

By way of contrast Fig. 4 shows the covariance for the GHM system using approximation preprocessing. In implementing approximation preprocessing we have rescaled the elements of the matrix prefilters in (5.3) by multiplying all elements by $\sqrt{3}$. Then, (5.2) and (5.1) become

$$v_{0,k}^{(0)} = \frac{3}{8\sqrt{2}}X_{2k} + \frac{10}{8\sqrt{2}}X_{2k+1} + \frac{3}{8\sqrt{2}}X_{2k+2} \text{ and } v_{0,k}^{(1)} = X_{2k+2}.$$

It then follows for example that

$$\text{var}\{v_{0,0}^{(0)}\} = (118/128)\sigma_\epsilon^2 \approx 0.922\sigma_\epsilon^2 \text{ and } \text{var}\{v_{0,0}^{(1)}\} = \sigma_\epsilon^2,$$

$$\text{cov}\{v_{0,0}^{(0)}, v_{0,0}^{(1)}\} = \frac{3}{8\sqrt{2}}\sigma_\epsilon^2 \approx 0.265\sigma_\epsilon^2 = \text{cov}\{v_{0,0}^{(1)}, v_{0,1}^{(0)}\},$$

$$\mathrm{cov}\{v_{0,0}^{(0)}, v_{0,1}^{(0)}\} = \frac{9}{128}\sigma_\epsilon^2 \approx 0.070\sigma_\epsilon^2,$$

etc., so that for $N = 4$ the covariance matrix of $[v_{0,0}^{(0)}, v_{0,0}^{(1)}, v_{0,1}^{(0)}, v_{0,1}^{(1)}]^T$ is given approximately by $\sigma_\epsilon^2 \mathcal{P}\mathcal{P}^T$, i.e., by

$$\sigma_\epsilon^2 \begin{bmatrix} 0.922 & 0.265 & 0.070 & 0 \\ 0.265 & 1 & 0.265 & 0 \\ 0.070 & 0.265 & 0.922 & 0.265 \\ 0 & 0 & 0.265 & 1.0 \end{bmatrix}.$$

The variances on the main diagonal are alternately $0.922\sigma_\epsilon^2$ and σ_ϵ^2, i.e., an 'average' variance of $0.961\sigma_\epsilon^2$, while the off-diagonal terms are relatively small; such a covariance matrix is not that far from a scaled identity matrix, so the covariance matrix after the GHM DMWT, $\mathrm{cov}\{\mathbf{Z}\}$, will not be too far from one either. Referring to Fig. 4 we see that the variances along the diagonal again increase with level, but there is virtually zero covariance just off the main diagonal, and the covariances between scales are very small. If we add the variances of the coefficients at all levels in Fig. 4, and divide by the total number of coefficients, we obtain the expected 'average' variance deflation factor of $c = 0.961$.

6.2 CL multiwavelet system

Let us look now at the CL system, firstly using repeated row preprocessing. Considering the case $N = 2$ the covariance matrix of $[v_{0,0}^{(0)}, v_{0,0}^{(1)}, v_{0,1}^{(0)}, v_{0,1}^{(1)}]^T$ is given by $\sigma_\epsilon^2 \mathcal{P}\mathcal{P}^T$, i.e., by

$$\sigma_\epsilon^2 \begin{bmatrix} 1 & 0 & 0 & 0 \\ 0 & 0 & 0 & 0 \\ 0 & 0 & 1 & 0 \\ 0 & 0 & 0 & 0 \end{bmatrix}. \tag{6.3}$$

While there are no off-diagonal terms, the variances on the main diagonal are alternately σ_ϵ^2 and zero, i.e., an 'average' variance of $0.5\sigma_\epsilon^2$. Such a covariance matrix is far from a scaled identity matrix. Fig. 5 shows the covariance $\mathrm{cov}\{\mathbf{Z}\}$ for the CL system using repeated row preprocessing. Fig. 6 shows various rows through the matrix; we see that at level 1, row 119, there is significant covariance just off the main diagonal, but on the adjacent line 120 there is little variance on the main diagonal due to the alternating variances on the inputs to the CL DMWT. With increasing level the covariance just off the main diagonal essentially disappears; see for example the main diagonal spikes on lines 29 and 30. There is notable covariance between all scales. One notable feature in Fig. 5 is the apparent sharp decrease in variance at level 2. This is due to the alternately low and high diagonal variance values at level 1 being thoroughly redistributed by the multifiltering which gives the level 2 coefficients.

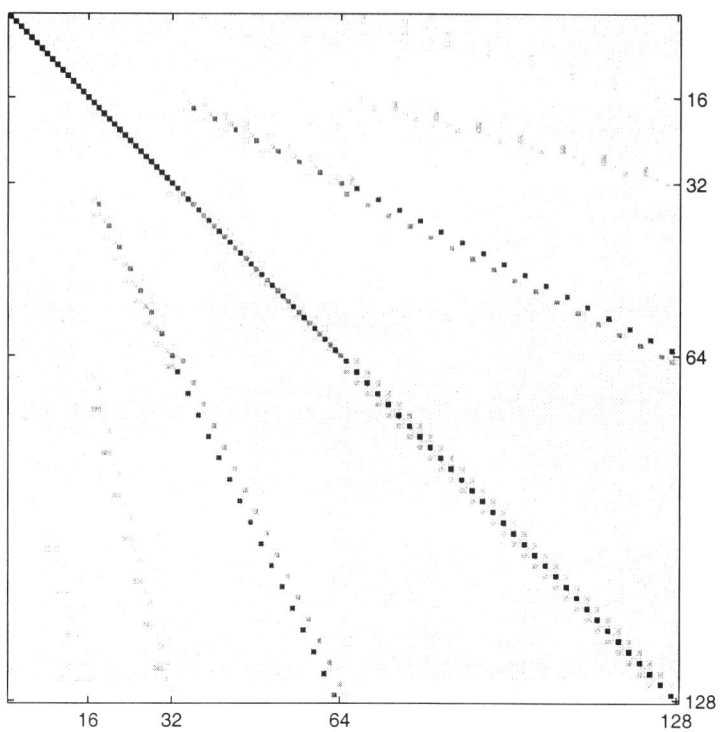

Figure 5: Grey-scale covariance matrix cov$\{\mathbf{Z}\}$ of size 128×128 for the CL system using repeated row preprocessing with $\sigma_\epsilon^2 = 1$.

If we add the variances of the coefficients at all levels in Fig. 5 and divide by the total number of coefficients, we obtain the expected 'average' variance deflation factor of $c = 0.5$.

For CL with preprocessing using the matrix prefilter given by (5.4) with $b = 1 + \sqrt{7}$,

$$v_{0,k}^{(0)} = \frac{1}{4}X_{2k} + \frac{1}{4}X_{2k+1} \text{ and } v_{0,k}^{(1)} = \frac{1}{1+\sqrt{7}}X_{2k} - \frac{1}{1+\sqrt{7}}X_{2k+1}.$$

It then follows for example that

$$\text{var}\{v_{0,k}^{(0)}\} = (2/16)\sigma_\epsilon^2 = 0.125\sigma_\epsilon^2 \text{ and } \text{var}\{v_{0,k}^{(1)}\} = [2/(1+\sqrt{7})^2]\sigma_\epsilon^2 \approx 0.15\sigma_\epsilon^2.$$

Since $\mathcal{P}\mathcal{P}^T$ is diagonal, all covariances are null, so that for $N = 4$ for example the covariance matrix of $[v_{0,0}^{(0)}, v_{0,0}^{(1)}, v_{0,1}^{(0)}, v_{0,1}^{(1)}]^T$ is given by $\sigma_\epsilon^2 \mathcal{P}\mathcal{P}^T$, i.e., approximately by

$$\sigma_\epsilon^2 \begin{bmatrix} 0.125 & 0 & 0 & 0 \\ 0 & 0.15 & 0 & 0 \\ 0 & 0 & 0.125 & 0 \\ 0 & 0 & 0 & 0.15 \end{bmatrix}.$$

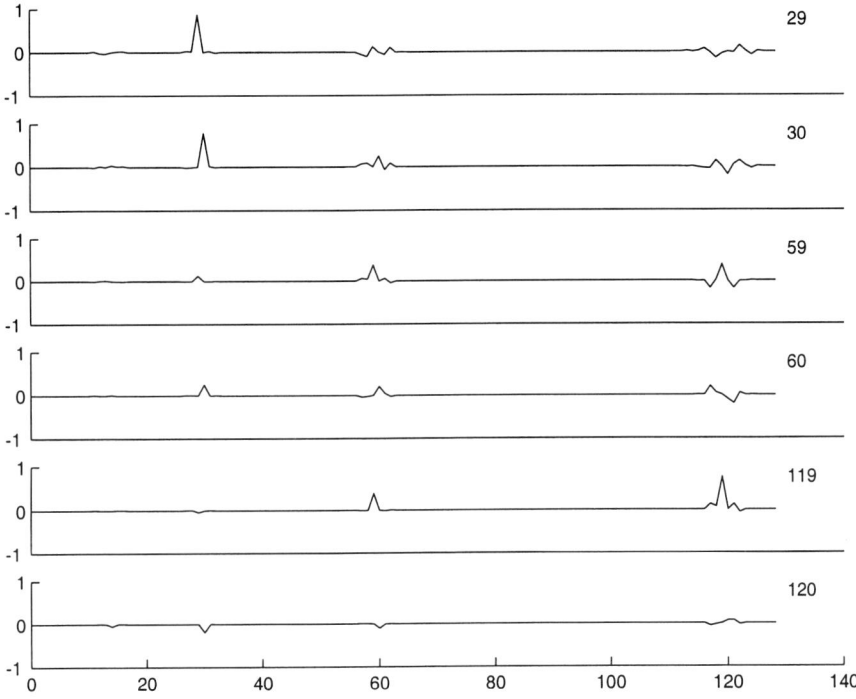

Figure 6: Various rows through the matrix.

The average variance is 0.138. The alternating-size diagonal (variance) elements are the only manifestation of a non-white noise as input to the CL DMWT. Following the CL DMWT the covariance matrix of Fig. 7 is obtained. As would be anticipated, this is close to a scaled identity matrix. At level 1 the alternating diagonal elements have been changed to 0.131 and 0.149, while at level 3 they are almost equal at 0.128 and 0.131.

If we add the variances of the coefficients at all levels in Fig. 7, and divide by the total number of coefficients, we obtain the expected 'average' variance deflation factor of $c = 0.138$.

6.3 Biorthogonal GHM and Hermite cubics multiwavelet system

The covariance matrix cov{**Z**} for the BiGHM system using repeated row preprocessing with $\alpha = 0$ was computed. The matrix $\sigma_\epsilon^2 \mathcal{P}\mathcal{P}^T$ is as for the CL system (e.g., equation (6.3)). The variances on the main diagonal again alternate in size. For example, at level 1 they are 0.16 and 3, and at level 3 they are 5.19 and 15.31.

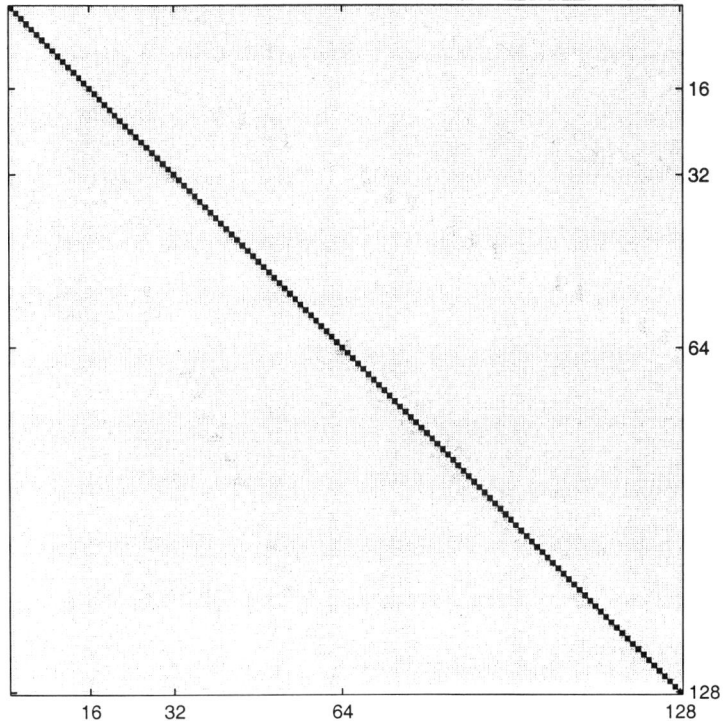

Figure 7: Grey-scale covariance matrix cov$\{\mathbf{Z}\}$ of size 128×128 for the CL system using approximation preprocessing with $\sigma_\epsilon^2 = 1$.

For BiGHM with preprocessing using the matrix prefilter given by (5.4) with $b = -6$, $\mathcal{PP}^T$ is diagonal and all covariances are null. The variances on the main diagonal of cov$\{\mathbf{Z}\}$ again alternate in size. For example, at level 1 they are 0.24 and 0.88, and at level 3 they are 1.02 and 2.84.

The covariance matrix cov$\{\mathbf{Z}\}$ for the BiHermite system using repeated row preprocessing with $\alpha = 0$ was also computed. At level 1 the alternating variances are 0.19 and 0.016, and at level 3 they are 1.81 and 1.63.

For BiHermite with preprocessing using the matrix prefilter given by (5.4) with $b = -3$, $\mathcal{PP}^T$ is diagonal and all covariances are null. The variances on the main diagonal of cov$\{\mathbf{Z}\}$ again alternate in size; at level 1 they are 0.055 and 0.033, and at level 3 they are 0.32 and 0.29.

These biorthogonal transforms are not orthonormal. Table 1 shows the ratio of output variance to input variance for the above biorthogonal transforms. The results are an average over 100 different input Gaussian white noise sequences of the lengths given, and are related to the maximum levels to which the transforms are taken. (Note that a given N means approximation preprocessing methods would start with two vectors of length $N/2$, and

N	max. level J	BiGHMr	BiGHMa	BiHr	BiHa	Bi9-7
256	5	4.059	0.960	0.805	0.188	1.032
	4	3.609	0.842	0.935	0.220	1.032
	3	2.704	0.636	1.178	0.271	1.029
	2	1.537	0.346	1.294	0.284	1.024
512	6	4.432	1.045	0.737	0.170	1.034
	5	4.105	0.975	0.806	0.189	1.029
	4	3.617	0.849	0.957	0.223	1.028
	3	2.711	0.633	1.171	0.271	1.026
	2	1.543	0.350	1.312	0.293	1.022
1024	7	4.579	1.089	0.696	0.161	1.029
	6	4.440	1.054	0.739	0.172	1.028
	5	4.211	0.994	0.811	0.189	1.027
	4	3.631	0.851	0.943	0.220	1.026
	3	2.710	0.631	1.195	0.276	1.024
	2	1.540	0.350	1.325	0.293	1.020
2048	8	4.709	1.113	0.686	0.158	1.028
	7	4.627	1.102	0.709	0.164	1.027
	6	4.458	1.057	0.745	0.173	1.027
	5	4.172	0.987	0.815	0.190	1.026
	4	3.626	0.852	0.958	0.225	1.025
	3	2.729	0.637	1.192	0.277	1.023
	2	1.540	0.350	1.313	0.291	1.020
4096	9	4.723	1.124	0.674	0.155	1.027
	8	4.721	1.122	0.687	0.158	1.026
	7	4.622	1.095	0.703	0.162	1.026
	6	4.462	1.056	0.745	0.173	1.027
	5	4.171	0.991	0.818	0.192	1.025
	4	3.638	0.855	0.967	0.226	1.025
	3	2.726	0.635	1.178	0.271	1.023
	2	1.539	0.348	1.319	0.292	1.020

Table 1: Variance inflation/deflation factors for biorthogonal transforms.

repeated row preprocessing would start with two vectors of length N. Also if the maximum level $J = j_0$, then j_0 steps of the cascade algorithm were used for repeated row methods and Bi9-7, and $j_0 - 1$ steps of the cascade were used for approximation methods, in order that the number of untouched scaling coefficients is the same.) For any particular transform, we see that the variance inflation/deflation factors are almost independent of sample size, but depend strongly on the maximum level to which the transform is taken.

6.4   Scalar Bi9-7 transform

The covariance matrix for the biorthogonal 9-7 transform is not far from the identity, meaning that this transform is not far from orthonormal. This is perhaps not too surprising in view of the discussion on 'Biorthogonal bases close to an orthonormal basis' in Daubechies (1992, Section 8.3.5). Table 1 shows that 1.026 is a good overall variance inflation factor for different sample sizes and maximum levels of the transform; hence, as expected, scale is nearly preserved ($c \approx 1$).

6.5   D4 and LA8 scalar transforms

Both these transforms are orthonormal, so that the covariance matrix is given simply by $\mathrm{cov}\{\mathbf{Z}\} = \sigma_\epsilon^2 \mathbf{I}_N$.

7   Thresholding

Starting with a signal $X_0, \ldots, X_{N-1}$ and referring to Fig. 1, we recall that with repeated row preprocessing $N' = N$ vectors $\mathbf{v}_{0,k}$ enter the pyramid algorithm, while with critically sampled preprocessing (such as approximation preprocessing) $N' = N/2$ vectors $\mathbf{v}_{0,k}$ enter the pyramid algorithm. After J steps of the pyramid algorithm of Fig. 1 we can write the wavelet coefficients in (6.1) in element-by-element form as

$$\mathbf{w}_{j,k} = \mathbf{s}_{j,k} + \mathbf{z}_{j,k}, \qquad j = 1, \ldots, J; \; k = 0, \ldots, N'_j - 1, \qquad (7.1)$$

where $N'_j = N'/2^j$, $\{\mathbf{w}_{j,k}\}$ are the coefficients corresponding to the transformed observations $\mathcal{MP}\mathbf{X}$, $\{\mathbf{s}_{j,k}\}$ are the coefficients corresponding to the transformed deterministic signal $\mathcal{MP}\mathbf{D}$, and $\{\mathbf{z}_{j,k}\}$ are the coefficients corresponding to the transformed stochastic noise $\mathcal{MP}\boldsymbol{\epsilon}$. There are also scaling coefficients $\mathbf{v}_{J,k}$, $k = 0, \ldots, N'_J - 1$. The idea of thresholding is to produce a denoised estimate using the following steps:

1. Apply preprocessing and the DMWT in one of the forms discussed out to level J.

2. Apply a thresholding procedure to the resulting empirical wavelet coefficients $\mathbf{w}_{j,k}$, $j = 1, \ldots, J$, $k = 0, \ldots, N'_j - 1$ (leaving the remaining empirical scaling coefficients $\mathbf{v}_{J,k}$, $k = 0, \ldots, N'_J - 1$ entirely alone as these carry low-frequency 'smooth' information).

3. Invert the DMWT and apply the postprocessor, producing a denoised estimate $\{\hat{X}_0, \ldots, \hat{X}_{N-1}\}$ of the input signal.

The thresholding step can be carried out as a scalar or vector operation, as discussed below.

7.1 Scalar thresholding

Suppose that we composite together the empirical wavelet coefficient vectors $\mathbf{w}_{j,k}, j = 1, \ldots, J; k = 0, \ldots, N'_j - 1$, and treat them as one scalar sequence of coefficients,

$$w_{1,0}^{(0)}, w_{1,0}^{(1)}, \ldots, w_{1,N_1'-1}^{(0)}, w_{1,N_1'-1}^{(1)}, \ldots, w_{J,0}^{(0)}, w_{J,0}^{(1)}, \ldots, w_{J,N_J'-1}^{(0)}, w_{J,N_J'-1}^{(1)}.$$

We know that *if* both preprocessing and transform were orthonormal, then white noise $\{\epsilon_k\}$ with mean zero and variance σ_ϵ^2 would be transformed to another scalar sequence of white noise coefficients with mean zero and variance σ_ϵ^2.

The idea of universal thresholding for such scalar sequences of white noise wavelet coefficients was introduced by Donoho and Johnstone (1994). The basic idea is that if the signal component is in fact zero, then with high probability the combination of (zero) signal plus noise should not exceed the threshold level. Suppose $\epsilon_1, \ldots, \epsilon_M$ is a sequence of independent and identically distributed Gaussian (normal) random variables with mean zero and variance σ_ϵ^2, then as $M \to \infty$,

$$P_M \equiv Pr\left[\max_i |\epsilon_i| > \sqrt{(2\sigma_\epsilon^2 \log M)}\right] \to 0.$$

This means that, asymptotically, if the signal component is in fact zero, then the probability of a 'false alarm' will tend to zero, so that the combination of (zero) signal plus noise will not exceed the threshold level, and will hence be set to zero. The scalar threshold is thus chosen as $T = T_s = \sqrt{(2\sigma_\epsilon^2 \log M)}$. The wavelet coefficients $w_{j,k}^{(i)}$ are thresholded using

$$\hat{w}_{j,k}^{(i)} = \delta(w_{j,k}^{(i)}, T_s), \quad i = 0, 1, \ 1 \leq j \leq J, \ k = 0, \ldots, N'_j - 1.$$

where $\delta(\cdot, \cdot)$ could be the hard-threshold rule:

$$\delta_H(w_{j,k}^{(i)}, T_s) = w_{j,k}^{(i)} \, \mathbf{1}\{|w_{j,k}^{(i)}| > T_s\}.$$

The inverse DMWT is then applied to the $\{\hat{\mathbf{w}}_{j,k}\}$ followed by the postprocessor.

The universal threshold is attractively simple, but is strictly suitable only when thresholding a Gaussian scalar sequence of white noise wavelet coefficients of mean zero and variance σ_ϵ^2. While the scalar DWT will not change the variance of input noise, we have seen that following preprocessing and the DMWT, the variances of the wavelet coefficients $w_{j,k}^{(0)}$ and $w_{j,k}^{(1)}$ can differ substantially, and we have defined 'average variances'; for example for GHM with repeated row preprocessing the noise variance σ_ϵ^2 is deflated by a factor $c = 0.75$ to $0.75\sigma_\epsilon^2$. Not only this, but for repeated row preprocessing $N' = 2N$. Hence, for GHM with repeated row preprocessing, and

no.	method	preprocessing	N'	c
1.	GHM	repeated row ($\alpha = 1/\sqrt{2}$)	$2N$	0.750
2.	GHM	approximation	N	0.961
3.	CL	repeated row ($\alpha = 0$)	$2N$	0.500
4.	CL	matrix (5.4), with $b = (1 + \sqrt{7})$	N	0.138
5.	Bi9-7*	none (scalar)	N	1.026
6.	D4	none (scalar)	N	1.0
7.	LA8	none (scalar)	N	1.0

*Note that for this non-orthogonal transform c will vary slightly with transform level and with N.

Table 2: Appropriate values of N' and c to use in the scalar threshold $T_s = \sqrt{(2\,c\,\sigma_\epsilon^2 \log N')}$ for orthonormal multiwavelet transforms, a scalar biorthogonal (nearly orthogonal) transform, and scalar orthonormal wavelet transforms.

ignoring covariances, $T_s = \sqrt{(2 \times 0.75\sigma_\epsilon^2 \log 2N)}$. This contrasts with for example GHM with approximation preprocessing, which is critically sampled, and for which the noise variance is essentially preserved ($0.96\sigma_\epsilon^2$), so that $T_s = \sqrt{(2\sigma_\epsilon^2 \log N)}$. Similarly, for the CL method with repeated row preprocessing, we used $T_s = \sqrt{(2 \times 0.5\sigma_\epsilon^2 \log 2N)}$, and for the CL method with matrix (approximation) preprocessing we used $T_s = \sqrt{(2 \times 0.138\sigma_\epsilon^2 \log N)}$; see Table 2.

We need to see how correlation in the wavelet coefficients due to non-orthogonal preprocessing damages the universal thresholding approach to denoising.

7.2   Vector thresholding

If the noise $\{\epsilon_k\}$ in (6.1) comes from zero-mean white Gaussian noise, then $\mathbf{z}_{j,k}$ in (7.1) has a multivariate Gaussian distribution with mean zero and covariance matrix given by the appropriate 2×2 block of the covariance matrix $\text{cov}\{\mathbf{Z}\} = \sigma_\epsilon^2 \mathcal{M}\mathcal{P}\mathcal{P}^T\mathcal{M}^T$. For $j = 1$ and referring to Fig. 2 in terms of row and column numbers, such 2×2 blocks are given by, for example, the following elements: $(128, 128), (128, 127), (127, 127), (127, 128)$ and $(126, 126), (126, 125)$, $(125, 125), (125, 126)$. At any level j all such 2×2 blocks are the same and hence we can write the 2×2 covariance matrix as Λ_j so that

$$\mathbf{z}_{j,k} \sim \mathcal{N}(\mathbf{0}, \Lambda_j).$$

In the absence of a signal component, Downie and Silverman (1998) pointed out that $\theta_{j,k} = \mathbf{w}_{j,k}^T \Lambda_j^{-1} \mathbf{w}_{j,k}$ has a χ_2^2 distribution. The vector wavelet coefficients $\mathbf{w}_{j,k}$ could be thresholded using

$$\hat{\mathbf{w}}_{j,k} = \delta(\mathbf{w}_{j,k}, T_v), \quad 1 \le j \le J, \ k = 0, \ldots, N'_j - 1,$$

where $\delta(\cdot, \cdot)$ could be the hard-threshold rule:

$$\delta_H(\mathbf{w}_{j,k}, T_v) = \mathbf{w}_{j,k} \, 1\{\theta_{j,k} > T_v\}.$$

A suitable vector threshold level T_v is required. The inverse DMWT is then applied to the $\{\hat{\mathbf{w}}_{j,k}\}$ followed by the postprocessor.

If $\eta_1, \ldots, \eta_M$ is a sequence of independent and identically distributed χ_2^2 random variables, then (e.g., Galambos, 1978) as $M \to \infty$

$$P_M \equiv Pr\left[\max_i \eta_i > 2\log M\right] \to 0.$$

Hence the vector threshold is chosen as $T_v = 2\log M$, as used by Downie and Silverman (1998). It is clear from our covariance matrices that the wavelet coefficients are generally correlated, to a greater or lesser extent, between scales, but this is not taken into account here. However, the vector threshold is at least scale-free, and need only be changed from $T_v = 2\log 2N$, for repeated row preprocessing, to $T_v = 2\log N$, for critical sampling.

7.3 Numerical simulations

For the reasons discussed above neither the scalar nor the vector thresholding methods is ideally matched to the properties of multiwavelet coefficients, although vector thresholding does at least build in the covariance structure of the $\mathbf{z}_{j,k}$ vectors. Also, multiwavelet theory was largely developed with compression properties in mind, rather than multiwavelet thresholding. In this section we discuss simulation results in which methods 1-7 listed in Table 2 were applied. For scalar thresholding T_s used values of N' and c as given in Table 2, while for vector thresholding T_v was set to $2\log 2N$ for repeated row preprocessing and to $2\log N$ for critical sampling. (For the biorthogonal multiwavelets methods the value of c depends on J and to a lesser extent, N, as in Table 1. This is less convenient in practice than using methods 1-7 of Table 2. In thresholding experiments, biorthogonal multiwavelets performed quite well, but for reasons of brevity details are not reported here.)

Four types of true (deterministic) signal were utilised: (i) the mixed signal used in Strela *et al.* (1999), and (ii) 'bumps,' (iii) 'blocks,' and (iv) 'Heavisine,' these latter three being standard test signals since their specification in Donoho and Johnstone (1994). The 'bumps' signal was also used by Downie and Silverman (1998). White Gaussian noise was added to create different noisy realizations. With the same definition of signal-to-noise (S/N) ratio as in Donoho and Johnstone (1994), i.e., the ratio of signal-to-noise standard deviations, we looked at S/N ratios of 2, 4 and 8 (variance ratios of 6, 12 and 18dB, respectively). For each signal type (i)-(iv) and for each S/N level, the following was done:

1. 100 realizations of the noisy signal were constructed. For (i) $N = 512$, and for (ii)-(iv), $N = 2048$, in line with previous studies.

2. Each realization was subjected in turn to each of the methods 1-7 in Table 2; for multiwavelet methods this consists of preprocessing, DMWT, hard thresholding, inverse DMWT, postprocessing, and for scalar methods this consists of DWT, hard thresholding, inverse DWT. For (i) $J = 4$ for repeated row preprocessing, and $J = 3$ for critical sampling; for (ii)-(iv) $J = 6$ for repeated row preprocessing and $J = 5$ for critical sampling. This leaves 32 pairs of coefficients untouched by thresholding in all cases, agreeing with Downie and Silverman (1998).

3. The sample root mean squared error (rmse) was computed for each realization and each method. This rmse is defined as $\sqrt{[\frac{1}{N}\sum_{t=0}^{N-1}(X_t - \hat{X}_t)^2]}$. It was also recorded whether or not the vector or scalar thresholding produced the smaller rmse for each realization and each multiwavelet method.

The average (over realizations) rmse results are shown in Fig. 8.

For the mixed signal (i), we see that for S/N=2 the best result is achieved using CL repeated row preprocessing with scalar thresholding; for S/N=4 the best result is achieved using GHM repeated row preprocessing with scalar thresholding, while for S/N=8 the best result corresponds again to GHM repeated row preprocessing, but with both scalar and vector thresholding producing essentially the same rmse.

For the 'bumps' signal (ii), with S/N=2 the best result is achieved using CL repeated row preprocessing with vector thresholding; for S/N=4 all the vector thresholding methods perform virtually identically, while for S/N=8 the best result corresponds to CL repeated row preprocessing, with vector thresholding. Note that for the first three methods the rmse differences between scalar and vector thresholding are quite small, and again the multiwavelet approach appears to offer most advantage over the scalar wavelet approach especially for lower S/N. The 'bumps' signal with S/N=4 was also analyzed by Downie and Silverman (1998, Table 2). Our rmse results agree with theirs for vector thresholding, but for scalar thresholding using GHM with repeated row preprocessing and with approximation preprocessing we obtained 0.0815 and 0.0840, respectively, compared to their 0.113 and 0.164. We believe our improved results arise due to our careful formulation of the form of the universal threshold.

For the 'blocks' signal (iii), the best result is achieved for all three S/N ratios using CL repeated row preprocessing, with vector thresholding.

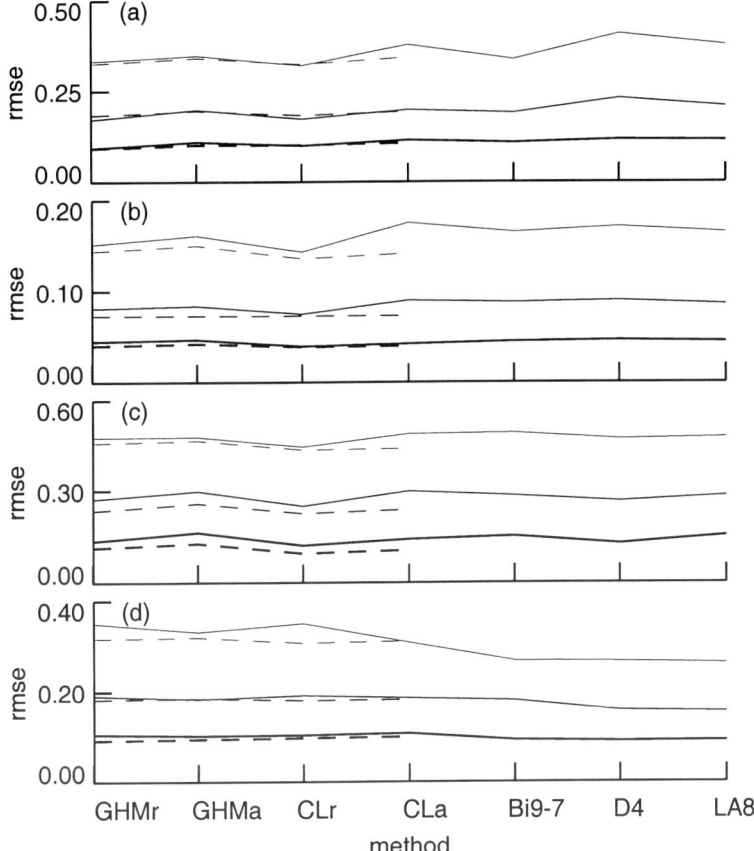

Figure 8: Average (over 100 realizations) rmse results for methods 1-7, for (a) mixed signal, (b) bumps, (c) blocks, and (d) Heavisine. Vector thresholding results are shown as dashed lines; signal-to-noise ratios are 8 (thick lines), 4 (medium) and 2 (thin).

For the Heavisine signal (iv), scalar wavelets give the best results especially for low S/N levels, with LA(8) best for S/N=2 and 4. All the scalar methods performed approximately equally well for S/N=8.

We thus see that multiwavelets outperform scalar wavelets for all models except Heavisine, and vector thresholding does not always outperform scalar thresholding. Generally repeated row preprocessing gives better results than matrix preprocessing. The average rmse of the (output) noise in the reconstructed signals across methods, for any S/N ratio, varies from about 1/4 to 2/3 (depending on the model) of the actual standard deviation of the (input) simulated noise; the smallest reduction in this intrinsic noise occurs for the mixed signal, and the largest for the Heavisine. Hence intrinsic noise is reduced

S/N	gain (dB)							
	Mixed	Bumps	Blocks	H'sine	Lenna	F'print	Boat	Bear
2	5.4	8.0	8.8	12.9	9.9	10.3	9.0	6.3
4	4.7	7.6	7.8	12.1	8.0	8.6	6.7	4.3
8	3.5	7.1	8.0	11.4	5.7	6.4	4.3	3.9

Table 3: Processing gains (dB) for different signals/images and S/N ratios. CL method with repeated row preprocessing and scalar thresholding used throughout. For image details see Section 8.

by practically useful amounts. CL repeated row preprocessing seems a good general method; only for the Heavisine signal and S/N=2 does it perform significantly worse than a competing method. Table 3 gives the processing gains for this method against S/N ratio and signal type, where processing gain is defined as

$$\text{gain} = 20 \log_{10} \left(\frac{\text{standard deviation of input noise}}{\text{average rmse in reconstruction}} \right) \text{dB}.$$

8 Two-dimensional signals

When dealing with 2D signals (images) we use the direct products of the scaling functions and wavelets. From the algorithmic point of view, the procedure is similar to the 1D case discussed above.

1. First we preprocess all rows, then all columns (of the previously row-preprocessed data). The amount of data stays unchanged if the matrix (critical sampling) preprocessing is used, but is quadrupled if repeated row preprocessing is used.

2. Next we perform the 2D wavelet cascade. During each step a multiwavelet filter bank (Fig. 1) is applied first to the rows, and then to the columns, of the low-low output from the previous step.

8.1 Algorithm illustrated

Let us describe this procedure in terms of matrix elements. We assume critical sampling preprocessing for ease of description, but the structure for repeated row preprocessing follows analogously. We start with an $N \times N$ matrix of elements

$$\begin{bmatrix} X_{0,0} & \cdots & X_{0,N-1} \\ \vdots & \ddots & \vdots \\ X_{N-1,0} & \cdots & X_{N-1,N-1} \end{bmatrix}.$$

After horizontal critical preprocessing the matrix is converted to

$$\begin{bmatrix} u_{0,0}^{(0)} & \cdots & u_{0,\frac{N}{2}-1}^{(0)} & u_{0,0}^{(1)} & \cdots & u_{0,\frac{N}{2}-1}^{(1)} \\ \vdots & \ddots & \vdots & \vdots & \ddots & \vdots \\ u_{N-1,0}^{(0)} & \cdots & u_{N-1,\frac{N}{2}-1}^{(0)} & u_{N-1,0}^{(1)} & \cdots & u_{N-1,\frac{N}{2}-1}^{(1)} \end{bmatrix},$$

where the vector of second components follows the vector of first components in each row. Next, vertical critical preprocessing is carried out to obtain

$$\mathbf{U} = \begin{bmatrix} u_{0,0}^{(0,0)} & \cdots & u_{0,\frac{N}{2}-1}^{(0,0)} & u_{0,0}^{(0,1)} & \cdots & u_{0,\frac{N}{2}-1}^{(0,1)} \\ \vdots & \ddots & \vdots & \vdots & \ddots & \vdots \\ u_{\frac{N}{2}-1,0}^{(0,0)} & \cdots & u_{\frac{N}{2}-1,\frac{N}{2}-1}^{(0,0)} & u_{\frac{N}{2}-1,0}^{(0,1)} & \cdots & u_{\frac{N}{2}-1,\frac{N}{2}-1}^{(0,1)} \\ u_{0,0}^{(1,0)} & \cdots & u_{0,\frac{N}{2}-1}^{(1,0)} & u_{0,0}^{(1,1)} & \cdots & u_{0,\frac{N}{2}-1}^{(1,1)} \\ \vdots & \ddots & \vdots & \vdots & \ddots & \vdots \\ u_{\frac{N}{2}-1,0}^{(1,0)} & \cdots & u_{\frac{N}{2}-1,\frac{N}{2}-1}^{(1,0)} & u_{\frac{N}{2}-1,0}^{(1,1)} & \cdots & u_{\frac{N}{2}-1,\frac{N}{2}-1}^{(1,1)} \end{bmatrix}.$$

The vector of second components follows the vector of first components in each column. For example the superscript $(0,1)$ indicates the first vector component is in the vertical direction and the second vector component is in the horizontal direction. This ends step 1 above.

Next we perform one step of the multiwavelet transform in the horizontal direction. Consider the first $N/2$ rows. For $n = 0, \ldots, (N/2) - 1$ we feed the multiwavelet transform the vectors $[u_{n,m}^{(0,0)}, u_{n,m}^{(0,1)}]^T$, $m = 0, \ldots, (N/2) - 1$. We thus obtain

$$\begin{bmatrix} v_{n,l}^{(0,0)} \\ v_{n,l}^{(0,1)} \end{bmatrix} = \sum_{m=0}^{\frac{N}{2}-1} \mathbf{G}_{m-2l} \begin{bmatrix} u_{n,m}^{(0,0)} \\ u_{n,m}^{(0,1)} \end{bmatrix}, \qquad \begin{bmatrix} w_{n,l}^{(0,0)} \\ w_{n,l}^{(0,1)} \end{bmatrix} = \sum_{m=0}^{\frac{N}{2}-1} \mathbf{H}_{m-2l} \begin{bmatrix} u_{n,m}^{(0,0)} \\ u_{n,m}^{(0,1)} \end{bmatrix},$$

for $l = 0, \ldots, (N/4) - 1$. For the next $N/2$ rows, for $n = 0, \ldots, (N/2) - 1$, we feed the multiwavelet transform the vectors $[u_{n,m}^{(1,0)}, u_{n,m}^{(1,1)}]^T$, $m = 0, \ldots, (N/2) - 1$. We thus obtain

$$\begin{bmatrix} v_{n,l}^{(1,0)} \\ v_{n,l}^{(1,1)} \end{bmatrix} = \sum_{m=0}^{\frac{N}{2}-1} \mathbf{G}_{m-2l} \begin{bmatrix} u_{n,m}^{(1,0)} \\ u_{n,m}^{(1,1)} \end{bmatrix}, \qquad \begin{bmatrix} w_{n,l}^{(1,0)} \\ w_{n,l}^{(1,1)} \end{bmatrix} = \sum_{m=0}^{\frac{N}{2}-1} \mathbf{H}_{m-2l} \begin{bmatrix} u_{n,m}^{(1,0)} \\ u_{n,m}^{(1,1)} \end{bmatrix},$$

for $l = 0, \ldots, (N/4) - 1$. The resulting matrix takes the form

$$
\begin{bmatrix}
v_{0,0}^{(0,0)} & \cdots & v_{0,0}^{(0,1)} & \cdots & w_{0,0}^{(0,0)} & \cdots & w_{0,0}^{(0,1)} & \cdots \\
v_{1,0}^{(0,0)} & \cdots & v_{1,0}^{(0,1)} & \cdots & w_{1,0}^{(0,0)} & \cdots & w_{1,0}^{(0,1)} & \cdots \\
\vdots & \ddots & \vdots & \ddots & \vdots & \ddots & \vdots & \ddots \\
v_{\frac{N}{2}-1,0}^{(0,0)} & \cdots & v_{\frac{N}{2}-1,0}^{(0,1)} & \cdots & w_{\frac{N}{2}-1,0}^{(0,0)} & \cdots & w_{\frac{N}{2}-1,0}^{(0,1)} & \cdots \\
v_{0,0}^{(1,0)} & \cdots & v_{0,0}^{(1,1)} & \cdots & w_{0,0}^{(1,0)} & \cdots & w_{0,0}^{(1,1)} & \cdots \\
v_{1,0}^{(1,0)} & \cdots & v_{1,0}^{(1,1)} & \cdots & w_{1,0}^{(1,0)} & \cdots & w_{1,0}^{(1,1)} & \cdots \\
\vdots & \ddots & \vdots & \ddots & \vdots & \ddots & \vdots & \ddots \\
v_{\frac{N}{2}-1,0}^{(1,0)} & \cdots & v_{\frac{N}{2}-1,0}^{(1,1)} & \cdots & w_{\frac{N}{2}-1,0}^{(1,0)} & \cdots & w_{\frac{N}{2}-1,0}^{(1,1)} & \cdots
\end{bmatrix},
$$

where we have only given the first element of each row sequence. Finally we perform one step of the multiwavelet transform in the vertical direction. Consider the first $N/4$ columns. For $l = 0, \ldots, (N/4) - 1$ we feed the multiwavelet transform the vectors $[v_{n,l}^{(0,0)}, v_{n,l}^{(1,0)}]^T$, $n = 0, \ldots, (N/2) - 1$ to obtain

$$
\begin{bmatrix} vv_{k,l}^{(0,0)} \\ vv_{k,l}^{(1,0)} \end{bmatrix} = \sum_{n=0}^{\frac{N}{2}-1} \mathbf{G}_{n-2k} \begin{bmatrix} v_{n,l}^{(0,0)} \\ v_{n,l}^{(1,0)} \end{bmatrix}, \qquad
\begin{bmatrix} wv_{k,l}^{(0,0)} \\ wv_{k,l}^{(1,0)} \end{bmatrix} = \sum_{n=0}^{\frac{N}{2}-1} \mathbf{H}_{n-2k} \begin{bmatrix} v_{n,l}^{(0,0)} \\ v_{n,l}^{(1,0)} \end{bmatrix},
$$

for $k = 0, \ldots, (N/4) - 1$. With respect to the next $N/4$ columns, for $l = 0, \ldots, (N/4) - 1$ we feed the multiwavelet transform the vectors $[v_{n,l}^{(0,1)}, v_{n,l}^{(1,1)}]^T$, $n = 0, \ldots, (N/2) - 1$, to obtain

$$
\begin{bmatrix} vv_{k,l}^{(0,1)} \\ vv_{k,l}^{(1,1)} \end{bmatrix} = \sum_{n=0}^{\frac{N}{2}-1} \mathbf{G}_{n-2k} \begin{bmatrix} v_{n,l}^{(0,1)} \\ v_{n,l}^{(1,1)} \end{bmatrix}, \qquad
\begin{bmatrix} wv_{k,l}^{(0,1)} \\ wv_{k,l}^{(1,1)} \end{bmatrix} = \sum_{n=0}^{\frac{N}{2}-1} \mathbf{H}_{n-2k} \begin{bmatrix} v_{n,l}^{(0,1)} \\ v_{n,l}^{(1,1)} \end{bmatrix},
$$

for $k = 0, \ldots, (N/4) - 1$. With respect to the third set of $N/4$ columns, for $l = 0, \ldots, (N/4) - 1$ we feed the multiwavelet transform the vectors $[w_{n,l}^{(0,0)}, w_{n,l}^{(1,0)}]^T$, $n = 0, \ldots, (N/2) - 1$, to obtain

$$
\begin{bmatrix} vw_{k,l}^{(0,0)} \\ vw_{k,l}^{(1,0)} \end{bmatrix} = \sum_{n=0}^{\frac{N}{2}-1} \mathbf{G}_{n-2k} \begin{bmatrix} w_{n,l}^{(0,0)} \\ w_{n,l}^{(1,0)} \end{bmatrix}, \qquad
\begin{bmatrix} ww_{k,l}^{(0,0)} \\ ww_{k,l}^{(1,0)} \end{bmatrix} = \sum_{n=0}^{\frac{N}{2}-1} \mathbf{H}_{n-2k} \begin{bmatrix} w_{n,l}^{(0,0)} \\ w_{n,l}^{(1,0)} \end{bmatrix},
$$

for $k = 0, \ldots, (N/4) - 1$. For the final $N/4$ columns, for $l = 0, \ldots, (N/4) - 1$ we feed the multiwavelet transform the vectors $[w_{n,l}^{(0,1)}, w_{n,l}^{(1,1)}]^T$, $n = 0, \ldots, (N/2) - 1$ to obtain

$$
\begin{bmatrix} vw_{k,l}^{(0,1)} \\ vw_{k,l}^{(1,1)} \end{bmatrix} = \sum_{n=0}^{\frac{N}{2}-1} \mathbf{G}_{n-2k} \begin{bmatrix} w_{n,l}^{(0,1)} \\ w_{n,l}^{(1,1)} \end{bmatrix}, \qquad
\begin{bmatrix} ww_{k,l}^{(0,1)} \\ wv_{k,l}^{(1,1)} \end{bmatrix} = \sum_{n=0}^{\frac{N}{2}-1} \mathbf{H}_{n-2k} \begin{bmatrix} w_{n,l}^{(0,1)} \\ w_{n,l}^{(1,1)} \end{bmatrix},
$$

for $k = 0, \ldots, (N/4) - 1$. The final matrix obtained takes the form

$$
\begin{bmatrix}
\begin{pmatrix}
\begin{array}{cccc}
vv^{(0,0)}_{0,0} & \cdots & vv^{(0,1)}_{0,0} & \cdots \\
vv^{(0,0)}_{1,0} & \cdots & vv^{(0,1)}_{1,0} & \cdots \\
\vdots & \ddots & \vdots & \ddots \\
vv^{(0,0)}_{\frac{N}{4}-1,0} & \cdots & vv^{(0,1)}_{\frac{N}{4}-1,0} & \cdots \\
vv^{(1,0)}_{0,0} & \cdots & vv^{(1,1)}_{0,0} & \cdots \\
vv^{(1,0)}_{1,0} & \cdots & vv^{(1,1)}_{1,0} & \cdots \\
\vdots & \ddots & \vdots & \ddots \\
vv^{(1,0)}_{\frac{N}{4}-1,0} & \cdots & vv^{(1,1)}_{\frac{N}{4}-1,0} & \cdots
\end{array}
\end{pmatrix}
&
\begin{pmatrix}
\begin{array}{cccc}
vw^{(0,0)}_{0,0} & \cdots & vw^{(0,1)}_{0,0} & \cdots \\
vw^{(0,0)}_{1,0} & \cdots & vw^{(0,1)}_{1,0} & \cdots \\
\vdots & \ddots & \vdots & \ddots \\
vw^{(0,0)}_{\frac{N}{4}-1,0} & \cdots & vw^{(0,1)}_{\frac{N}{4}-1,0} & \cdots \\
vw^{(1,0)}_{0,0} & \cdots & vw^{(1,1)}_{0,0} & \cdots \\
vw^{(1,0)}_{1,0} & \cdots & vw^{(1,1)}_{1,0} & \cdots \\
\vdots & \ddots & \vdots & \ddots \\
vw^{(1,0)}_{\frac{N}{4}-1,0} & \cdots & vw^{(1,1)}_{\frac{N}{4}-1,0} & \cdots
\end{array}
\end{pmatrix}
\\[2em]
\begin{pmatrix}
\begin{array}{cccc}
wv^{(0,0)}_{0,0} & \cdots & wv^{(0,1)}_{0,0} & \cdots \\
wv^{(0,0)}_{1,0} & \cdots & wv^{(0,1)}_{1,0} & \cdots \\
\vdots & \ddots & \vdots & \ddots \\
wv^{(0,0)}_{\frac{N}{4}-1,0} & \cdots & wv^{(0,1)}_{\frac{N}{4}-1,0} & \cdots \\
wv^{(1,0)}_{0,0} & \cdots & wv^{(1,1)}_{0,0} & \cdots \\
wv^{(1,0)}_{1,0} & \cdots & wv^{(1,1)}_{1,0} & \cdots \\
\vdots & \ddots & \vdots & \ddots \\
wv^{(1,0)}_{\frac{N}{4}-1,0} & \cdots & wv^{(1,1)}_{\frac{N}{4}-1,0} & \cdots
\end{array}
\end{pmatrix}
&
\begin{pmatrix}
\begin{array}{cccc}
ww^{(0,0)}_{0,0} & \cdots & ww^{(0,1)}_{0,0} & \cdots \\
ww^{(0,0)}_{1,0} & \cdots & ww^{(0,1)}_{1,0} & \cdots \\
\vdots & \ddots & \vdots & \ddots \\
ww^{(0,0)}_{\frac{N}{4}-1,0} & \cdots & ww^{(0,1)}_{\frac{N}{4}-1,0} & \cdots \\
ww^{(1,0)}_{0,0} & \cdots & ww^{(1,1)}_{0,0} & \cdots \\
ww^{(1,0)}_{1,0} & \cdots & ww^{(1,1)}_{1,0} & \cdots \\
\vdots & \ddots & \vdots & \ddots \\
ww^{(1,0)}_{\frac{N}{4}-1,0} & \cdots & ww^{(1,1)}_{\frac{N}{4}-1,0} & \cdots
\end{array}
\end{pmatrix}
\end{bmatrix},
$$

where we have again only given the first element of each row sequence. The full matrix has been split into four submatrices which may be denoted

$$
\begin{bmatrix} \mathbf{GG} & \mathbf{GH} \\ \mathbf{HG} & \mathbf{HH} \end{bmatrix},
$$

where, e.g., '**GH**' denotes application of the '**H**' filter coefficients in the horizontal direction, followed by application of the '**G**' filter coefficients in the vertical direction. These four submatrices are each further subdivided into four matrices, i.e., those with superscripts $(0,0), (0,1), (1,0)$ and $(1,1)$. From the original $N \times N$ array we thus obtain 16 submatrices of sizes $N/4 \times N/4$. Examples of this one-step processing are given in Figs. 9 and 10 for GHM and CL, respectively, both with approximation preprocessing. The image is the well-known photo of Lenna.

In the CL one-step plot almost all information is concentrated in one block of **GG** with superscripts $(0,0)$ (in the scalar case we would expect a picture similar to the **GG** block after one step of the cascade algorithm). This reflects the fact that for CL one scaling function is symmetric and the other is anti-symmetric. The $(0,0)$ block of the **GG** matrix contains the inner products

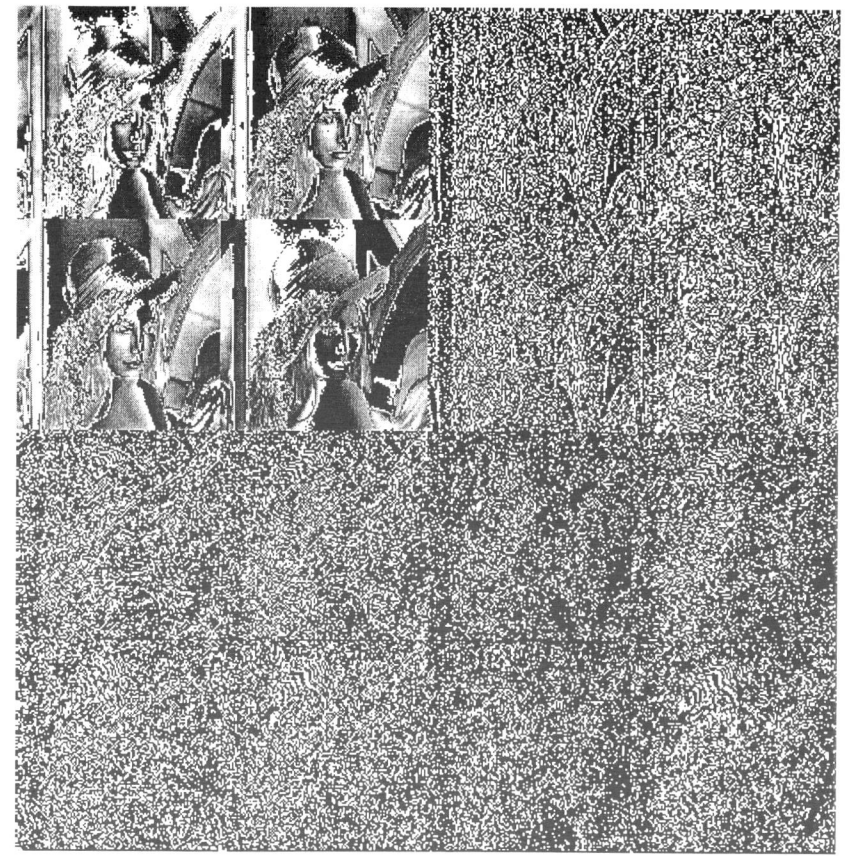

Figure 9: Result of one step of the 2D wavelet cascade algorithm for GHM with approximation preprocessing.

of the 2D image function $F(x, y)$, say, with the product of the shifted scaling function $\phi_1(x)\phi_1(y)$ (see (2.9) for the 1D case):

$$vv_{k,l}^{(0,0)} = \int\int F(x, y)\phi_1(x - l)\phi_1(y - k)dxdy.$$

Analogously,

$$vv_{k,l}^{(1,0)} = \int\int F(x, y)\phi_1(x - l)\phi_2(y - k)dxdy,$$

$$vv_{k,l}^{(0,1)} = \int\int F(x, y)\phi_2(x - l)\phi_1(y - k)dxdy,$$

$$vv_{k,l}^{(1,1)} = \int\int F(x, y)\phi_2(x - l)\phi_2(y - k)dxdy.$$

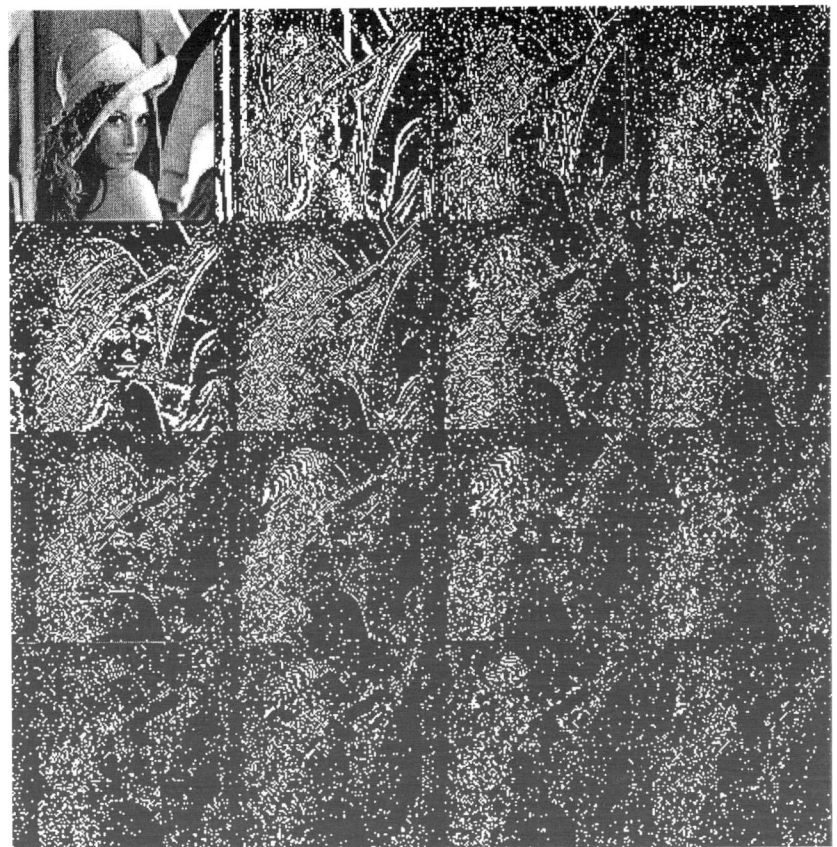

Figure 10: Result of one step of the 2D wavelet cascade algorithm for CL with approximation preprocessing.

Here ϕ_1 is the symmetric scaling function of the CL pair, and ϕ_2 is the antisymmetric one. As is well known, the inner product of the signal with a symmetric function corresponds to low-pass filtering, while the inner product with an antisymmetric function corresponds to high-pass filtering. This means that the $(0,0)$ block contains low-pass data in both directions. On the other hand, $(0,1),(1,0)$ and $(0,0)$ blocks correspond to high-pass filtering in one of the directions and contain mostly the information about edges. Nevertheless all four parts of $\mathbf{GG}$ come from the inner products with the components of the 2D multiscaling function and should be considered as the result of *low-pass multifiltering*.

In the GHM one-step plot all four submatrices of $\mathbf{GG}$ contribute in a similar fashion, since both scaling functions are symmetric, and hence $(0,0),(0,1)$, $(1,0)$ and $(1,1)$ coefficients are all a result of low-pass filtering in both directions.

From the structures of the submatrices we find that

$$
\begin{bmatrix} vv_{k,l}^{(0,0)} \\ vv_{k,l}^{(0,1)} \\ vv_{k,l}^{(1,0)} \\ vv_{k,l}^{(1,1)} \end{bmatrix} = \sum_{n=0}^{\frac{N}{2}-1} \sum_{m=0}^{\frac{N}{2}-1} \mathbf{G}_{n-2k} \otimes \mathbf{G}_{m-2l} \begin{bmatrix} u_{n,m}^{(0,0)} \\ u_{n,m}^{(0,1)} \\ u_{n,m}^{(1,0)} \\ u_{n,m}^{(1,1)} \end{bmatrix},
$$

$$
\begin{bmatrix} vw_{k,l}^{(0,0)} \\ vw_{k,l}^{(0,1)} \\ vw_{k,l}^{(1,0)} \\ vw_{k,l}^{(1,1)} \end{bmatrix} = \sum_{n=0}^{\frac{N}{2}-1} \sum_{m=0}^{\frac{N}{2}-1} \mathbf{G}_{n-2k} \otimes \mathbf{H}_{m-2l} \begin{bmatrix} u_{n,m}^{(0,0)} \\ u_{n,m}^{(0,1)} \\ u_{n,m}^{(1,0)} \\ u_{n,m}^{(1,1)} \end{bmatrix},
$$

$$
\begin{bmatrix} wv_{k,l}^{(0,0)} \\ wv_{k,l}^{(0,1)} \\ wv_{k,l}^{(1,0)} \\ wv_{k,l}^{(1,1)} \end{bmatrix} = \sum_{n=0}^{\frac{N}{2}-1} \sum_{m=0}^{\frac{N}{2}-1} \mathbf{H}_{n-2k} \otimes \mathbf{G}_{m-2l} \begin{bmatrix} u_{n,m}^{(0,0)} \\ u_{n,m}^{(0,1)} \\ u_{n,m}^{(1,0)} \\ u_{n,m}^{(1,1)} \end{bmatrix},
$$

$$
\begin{bmatrix} ww_{k,l}^{(0,0)} \\ ww_{k,l}^{(0,1)} \\ ww_{k,l}^{(1,0)} \\ ww_{k,l}^{(1,1)} \end{bmatrix} = \sum_{n=0}^{\frac{N}{2}-1} \sum_{m=0}^{\frac{N}{2}-1} \mathbf{H}_{n-2k} \otimes \mathbf{H}_{m-2l} \begin{bmatrix} u_{n,m}^{(0,0)} \\ u_{n,m}^{(0,1)} \\ u_{n,m}^{(1,0)} \\ u_{n,m}^{(1,1)} \end{bmatrix},
$$

where $\otimes$ denotes matrix Kronecker product, and $0 \le n, m \le (N/2) - 1$ and $0 \le k, l \le (N/4) - 1$.

The next step of the algorithm is to apply the multiwavelet transform horizontally and then vertically to the submatrix $\mathbf{GG}$, which has exactly the same layout as $\mathbf{U}$, except that the dimension of the matrix is halved. The process then continues in cascade fashion.

Initial data can be reconstructed exactly from the multiwavelet coefficients. Reconstruction starts with the submatrix $\begin{bmatrix} \mathbf{GG}^J & \mathbf{GH}^J \\ \mathbf{HG}^J & \mathbf{HH}^J \end{bmatrix}$, of the highest decomposition level J, and employs formulas of the type (2.12) first in the vertical and then in the horizontal direction in order to get the low-pass matrix $\mathbf{GG}^{J-1}$ of the next level. The process then continues until the matrix $\mathbf{U}$ is reconstructed. Finally, the initial data can be recovered from $\mathbf{U}$ by postprocessing vertically and then horizontally.

8.2   Image denoising

Taking into account the results of our 1D experiments, we decided to restrict ourselves to orthogonal multifilters for the denoising of images. In this case, the only change in variance of the input noise is due to the preprocessing.

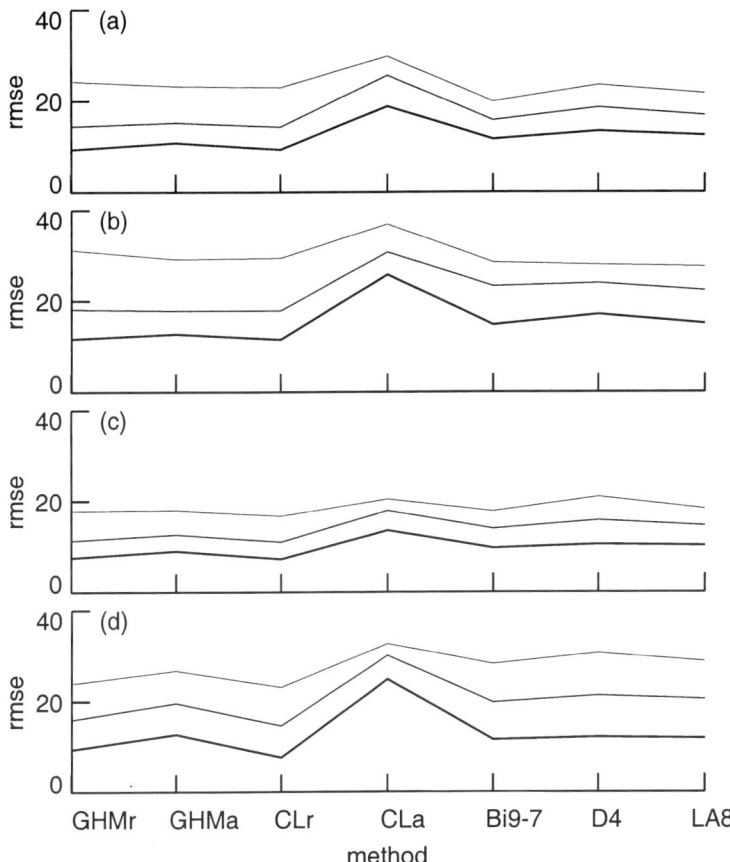

Figure 11: Average (over 100 realizations) rmse results for methods 1-7 of Table 2, for (a) Lenna, (b) a fingerprint, (c) the ship *Picardie*, and (d) cartoon bear. Scalar thresholding used throughout; signal-to-noise ratios are 8 (thick lines), 4 (medium) and 2 (thin).

Analogously to the 1D case, for the thresholding we make use of the 'average' variance of the preprocessed and transformed coefficients; the variance deflation/inflation factor is the square of that obtained in the 1D case.

To see this, consider an $N \times N$ image where each cell has noise with variance σ_ϵ^2. Then, for example, for 2D GHM repeated row preprocessing, which uses $\alpha = 1/\sqrt{2}$, we firstly preprocess all rows, thus producing an unaltered $N \times N$ image, A say, and another scaled $N \times N$ image 'to its right,' B say, where each cell has been scaled by $1/\sqrt{2}$ relative to A. Next we preprocess all the $2N$ columns in A and B. This gives four $N \times N$ images, A left unaltered, plus a scaled version 'below it,' C say, where each cell has been scaled by $1/\sqrt{2}$ relative to A; the image B is left unaltered, while 'below it' is created a scaled version of B, where each cell has been scaled by $1/\sqrt{2} \times 1/\sqrt{2} = 1/2$ relative to A. We thus have N^2 entries with noise variance σ_ϵ^2, $2N^2$ entries with variance $\sigma_\epsilon^2/2$, and N^2 entries with variance $\sigma_\epsilon^2/4$. The average variance is then $(9/16)\sigma_\epsilon^2$, and $c = 9/16$ is the square of the factor $(3/4)$ obtained in the 1D case (e.g., Table 2). For 2D CL repeated row preprocessing, which uses $\alpha = 0$, only N^2 coefficients of the $4N^2$ produced are non-zero, and these have variance σ_ϵ^2, so that the average variance is $\sigma_\epsilon^2/4$, and $c = 1/4$ is the square of the value of $1/2$ obtained in the 1D case. For GHM or CL with matrix (critical sampling) preprocessing, the same rescaling takes place in the row direction and column direction, so that the 1D variance deflation/inflation factor c in Table 2 becomes squared in the 2D case.

The performance of methods 1-7 of Table 2 on images was tested with four different (128×128)-cell images. In implementing the scalar thresholding for these images, for which $N = 128$, N' and c in Table 2 are replaced by their squares, as explained above. The four images used were (a) of Lenna again, (b) a fingerprint, (c) a photo of part of a ship including its name, *Picardie*, and (d) a textural image containing part of a cartoon bear. For each image, 100 realizations of the noisy image were constructed, for S/N ratios of 2, 4 and 8. After preprocessing, DMWT, hard thresholding, inverse DMWT, postprocessing (and, for scalar methods, DWT, hard thresholding, inverse DWT), the rmse was computed for each method and each realization.

The average rmse over realizations was computed, and is shown in Fig. 11.

For the Lenna image (a), the best result is achieved using CL repeated row preprocessing for S/N=8 and 4, but for S/N=2 Bi 9-7 was best; however in this case, CL repeated row preprocessing was still the best of the multiwavelet methods. Examples of the processing are shown in Figs. 12–13 for S/N=4 and 2, respectively, where CL repeated row preprocessing is seen visually to perform well, and would probably be felt to have performed *visually* better than Bi 9-7 for S/N=2 in this example.

For the fingerprint image (b), CL repeated row preprocessing is best for S/N=8, is tie-best with GHM approximation preprocessing for S/N=4, but for S/N=2, LA(8) performed best. Examples of the processing are shown in

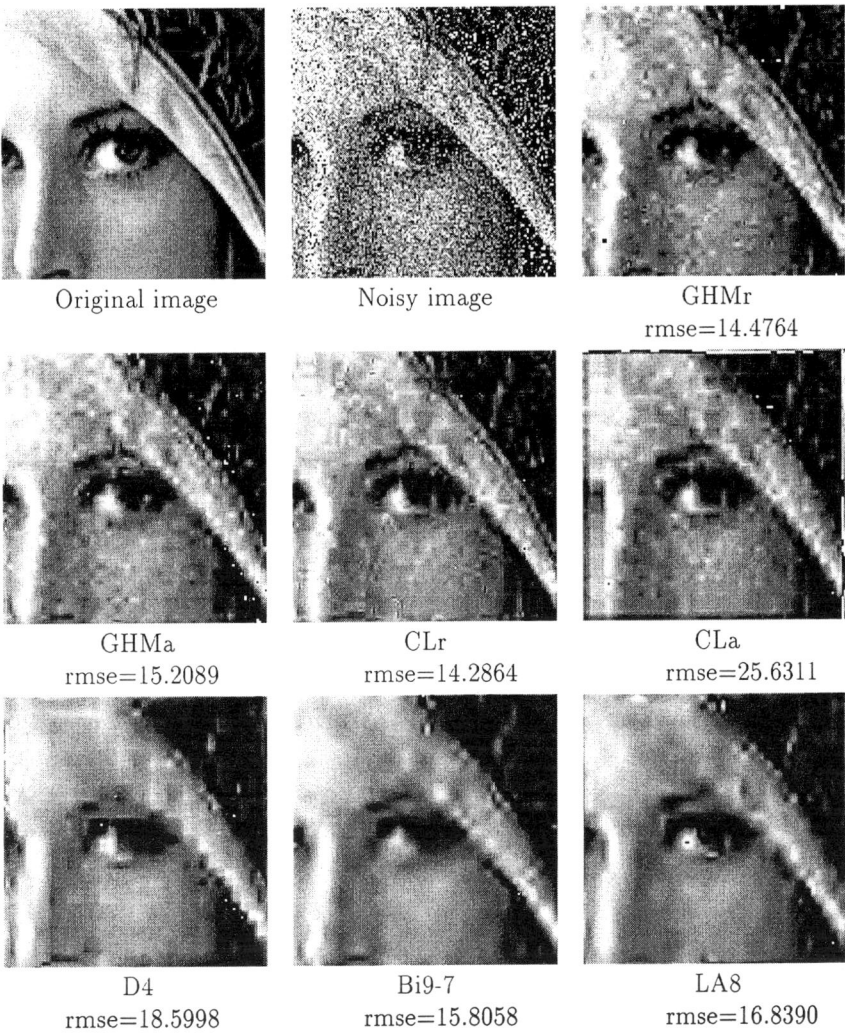

Figure 12: Example of 2D processing methods applied to Lenna image, for S/N=4.

Figs. 14–15 for S/N=4 and 2, respectively, where CL repeated row preprocessing is seen visually to perform well for all S/N ratios, and is visually better than LA(8) for S/N=2 in this example.

For the ship image (c), CL repeated row preprocessing is best for all S/N ratios. Processing examples are given in Figs. 16–17 for for S/N=4 and 2. Some letters of the name are interpretable at S/N=4 for GHM and CL repeated row preprocessing; when the S/N is as low as 2, these same methods preserve

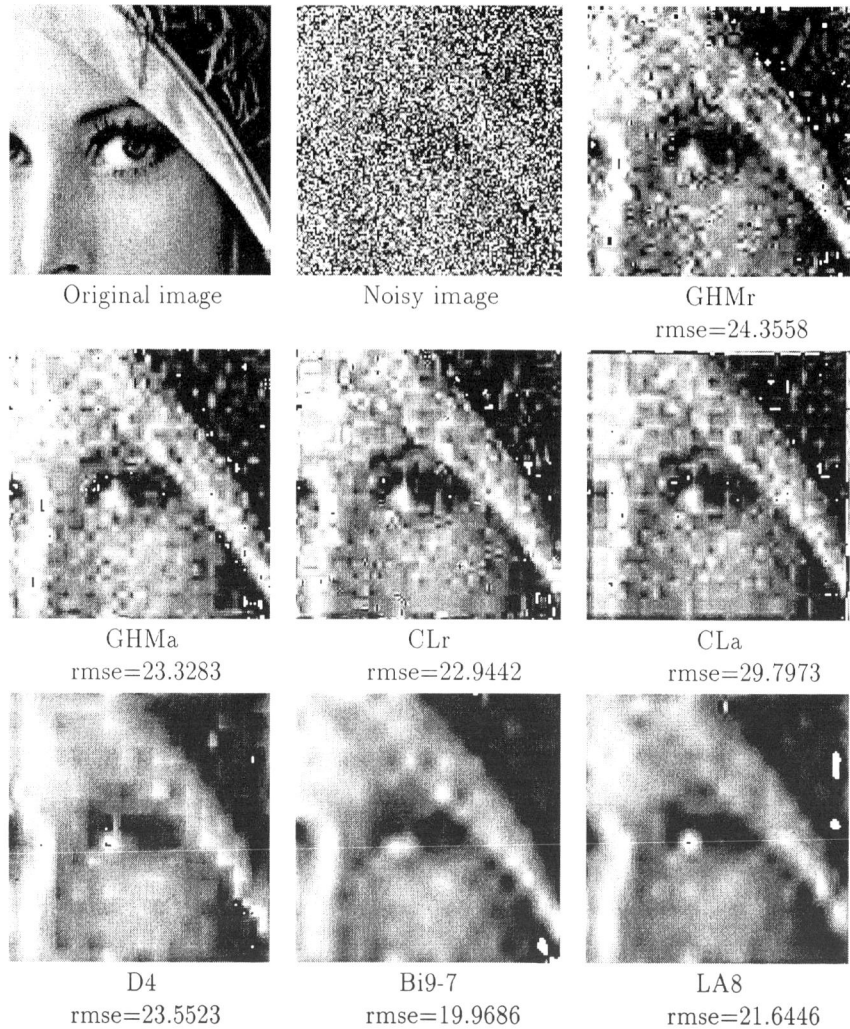

Original image   Noisy image   GHMr
                                 rmse=24.3558

GHMa   CLr   CLa
rmse=23.3283   rmse=22.9442   rmse=29.7973

D4   Bi9-7   LA8
rmse=23.5523   rmse=19.9686   rmse=21.6446

Figure 13: Example of 2D processing methods applied to Lenna image, for S/N=2.

the main characteristics of the image, but the name is unreadable.

For the textural cartoon bear image (d), CL repeated row preprocessing is best for all S/N ratios. GHM repeated row preprocessing does nearly as well. Processing examples are given in Figs. 18–19 for S/N=4 and 2. GHM and CL repeated row preprocessing do well in preserving the speckle texture in the upper right of the image, and in preserving the continuity of lines, even for S/N=2.

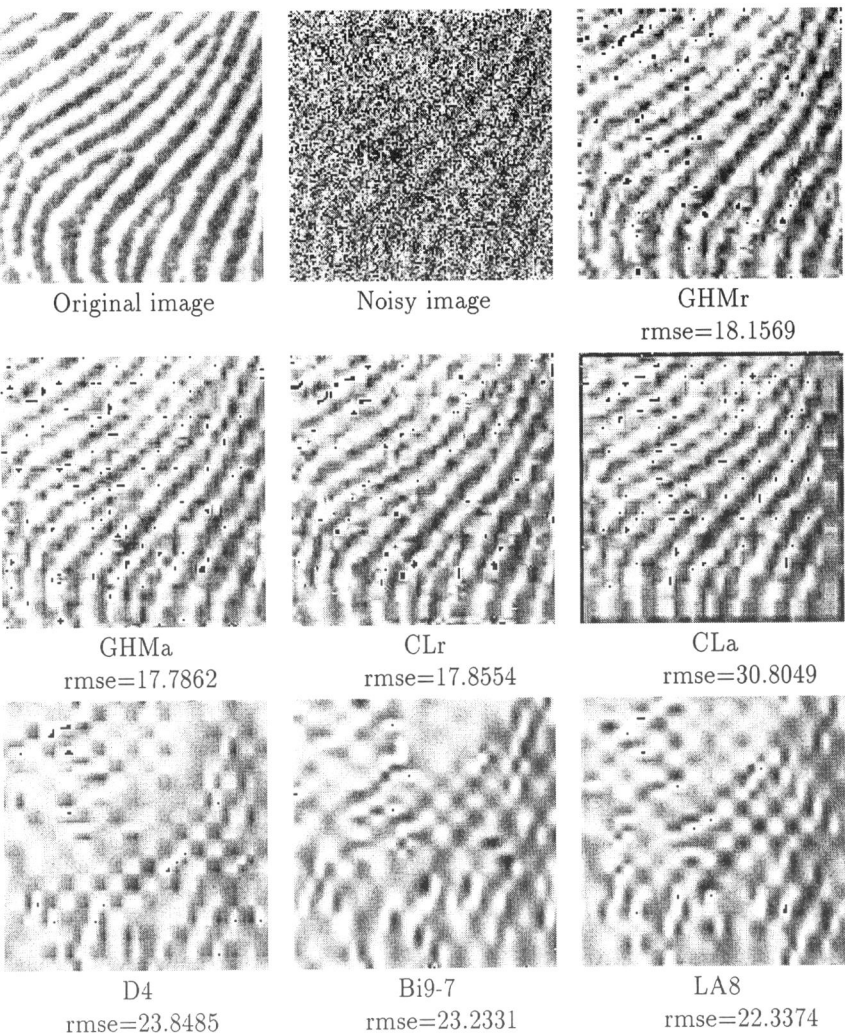

Figure 14: Example of 2D processing methods applied to fingerprint image, for S/N=4.

Multiwavelets generally outperform scalar wavelets for image denoising for all models; only for Lenna and fingerprints with S/N=2 do scalar wavelets perform best. The rmse obtained for the best method varies (over image type) between 45 and 65% of the actual standard deviation of the simulated noise for S/N=8, between 35 and 60% for S/N=4, and between 25 and 50% for S/N=2. Again we see intrinsic noise can be reduced by practically useful amounts. As for 1D signal processing, CL repeated row preprocessing seems

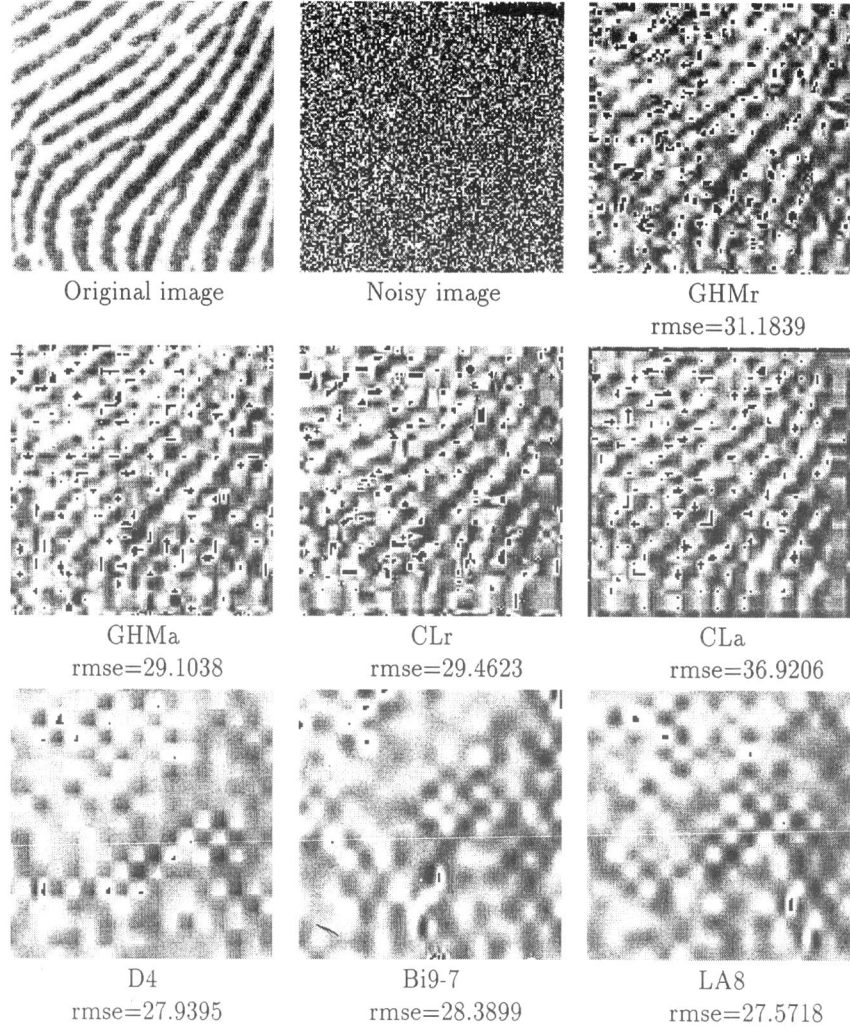

Figure 15: Example of 2D processing methods applied to fingerprint image, for S/N=2.

a good general method; Table 3 gives the processing gains for this method against S/N ratio and image type.

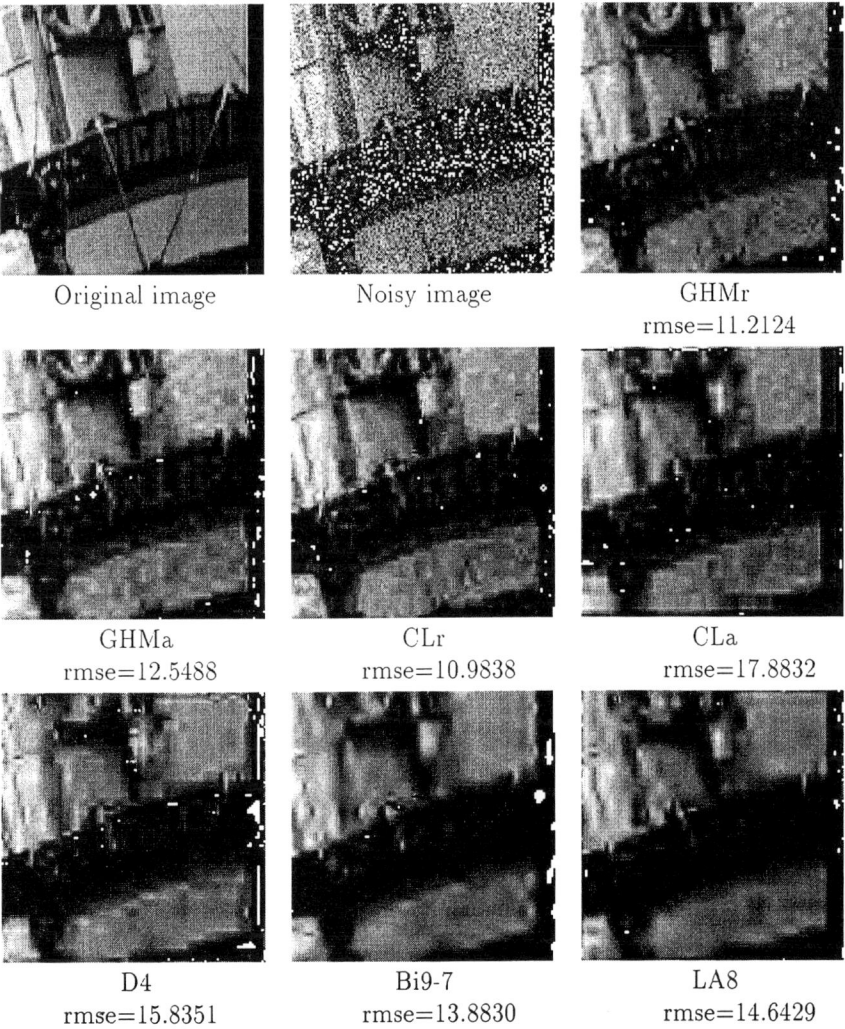

Original image	Noisy image	GHMr rmse=11.2124
GHMa rmse=12.5488	CLr rmse=10.9838	CLa rmse=17.8832
D4 rmse=15.8351	Bi9-7 rmse=13.8830	LA8 rmse=14.6429

Figure 16: Example of 2D processing methods applied to *Picardie* boat image, for S/N=4.

9 Concluding remarks

In this article we studied the covariance structure of 1D scalar orthogonal, scalar biorthogonal and multiple orthogonal and multiple biorthogonal wavelet transforms with appropriate preprocessing. This led to a careful formulation of the universal threshold for scalar thresholding. Vector thresholding was also considered.

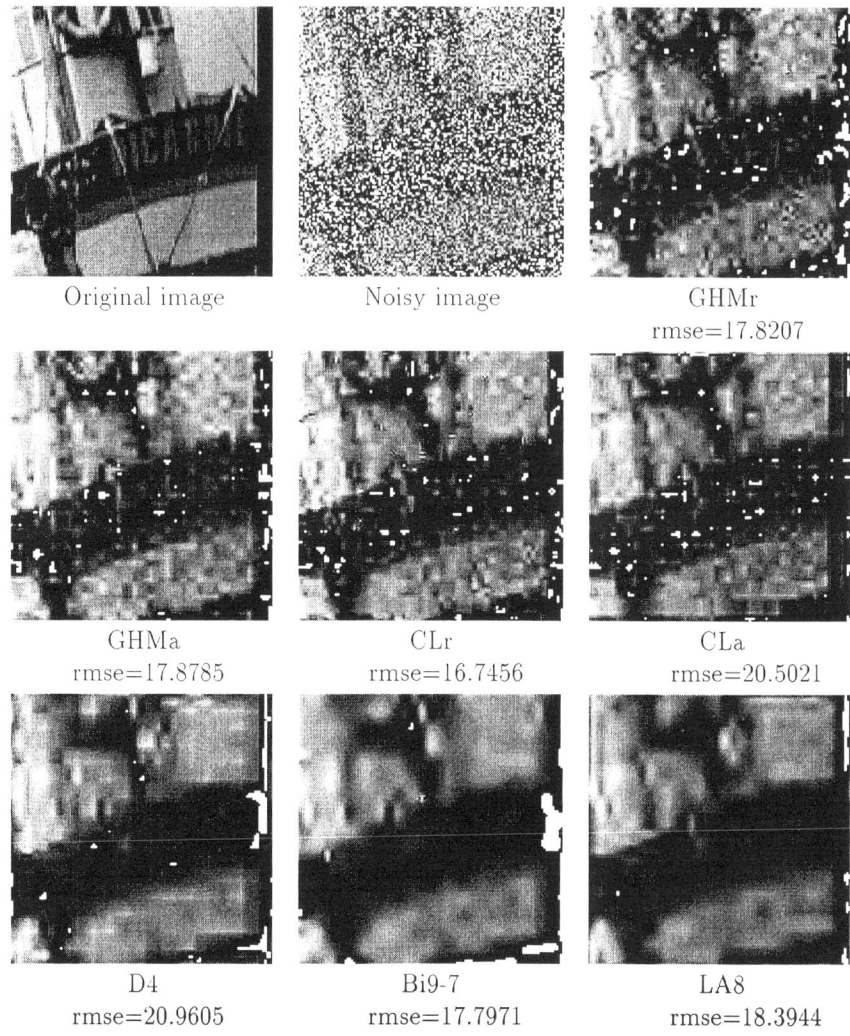

Figure 17: Example of 2D processing methods applied to *Picardie* boat image, for S/N=2.

Multiwavelets outperform scalar wavelets for three out of four noisy 1D test signals. Vector thresholding does not always outperform scalar thresholding. Multiwavelets generally outperform scalar wavelets for image denoising for all four noisy 2D test images, and the results are visually very impressive. Chui–Lian scaling functions and wavelets combined with repeated row preprocessing appear to be a good general method. For both 1D and 2D cases, the reconstructed signals derived from such a good general method demonstrate much

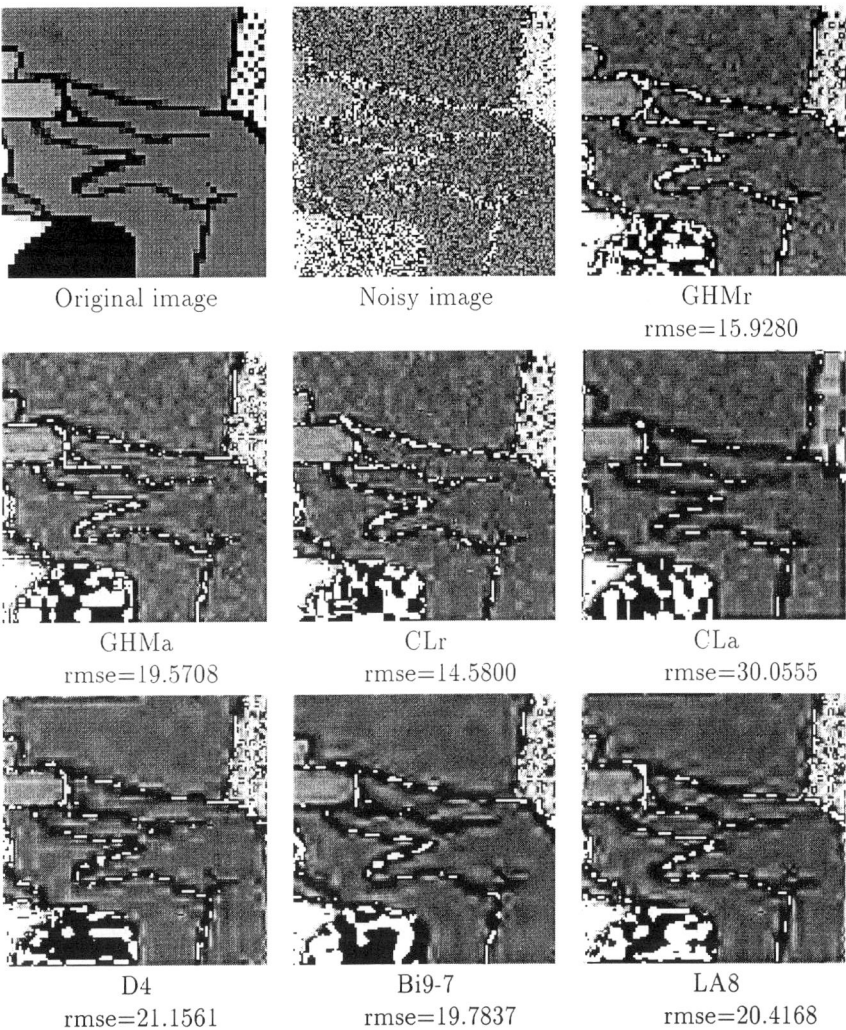

Original image	Noisy image	GHMr rmse=15.9280
GHMa rmse=19.5708	CLr rmse=14.5800	CLa rmse=30.0555
D4 rmse=21.1561	Bi9-7 rmse=19.7837	LA8 rmse=20.4168

Figure 18: Example of 2D processing methods applied to cartoon bear image, for S/N=4.

reduced noise levels – an average noise reduction gain of 7.5dB is achieved over test examples. Still better results could be achievable using an undecimated multiple wavelet transform due to its shift invariance.

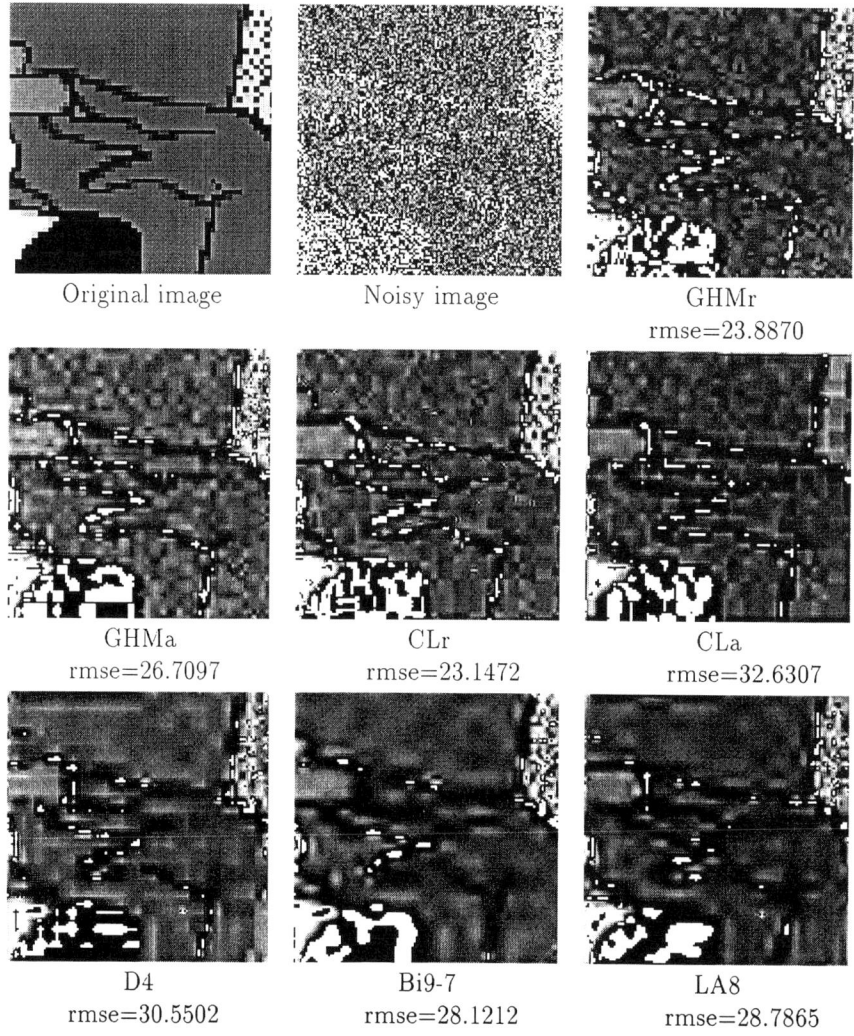

Figure 19: Example of 2D processing methods applied to cartoon bear image, for S/N=2.

Acknowledgements

This work was done while Vasily Strela was at Imperial College on EPSRC grant GR/L11182. The authors are grateful for this support. This article is based on a presentation on 7th December 1998 at the Data Analysis Workshop, part of the programme on Nonlinear and Nonstationary Signal Processing at the Isaac Newton Institute for Mathematical Sciences, Cambridge, UK.

References

Chui, C. K. and Lian J. A. (1996) 'A study of orthonormal multi-wavelets,' *Applied Numerical Mathematics* **20**, 273–98.

Cotronei, M., Montefusco, L. B. and Puccio, L. (1998) 'Multiwavelet analysis and signal processing,' *IEEE Transactions on Circuits and Systems II: Analog and Digital Signal Processing* **45**, 970–87.

Daubechies, I. (1988) 'Orthonormal bases of compactly supported wavelets,' *Communications in Pure and Applied Mathematics* **41**, 909–96.

Daubechies, I. (1992) *Ten Lectures on Wavelets*, Philadelphia: SIAM.

Donoho, D. L. and Johnstone, I. M. (1994) 'Ideal spatial adaptation by wavelet shrinkage,' *Biometrika* **81**, 425–55.

Donoho, D. L. and Johnstone, I. M. (1995) 'Adapting to unknown smoothness via wavelet shrinkage,' *Journal of the American Statistical Association* **90**, 1200–24.

Downie, T. R. and Silverman, B. W. (1998) 'The discrete multiple wavelet transform and thresholding methods,' *IEEE Transactions on Signal Processing* **46**, 2558–61.

Galambos, J. (1978) *The Asymptotic Theory of Extreme Order Statistics*, New York: Wiley.

Geronimo, J., Hardin, D. and Massopust P. R. (1994) 'Fractal functions and wavelet expansions based on several functions,' *Journal of Approximation Theory* **78**, 373–401.

Goodman, T. N. T. and Lee, S. L. (1994) 'Wavelets of multiplicity r,' *Transactions of the American Mathematical Society* **342**, 307–24.

Heil, C., Strang, G. and Strela, V. (1996) 'Approximation by translates of refinable functions,' *Numerische Mathematik* **73**, 75–94.

Strela, V. (1998) 'A note on construction of biorthogonal multi-scaling functions,' in *Contemporary Mathematics: Wavelets, Multiwavelets and their Applications*, edited by A. Aldoubi and E. B. Lin, Providence: AMS, 149–157.

Strela, V. and Strang, G. (1995) 'Finite element multiwavelets,' in *Approximation Theory, Wavelets and Applications*, edited by S. P. Singh, Kluwer, 485–96.

Strela, V., Heller, P. N., Strang, G., Topiwala, P. and Heil, C. (1999) 'The application of multiwavelet filter banks to signal and image processing,' *IEEE Transactions on Image Processing* **8**, 548–63.

Xia, X.-G., Geronimo, J. S., Hardin, D. P. and Suter, B.W. (1996) 'Design of prefilters for discrete multiwavelet transforms,' *IEEE Transactions on Signal Processing* **44**, 25–35.

Wavestrapping Time Series: Adaptive Wavelet-Based Bootstrapping

D. B. Percival, S. Sardy and A. C. Davison

1 Introduction

Suppose we observe a time series that can be regarded as a realization of a portion $X_0, X_1, \ldots, X_{N-1}$ of a real-valued zero mean Gaussian stationary process $\{X_t\}$ with autocovariance sequence (ACVS) $s_{X,\tau} \equiv \text{cov}\{X_t, X_{t+\tau}\}$. Suppose also that we compute a statistic based upon our time series, e.g., the sample autocorrelation for unit lag:

$$\hat{\rho}_{X,1} \equiv \frac{\sum_{t=0}^{N-2} X_t X_{t+1}}{\sum_{t=0}^{N-1} X_t^2}. \tag{1.1}$$

To thoroughly assess the quality of $\hat{\rho}_{X,1}$ as an estimator of the corresponding population quantity $\rho_{X,1} \equiv s_{X,1}/s_{X,0}$, we need to know the distribution of $\hat{\rho}_{X,1}$; however, calculating the exact distribution of a statistic of a time series can be very difficult, so it is of interest to find reasonable approximations. If our time series were a white noise process (i.e., a sample of uncorrelated random variables (RVs), which — because of the Gaussian assumption — yields independent and identically distributed (IID) RVs), we could make use of two quite different approximations. The first approximation is based on large sample theory, which says that, as $N \to \infty$, $\hat{\rho}_{X,1}$ is approximately normally distributed with mean zero and variance $1/N$ (Bartlett, 1946; Priestley, 1981, Equation (5.3.39)). The second approximation is based on bootstrapping (Efron and Tibshirani, 1993; Davison and Hinkley, 1997). Here we randomly sample with replacement from the original time series to create a new series of N values, for which we then compute the unit lag sample autocorrelation, say, $\hat{\rho}_{X,1}^{(1)}$. If we repeat this procedure M times to obtain $\hat{\rho}_{X,1}^{(1)}, \hat{\rho}_{X,1}^{(2)}, \ldots, \hat{\rho}_{X,1}^{(M)}$, we can use the sample distribution of these M bootstrap estimates as an approximation to the unknown distribution of $\hat{\rho}_{X,1}$. While the large sample distribution is obviously faster to compute than the bootstrap distribution for the case of $\hat{\rho}_{X,1}$, a major advantage of the bootstrap approximation is its adaptability to other statistics of interest, for which a significant amount of research might be required to work out the large sample distribution.

More generally, if $\{X_t\}$ is not necessarily white noise, we must reconsider both the large sample and bootstrap approximations to the distribution of $\hat{\rho}_{X,1}$. Under an assumption that the ACVS damps down to zero 'rapidly,' large sample approximations to the distribution of $\hat{\rho}_{X,1}$ have been worked out, but are unappealing. In particular, if we let $\rho_{X,\tau} \equiv s_{X,\tau}/s_{X,0}$ denote the τth element of the autocorrelation sequence (ACS), then $N^{1/2}(\hat\rho_{X,\tau} - \rho_{X,\tau})$ converges in distribution to a Gaussian distribution with mean zero and variance (Priestley, 1981, (5.3.37))

$$\sum_{\tau=-\infty}^{\infty} \left\{ \rho_{X,\tau}^2(1 + 2\rho_{X,1}^2) + \rho_{X,\tau+1}\rho_{X,\tau-1} - 4\rho_{X,1}\rho_{X,\tau}\rho_{X,\tau-1} \right\}.$$

This expression depends upon the entire ACS, which is typically unknown in practice. Under the same assumption of rapid decorrelation, several variations on the bootstrap procedure have been proposed that provide good approximations to the distribution of $\hat\rho_{X,1}$ and related statistics (see Davison and Hinkley, 1997, Chapter 8, and §3 below). On the other hand, if the ACVS does not damp down rapidly but rather exhibits 'long memory' (see §2), then the large sample theory for $\hat\rho_{X,1}$ is currently incomplete, and standard bootstrapping procedures are known not to work very well. Stationary long memory processes (LMPs) are becoming increasingly important as models for a wide range of time series (Beran, 1994), so it is of interest to have decent approximations for the distribution of $\hat\rho_{X,1}$ and related statistics that allow for such processes.

We propose here 'wavestrapping' as an adaptive wavelet-based scheme for bootstrapping certain statistics for time series that can be modeled by stationary processes with either rapidly decaying ('short memory') or long memory ACVSs. The basis for this methodology is the work of Flandrin (1992), Wornell (1995) and McCoy and Walden (1996), who show that the discrete wavelet transform (DWT) approximately decorrelates long memory processes. We demonstrate that, by applying the bootstrap in the wavelet domain, we can approximate the distribution of $\hat\rho_{X,1}$ reasonably well for long memory processes. When applied to certain short memory processes, this DWT-based scheme is not as successful, a result that can be attributed to the fact that the DWT need not be an adequate decorrelating transform for such processes; however, in such cases, a generalization of the DWT-based on discrete wavelet packet transforms (DWPTs) can yield an acceptable decorrelating transform. We propose a procedure for adaptively selecting a decorrelating transform for a given time series that involves a 'top-down' search of a collection of DWPTs with the help of white noise tests.

The remainder of this article is organized as follows. We first review short and long memory models for time series (§2) and current approaches for bootstrapping time series (§3), after which we discuss the basic ideas behind the DWT (§4). Because the DWT acts as a decorrelating transform for long memory processes, we can use it to define a bootstrapping scheme. We demonstrate the effectiveness of this scheme via Monte Carlo experiments (§5). We then consider why DWT-based bootstrapping does not work well for certain short memory processes and why wavestrapping can correct this deficiency (§6).

We demonstrate via Monte Carlo experiments that wavestrapping works reasonably well for both short and long memory processes (§7). We then give examples of wavestrapping (§8), including one involving two time series of interest in atmospheric science, and we conclude with a discussion of directions for future research (§9).

2   Models for stationary time series

In this article we concentrate on time series that can be modeled as a stationary process $\{X_t\}$ with an ACVS $s_{X,\tau}$ and spectral density function (SDF) $S_X(\cdot)$ related by

$$s_{X,\tau} = \int_{-1/2}^{1/2} e^{i2\pi f\tau} S_X(f)\, df, \quad \tau = \dots, -1, 0, 1, \dots.$$

Let $\{\epsilon_t\}$ be a Gaussian white noise process with mean zero and variance σ_ϵ^2. Two simple models that fit into the above framework are the first-order autoregressive model (AR(1)) $X_t = \phi X_{t-1} + \epsilon_t$ with $|\phi| < 1$, for which

$$s_{X,\tau} = \frac{\phi^{|\tau|}\sigma_\epsilon^2}{1 - \phi^2} \text{ and } S_X(f) = \frac{\sigma_\epsilon^2}{|1 - \phi e^{-i2\pi f}|^2},$$

and the first-order moving average model (MA(1)) $X_t = \epsilon_t - \theta\epsilon_{t-1}$, for which

$$s_{X,\tau} = \begin{cases} (1 + \theta^2)\sigma_\epsilon^2, & \tau = 0, \\ -\theta\sigma_\epsilon^2, & \tau = \pm 1,; \\ 0, & \text{otherwise.} \end{cases} \quad \text{and } S_X(f) = \sigma_\epsilon^2 |1 - \theta e^{-i2\pi f}|^2;$$

Both these models have ACVSs that rapidly decay to zero: in the case of the AR(1) model, the rate of decay is exponential, whereas the MA(1) ACVS is identically zero for all lags $|\tau| \geq 2$. Because of this rapid decorrelation with increasing τ, the AR(1) and MA(1) models are sometimes said to have 'short memory.'

As an example of a simple model exhibiting long memory, let us consider a stationary Gaussian fractionally differenced (FD) process $\{X_t\}$ (Granger and Joyeux, 1980; Hosking, 1981; Beran, 1994). In terms of the white noise process $\{\epsilon_t\}$, we can represent an FD process as an infinite order MA process, namely,

$$X_t = \sum_{k=0}^{\infty} \frac{\Gamma(k + \delta)}{\Gamma(k + 1)\Gamma(\delta)} \epsilon_{t-k},$$

where $-\frac{1}{2} < \delta < \frac{1}{2}$. The SDF for this process is given by

$$S_X(f) = \sigma_\epsilon^2 |2\sin(\pi f)|^{-2\delta},$$

while its ACVS can be obtained using

$$s_{X,\tau} = s_{X,\tau-1}\frac{\tau+\delta-1}{\tau-\delta}, \quad \tau = 1, 2, \ldots, \quad \text{with } s_{X,0} = \frac{\sigma_\epsilon^2 \Gamma(1-2\delta)}{\Gamma^2(1-\delta)}$$

(for $\tau < 0$, we have $s_{X,\tau} = s_{X,-\tau}$). When $0 < \delta < \frac{1}{2}$, the SDF has a pole at zero, in which case the process exhibits slowly decaying autocovariances because we have, for some $C_s > 0$,

$$\lim_{\tau\to\infty} \frac{s_{X,\tau}}{C_s \tau^{2\delta-1}} = 1;$$

i.e., the ACVS decays at a slower (hyperbolic) rate than for the AR(1) and MA(1) models.

3   Current approaches for bootstrapping time series

Existing procedures for bootstrapping time series can be divided into those which resample in the time and the frequency domains. In this section we review them; see Davison and Hinkley (1997, Chapter 8) and Bühlmann (1999) for details and further references.

3.1   Time domain

3.1.1   Residual bootstrap

When it is credible that $X_0, \ldots, X_{N-1}$ result from a model for which residuals can be identified, a form of model-based resampling may be applied. For example, if the series has AR(p) representation

$$X_t = \sum_{u=1}^p \phi_u X_{t-u} + \epsilon_t, \tag{3.1}$$

where $\{\epsilon_t\}$ is a white noise process, we can use the estimated coefficients $\hat\phi_u$ to determine residuals

$$r_t = X_t - \sum_{u=1}^p \hat\phi_u X_{t-u}.$$

We then generate bootstrap series according to (3.1), but with the ϕ_u replaced by their estimates, and with $\{\epsilon_t\}$ replaced with a white noise process generated by sampling independently with replacement from the residuals r_t, ideally centered and scaled to have the same mean and variance as the $\{\epsilon_t\}$. Under suitable conditions the properties of statistics constructed from the bootstrap series will mimic repeated sampling properties of statistics constructed from the original series. This procedure has the drawback that a specific model must be fitted and used for the resampling, and it will fail if that model is incorrect.

In practice the model fitted is generally selected from the data. For example, the p in (3.1) is often selected by minimizing a model selection criterion such as AIC. This corresponds to the sieve bootstrap, whose philosophy is that a wide class of models should be compared, with the best-fitting model chosen for the bootstrap. This can greatly improve on the simplistic approach in which the model is fixed, but its performance depends heavily on the adequacy of the model class chosen.

3.1.2   Block bootstrap

A nonparametric time domain approach is block resampling (Künsch, 1989). The motivation is that for many purposes the dominant property of a time series is its short-range dependence, which may (largely) be preserved by re-sampling blocks of consecutive observations. The simplest example is the sample autocorrelation for unit lag, $\hat{\rho}_{X,1}$, which is (almost) the solution of

$$\sum_{t=0}^{N-2} X_t(X_{t+1} - \rho X_t) = 0$$

and depends only on the marginal distribution of successive pairs of observations. The idea is to rewrite the original series as the bivariate time series

$$Y_0, Y_1, \ldots, Y_{N-2} = \begin{pmatrix} X_0 \\ X_1 \end{pmatrix}, \begin{pmatrix} X_1 \\ X_2 \end{pmatrix}, \ldots, \begin{pmatrix} X_{N-2} \\ X_{N-1} \end{pmatrix},$$

rewrite the algorithm that computes $\hat{\rho}_{X,1}$ as a function of the Ys, resample b blocks of l consecutive Ys, where $bl = N$, and compute the statistic from the resampled data. This preserves dependence between the Xs from which the statistic is calculated and can give excellent results for short-range dependent series. Its main drawbacks are twofold. First, it is not automatic because consideration has to be given to the rewriting of the statistic. Second, not every statistic can be written as a function of short blocks of data. Both drawbacks are non-trivial, and a simplified approach is usually applied in practice.

The simplification is to construct a new series by concatenating resampled blocks of $X_0, \ldots, X_{N-1}$, but this generally yields resampled series that are much less dependent than the original data. A drawback of this simple approach is the choice of block length, about which little is known of practical use; somewhat discouragingly, this is analogous to the choice of bandwidth in smoothing problems. Moreover, the method requires that certain sums of ACVS elements be bounded and so fails entirely for LMPs.

3.2 Bootstrapping in the frequency domain

A quite different approach is based on the Fourier transform for stationary processes (Priestley, 1981). Let

$$\widetilde{X}_k = \sum_{t=0}^{N-1} X_t e^{-i2\pi kt/N}, \quad k = 0, \ldots, N-1,$$

be the discrete Fourier transform (DFT) of the time series. The sequence $\widetilde{X}_0, \ldots, \widetilde{X}_{N-1}$ comprises the empirical Fourier transform of the data; the periodogram has elements $N^{-1}|\widetilde{X}_k|^2$, which summarize frequency information in the series. Under suitable conditions and as $N \to \infty$, the real and imaginary parts of the $\widetilde{X}_k$ are distributed like a sample of independent normal variables, with means zero and variances $NS_X(k/N)/2$. One implication of this is that the phase and modulus of each $\widetilde{X}_k$ are independent. A second is that the $N^{-1}|\widetilde{X}_k|^2$ are asymptotically independent with scaled chi-squared distributions. Both properties suggest possible resampling schemes.

3.2.1 Phase scrambling

The independence of the phase and modulus of the $\widetilde{X}_k$ suggests that a resampled series with the same periodogram can be made by generating phases but keeping moduli fixed. To be specific, let $U_0, \ldots, U_{N-1}$ be independent variables uniform on $(0, 2\pi)$ and set

$$\widetilde{X}_k^* = 2^{-1/2}\left\{e^{-iU_k}\widetilde{X}_k + \overline{e^{-iU_{N-k}}\widetilde{X}_{N-k}}\right\}, \quad k = 0, \ldots, N-1,$$

where the overbar denotes complex conjugate. Then the inverse Fourier transform of $\widetilde{X}_0^*, \ldots, \widetilde{X}_{N-1}^*$ is a series with the same periodogram as $X_0, \ldots, X_{N-1}$ but randomized phases. Unfortunately this resampling scheme and its variants apply to a very limited range of statistics, because they mimic only second-order properties of the original data. Moreover variability is underestimated because this resampling scheme fixes the periodogram, unlike for the original series whose periodogram is random, and statistics that can be computed from the periodogram, such as $\hat{\rho}_{X,1}$, display no variation across resamples.

3.2.2 Bootstrapping the periodogram

Another frequency domain approach potentially suitable for statistics that can be computed from the periodogram stems from the observation that the $N^{-1}|\widetilde{X}_k|^2$, $k = 1, \ldots, N-1$, have independent exponential distributions with means $S_X(k/N)$ as $N \to \infty$. This suggests using an estimate $\hat{S}_X(k/N)$ of the SDF to make residuals $r_t' = N^{-1}|\widetilde{X}_k|^2/\hat{S}_X(k/N)$, $k = 1, \ldots, N-1$, which are then resampled and merged with the estimate to give a new periodogram with elements $S_X(k/N)r_t'^*$, where the $r_t'^*$ are a random sample taken with replacement from the r'_t. The motivation is that the $N^{-1}|\tilde{X}_k|^2/S_X(k/N)$ form a random sample from the exponential distribution, and the hope is that the r'_t are (almost) such a sample also.

If the SDF $S_X(f)$ is known apart from the values of a few parameters, this approach is essentially model-based, and will share the good and bad aspects of the schemes discussed in §3.1.1. If $S_X(f)$ is estimated nonparametrically, for example by a kernel method, then a bandwidth must be chosen. It turns out that three bandwidths are needed if the bootstrap is to work, one for the original estimate, a smaller one to estimate the residuals consistently and a larger one to give an estimate to which the resampled residuals should be added. Unfortunately the literature contains little theoretical guidance about how they should be chosen, while the numerical evidence is scant and equivocal. Hence although this method has the appeal of not involving the construction of a bootstrap series, it cannot yet be recommended for general use, even for statistics that depend only on the periodogram. In any case it may not be applied to other statistics.

4 The discrete wavelet transform

The discrete wavelet transform (DWT) is an orthonormal transform $\mathcal{W}$ that takes a time series $\mathbf{X} = [X_0, X_1, \ldots, X_{N-1}]^T$ and yields a vector of N DWT coefficients $\mathbf{W} \equiv \mathcal{W}\mathbf{X}$. The orthonormality condition $\mathcal{W}^T\mathcal{W} = I_N$ implies that we can reconstruct the time series from its DWT coefficients via $\mathbf{X} = \mathcal{W}^T\mathbf{W}$, so $\mathbf{W}$ is fully equivalent to $\mathbf{X}$. Under the assumption that N is an integer multiple of 2^{J_0}, where J_0 is an integer denoting the number of levels in the DWT, we can partition the DWT coefficient vector into subvectors:

$$\mathbf{W} = [\mathbf{W}_1^T, \mathbf{W}_2^T, \ldots, \mathbf{W}_{J_0}^T, \mathbf{V}_{J_0}^T]^T.$$

The subvector $\mathbf{W}_j$ contains $N_j \equiv N/2^j$ wavelet coefficients associated with scale $\tau_j \equiv 2^{j-1}$, whereas $\mathbf{V}_{J_0}$ contains $N/2^{J_0}$ scaling coefficients associated with scale $\lambda_{J_0} \equiv 2^{J_0}$. To see what we mean by scale and what the wavelet and scaling coefficients are telling us about the time series, let us define an average of λ contiguous time series values ending with index t as

$$\overline{X}_t(\lambda) \equiv \frac{1}{\lambda}\sum_{l=0}^{\lambda-1} X_{t-l}.$$

We define the scale associated with this average to be λ. With this definition, let us consider the special case of the Haar DWT, for which the DWT coefficients have the form

$$W_{j,n} \propto \overline{X}_{\lambda_j(n+1)-1}(\tau_j) - \overline{X}_{\lambda_j(n+1)-1-\tau_j}(\tau_j) \text{ and } V_{J_0,n} \propto \overline{X}_{\lambda_{J_0}(n+1)-1}(\lambda_j),$$

where $W_{j,n}$ and $V_{J_0,n}$ are the nth elements of, respectively, $\mathbf{W}_j$ and $\mathbf{V}_{J_0}$. Note that the Haar wavelet coefficients $W_{j,n}$ are proportional to first differences of adjacent averages over scale τ_j, whereas the Haar scaling coefficients $V_{J_0,n}$ are proportional to averages over scale λ_{J_0}. This same pattern holds for DWTs other than the Haar, in that we can regard the wavelet coefficients as being proportional to (higher order) differences of (weighted) averages over scale τ_j, and the scaling coefficients as being proportional to (weighted) averages over scale λ_{J_0}.

We can formally describe the DWT in terms of wavelet and scaling filters as follows. Let $\mathbf{h}_1 = [h_{1,0}, \ldots, h_{1,L-1}, 0 \ldots, 0]^T$ be a vector of length N whose first $L < N$ elements are the unit level wavelet filter coefficients for a Daubechies compactly supported wavelet (see Daubechies, 1992, Chapter 6). For example, the Haar wavelet filter has $L = 2$ coefficients, namely, $h_{1,0} = \frac{1}{\sqrt{2}}$ and $h_{1,1} = -\frac{1}{\sqrt{2}}$. Let $H_{1,k}, k = 0, \ldots, N - 1$, be the DFT of $\mathbf{h}_1$. In the Haar case, we have $H_{1,k} = (1 - e^{-i2\pi k/N})/\sqrt{2}$. Let $\mathbf{g}_1 = [g_{1,0}, \ldots, g_{1,L-1}, 0, \ldots, 0]^T$ be a vector of length N containing the zero padded scaling filter coefficients for unit level, defined via $g_{1,l} = (-1)^{l+1} h_{1,L-1-l}$ for $l = 0, \ldots, L - 1$, and let $G_{1,k}$ denote its DFT. Like the Haar wavelet filter, the Haar scaling filter has two non-zero elements, namely, $g_{1,0} = g_{1,1} = \frac{1}{\sqrt{2}}$, and its DFT is $G_{1,k} = (1 + e^{-i2\pi k/N})/\sqrt{2}$. The level j wavelet filter is given by the elements of the vector $\mathbf{h}_j$, which is the inverse DFT of

$$H_{j,k} = H_{1,2^{j-1}k \bmod N} \prod_{l=0}^{j-2} G_{1,2^l k \bmod N}, \quad k = 0, \ldots, N - 1.$$

When $N > L_j = (2^j - 1)(L - 1) + 1$, the last $N - L_j$ elements of $\mathbf{h}_j$ are zero, so the jth wavelet filter $\mathbf{h}_j$ has no more than L_j non-zero elements. In the Haar case, we have $L_j = 2^j$, and, when $N > 2^j$, the elements of $\mathbf{h}_j$ are

$$h_{j,l} = \begin{cases} 1/2^{j/2}, & l = 0, \ldots, 2^{j-1} - 1, \\ -1/2^{j/2}, & l = 2^{j-1}, \ldots, 2^j - 1, \text{ and} \\ 0, & l = 2^j, \ldots, N - 1. \end{cases}$$

Similarly, the level J_0 scaling filter is contained in $\mathbf{g}_{J_0}$, whose elements are the inverse DFT of

$$G_{J_0,k} = \prod_{l=0}^{J_0-1} G_{1,2^l k \bmod N}, \quad k = 0, \ldots, N - 1.$$

The elements of the Haar $\mathbf{g}_{J_0}$ are

$$g_{J_0,l} = \begin{cases} 1/2^{J_0/2}, & l = 0, \ldots, 2^{J_0} - 1, \text{ and} \\ 0, & l = 2^{J_0}, \ldots, N - 1. \end{cases}$$

To obtain the jth level wavelet coefficients, we filter $\mathbf{X}$ using $\mathbf{h}_j$ and sub-sample every 2^jth value from the filter output:

$$W_{j,n} = \sum_{l=0}^{\min(L_j,N)-1} h_{j,l} X_{2^j(n+1)-1-l \bmod N}, \quad n = 0, \ldots, N_j - 1; \quad (4.1)$$

an analogous expression yields the J_0th level scaling coefficients. In the Haar case we can write

$$W_{j,n} = \frac{1}{2^{j/2}} \sum_{l=0}^{2^{j-1}-1} X_{2^j(n+1)-1-l} - \frac{1}{2^{j/2}} \sum_{l=2^{j-1}}^{2^j-1} X_{2^j(n+1)-1-l}.$$

This example is atypical in that we do not need to use the 'modulo N' operation. For wavelet filters such that $L > 2$, the wavelet coefficients are obtained by treating the time series as if it were circular (i.e., as if it were a periodic sequence with period N). This assumption is problematic and yields a certain number of 'boundary' coefficients whose statistical properties differ from coefficients unaffected by circularity (the number of boundary coefficients on any given level is no more than $L - 2$, which is consistent with there being no such coefficients in the Haar case). In practice the DWT coefficients are not computed directly via (4.1) but rather via an elegant pyramid algorithm (Mallat, 1989) that filters $\mathbf{X}$ using $\mathbf{h}_1$ and $\mathbf{g}_1$, retains the odd-indexed values of the wavelet filter output as the unit level wavelet coefficients and then repeats this process with $\mathbf{X}$ replaced by the odd-indexed values of the scaling filter output to obtain the level $j = 2$ wavelet coefficients and so forth.

5 DWT-based bootstrapping

The idea behind DWT-based bootstrapping is to make use of the fact that, for FD and certain other stationary processes, the DWT acts as a decorrelating transform for a time series; i.e., whereas the time series itself can exhibit a high degree of autocorrelation, its DWT coefficients can — to a reasonable approximation — be regarded as uncorrelated. To quantify this decorrelation effect, we first note that, if we ignore boundary coefficients, then within a given level we have

$$\operatorname{cov}\{W_{j,n}, W_{j,n+\tau}\} = \sum_{m=-(L_j-1)}^{L_j-1} s_{X,2^j\tau+m} \sum_{l=0}^{L_j-|m|-1} h_{j,l} h_{j,l+|m|}. \quad (5.1)$$

We can use the above to compute the unit lag correlations $\operatorname{corr}\{W_{j,t}, W_{j,t+1}\}$ for, e.g., an FD process with $\delta = 0.45$. Table 5.1 lists these correlations for the Haar, D(4) and LA(8) wavelet filters and scales 1, 2, 4 and 8; here 'D(4)' and 'LA(8)' refer to the Daubechies extremal phase filter with four non-zero coefficients and to her least asymmetric filter with eight coefficients (Daubechies,

Scale	Haar	D(4)	LA(8)
1	−0.0626	−0.0797	−0.0767
2	−0.0947	−0.1320	−0.1356
4	−0.1133	−0.1511	−0.1501
8	−0.1211	−0.1559	−0.1535

Table 5.1: Lag 1 autocorrelations for wavelet coefficients of scales 1, 2, 4, and 8 for an FD process with $\delta = 0.45$ using the Haar, D(4) and LA(8) wavelet filters.

	Haar			D(4)			LA(8)		
Scale	2	4	8	2	4	8	2	4	8
1	0.13	0.17	0.14	0.09	0.09	0.04	0.06	0.03	0.00
2		0.17	0.21		0.12	0.11		0.08	0.03
4			0.18			0.13			0.08

Table 5.2: Maximum absolute cross-correlations for wavelet coefficients between scales for an FD process with $\delta = 0.45$ using the Haar, D(4) and LA(8) wavelet filters.

1992). Note that these correlations are all negative, with departures from zero increasing somewhat as j increases. For larger lags, computations indicate that the autocorrelation damps down roughly as dictated by an AR(1) model, i.e., $\text{corr}\{W_{j,t}, W_{j,t+\tau}\} \approx (\text{corr}\{W_{j,t}, W_{j,t+1}\})^{|\tau|}$. To quantify correlation between different levels, we note that (again ignoring boundary coefficients)

$$\text{cov}\{W_{j,n}, W_{j',n'}\} = \sum_{l=0}^{L_j-1} \sum_{l'=0}^{L_{j'}-1} h_{j,l} h_{j',l'} s_{X, 2^j(n+1)-l-2^{j'}(n'+1)+l'}. \tag{5.2}$$

For the same FD process as before, Table 5.2 lists $\max_{n,n'} |\text{corr}\{W_{j,n}, W_{j',n'}\}|$ for $1 \le j < j' \le 4$. We can deduce from these two tables that, while the unit lag correlations within levels are somewhat larger for the D(4) and LA(8) wavelets than for the Haar, wavelets of greater width than the Haar lead to a decrease in maximum absolute correlation between levels.

We can gain additional insight into the decorrelation properties of the DWT by noting the frequency domain equivalent of (5.2), namely,

$$\text{cov}\{W_{j,n}, W_{j',n'}\} = \int_{-1/2}^{1/2} e^{i2\pi f(2^j(n+1)-2^{j'}(n'+1))} H_j(f) H_{j'}^*(f) S_X(f) \, df, \tag{5.3}$$

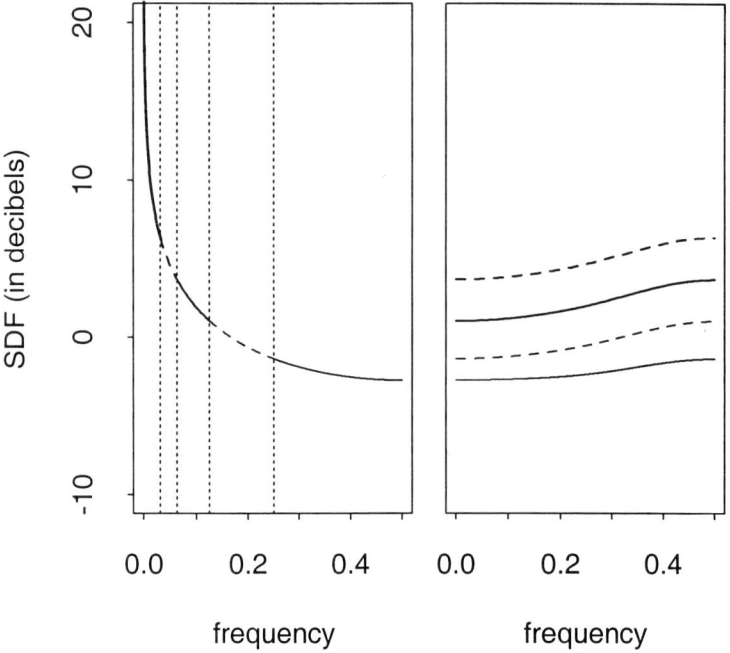

Figure 5.1: SDFs for an FD process with $\delta = 0.45$ (left-hand plot) and for the corresponding nonboundary LA(8) wavelet coefficients in $\mathbf{W}_j$, $j = 1, 2, 3, 4$ (bottom to top curves in the right-hand plot). The vertical dotted lines mark the beginning of the nominal pass-bands $[\frac{1}{2^{j+1}}, \frac{1}{2^j}]$ for $\mathbf{W}_j$.

where $H_j(\cdot)$ is the transfer function for the jth level wavelet filter:

$$H_j(f) \equiv \sum_{l=0}^{L_j-1} h_{j,l} e^{-i2\pi f l}.$$

When $j = j'$ and $n' = n + \tau$, we obtain the frequency domain equivalent of (5.1):

$$\operatorname{cov}\{W_{j,n}, W_{j,n+\tau}\} = \int_{-1/2}^{1/2} e^{i2^{j+1}\pi f \tau} \mathcal{H}_j(f) S_X(f) \, df, \qquad (5.4)$$

where $\mathcal{H}_j(f) \equiv |H_j(f)|^2$. A jth level wavelet filter has a nominal pass-band given by $|f| \in [\frac{1}{2^{j+1}}, \frac{1}{2^j}]$. We can thus argue that the above should be approximately zero for $\tau \neq 0$ when $S_X(\cdot)$ is approximately constant over this pass-band. An alternative formulation is to note that

$$\text{cov}\{W_{j,n}, W_{j,n+\tau}\} = \int_{-1/2}^{1/2} e^{i2\pi f \tau} S_j(f)\, df,$$

where

$$S_j(f) \equiv \frac{1}{2^j} \sum_{k=0}^{2^j-1} \mathcal{H}_j(\tfrac{f+k}{2^j}) S_X(\tfrac{f+k}{2^j}); \qquad (5.5)$$

i.e., if we ignore boundary coefficients, then $W_{j,n}$ can be regarded as a portion of a stationary process with SDF $S_j(\cdot)$, so $W_{j,n}$ will be approximately white noise if $S_j(\cdot)$ is approximately constant. As an example, the left-hand panel of Figure 5.1 shows the SDF $S_X(\cdot)$ on a decibel (dB) scale for an FD process with $\delta = 0.45$. The right-hand panel shows, from bottom to top, $S_j(f), j = 1, \ldots, 4$, based upon an LA(8) wavelet. We see that, in contrast to $S_X(\cdot)$, the SDFs for the wavelet coefficients have a quite limited range of variation (less than 3 dB, i.e., a factor of 2). We can also see why the DWT is so well suited for SDFs for FD processes: as $S_X(f)$ diverges to infinity with decreasing f, the widths of the nominal pass-bands decrease commensurately so that $S_X(\cdot)$ does not vary much over any given pass-band.

Finally, let us consider the exact covariance matrix $\Sigma_\mathbf{W}$ for all the DWT coefficients $\mathbf{W}$ (this allows us to examine covariances involving the scaling and boundary wavelet coefficients). Let $\Sigma_\mathbf{X}$ be the covariance matrix for $\mathbf{X}$ (because of stationarity, its (j, k)th element is $s_{X,j-k}$). Since $\mathbf{W} = \mathcal{W}\mathbf{X}$, we have $\Sigma_\mathbf{W} = \mathcal{W}\Sigma_\mathbf{X}\mathcal{W}^T$ (note that the elements of $\mathcal{W}$ can be deduced from (4.1)). The top row of Figure 5.2 depicts the corresponding correlation matrix for level $J_0 = 6$ Haar, D(4) and LA(8) DWTs when $\mathbf{X}$ consists of a portion of size $N = 256$ from an FD process with $\delta = 0.45$. These plots show a grey-scale coding of the magnitudes of the elements of the correlation matrices after setting the diagonal elements to zero (these elements are unity by definition and dominate the off-diagonal elements). Let us focus first on the Haar case (upper left-hand corner). The dotted vertical and horizontal lines delineate the portions of the correlation matrix involving the DWT coefficients of different scales. As we go from the upper left-hand to lower right-hand corners, we pass along the diagonals of square submatrices that involve correlations within a given scale. The faint diagonal within each of these submatrices is primarily due to the lag 1 autocorrelations (see the values in the second column of Table 5.1, which can be used to gauge the magnitudes depicted in the plot). The faint lines going between opposite corners of the off-diagonal (nonsquare) submatrices are due to correlations between scales (Table 5.2 lists the largest such magnitudes). If we compare this plot with the corresponding plots for the D(4) and LA(8) DWTs, we see that the square submatrices are roughly the same (indicating that the autocorrelations within a scale are similar for the three wavelets), but that the lines going between opposite corners of the nonsquare submatrices are fainter (indicating a decrease in correlations between scales

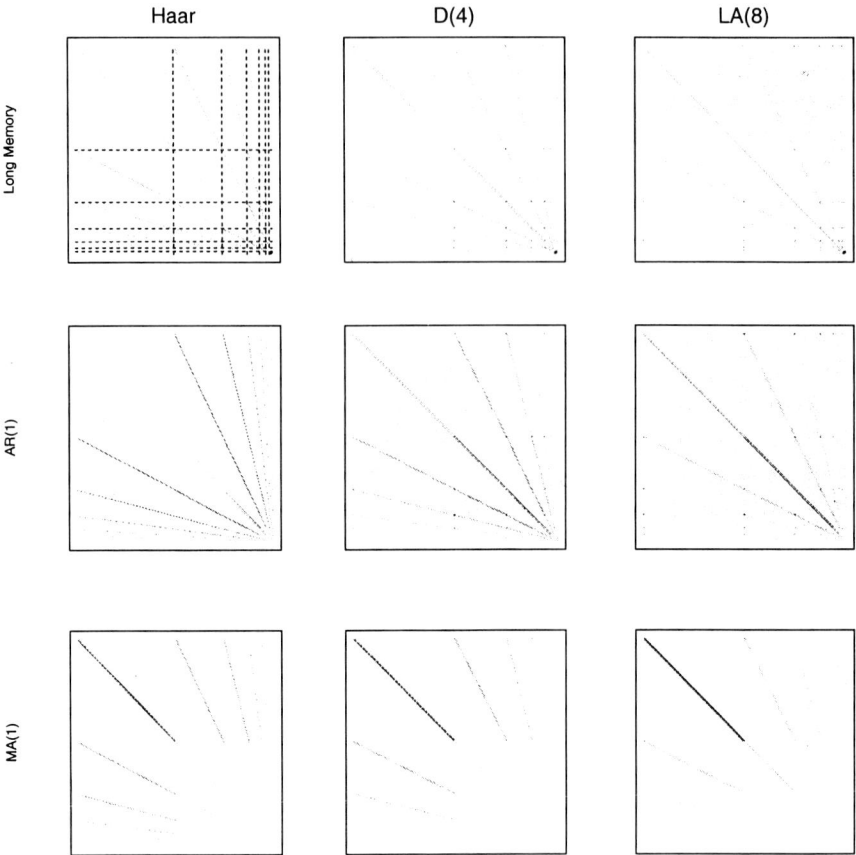

Figure 5.2: Correlation matrices for DWT coefficients **W** when $N = 256$ (see text for details).

as L increases). Additionally, there are some dark points in these latter two plots that tend to line up horizontally and vertically. These are attributable to the scaling and boundary wavelet coefficients and are seen to be relatively few in number (the Haar DWT is free of boundary coefficients). Finally the dark spot in the lower right-hand corner of all three plots is due to the four scaling coefficients, which are highly autocorrelated for an FD process. The overall impression that the top row of plots gives is that the three DWTs do a credible job of decorrelating the highly autocorrelated FD process.

We can thus bootstrap a time series via its DWT using the following steps.

1. Given a time series **X** of length 2^J, compute a level $J_0 = J - 2$ DWT

to obtain the wavelet coefficient vectors $\mathbf{W}_1, \ldots, \mathbf{W}_{J_0}$ and the scaling coefficient vector $\mathbf{V}_{J_0}$. This recipe for setting J_0 yields four coefficients each in $\mathbf{W}_{J_0}$ and $\mathbf{V}_{J_0}$ — decreasing J_0 has the effect of giving us more coefficients in $\mathbf{W}_{J_0}$ from which to bootstrap, but at the price of having more scaling coefficients. As noted before, these are highly correlated for an FD process, and hence are not amenable to bootstrapping.

2. Randomly sample with replacement N_j times from $\mathbf{W}_j$ to create the similarly dimensioned bootstrapped vectors $\mathbf{W}_j^{(b)}$, $j = 1, \ldots, J_0$; likewise, create a bootstrapped vector of scaling coefficients $\mathbf{V}_{J_0}^{(b)}$. While theoretical justification for resampling the scaling coefficients is lacking, we obtained somewhat better results in the Monte Carlo experiments discussed below by doing so — see §9 for further discussion.

3. Apply the inverse DWT to $\mathbf{W}_1^{(b)}, \ldots, \mathbf{W}_{J_0}^{(b)}$ and $\mathbf{V}_{J_0}^{(b)}$ to obtain the bootstrapped time series $\mathbf{X}^{(b)}$, for which we can then compute our statistic of interest, i.e., the unit lag sample autocorrelation $\hat{\rho}_{X,1}^{(b)}$ obtained from (1.1) with X_t replaced by $X_t^{(b)}$.

By repeating the above over and over again, we can build up a sample distribution of bootstrapped autocorrelations, which we use as a surrogate for the distribution of the actual sample autocorrelation.

Let us comment on a variation of the above scheme. As noted before, the DWT treats a time series as if it were circular. This aspect of the DWT can be problematic for a long memory series, for which there can be a large discrepancy between X_0 and X_{N-1}. Greenhall *et al.* (1999) provide evidence that an effective way to get around this difficulty for long memory processes is to replace $\mathbf{X}$ by a series of length $2N$ created by tacking on a time-reversed version of $\mathbf{X}$ to itself:

$$\mathbf{X}_{(c)} \equiv [X_0, X_1, \ldots, X_{N-2}, X_{N-1}, X_{N-1}, X_{N-2}, \ldots, X_1, X_0]^T.$$

We then use the DWT of this circularized series to form our bootstrapped series $\mathbf{X}_{(c)}^{(b)}$, from which we extract the first N elements to compute the sample autocorrelation. We refer to using the DWT on $\mathbf{X}_{(c)}$ rather than $\mathbf{X}$ as using reflection — rather than periodic — boundary conditions.

To see how well DWT-based bootstrapping works, see Table 7.1, which reports the results of a Monte Carlo study described in detail in §7. The bottom quarter of this table shows how well the standard deviations of the DWT-based bootstrapped $\hat{\rho}_{X,1}^{(b)}$ (under the column labeled 'DWT') compare with the actual standard deviation for $\hat{\rho}_{X,1}$ (under 'True'). Here we looked at time series of length $N = 128$ and 1024 that are realizations of an FD process with $\delta = 0.45$, and we used the LA(8) DWT with both periodic and reflection boundary conditions. We also report results for the block bootstrap (under the 'Block' column). We see that, while the DWT-based bootstrap tends to underestimate the true standard deviation by about 15% and 10% using, respectively, periodic and reflection boundary conditions, it is an improvement on the block bootstrap, which underestimates by about 30%.

6   Wavestrapping time series

While DWT-based bootstrapping works reasonably well for long memory FD processes, the question arises as to whether we can expect it to be useful for other processes. As simple examples, let us consider realizations of the AR(1) process $X_t = 0.9X_{t-1} + \epsilon_t$ and the MA(1) process $X_t = \epsilon_t + 0.99\epsilon_{t-1}$. The correlation matrices for the DWT coefficients $\mathbf{W}$ are shown in the middle and bottom rows of Figure 5.2 for, respectively, AR(1) and MA(1) series of length $N = 256$. When compared to the FD case in the first row, we see higher levels of correlation, particularly within scale $j = 1$ for the MA(1) process. Figure 6.1 shows the SDF for this process, along with the SDFs $S_j(\cdot)$ for the nonboundary LA(8) wavelet coefficients in $\mathbf{W}_1, \ldots, \mathbf{W}_4$. The MA(1) SDF has considerable variation within the nominal pass-band $[\frac{1}{4}, \frac{1}{2}]$ for $\mathbf{W}_1$, which leads to $S_1(\cdot)$ being a poor approximation to white noise, thus explaining the high levels of correlation within scale $j = 1$.

Let us attempt to correct the poor decorrelating properties of the DWT in cases like the MA(1) process by considering a generalization of the DWT based upon adaptively picking out a transform from a level J_0 wavelet packet (WP) table (details on how to compute WP tables can be found in, e.g., Wickerhauser, 1994, Bruce and Gao, 1996, and Percival and Walden, 2000). Figure 6.2 shows an example of such a table. The jth row of the table is composed of 2^j vectors $\mathbf{W}_{j,n}$, $n = 0, \ldots, 2^j - 1$. Each vector has $N_j = N/2^j$ elements, and collectively all 2^j vectors form the coefficients for a jth level discrete wavelet packet transform (DWPT). Like the DWT, a DWPT is an orthonormal transform from which we can recover $\mathbf{X}$; moreover, the transform can be formulated as filtering operations involving a wavelet and a scaling filter. The filter that yields the nth subvector has a nominal pass-band given by $[\frac{n}{2^{j+1}}, \frac{n+1}{2^{j+1}}]$. When taken together, the 2^j pass-bands partition the interval $[0, \frac{1}{2}]$ into 2^j intervals of equal length. Figure 6.2 shows the subvectors $\mathbf{W}_{j,n}$ for levels $j = 1, 2$ and 3 enclosed by rectangles spanning the nominal pass-bands. For convenience, we define $\mathbf{W}_{0,0} = \mathbf{X}$ so that the time series itself is associated with a 'zeroth level' DWPT (i.e., the identity transform) covering the entire frequency band.

The collection of DWPT coefficients for levels $j = 0, \ldots, J_0$ forms a level J_0 WP table. The vertical stacking of coefficients in the figure tells us how coefficients from DWPTs of different levels are related: given a subvector $\mathbf{W}_{j,n}$ of level j, we obtain the subvectors $\mathbf{W}_{j+1,2n}$ and $\mathbf{W}_{j+1,2n+1}$ of level $j + 1$ via an orthonormal transform (one subvector is formed using the wavelet

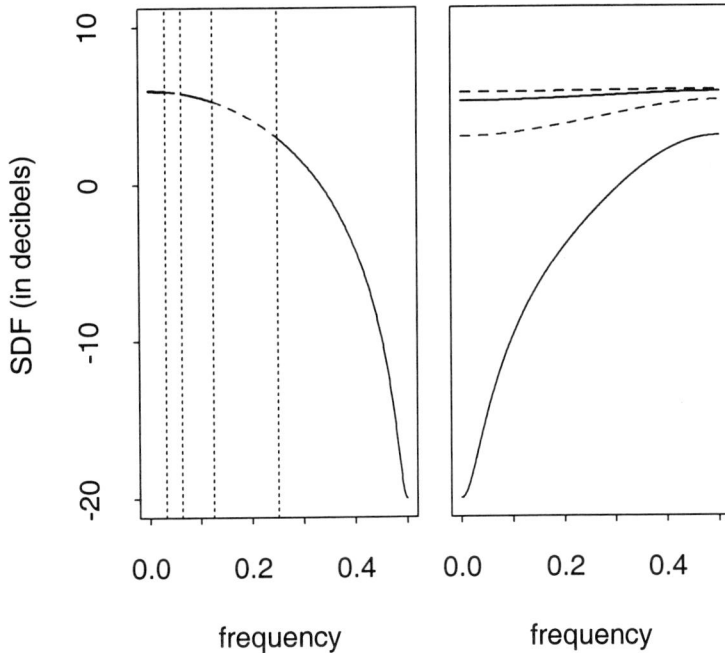

Figure 6.1: As in Figure 5.1, but now for an MA(1) process with $\theta = -0.99$.

filter, and the other, the scaling filter, but the order in which these get used depends upon n). Thus, as depicted in the table, we can obtain $\mathbf{W}_{3,4}$ and $\mathbf{W}_{3,5}$ via an orthonormal transform of $\mathbf{W}_{2,2}$. We can extract a large number of different orthonormal transforms from a WP table. For example, if we start with a jth level DWPT, we can obtain 2^{2^j} different transforms by choosing either to keep each $\mathbf{W}_{j,n}$ or to transform it into the two subvectors $\mathbf{W}_{j+1,2n}$ and $\mathbf{W}_{j+1,2n+1}$. We can obtain even more transforms by keeping or splitting across more than two levels. In fact a DWT of level J_0 is one such transform, consisting of $\mathbf{W}_1 = \mathbf{W}_{1,1}$, $\mathbf{W}_2 = \mathbf{W}_{2,1}$, ..., $\mathbf{W}_{J_0} = \mathbf{W}_{J_0,1}$ and $\mathbf{V}_{J_0} = \mathbf{W}_{J_0,0}$.

With so many different transforms at our disposal, a careful selection of coefficients from the WP table can lead to an orthonormal transform that partitions the frequency interval $[0, \frac{1}{2}]$ into subintervals such that, within each subinterval, the SDF for $\mathbf{X}$ does not vary much. Given knowledge of the SDF $S_X(\cdot)$ and a stopping level J_0, we can adaptively select a transform by starting with $\mathbf{W}_{0,0} = \mathbf{X}$ and recursively applying the following simple rule. If the level of $\mathbf{W}_{j,n}$ is J_0, we retain it; otherwise, we consider the SDF associated with the nonboundary coefficients in $\mathbf{W}_{j,n}$ (this SDF can be computed via an equation

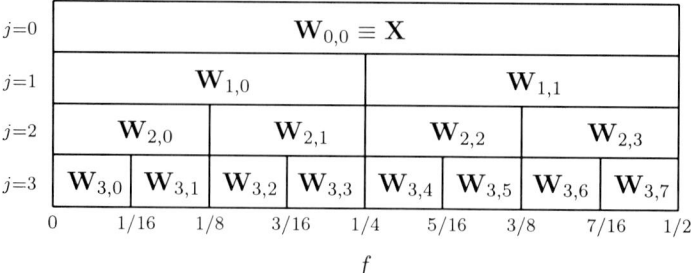

Figure 6.2: Wavelet packet table of order $J_0 = 3$ and associated pass-bands. The jth row of the table contains the subvectors $\mathbf{W}_{j,n}$ of a jth level DWPT.

analogous to (5.5)). If the SDF varies no more than, say, 3 dB, then we retain $\mathbf{W}_{j,n}$; otherwise, we replace $\mathbf{W}_{j,n}$ by $\mathbf{W}_{j+1,2n}$ and $\mathbf{W}_{j+1,2n+1}$, and then apply the simple rule to both of these vectors. Figure 6.3 shows a transform from a fourth level WP table that is adapted to the MA(1) process.

In practice, of course, we do not know $S_X(\cdot)$, so we propose to replace the 3 dB criterion with a statistical test for the null hypothesis that the values in $\mathbf{W}_{j,n}$ are a sample from a white noise process (there are a number of appropriate test statistics in the literature, two of which we describe briefly in §6.1 below). We can now outline the steps needed to create 'wavestrap' samples of a time series.

1. Given a time series $\mathbf{X}$ of length 2^J, compute a level $J_0 = J - 2$ WP table. Enter step 2 with starting values $j = n = 0$ and $\mathbf{W}_{0,0} \equiv \mathbf{X}$.

2. If $j = J_0$, retain $\mathbf{W}_{j,n}$; if $j < J_0$, perform a white noise test on $\mathbf{W}_{j,n}$ using one of the test statistics given in §6.1. If we fail to reject the null hypothesis, then retain $\mathbf{W}_{j,n}$; if we reject, then discard $\mathbf{W}_{j,n}$ (after transforming it into $\mathbf{W}_{j+1,2n}$ and $\mathbf{W}_{j+1,2n+1}$), and repeat this step twice again, once on $\mathbf{W}_{j+1,2n}$, and once on $\mathbf{W}_{j+1,2n+1}$.

3. The desired adaptively chosen transform consists of all the subvectors that are retained after step 2 has been applied as many times as needed. Randomly sample (with replacement) from each of the subvectors in the transform to create the similarly dimensioned wavestrapped subvectors.

4. Apply the inverse of the adaptively chosen transform to the wavestrapped subvectors to obtain the wavestrapped time series, for which we can then compute, e.g., a unit lag sample autocorrelation.

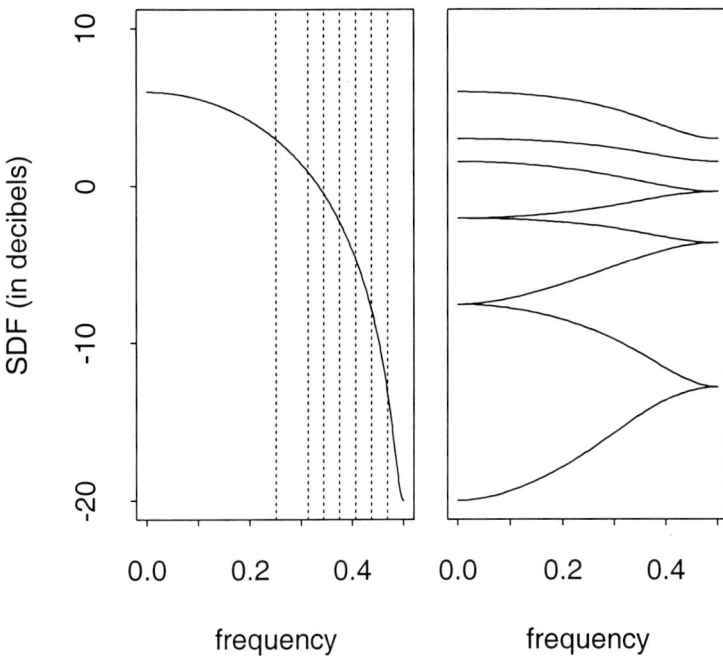

Figure 6.3: Transform selected from an LA(8) WP table of level $J_0 = 4$ that converts the MA(1) process with $\theta = -0.99$ into approximately uncorrelated coefficients. The transform consists of the eight subvectors $\mathbf{W}_{1,0}$, $\mathbf{W}_{3,4}$, $\mathbf{W}_{4,10}, \dots, \mathbf{W}_{4,15}$, which partition $[0, \frac{1}{2}]$ into the nominal pass-bands shown by the vertical lines in the left-hand plot (the solid curve is the SDF for the MA(1) process). The corresponding SDFs for the subvectors are shown from top to bottom in the right-hand plot. The first five of these SDFs have variations less than 3 dB, while the SDFs for $\mathbf{W}_{4,13}$, $\mathbf{W}_{4,14}$ and $\mathbf{W}_{4,15}$ vary by, respectively, 3.9, 5.3 and 7.2 dB. If we were to increase the level to $J_0 = 6$, these three subvectors would be replaced by three $j = 5$ level subvectors $\mathbf{W}_{5,26}, \mathbf{W}_{5,27}, \mathbf{W}_{5,28}$ and six $j = 6$ level subvectors $\mathbf{W}_{6,58}, \dots, \mathbf{W}_{6,63}$, all of whose SDFs vary by less than 3 dB.

As was the case for the DWT-based bootstrap, we repeat the last two steps above over and over again to build up a sample distribution of wavestrapped autocorrelations.

Figure 6.4 shows the correlation matrices for the wavestrap transforms picked out for a single realization of the same three processes considered in

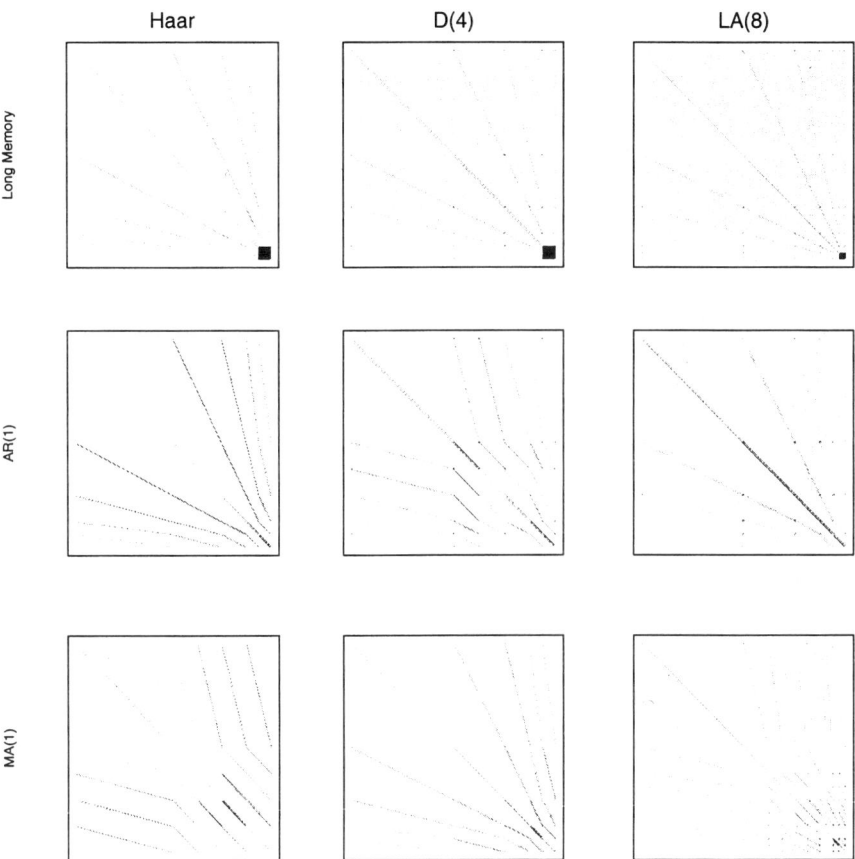

Figure 6.4: Correlation matrices for adaptively selected transforms based upon a single realization of an FD process with $\delta = 0.45$ (top row), an AR(1) process with $\phi = 0.9$ (middle) and an MA(1) process with $\theta = -0.99$ (bottom); $N = 256$ (see text for details).

Figure 5.2 for DWT-based bootstrapping. A comparison of the largest correlations in the corresponding plots of these two figures shows that, by this measure, wavestrapping does better than the DWT-based procedure for the MA(1) process, is about the same for the AR(1) process, and, not unexpectedly, is worse for the FD process, which is well matched to the DWT. The dark squares in the lower right-hand corners in the top row are due to $\mathbf{W}_{j,0}$, which are highly correlated for an FD process.

6.1 White noise tests

Here we briefly describe two well-known test statistics that can be used to assess the null hypothesis that the WP coefficients $\mathbf{W}_{j,n}$ are a sample from a white noise process.

6.1.1 Portmanteau tests

The portmanteau test is designed to see if the sample autocorrelation estimates of $\mathbf{W}_{j,n}$ for lags $\tau = 1, \ldots, K$ are jointly consistent with a hypothesis of zero mean white noise, where K is taken to be relatively small in relation to the number of elements N_j in $\mathbf{W}_{j,n}$ (in the Monte Carlo simulations, we have set $K = \max\{2, \min\{20, N_j/10\}\}$, which is in keeping with recommendations in the literature). For $0 < \tau < N_j$, we define the sample autocorrelation to be

$$\hat{\rho}_{j,n,\tau} = \frac{\sum_{t=0}^{N_j-1-\tau} W_{j,n,t} W_{j,n,t+\tau}}{\sum_{t=0}^{N_j-1} W_{j,n,t}^2}.$$

There are three variations on the portmanteau test in the literature. The Box–Pierce test statistic and Ljung–Box–Pierce test statistic are respectively

$$Q_{j,n} = N_j \sum_{\tau=1}^{K} \hat{\rho}_{j,n,\tau}^2 \quad \text{and} \quad \tilde{Q}_{j,n} = N_j(N_j+2) \sum_{\tau=1}^{K} \frac{\hat{\rho}_{j,n,\tau}^2}{N_j - \tau}$$

(Box and Pierce, 1970; Ljung and Box, 1978). For either test statistic, we reject the null hypothesis of white noise at significance level α when the statistic exceeds the $(1 - \alpha) \times 100\%$ percentage point $Q_K(1 - \alpha)$ for the chi-square distribution with K degrees of freedom. The third variation (McLeod and Li, 1983; Brockwell and Davis, 1991, §9.4) is to use the Ljung–Box–Pierce test on the sample autocorrelations for the squares of $W_{j,n,t}$, namely,

$$\hat{\rho}_{j,n,\tau}^{[2]} = \frac{\sum_{t=0}^{N_j-1-\tau} (W_{j,n,t}^2 - \overline{W^2}_{j,n})(W_{j,n,t+\tau}^2 - \overline{W^2}_{j,n})}{\sum_{t=0}^{N_j-1} (W_{j,n,t}^2 - \overline{W^2}_{j,n})^2},$$

where $\overline{W^2}_{j,n}$ is the sample mean of the squares of the elements of $\mathbf{W}_{j,n}$.

6.1.2 Cumulative periodogram test

Let $|\widetilde{W}_{j,n,k}|^2$ be the squared modulus of the DFT of $\mathbf{W}_{j,n}$ at the Fourier frequency $f_k \equiv k/N_j$. Based upon the $M_j \equiv \frac{N_j}{2} - 1$ frequencies satisfying $0 < f_k < 1/2$, we form the normalized cumulative periodogram

$$\mathcal{P}_l \equiv \frac{\sum_{k=1}^{l} |\widetilde{W}_{j,n,k}|^2}{\sum_{k=1}^{M_j} |\widetilde{W}_{j,n,k}|^2}, \qquad l = 1, \ldots, M_j.$$

We then compute the test statistic $D \equiv \max \{D^+, D^-\}$, where

$$D^+ \equiv \max_{1 \leq l \leq M_j - 1} \left(\frac{l}{M_j - 1} - \mathcal{P}_l \right) \text{ and } D^- \equiv \max_{1 \leq l \leq M_j - 1} \left(\mathcal{P}_l - \frac{l - 1}{M_j - 1} \right).$$

We reject the null hypothesis of white noise at the α level of significance if D exceeds the upper $\alpha \times 100\%$ percentage point $D(\alpha)$ for D under the null hypothesis. A simple approximation for $D(\alpha)$ is given by

$$\tilde{D}(\alpha) \equiv \frac{C(\alpha)}{(M_j - 1)^{1/2} + 0.12 + \frac{0.11}{(M_j - 1)^{1/2}}},$$

where $C(0.05) = 1.358$ (Stephens, 1974).

7 Simulation study

Here we report on a Monte Čarlo experiment that we conducted to see how well the DWT-based bootstrap, wavestrapping and the block bootstrap do at assessing the standard deviation of the unit lag sample autocorrelation $\hat{\rho}_{X,1}$ for Gaussian white noise and the three nonwhite processes considered in previous sections, namely, an FD process with $\delta = 0.45$, an AR(1) process with $\phi = 0.9$ and an MA(1) process with $\theta = -0.99$. We used a pseudo-random number generator of uncorrelated Gaussian deviates ϵ_t with mean zero and unit variance to simulate the white noise process. Using the same generator, we can simulate

- AR(1) time series by setting $X_0 = (\frac{1}{1-0.9^2})^{1/2} \epsilon_0$ and $X_t = 0.9 X_{t-1} + \epsilon_t$, $t = 1, \ldots, N - 1$ (Kay, 1981);

- MA(1) series by using the process definition $X_t = \epsilon_t + 0.99 \epsilon_{t-1}$, $t = 0, \ldots, N - 1$; and

- FD series by using the Davies–Harte algorithm (Davies and Harte, 1987; Wood and Chan, 1996).

In order to better match what must be done in practice, we act as if we do not know the process mean and recenter each simulated series by subtracting off its sample mean $\overline{X}$ prior to any other computations; i.e, we compute $\hat{\rho}_{X,1}$ by replacing X_t with $X_t - \overline{X}$ in (1.1), and we do all bootstrapping and wavestrapping operations on the latter also. We considered two sample sizes, namely, $N = 128$ and $N = 1024$. For each sample size and each process, we used 10,000 simulated time series to determine the 'true' standard deviation of $\hat{\rho}_{X,1}$. The final column of Table 7.1 gives these 'true' values, multiplied by 100. We used the LA(8) wavelet throughout with $J_0 = J - 2$ for sample size $N = 2^J$. For the white noise test needed by wavestrapping, we considered all

Process	Boundary	DWT	Wavestrap Port	Wavestrap Pgrm	Block	True
WN						
$N = 128$	periodic	8.2	8.7	8.8	8.1	8.7
	reflection	8.3	8.6	8.7		
$N = 1024$	periodic	3.1	3.1	3.1	3.0	3.1
	reflection	3.2	3.2	3.1		
AR(1)						
$N = 128$	periodic	5.7	5.2	5.1	5.4	4.8
	reflection	5.5	5.1	5.4		
$N = 1024$	periodic	1.6	1.5	1.5	1.5	1.4
	reflection	1.6	1.5	1.5		
MA(1)						
$N = 128$	periodic	7.1	6.8	6.8	6.5	6.3
	reflection	7.0	6.8	6.6		
$N = 1024$	periodic	2.6	2.4	2.3	2.2	2.2
	reflection	2.6	2.4	2.4		
FD						
$N = 128$	periodic	9.4	8.3	8.5	7.7	10.7
	reflection	9.9	8.8	9.6		
$N = 1024$	periodic	4.4	4.2	4.2	3.4	5.3
	reflection	4.7	4.5	4.7		

Table 7.1: Standard deviations ($\times 100$) of unit lag sample autocorrelations as assessed by DWT-based bootstrapping, wavestrapping with the Ljung–Box–Pierce portmanteau test statistic, wavestrapping with the cumulative periodogram test statistic, and the block bootstrap, along with the 'true' standard deviation as determined by 10,000 simulated series. Independent replications indicate that the standard error for all numbers reported above is roughly 0.1.

three variations on the portmanteau test mentioned in §6.1.1 and the cumulative periodogram test of §6.1.2; however, we obtained nearly identical results for the portmanteau tests, so we only report the Ljung–Box–Pierce statistic in Table 7.1. We set the block size of the block bootstrap to be $\sqrt{N}$. For each variation on bootstrapping and wavestrapping, we generated 50 different simulated series, and we then created 100 boot/wavestrapped series for each of these, from which we can then compute an estimate of the standard deviation for $\hat{\rho}_{X,1}$ (see §9 for comments on the bias in the distribution of $\hat{\rho}_{X,1}$).

Let us examine the $N = 1024$ results given in Table 7.1. For this sample size there is little difference between any of the techniques for the three short memory processes, the one exception being that DWT-based bootstrapping

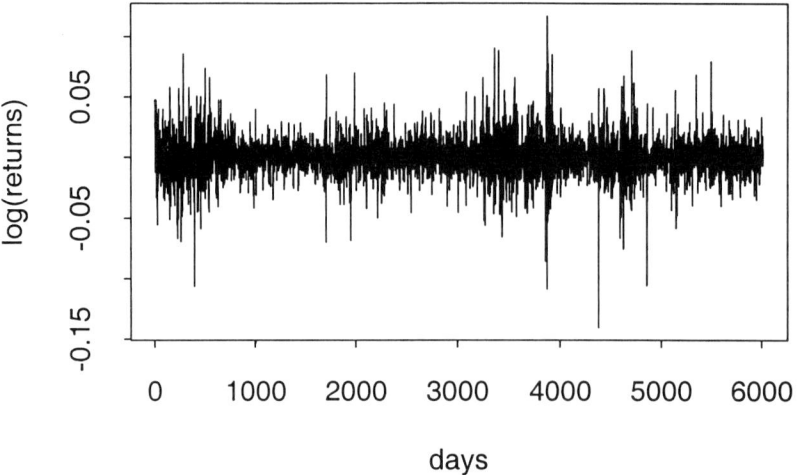

Figure 8.1: Log of daily returns on BMW share prices (1973–96).

is somewhat inferior for the MA(1) process (in view of Figure 6.1 this is not unexpected). The choice of boundary conditions or white noise test makes little difference for these processes, and the wavestrapped standard deviation is quite close to the true value. On the other hand, block bootstrapping is inferior to the other techniques for the FD process. Reflection boundary conditions work better with both DWT-based bootstrapping and wavestrapping, and the cumulative periodogram test statistic is better with the latter than the portmanteau statistic. While DWT-based bootstrapping and wavestrapping yield similar results, they both underestimate the true standard deviation by about 10%.

When we decrease the sample size to $N = 128$, there is more disparity among the four methods. Wavestrapping outperforms the DWT-based bootstrap for the three short memory processes, but the converse is true for the FD process. With the exception of the MA(1) process, wavestrapping also does better than the block bootstrap. Finally, we note that the cumulative periodogram test and reflection boundary conditions generally do better with wavestrapping than the portmanteau test and periodic boundary conditions.

8   Applications

We now apply our methodology to a series X_t of $N = 6016$ daily log returns on BMW share prices between 1973 and 1996; see Figure 8.1. This time series is actually irregularly sampled because no trading takes place on weekends

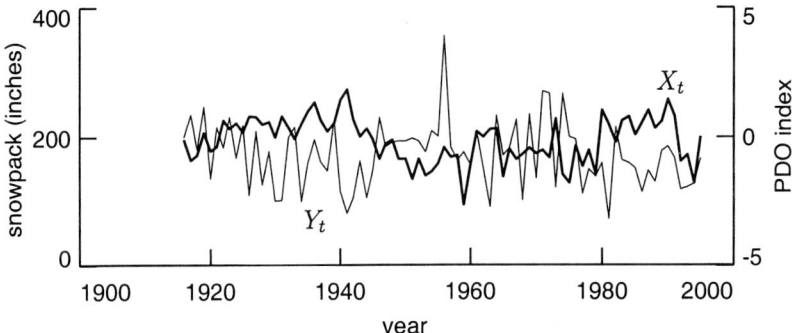

Figure 8.2: Pacific decadal oscillation (PDO) index X_t (thick curve, right-hand axis) and March 15 snow depth at Paradise Ranger Station (1600 meters above sea level) on Mt Rainer Y_t (thin curve, left-hand axis). Both time series have one value per year from 1916 to 1995.

and holidays, but we ignore these gaps and treat the data as a regularly sampled series. The unit lag sample autocorrelation is small, $\hat{\rho}_{X,1} \doteq 0.081$. If we applied the standard large sample theory appropriate for Gaussian white noise, we would attach to this estimate a standard error of $1/\sqrt{N} \doteq 0.013$. In fact the Gaussian assumption is suspect, and the data are better modeled by a t distribution with 3.9 degrees of freedom. Taken at face value, however, the standard error tells us that although small, $\hat{\rho}_{X,1}$ is significantly different from zero; this could presumably be exploited by traders. When we apply the block bootstrap with blocks of length 30, 50, 100, 200 and 500, the standard errors are 0.012, 0.012, 0.014, 0.016 and 0.015, while the DWT-based bootstrap and the wavestrap give 0.023 and 0.020. Though all are larger than the value 0.013 for Gaussian white noise, these confirm the presence of autocorrelation. Simulation using blocks of t_4 innovations with variances 1, 4, 9 and 16, to give the type of stochastic volatility seen in Figure 8.1, gives values of $\hat{\rho}_{X,1}$ whose standard error is 0.02. It seems that the DWT and the wavestrap are able to reproduce this, but that the block bootstrap is not.

As a second example, let us show how wavestrapping can help assess the significance of the sample cross-correlation between two time series. This statistic is often used in the physical sciences as a first step in investigating possible relationships between two series. Figure 8.2 shows two annual time series of interest in atmospheric science, namely, the Pacific decadal oscillation (PDO) index X_t (thick curve) and springtime snow depth Y_t at a location in the Washington Cascade Mountains (thin). The PDO index (Mantua *et al.*, 1997) is the leading principal component of sea-surface temperature over the extratropical north Pacific ocean and has been implicated as a major source

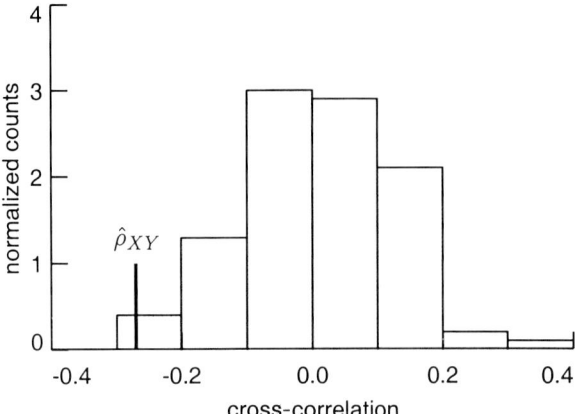

Figure 8.3: Histogram of wavestrapped cross-correlations to assess significance of $\hat{\rho}_{XY}$ computed for the two time series in Figure 8.2.

of interannual variability in temperature and precipitation in western North America. When the PDO index is positive (corresponding to cold sea-surface temperatures in the central Pacific Ocean and warm temperatures off the west coast of North America), an observational record of nearly a century suggests that the mean winter time temperature tends to be high, while precipitation tends to be low. The data shown in Figure 8.2 support this statement, as does the fact that the sample correlation coefficient between X_t and Y_t is negative:

$$\hat{\rho}_{XY} \equiv \frac{\sum_{t=0}^{N-1}(X_t - \overline{X})(Y_t - \overline{Y})}{\left[\sum_{t=0}^{N-1}(X_t - \overline{X})^2 \sum_{t=0}^{N-1}(Y_t - \overline{Y})^2\right]^{1/2}} \doteq -0.27.$$

Were we to assume that all 80 observations were independent and normally distributed, confidence limits based upon large sample statistical theory would declare $\hat{\rho}_{XY}$ to be significantly different from zero at more than the 95% level. However, since both the PDO index and (to a lesser extent) snowpack have considerable year-to-year correlation and since both time series are short, we need another way of ascertaining if the sample cross-correlation is significantly different from zero.

We can address the question of the significance of $\hat{\rho}_{XY}$ by wavestrapping X_t and Y_t separately. The resulting wavestrapped series should be approximately pairwise uncorrelated because any relationship between the two series will be destroyed by resampling separately. The values of $\hat{\rho}_{XY}$ over many wavestrapped series will have a distribution reflecting a null hypothesis of $\rho_{XY} = 0$. We can thus assess the significance of an observed $\hat{\rho}_{XY}$ by comparing it to the distribution of the corresponding $\hat{\rho}_{XY}$ for the wavestrapped series. To do so, we generated 1000 wavestrap samples for each of the two series in Figure 8.2 (we used a WP table of level $J_0 = 3$ based upon an LA(8) wavelet). The resulting histogram of wavestrapped $\hat{\rho}_{XY}$ is shown in Figure 8.3, along with a vertical line marking the observed cross-correlation estimate (-0.27). Based upon this test, we can conclude that $\hat{\rho}_{XY}$ is significantly different from zero with an observed significance around 1%. This result agrees well with an 'equivalent degrees of freedom' assessment of the significance of $\hat{\rho}_{XY}$ based on a t-test with 50 degrees of freedom as described in Bretherton *et al.* (1990).

9   Concluding remarks

We have demonstrated that, for statistics such as the sample ACS, wavestrapping is competitive with existing bootstrap methodology for short memory processes and offers an improvement for long memory processes. While these results are promising, there is considerable work to be done to put wavestrapping on a sound theoretical foundation. Questions that need to be addressed include the following.

1. For what kinds of statistics and processes can we expect wavestrapping to yield either a reasonable approximate distribution or a reasonable approximation to certain aspects of that distribution? For example, Monte Carlo experiments indicate that, whereas the standard deviation of the wavestrap distribution for $\hat{\rho}_{X,1}$ is a good approximation to the actual standard deviation, the same cannot be said for the bias.

2. What are the asymptotic properties of wavestrapping? Although we are really interested in small to moderate sample sizes, it would be of interest to know what conditions are needed for wavestrapping to be a consistent estimator.

3. Can wavestrapping handle non-Gaussian and/or nonlinear processes? The Gaussian assumption is a convenient starting point, but real-world applications dictate that we move beyond it.

4. Can we offer better guidance on the subjective aspects of wavestrap, namely, choice of wavelet and level J_0? The LA(8) wavelet and picking $J_0 = J - 2$ gave good results in our Monte Carlo study, but it is not clear if these would be good choices for other statistics and processes.

Even for statistics such as the sample ACS, there is room for improving the performance of wavestrap, particularly for long memory processes, where it tends to underestimate the variability in the sample autocorrelation. Two possible improvements to our work would be to combine wavestrapping with a parametric approach and to use a different procedure for picking out a decorrelating transform from a WP table. Let us close by briefly commenting on why we feel these to be worth studying.

If we compare the wavestrapping results in Table 7.1 for the AR and FD processes when $N = 1024$, we see somewhat better estimation of the true standard deviation for the AR case (a 7% overestimate as compared to an 11% underestimate in the FD case). If we focus on the DWT, we find that the between/across scale correlations of the wavelet coefficients for the AR and FD processes are quite similar to each other (in fact the correlations in the AR case are a little larger in magnitude). There is a big difference, however, in the properties of their scaling coefficients: for the AR process, the scaling coefficients are reasonably close to white noise (because the SDF flattens out as $f \to 0$), whereas they have a long memory structure for the FD process. This suggests that the underestimation of variability in the FD case is attributable to the scaling coefficients (note that any orthonormal transform we pick from a WP table must include one subvector corresponding to the scaling coefficients from a DWT of some level $J' \leq J_0$). One way to account for the correlation in the scaling coefficients would be to use a parametric bootstrap. If we set J_0 to, say, $J - 4$ rather than our standard choice of $J - 2$, we would have at least 16 scaling coefficients, which would be enough to entertain an AR(1) model. Although an AR(1) model is not a perfect match to the correlation properties of the scaling coefficients for an FD process, this simple model is capable of approximating the correlation structure over limited number of lags, which is really all we require. In addition, a study of the SDFs on the right-hand sides of Figures 5.1 and 6.3 suggests that the remaining correlation structure in wavelet and WP coefficients might be well modeled by fitting an AR(1) process to each set of coefficients and then bootstrapping with respect to the fitted models. Limited tests suggest that this is a promising idea.

Finally, with regard to picking a decorrelating transform from a WP table, wavestrapping does its search through the table in a 'top-down' manner, so the obvious alternative to consider is a 'bottom-up' approach. A well-known example of such an approach is the 'best basis' algorithm (Coifman and Wickerhauser, 1992), which selects between a 'parent' node $\mathbf{W}_{j,n}$ and its 'children' $\mathbf{W}_{j+1,2n}$ and $\mathbf{W}_{j+1,2n+1}$ based upon a cost functional. To see why this algorithm leads to a decorrelating transform, let $j = 1$ and $n = 0$ for simplicity, and suppose that the nonboundary WP coefficients in the parent node have the following SDF:

$$S_{1,0}(f) = \begin{cases} \sigma_{2,0}^2, & 0 \leq |f| \leq 1/4, \\ \sigma_{2,1}^2, & 1/4 \leq |f| \leq 1/2. \end{cases}$$

The variance for a process with this SDF is $\sigma_{1,0}^2 = \frac{1}{2}(\sigma_{2,0}^2 + \sigma_{2,1}^2)$. If we assume for simplicity that the scaling and wavelet filters are perfect high- and low-pass filters, the SDFs of the nonboundary WP coefficients in $\mathbf{W}_{2,0}$ and $\mathbf{W}_{2,1}$ are given by, respectively, $S_{2,0}(f) = \sigma_{2,0}^2$ and $S_{2,1}(f) = \sigma_{2,1}^2$ for $-1/2 \leq f \leq 1/2$;

i.e., both are white noise processes with variances given by, respectively, $\sigma_{2,0}^2$ and $\sigma_{2,1}^2$ (note that $S_{1,0}(\cdot)$ is not a white noise SDF unless $\sigma_{2,0}^2 = \sigma_{2,1}^2$). If we assume Gaussianity and, e.g., the L_1 cost functional, then the costs of each coefficient $W_{1,0,t}$ in $\mathbf{W}_{1,0}$ and of each coefficient $W_{2,m,t}$ in $\mathbf{W}_{2,m}$ are, respectively,

$$E\{|W_{1,0,t}|\} = \left(\frac{\sigma_{2,0}^2 + \sigma_{2,1}^2}{2\pi}\right)^{1/2} \text{ and } E\{|W_{2,m,t}|\} = \left(\frac{2\sigma_{2,m}^2}{\pi}\right)^{1/2}.$$

Since there are N_1 coefficients in $\mathbf{W}_{1,0}$ and N_2 coefficients in each of $\mathbf{W}_{2,0}$ and $\mathbf{W}_{2,1}$, the total expected costs of the parent node and its children are thus, respectively,

$$C_1 \equiv N_1 \left(\frac{\sigma_{2,0}^2 + \sigma_{2,1}^2}{2\pi}\right)^{1/2} \text{ and } C_2 \equiv N_2 \left[\left(\frac{2\sigma_{2,0}^2}{\pi}\right)^{1/2} + \left(\frac{2\sigma_{2,1}^2}{\pi}\right)^{1/2}\right].$$

It is an easy exercise to verify that $C_2 \leq C_1$ always, with equality occurring if and only if $\sigma_{2,0}^2 = \sigma_{2,1}^2$ (i.e., the nonboundary coefficients in $\mathbf{W}_{1,0}$ are white noise). Since the best basis algorithm works by making a comparison such as the above on each node, we can argue that this algorithm will tend to pick out a decorrelating transform. Tests to date, however, indicate that best basis picks out too many small groups of coefficients (not ideal for bootstrapping), so we are currently exploring ways of 'pruning' back the best basis transform.

10 Acknowledgments

We would like to thank Bill Fitzgerald, Richard Smith, Andrew Walden and Peter Young for the considerable efforts that they put into organizing the program on Nonlinear and Nonstationary Signal Processing at the Isaac Newton Institute. We would also like to thank Chris Bretherton (Department of Atmospheric Sciences, University of Washington) for providing us with the snowpack/PDO time series and for discussions concerning them. This work was supported in part by the Swiss National Science Foundation (Sardy and Davison) and by the US National Science Foundation (Percival).

References

Bartlett, M.S. (1946) 'On the theoretical specification of sampling properties of auto-correlated time series,' *Supplement to the Journal of the Royal Statistical Society* **8**, 27–41.

Beran, J. (1994) *Statistics for Long-Memory Processes*, New York: Chapman & Hall.

Box, G.E.P., Pierce, D.A. (1970) 'Distribution of residual autocorrelations in autoregressive-integrated moving average time series models,' *Journal of the American Statistical Association* **65**, 1509–1526.

Bretherton, C.S., Widmann, M., Dymnikov, V.P., Wallace, J.M., Bladé, I. (1999) 'The effective number of spatial degrees of freedom of a time-varying field,' *Journal of Climate* **12**, 1990–2009.

Brockwell, P.J., Davis, R.A. (1991) *Time Series: Theory and Methods* (Second Edition), New York: Springer.

Bruce, A.G., Gao, H.–Y. (1996) *Applied Wavelet Analysis with S-PLUS*, New York: Springer.

Bühlmann, P. (1999) 'Bootstrapping time series,' *Bulletin of the 52nd Session of the International Statistical Institute* **1**, 201–204.

Coifman, R.R., Wickerhauser, M.V. (1992) 'Entropy-based algorithms for best basis selection,' *IEEE Transactions on Information Theory* **38**, 713–718.

Daubechies, I. (1992) *Ten Lectures on Wavelets*, Philadelphia: SIAM.

Davies, R.B., Harte, D.S. (1987) 'Tests for Hurst effect,' *Biometrika* **74**, 95–101.

Davison, A.C., Hinkley, D.V. (1997) *Bootstrap Methods and their Application*, Cambridge, UK: Cambridge University Press.

Efron, B., Tibshirani, R.J. (1993) *An Introduction to the Bootstrap*, New York: Chapman & Hall.

Flandrin, P. (1992) 'Wavelet analysis and synthesis of fractional Brownian motion,' *IEEE Transactions on Information Theory* **38**, 910–917.

Granger, C.W.J., Joyeux, R. (1980) 'An introduction to long-memory time series models and fractional differencing,' *Journal of Time Series Analysis* **1**, 15–30.

Greenhall, C.A., Howe, D.A., Percival, D.B. (1999) 'Total variance, an estimator of long-term frequency stability,' *IEEE Transactions on Ultrasonics, Ferroelectrics, and Frequency Control* **46**, 1183–1191.

Hosking, J.R.M. (1981) 'Fractional differencing,' *Biometrika* **68**, 165–176.

Kay, S.M. (1981) 'Efficient generation of colored noise,' *Proceedings of the IEEE* **69**, 480–481.

Künsch, H.R. (1989) 'The jackknife and the bootstrap for general stationary observations,' *Annals of Statistics* **17**, 1217–1241.

Ljung, G.M., Box, G.E.P. (1978) 'On a measure of lack of fit in time series models,' *Biometrika* **65**, 297–303.

Mallat, S.G. (1989) 'A theory for multiresolution signal decomposition: the wavelet representation,' *IEEE Transactions on Pattern Analysis and Machine Intelligence* **11**, 674–693.

Mantua, N.J., Hare, S.R., Zhang, Y., Wallace, J.M., Francis, R.C. (1997) 'A Pacific interdecadal climate oscillation with impacts on salmon production,' *Bulletin of the American Meteorological Society* **78**, 1069–1079.

McCoy, E.J., Walden, A.T. (1996) 'Wavelet analysis and synthesis of stationary long-memory processes,' *Journal of Computational and Graphical Statistics* **5**, 26–56.

McLeod, A.I., Li, W.K. (1983) 'Diagnostic checking ARMA time series models using squared-residual autocorrelations,' *Journal of Time Series Analysis* **4**, 269–273.

Percival, D.B., Walden, A.T. (2000) *Wavelet Methods for Time Series Analysis*, Cambridge, UK: Cambridge University Press.

Priestley, M.B. (1981) *Spectral Analysis and Time Series*, London: Academic Press.

Stephens, M.A. (1974) 'EDF statistics for goodness of fit and some comparisions,' *Journal of the American Statistical Association* **69**, 730–737.

Wickerhauser, M.V. (1994) *Adapted Wavelet Analysis from Theory to Software*, Wellesley, MA: A.K. Peters.

Wood, A.T.A., Chan, G. (1996) 'Simulation of stationary Gaussian processes in $[0, 1]^d$,' *Journal of Computational and Graphical Statistics* **3**, 3, 409–432.

Wornell, G.W. (1995) *Signal Processing with Fractals: A Wavelet Based Approach*, Englewood Cliffs, NJ: Prentice-Hall.